# Lecture Notes in Computer Science 14834

The series Lecture Notes in Computer Science (LNCS), including its subseries Lecture Notes in Artificial Intelligence (LNAI) and Lecture Notes in Bioinformatics (LNBI), has established itself as a medium for the publication of new developments in computer science and information technology research, teaching, and education.

LNCS enjoys close cooperation with the computer science R & D community, the series counts many renowned academics among its volume editors and paper authors, and collaborates with prestigious societies. Its mission is to serve this international community by providing an invaluable service, mainly focused on the publication of conference and workshop proceedings and postproceedings. LNCS commenced publication in 1973.

Leonardo Franco · Clélia de Mulatier ·
Maciej Paszynski · Valeria V. Krzhizhanovskaya ·
Jack J. Dongarra · Peter M. A. Sloot
Editors

# Computational Science – ICCS 2024

24th International Conference
Malaga, Spain, July 2–4, 2024
Proceedings, Part III

 Springer

*Editors*
Leonardo Franco (iD)
University of Malaga
Malaga, Spain

Maciej Paszynski (iD)
AGH University of Science and Technology
Krakow, Poland

Jack J. Dongarra (iD)
University of Tennessee
Knoxville, TN, USA

Clélia de Mulatier (iD)
University of Amsterdam
Amsterdam, The Netherlands

Valeria V. Krzhizhanovskaya (iD)
University of Amsterdam
Amsterdam, The Netherlands

Peter M. A. Sloot (iD)
University of Amsterdam
Amsterdam, The Netherlands

ISSN 0302-9743          ISSN 1611-3349 (electronic)
Lecture Notes in Computer Science
ISBN 978-3-031-63758-2      ISBN 978-3-031-63759-9 (eBook)
https://doi.org/10.1007/978-3-031-63759-9

This Springer imprint is published by the registered company Springer Nature Switzerland AG
The registered company address is: Gewerbestrasse 11, 6330 Cham, Switzerland

If disposing of this product, please recycle the paper.

# Preface

Welcome to the proceedings of the 24th International Conference on Computational Science (https://www.iccs-meeting.org/iccs2024/), held on July 2–4, 2024 at the University of Málaga, Spain.

In keeping with the new normal of our times, ICCS featured both in-person and online sessions. Although the challenges of such a hybrid format are manifold, we have always tried our best to keep the ICCS community as dynamic, creative, and productive as possible. We are proud to present the proceedings you are reading as a result.

ICCS 2024 was jointly organized by the University of Málaga, the University of Amsterdam, and the University of Tennessee.

Facing the Mediterranean in Spain's Costa del Sol, Málaga is the country's sixth-largest city, and a major hub for finance, tourism, and technology in the region.

The University of Málaga (Universidad de Málaga, UMA) is a modern, public university, offering 63 degrees and 120 postgraduate degrees. Close to 40,000 students study at UMA, taught by 2500 lecturers, distributed over 81 departments and 19 centers. The UMA has 278 research groups, which are involved in 80 national projects and 30 European and international projects. ICCS took place at the Teatinos Campus, home to the School of Computer Science and Engineering (ETSI Informática), which is a pioneer in its field and offers the widest range of IT-related subjects in the region of Andalusia.

The International Conference on Computational Science is an annual conference that brings together researchers and scientists from mathematics and computer science as basic computing disciplines, as well as researchers from various application areas who are pioneering computational methods in sciences such as physics, chemistry, life sciences, engineering, arts, and the humanities, to discuss problems and solutions in the area, identify new issues, and shape future directions for research.

The ICCS proceedings series have become a primary intellectual resource for computational science researchers, defining and advancing the state of the art in this field.

We are proud to note that this 24th edition, with 17 tracks (16 thematic tracks and one main track) and close to 300 participants, has kept to the tradition and high standards of previous editions.

The theme for 2024, "Computational Science: Guiding the Way Towards a Sustainable Society", highlights the role of Computational Science in assisting multidisciplinary research on sustainable solutions. This conference was a unique event focusing on recent developments in scalable scientific algorithms; advanced software tools; computational grids; advanced numerical methods; and novel application areas. These innovative novel models, algorithms, and tools drive new science through efficient application in physical systems, computational and systems biology, environmental systems, finance, and others.

ICCS is well known for its excellent lineup of keynote speakers. The keynotes for 2024 were:

- David Abramson, University of Queensland, Australia
- Manuel Castro Díaz, University of Málaga, Spain
- Jiří Mikyška, Czech Technical University in Prague, Czechia
- Takemasa Miyoshi, RIKEN, Japan
- Coral Calero Muñoz, University of Castilla-La Mancha, Spain
- Petra Ritter, Berlin Institute of Health & Charité University Hospital Berlin, Germany

This year we had 430 submissions (152 to the main track and 278 to the thematic tracks). In the main track, 51 full papers were accepted (33.5%); in the thematic tracks, 104 full papers (37.4%). The higher acceptance rate in the thematic tracks is explained by their particular nature, whereby track organizers personally invite many experts in the field to participate. Each submission received at least 2 single-blind reviews (2.6 reviews per paper on average).

ICCS relies strongly on our thematic track organizers' vital contributions to attract high-quality papers in many subject areas. We would like to thank all committee members from the main and thematic tracks for their contribution to ensuring a high standard for the accepted papers. We would also like to thank Springer, Elsevier, and Intellegibilis for their support. Finally, we appreciate all the local organizing committee members for their hard work in preparing this conference.

We hope the attendees enjoyed the conference, whether virtually or in person.

July 2024

Leonardo Franco
Clélia de Mulatier
Maciej Paszynski
Valeria V. Krzhizhanovskaya
Jack J. Dongarra
Peter M. A. Sloot

# Organization

## Conference Chairs

### General Chair

Valeria Krzhizhanovskaya        University of Amsterdam, The Netherlands

### Main Track Chair

Clélia de Mulatier        University of Amsterdam, The Netherlands

### Thematic Tracks Chair

Maciej Paszynski        AGH University of Krakow, Poland

### Thematic Tracks Vice Chair

Michael Harold Lees        University of Amsterdam, The Netherlands

### Scientific Chairs

Peter M. A. Sloot        University of Amsterdam, The Netherlands
Jack Dongarra        University of Tennessee, USA

### Local Organizing Committee

Leonardo Franco (Chair)        University of Malaga, Spain
Francisco Ortega-Zamorano        University of Malaga, Spain
Francisco J. Moreno-Barea        University of Malaga, Spain
José L. Subirats-Contreras        University of Malaga, Spain

## Thematic Tracks and Organizers

## Advances in High-Performance Computational Earth Sciences: Numerical Methods, Frameworks & Applications (IHPCES)

| | |
|---|---|
| Takashi Shimokawabe | University of Tokyo, Japan |
| Kohei Fujita | University of Tokyo, Japan |
| Dominik Bartuschat | FAU Erlangen-Nürnberg, Germany |

## Artificial Intelligence and High-Performance Computing for Advanced Simulations (AIHPC4AS)

| | |
|---|---|
| Maciej Paszynski | AGH University of Krakow, Poland |

## Biomedical and Bioinformatics Challenges for Computer Science (BBC)

| | |
|---|---|
| Mario Cannataro | University Magna Graecia of Catanzaro, Italy |
| Giuseppe Agapito | University Magna Graecia of Catanzaro, Italy |
| Mauro Castelli | Universidade Nova de Lisboa, Portugal |
| Riccardo Dondi | University of Bergamo, Italy |
| Rodrigo Weber dos Santos | Federal University of Juiz de Fora, Brazil |
| Italo Zoppis | University of Milano-Bicocca, Italy |

## Computational Diplomacy and Policy (CoDiP)

| | |
|---|---|
| Roland Bouffanais | University of Geneva, Switzerland |
| Michael Lees | University of Amsterdam, The Netherlands |
| Brian Castellani | Durham University, UK |

## Computational Health (CompHealth)

| | |
|---|---|
| Sergey Kovalchuk | Huawei, Russia |
| Georgiy Bobashev | RTI International, USA |
| Anastasia Angelopoulou | University of Westminster, UK |
| Jude Hemanth | Karunya University, India |

# Computational Optimization, Modelling, and Simulation (COMS)

| | |
|---|---|
| Xin-She Yang | Middlesex University London, UK |
| Slawomir Koziel | Reykjavik University, Iceland |
| Leifur Leifsson | Purdue University, USA |

# Generative AI and Large Language Models (LLMs) in Advancing Computational Medicine (CMGAI)

| | |
|---|---|
| Ahmed Abdeen Hamed | State University of New York at Binghamton, USA |
| Qiao Jin | National Institutes of Health, USA |
| Xindong Wu | Hefei University of Technology, China |
| Byung Lee | University of Vermont, USA |
| Zhiyong Lu | National Institutes of Health, USA |
| Karin Verspoor | RMIT University, Australia |
| Christopher Savoie | Zapata AI, USA |

# Machine Learning and Data Assimilation for Dynamical Systems (MLDADS)

| | |
|---|---|
| Rossella Arcucci | Imperial College London, UK |
| Cesar Quilodran-Casas | Imperial College London, UK |

# Multiscale Modelling and Simulation (MMS)

| | |
|---|---|
| Derek Groen | Brunel University London, UK |
| Diana Suleimenova | Brunel University London, UK |

# Network Models and Analysis: From Foundations to Artificial Intelligence (NMAI)

| | |
|---|---|
| Marianna Milano | Università Magna Graecia of Catanzaro, Italy |
| Giuseppe Agapito | University Magna Graecia of Catanzaro, Italy |
| Pietro Cinaglia | University Magna Graecia of Catanzaro, Italy |
| Chiara Zucco | University Magna Graecia of Catanzaro, Italy |

## Numerical Algorithms and Computer Arithmetic for Computational Science (NACA)

Pawel Gepner                Warsaw Technical University, Poland
Ewa Deelman                 University of Southern California, Marina del
                            Rey, USA
Hatem Ltaief                KAUST, Saudi Arabia

## Quantum Computing (QCW)

Katarzyna Rycerz            AGH University of Krakow, Poland
Marian Bubak                Sano and AGH University of Krakow, Poland

## Simulations of Flow and Transport: Modeling, Algorithms, and Computation (SOFTMAC)

Shuyu Sun                   King Abdullah University of Science and
                            Technology, Saudi Arabia
Jingfa Li                   Beijing Institute of Petrochemical Technology,
                            China
James Liu                   Colorado State University, USA

## Smart Systems: Bringing Together Computer Vision, Sensor Networks and Artificial Intelligence (SmartSys)

Pedro Cardoso               University of Algarve, Portugal
João Rodrigues              University of Algarve, Portugal
Jânio Monteiro              University of Algarve, Portugal
Roberto Lam                 University of Algarve, Portugal

## Solving Problems with Uncertainties (SPU)

Vassil Alexandrov           Hartree Centre – STFC, UK
Aneta Karaivanova           IICT – Bulgarian Academy of Science, Bulgaria

# Teaching Computational Science (WTCS)

Evguenia Alexandrova            Hartree Centre – STFC, UK
Tseden Taddese                  UK Research and Innovation, UK

# Reviewers

Ahmed Abdelgawad                Central Michigan University, USA
Samaneh Abolpour Mofrad         Imperial College London, UK
Tesfamariam Mulugeta Abuhay     Queen's University, Canada
Giuseppe Agapito                University of Catanzaro, Italy
Elisabete Alberdi               University of the Basque Country, Spain
Luis Alexandre                  UBI and NOVA LINCS, Portugal
Vassil Alexandrov               Hartree Centre – STFC, UK
Evguenia Alexandrova            Hartree Centre – STFC, UK
Julen Alvarez-Aramberri         Basque Center for Applied Mathematics, Spain
Domingos Alves                  Ribeirão Preto Medical School, University of São
                                Paulo, Brazil
Sergey Alyaev                   NORCE, Norway
Anastasia Anagnostou            Brunel University London, UK
Anastasia Angelopoulou          University of Westminster, UK
Rossella Arcucci                Imperial College London, UK
Emanouil Atanasov               IICT – Bulgarian Academy of Sciences, Bulgaria
Krzysztof Banaś                 AGH University of Krakow, Poland
Luca Barillaro                  Magna Graecia University of Catanzaro, Italy
Dominik Bartuschat              FAU Erlangen-Nürnberg, Germany
Pouria Behnodfaur               Curtin University, Australia
Jörn Behrens                    University of Hamburg, Germany
Adrian Bekasiewicz              Gdansk University of Technology, Poland
Gebrail Bekdas                  Istanbul University, Turkey
Mehmet Ali Belen                Iskenderun Technical University, Turkey
Stefano Beretta                 San Raffaele Telethon Institute for Gene Therapy,
                                Italy
Anabela Moreira Bernardino      Polytechnic Institute of Leiria, Portugal
Eugénia Bernardino              Polytechnic Institute of Leiria, Portugal
Daniel Berrar                   Tokyo Institute of Technology, Japan
Piotr Biskupski                 IBM, Poland
Georgiy Bobashev                RTI International, USA
Carlos Bordons                  University of Seville, Spain
Bartosz Bosak                   PSNC, Poland
Lorella Bottino                 University Magna Graecia of Catanzaro, Italy

| Roland Bouffanais | University of Geneva, Switzerland |
| Marian Bubak | Sano and AGH University of Krakow, Poland |
| Aleksander Byrski | AGH University of Krakow, Poland |
| Cristiano Cabrita | Universidade do Algarve, Portugal |
| Xing Cai | Simula Research Laboratory, Norway |
| Carlos Calafate | Universitat Politècnica de València, Spain |
| Victor Calo | Curtin University, Australia |
| Mario Cannataro | University Magna Graecia of Catanzaro, Italy |
| Karol Capała | AGH University of Krakow, Poland |
| Pedro J. S. Cardoso | Universidade do Algarve, Portugal |
| Eddy Caron | ENS-Lyon/Inria/LIP, France |
| Stefano Casarin | Houston Methodist Hospital, USA |
| Brian Castellani | Durham University, UK |
| Mauro Castelli | Universidade Nova de Lisboa, Portugal |
| Nicholas Chancellor | Durham University, UK |
| Thierry Chaussalet | University of Westminster, UK |
| Sibo Cheng | Imperial College London, UK |
| Lock-Yue Chew | Nanyang Technological University, Singapore |
| Pastrello Chiara | Krembil Research Institute, Canada |
| Su-Fong Chien | MIMOS Berhad, Malaysia |
| Marta Chinnici | enea, Italy |
| Bastien Chopard | University of Geneva, Switzerland |
| Maciej Ciesielski | University of Massachusetts, USA |
| Pietro Cinaglia | University of Catanzaro, Italy |
| Noelia Correia | Universidade do Algarve, Portugal |
| Adriano Cortes | University of Rio de Janeiro, Brazil |
| Ana Cortes | Universitat Autònoma de Barcelona, Spain |
| Enrique Costa-Montenegro | Universidad de Vigo, Spain |
| David Coster | Max Planck Institute for Plasma Physics, Germany |
| Carlos Cotta | University of Málaga, Spain |
| Peter Coveney | University College London, UK |
| Alex Crimi | AGH University of Krakow, Poland |
| Daan Crommelin | CWI Amsterdam, The Netherlands |
| Attila Csikasz-Nagy | King's College London, UK/Pázmány Péter Catholic University, Hungary |
| Javier Cuenca | University of Murcia, Spain |
| António Cunha | UTAD, Portugal |
| Pawel Czarnul | Gdansk University of Technology, Poland |
| Pasqua D'Ambra | IAC-CNR, Italy |
| Alberto D'Onofrio | University of Trieste, Italy |
| Lisandro Dalcin | KAUST, Saudi Arabia |

| | |
|---|---|
| Bhaskar Dasgupta | University of Illinois at Chicago, USA |
| Clélia de Mulatier | University of Amsterdam, The Netherlands |
| Ewa Deelman | University of Southern California, Marina del Rey, USA |
| Quanling Deng | Australian National University, Australia |
| Eric Dignum | University of Amsterdam, The Netherlands |
| Riccardo Dondi | University of Bergamo, Italy |
| Rafal Drezewski | AGH University of Krakow, Poland |
| Simon Driscoll | University of Reading, UK |
| Hans du Buf | University of the Algarve, Portugal |
| Vitor Duarte | Universidade NOVA de Lisboa, Portugal |
| Jacek Długopolski | AGH University of Krakow, Poland |
| Wouter Edeling | Vrije Universiteit Amsterdam, The Netherlands |
| Nahid Emad | University of Paris Saclay, France |
| Christian Engelmann | ORNL, USA |
| August Ernstsson | Linköping University, Sweden |
| Aniello Esposito | Hewlett Packard Enterprise, Switzerland |
| Roberto R. Expósito | Universidade da Coruna, Spain |
| Hongwei Fan | Imperial College London, UK |
| Tamer Fandy | University of Charleston, USA |
| Giuseppe Fedele | University of Calabria, Italy |
| Christos Filelis-Papadopoulos | Democritus University of Thrace, Greece |
| Alberto Freitas | University of Porto, Portugal |
| Ruy Freitas Reis | Universidade Federal de Juiz de Fora, Brazil |
| Kohei Fujita | University of Tokyo, Japan |
| Takeshi Fukaya | Hokkaido University, Japan |
| Wlodzimierz Funika | AGH University of Krakow, Poland |
| Takashi Furumura | University of Tokyo, Japan |
| Teresa Galvão | University of Porto, Portugal |
| Luis Garcia-Castillo | Carlos III University of Madrid, Spain |
| Bartłomiej Gardas | Institute of Theoretical and Applied Informatics, Polish Academy of Sciences, Poland |
| Victoria Garibay | University of Amsterdam, The Netherlands |
| Frédéric Gava | Paris-East Créteil University, France |
| Piotr Gawron | Nicolaus Copernicus Astronomical Centre, Polish Academy of Sciences, Poland |
| Bernhard Geiger | Know-Center GmbH, Austria |
| Pawel Gepner | Warsaw Technical University, Poland |
| Alex Gerbessiotis | NJIT, USA |
| Maziar Ghorbani | Brunel University London, UK |
| Konstantinos Giannoutakis | University of Macedonia, Greece |
| Alfonso Gijón | University of Granada, Spain |

| Jorge González-Domínguez | Universidade da Coruña, Spain |
| Alexandrino Gonçalves | CIIC – ESTG – Polytechnic University of Leiria, Portugal |
| Yuriy Gorbachev | Soft-Impact LLC, Russia |
| Pawel Gorecki | University of Warsaw, Poland |
| Michael Gowanlock | Northern Arizona University, USA |
| George Gravvanis | Democritus University of Thrace, Greece |
| Derek Groen | Brunel University London, UK |
| Loïc Guégan | UiT the Arctic University of Norway, Norway |
| Tobias Guggemos | University of Vienna, Austria |
| Serge Guillas | University College London, UK |
| Manish Gupta | Harish-Chandra Research Institute, India |
| Piotr Gurgul | SnapChat, Switzerland |
| Oscar Gustafsson | Linköping University, Sweden |
| Ahmed Abdeen Hamed | State University of New York at Binghamton, USA |
| Laura Harbach | Brunel University London, UK |
| Agus Hartoyo | TU Kaiserslautern, Germany |
| Ali Hashemian | Basque Center for Applied Mathematics, Spain |
| Mohamed Hassan | Virginia Tech, USA |
| Alexander Heinecke | Intel Parallel Computing Lab, USA |
| Jude Hemanth | Karunya University, India |
| Aochi Hideo | BRGM, France |
| Alfons Hoekstra | University of Amsterdam, The Netherlands |
| George Holt | UK Research and Innovation, UK |
| Maximilian Höb | Leibniz-Rechenzentrum der Bayerischen Akademie der Wissenschaften, Germany |
| Huda Ibeid | Intel Corporation, USA |
| Alireza Jahani | Brunel University London, UK |
| Jiří Jaroš | Brno University of Technology, Czechia |
| Qiao Jin | National Institutes of Health, USA |
| Zhong Jin | Computer Network Information Center, Chinese Academy of Sciences, China |
| David Johnson | Uppsala University, Sweden |
| Eleda Johnson | Imperial College London, UK |
| Piotr Kalita | Jagiellonian University, Poland |
| Drona Kandhai | University of Amsterdam, The Netherlands |
| Aneta Karaivanova | IICT-Bulgarian Academy of Science, Bulgaria |
| Sven Karbach | University of Amsterdam, The Netherlands |
| Takahiro Katagiri | Nagoya University, Japan |
| Haruo Kobayashi | Gunma University, Japan |
| Marcel Koch | KIT, Germany |

| Harald Koestler | University of Erlangen-Nuremberg, Germany |
| Georgy Kopanitsa | Tomsk Polytechnic University, Russia |
| Sotiris Kotsiantis | University of Patras, Greece |
| Remous-Aris Koutsiamanis | IMT Atlantique/DAPI, STACK (LS2N/Inria), France |
| Sergey Kovalchuk | Huawei, Russia |
| Slawomir Koziel | Reykjavik University, Iceland |
| Ronald Kriemann | MPI MIS Leipzig, Germany |
| Valeria Krzhizhanovskaya | University of Amsterdam, The Netherlands |
| Sebastian Kuckuk | Friedrich-Alexander-Universität Erlangen-Nürnberg, Germany |
| Michael Kuhn | Otto von Guericke University Magdeburg, Germany |
| Ryszard Kukulski | Institute of Theoretical and Applied Informatics, Polish Academy of Sciences, Poland |
| Krzysztof Kurowski | PSNC, Poland |
| Marcin Kuta | AGH University of Krakow, Poland |
| Marcin Łoś | AGH University of Krakow, Poland |
| Roberto Lam | Universidade do Algarve, Portugal |
| Tomasz Lamża | ACK Cyfronet, Poland |
| Ilaria Lazzaro | Università degli studi Magna Graecia di Catanzaro, Italy |
| Paola Lecca | Free University of Bozen-Bolzano, Italy |
| Byung Lee | University of Vermont, USA |
| Mike Lees | University of Amsterdam, The Netherlands |
| Leifur Leifsson | Purdue University, USA |
| Kenneth Leiter | U.S. Army Research Laboratory, USA |
| Paulina Lewandowska | IT4Innovations National Supercomputing Center, Czechia |
| Jingfa Li | Beijing Institute of Petrochemical Technology, China |
| Siyi Li | Imperial College London, UK |
| Che Liu | Imperial College London, UK |
| James Liu | Colorado State University, USA |
| Zhao Liu | National Supercomputing Center in Wuxi, China |
| Marcelo Lobosco | UFJF, Brazil |
| Jay F. Lofstead | Sandia National Laboratories, USA |
| Chu Kiong Loo | University of Malaya, Malaysia |
| Stephane Louise | CEA, LIST, France |
| Frédéric Loulergue | University of Orléans, INSA CVL, LIFO EA 4022, France |
| Hatem Ltaief | KAUST, Saudi Arabia |
| Zhiyong Lu | National Institutes of Health, USA |

| | |
|---|---|
| Fernando Nobrega Santos | University of Amsterdam, The Netherlands |
| Joseph O'Connor | University of Edinburgh, UK |
| Frederike Oetker | University of Amsterdam, The Netherlands |
| Arianna Olivelli | Imperial College London, UK |
| Ángel Omella | Basque Center for Applied Mathematics, Spain |
| Kenji Ono | Kyushu University, Japan |
| Hiroyuki Ootomo | Tokyo Institute of Technology, Japan |
| Eneko Osaba | TECNALIA Research & Innovation, Spain |
| George Papadimitriou | University of Southern California, USA |
| Nikela Papadopoulou | University of Glasgow, UK |
| Marcin Paprzycki | IBS PAN and WSM, Poland |
| David Pardo | Basque Center for Applied Mathematics, Spain |
| Anna Paszynska | Jagiellonian University, Poland |
| Maciej Paszynski | AGH University of Krakow, Poland |
| Łukasz Pawela | Institute of Theoretical and Applied Informatics, Polish Academy of Sciences, Poland |
| Giulia Pederzani | Universiteit van Amsterdam, The Netherlands |
| Alberto Perez de Alba Ortiz | University of Amsterdam, The Netherlands |
| Dana Petcu | West University of Timisoara, Romania |
| Beáta Petrovski | University of Oslo, Norway |
| Frank Phillipson | TNO, The Netherlands |
| Eugenio Piasini | International School for Advanced Studies (SISSA), Italy |
| Juan C. Pichel | Universidade de Santiago de Compostela, Spain |
| Anna Pietrenko-Dabrowska | Gdansk University of Technology, Poland |
| Armando Pinho | University of Aveiro, Portugal |
| Pietro Pinoli | Politecnico di Milano, Italy |
| Yuri Pirola | Università degli Studi di Milano-Bicocca, Italy |
| Ollie Pitts | Imperial College London, UK |
| Robert Platt | Imperial College London, UK |
| Dirk Pleiter | KTH/Forschungszentrum Jülich, Germany |
| Paweł Poczekajło | Koszalin University of Technology, Poland |
| Cristina Portalés Ricart | Universidad de Valencia, Spain |
| Simon Portegies Zwart | Leiden University, The Netherlands |
| Anna Procopio | Università Magna Graecia di Catanzaro, Italy |
| Ela Pustulka-Hunt | FHNW Olten, Switzerland |
| Marcin Płodzień | ICFO, Spain |
| Ubaid Qadri | Hartree Centre – STFC, UK |
| Rick Quax | University of Amsterdam, The Netherlands |
| Cesar Quilodran Casas | Imperial College London, UK |
| Andrianirina Rakotoharisoa | Imperial College London, UK |
| Celia Ramos | University of the Algarve, Portugal |

| Robin Richardson | Netherlands eScience Center, The Netherlands |
| Sophie Robert | University of Orléans, France |
| João Rodrigues | Universidade do Algarve, Portugal |
| Daniel Rodriguez | University of Alcalá, Spain |
| Marcin Rogowski | Saudi Aramco, Saudi Arabia |
| Sergio Rojas | Pontifical Catholic University of Valparaiso, Chile |
| Diego Romano | ICAR-CNR, Italy |
| Albert Romkes | South Dakota School of Mines and Technology, USA |
| Juan Ruiz | University of Buenos Aires, Argentina |
| Tomasz Rybotycki | IBS PAN, CAMK PAN, AGH, Poland |
| Katarzyna Rycerz | AGH University of Krakow, Poland |
| Grażyna Ślusarczyk | Jagiellonian University, Poland |
| Emre Sahin | Science and Technology Facilities Council, UK |
| Ozlem Salehi | Özyeğin University, Turkey |
| Ayşin Sancı | Altinay, Turkey |
| Christopher Savoie | Zapata Computing, USA |
| Ileana Scarpino | University "Magna Graecia" of Catanzaro, Italy |
| Robert Schaefer | AGH University of Krakow, Poland |
| Ulf D. Schiller | University of Delaware, USA |
| Bertil Schmidt | University of Mainz, Germany |
| Karen Scholz | Fraunhofer MEVIS, Germany |
| Martin Schreiber | Université Grenoble Alpes, France |
| Paulina Sepúlveda-Salas | Pontifical Catholic University of Valparaiso, Chile |
| Marzia Settino | Università Magna Graecia di Catanzaro, Italy |
| Mostafa Shahriari | Basque Center for Applied Mathematics, Spain |
| Takashi Shimokawabe | University of Tokyo, Japan |
| Alexander Shukhman | Orenburg State University, Russia |
| Marcin Sieniek | Google, USA |
| Joaquim Silva | Nova School of Science and Technology – NOVA LINCS, Portugal |
| Mateusz Sitko | AGH University of Krakow, Poland |
| Haozhen Situ | South China Agricultural University, China |
| Leszek Siwik | AGH University of Krakow, Poland |
| Peter Sloot | University of Amsterdam, The Netherlands |
| Oskar Slowik | Center for Theoretical Physics PAS, Poland |
| Sucha Smanchat | King Mongkut's University of Technology North Bangkok, Thailand |
| Alexander Smirnovsky | SPbPU, Russia |
| Maciej Smołka | AGH University of Krakow, Poland |
| Isabel Sofia | Instituto Politécnico de Beja, Portugal |
| Robert Staszewski | University College Dublin, Ireland |

| | |
|---|---|
| Magdalena Stobińska | University of Warsaw, Poland |
| Tomasz Stopa | IBM, Poland |
| Achim Streit | KIT, Germany |
| Barbara Strug | Jagiellonian University, Poland |
| Diana Suleimenova | Brunel University London, UK |
| Shuyu Sun | King Abdullah University of Science and Technology, Saudi Arabia |
| Martin Swain | Aberystwyth University, UK |
| Renata G. Słota | AGH University of Krakow, Poland |
| Tseden Taddese | UK Research and Innovation, UK |
| Ryszard Tadeusiewicz | AGH University of Krakow, Poland |
| Claude Tadonki | Mines ParisTech/CRI – Centre de Recherche en Informatique, France |
| Daisuke Takahashi | University of Tsukuba, Japan |
| Osamu Tatebe | University of Tsukuba, Japan |
| Michela Taufer | University of Tennessee, USA |
| Andrei Tchernykh | CICESE, Mexico |
| Kasim Terzic | University of St Andrews, UK |
| Jannis Teunissen | KU Leuven, Belgium |
| Sue Thorne | Hartree Centre – STFC, UK |
| Ed Threlfall | United Kingdom Atomic Energy Authority, UK |
| Vinod Tipparaju | AMD, USA |
| Pawel Topa | AGH University of Krakow, Poland |
| Paolo Trunfio | University of Calabria, Italy |
| Ola Tørudbakken | Meta, Norway |
| Carlos Uriarte | University of the Basque Country, BCAM – Basque Center for Applied Mathematics, Spain |
| Eirik Valseth | University of Life Sciences & Simula, Norway |
| Rein van den Boomgaard | University of Amsterdam, The Netherlands |
| Vítor V. Vasconcelos | University of Amsterdam, The Netherlands |
| Aleksandra Vatian | ITMO University, Russia |
| Francesc Verdugo | Vrije Universiteit Amsterdam, The Netherlands |
| Karin Verspoor | RMIT University, Australia |
| Salvatore Vitabile | University of Palermo, Italy |
| Milana Vuckovic | European Centre for Medium-Range Weather Forecasts, UK |
| Kun Wang | Imperial College London, UK |
| Peng Wang | NVIDIA, China |
| Rodrigo Weber dos Santos | Federal University of Juiz de Fora, Brazil |
| Markus Wenzel | Fraunhofer Institute for Digital Medicine MEVIS, Germany |

# Contents – Part III

## Advances in High-Performance Computational Earth Sciences: Numerical Methods, Frameworks and Applications

## Artificial Intelligence and High-Performance Computing for Advanced Simulations

# ICCS 2024 Main Track Short Papers

# Multitaper-Based Post-processing of Compact Antenna Responses Obtained in Non-anechoic Conditions

Mariusz Dzwonkowski[1,2] (ID), Adrian Bekasiewicz[1](✉) (ID), and Slawomir Koziel[1,3] (ID)

[1] Faculty of Electronics, Telecommunications and Informatics,
Gdansk University of Technology, Narutowicza 11/12, 80-233 Gdansk, Poland
bekasiewicz@ru.is
[2] Department of Radiology Informatics and Statistics, Faculty of Health Sciences,
Medical University of Gdansk, Tuwima 15, 80-210 Gdansk, Poland
[3] Department of Engineering, Reykjavik University, Menntavegur 1, 102 Reykjavík, Iceland

**Abstract.** The process of developing antenna structures typically involves prototype measurements. While accurate validation of far-field performance can be performed in dedicated facilities like anechoic chambers, high cost of construction and maintenance might not justify their use for teaching, or low-budget research scenarios. Non-anechoic experiments provide a cost-effective alternative, however the performance metrics obtained in such conditions require appropriate correction. In this paper, we consider a multitaper approach for post-processing antenna far-field characteristics measured in challenging, non-anechoic environments. The discussed algorithm enhances one-shot measurements to enable extraction of line-of-sight responses while attenuating interferences from multi-path propagation and the noise from external sources of electromagnetic radiation. The performance of the considered method has been demonstrated in uncontrolled conditions using a compact spline-based monopole. Furthermore, the approach has been favorably validated against the state-of-the-art techniques from the literature.

**Keywords:** Antenna calibration · data post-processing · non-anechoic measurements · radiation pattern · multitaper

## 1 Introduction

Antennas are traditionally validated in costly, specialized laboratories where strict control over the propagation environment is maintained in order to mitigate its effects on the measured performance of the prototype [1–4]. However, for budget-sensitive applications like training of students, underfunded research, or rapid iterative development conducted by start-up companies the use or lease of professional facilities might be economically impractical. Instead, the tests can be performed in in-door or outdoor conditions not intended for precise far-field experiments such as offices, hallways, courtyards, or parks [5]. Despite potential cost savings, the results obtained in such uncontrolled propagation conditions are of little to no use for interpretation of the antenna under test (AUT) real-world performance [5–7].

L. Franco et al. (Eds.): ICCS 2024, LNCS 14834, pp. 3–10, 2024.
https://doi.org/10.1007/978-3-031-63759-9_1

The challenges related to insufficient fidelity of non-anechoic measurements can be mitigated using appropriate post-processing methods. The latter ones belong to two main categories that include: (i) decomposition of transmitted signals and (ii) characterization of propagation environments [5–18]. Signal decomposition techniques boil down to extraction of the Line-of-Sight (LoS) transmission between the reference antenna (RA) and AUT based on time, or frequency domain analyses that employ suitable windowing kernels, or a truncated composition of basis functions [5–14]. The second group of methods focuses on extracting the effects of propagation environment on the RA-AUT system performance (e.g., based on comparative measurements conducted within the professional test sites) [15–17]. Their practical implementations include analyzing the AUT in multiple spatially separated locations [17], determining the noise floor [15], or estimating equivalent currents on a hull enclosing the radiator [18].

Unfortunately, the existing approaches face challenges that limit their applicability to measurements in truly uncontrolled environments. Issues include not only cognitive and problem-specific setup of correction parameters, but also validation using electrically large, high-gain (and hence featuring increased signal-to-noise ratio) antennas [16, 17]. Finally, the state-of-the-art methods are predominantly demonstrated in favorable propagation conditions compared to standard office rooms, such as anechoic chambers (ACs), semi-ACs, or even based on electromagnetic (EM) simulation environments [10–14]. From this perspective, the problem concerning reliable measurements of small antennas in non-anechoic conditions remains unsolved.

In this work, a framework for automatic correction of measurements performed in uncontrolled environments is considered. The method modifies the noise-distorted responses obtained in non-anechoic conditions using a series of locally optimal discrete prolate spheroidal sequences (DPSS) to enhance the relevant fraction of the RA-AUT impulse response while suppressing the noise and interferences. The post-processing is performed as a convex combination of the tapered responses. The performance of the algorithm is demonstrated based on four experiments conducted in a non-anechoic test site (a standard office room) and oriented towards evaluating far-field characteristics of a geometrically small spline-based monopole. The discussed method outperforms other window-function-based correction schemes while eliminating the need for expert knowledge in the course of algorithm setup. Potential applications of the considered antenna include in-door localization, or radiology with emphasis on microwave imaging devices dedicated to breast cancer detection [19, 20].

## 2  Methodology

In this section, a multitaper-based post-processing is discussed, starting with a formulation of the problem related to refining non-anechoic measurements. Subsequently, the correction procedure utilizing DPSS taper functions is outlined, followed by an explanation of the parametric tuning method applied to the considered framework.

### 2.1  Problem Formulation

Assume the set of uncorrected transmission responses (distorted by EM noise and multipath interferences) of the RA-AUT system is denoted as $R_u(\omega, \theta)$. The parameters $\omega$

$= [\omega_1 \dots \omega_k \dots \omega_K]^T$ and $\boldsymbol{\theta} = [\theta_1 \dots \theta_a \dots \theta_A]^T$ refer to frequency sweep around $f_0 = (\omega_K - \omega_1)/2$ and angular rotation of AUT w.r.t. RA, respectively. The assumed bandwidth around $f_0$ is $B = \omega_K - \omega_1$. The multitaper post-processing method entails spectral analysis of $\boldsymbol{R}_u(\omega, \boldsymbol{\theta})$ to obtain $\boldsymbol{R}_c^* = \boldsymbol{R}_c^*(f_0, \boldsymbol{\theta})$, i.e. the refined response of AUT at the frequency of interest $f_0$. The vector $\boldsymbol{R}_c^*$ can be interpreted as an estimate of the measurements carried out in an anechoic environment [1].

## 2.2 Multitaper Post-processing

The post-processing relies on spectral analysis of the signal, incorporating multiple modifications of the uncorrected signals using a composition of mutually orthogonal taper functions, i.e. DPSSs. The correction process is as follows [7].

Assuming a specific angle $\theta_a$ for RA-AUT system, let the time-domain response be denoted as $\boldsymbol{T}_u = \boldsymbol{T}_u(t, \theta_a) = F^{-1}(\boldsymbol{R}_u, N)$ and extracted via an inverse Fourier transform $F^{-1}(\cdot)$ with $N = 2^{\lceil log_2 K \rceil + 3}$ points from $\boldsymbol{R}_u = \boldsymbol{R}_u(\omega, \theta_a)$, where $\lceil \cdot \rceil$ denotes rounding up to the nearest integer. Additionally, let $t = [t_1, \dots, t_N]^T = \partial t \cdot \boldsymbol{M}$ be the sweep in time for $\partial t = (\omega_K - \omega_1)^{-1} \cdot (K-1)/(N-1)$ and $\boldsymbol{M} = [-N/2, \dots, N/2-2, N/2-1]^T$. Efficient noise isolation from the relevant fraction of the RA-AUT transmission can be achieved through multiple tapering of the segmented $\boldsymbol{T}_u$ with partial overlap of its subsequent intervals (segments) [21]. To segment $\boldsymbol{T}_u$ with an overlap, let $\boldsymbol{T}_s = \boldsymbol{T}_u(t_s, \theta_a)$ be the $i$-th segment of $\boldsymbol{T}_u$, such that $t_s = \partial t \cdot \boldsymbol{M}_s$, where $\boldsymbol{M}_s = \boldsymbol{M}_s(i) = [-N/2 + (i-1) \cdot s, \dots, -N/2 + (i-1) \cdot s + n-1]^T$. The parameter $i = 1, 2, \dots, \lfloor (N-n+s)/s \rfloor$ (where $\lfloor \cdot \rfloor$ indicates rounding down to the nearest integer) is determined by segment length $n = 1, 2, \dots, N-1$ and step length $s = 1, 2, \dots, n$, both defined in points. To achieve a reasonable $\boldsymbol{T}_u$ segmentation with a desirable overlap of consecutive $\boldsymbol{T}_u$ fractions, assume $1 \le s \le n \le N/2$.

The multiple tapering of segments from $\boldsymbol{T}_u$ requires the prior formulation of the specific DPSS functions. These can be obtained by finding all $n$-point finite energy sequences (eigenvectors) that maximize the spectral concentration ratio (corresponding eigenvalues) for the selected $2W$ bandwidth, where $W < 0.5\partial t^{-1}$ [21, 22]. The procedure of finding all DPSS components $\boldsymbol{T}_\kappa(t_s, w)$ for the $i$-th segment of $\boldsymbol{T}_u$ can be depicted as:

$$\sum_{\rho=0}^{n-1} \frac{\sin(2\pi(\gamma - \rho) \cdot W)}{\pi(\gamma - \rho)} \cdot \boldsymbol{T}_\kappa(t_s(\rho), w) = \lambda_w \cdot \boldsymbol{T}_\kappa(t_s(\gamma), w) \quad (1)$$

where $\gamma = 0, 1, \dots, n-1$. Parameter $w = 0, 1, \dots, n-1$ is the order of the identified $\boldsymbol{T}_\kappa(t_s, w)$ sequence for the spectral concentration ratio $\lambda_w$. Note that sequences of lower order have higher concentration ratios. The optimal number of DPSS tapers (with high enough $\lambda_w$) can be denoted as $w_{opt} = \lfloor 2t_{HB} \rfloor - 1$, where $t_{HB} = n \cdot \partial t \cdot W < n/2$ is a time-half-bandwidth product [22].

The identification of the DPSS functions enables modification of the segments from $\boldsymbol{T}_u$ using the obtained set of tapers, where each segment is modified using $w_{opt}$ functions. Subsequently, the set of $n$-point Fourier transforms of the individual tapered events are computed, and their convex combination, using $\lambda_w$ weights, is calculated to extract the frequency-domain response for each segment. Assuming the $i$-th segment $\boldsymbol{T}_s$, this task can be formulated as follows:

$$R_{cs}(\mathbf{\Omega}_s, \theta_a) = \sum_{w=0}^{w_{opt}-1} F(T_s \circ T_\kappa(t_s, w), n) \cdot \frac{\lambda_w}{\sum_{w'=0}^{w_{opt}-1} \lambda_{w'}} \qquad (2)$$

where $\circ$ is the component-wise multiplication. The vector $R_{cs}(\mathbf{\Omega}_s, \theta_a)$ represents the frequency-domain response for the $i$-th segment $T_s$, such that $\mathbf{\Omega}_s = \partial\omega \cdot M_s$ and $\partial\omega = (t_N - t_1)^{-1}$. Upon calculation, the frequency responses for all subsequent segments of $T_u$ can be concatenated (with an applied arithmetic mean for overlapping points) to yield the overall frequency-domain response $R_c(\mathbf{\Omega}, \theta_a)$. Here, $\mathbf{\Omega} = \partial\omega \cdot M_\omega$, and $M_\omega = [-N/2, ..., -N/2 + (\lfloor(N-n+s)/s\rfloor-1) \cdot s + n-1]^T$. The corrected response of the RA-AUT system in non-anechoic conditions, i.e. $R_c(f_0, \theta_a)$, is obtained at $f_0 \in \mathbf{\Omega}$ and $\theta_a$ angle. This procedure is repeated for all $\theta_a$ angles to determine $R_c(f_0, \theta)$.

### 2.3  Optimization of Multitaper Functions

The discussed multitaper post-processing is optimized using two parameters, i.e., segment length $n$ and step length $s$, both crucial for ensuring high correction performance when using DPSS functions. The optimal values for $n$ and $s$ (for the assumed $t_{HB} = 4$) are determined based on evaluations of a scalar objective function [7]:

$$U(x) = \sum \left(R_c(x) - \alpha R_f(f_0, \theta)\right) \qquad (3)$$

where $R_c(x)$ is the corrected response $R_c(f_0, \theta)$ obtained by utilizing the DPSS tapers in accordance to the vector of setup parameters $x = [x_1\ x_2]^T = [n\ s]^T$. $R_f = R_f(f_0, \theta)$ is the reference performance figure (here, the radiation pattern) obtained from simulations of the AUT EM model. The factor $\alpha = (R_f^T R_f)^{-1} R_f^T R_c(x)$ offers an analytical solution for a curve fitting problem that minimizes the difference between $R_c(x)$ and $R_f$ through appropriate scaling of the latter. The DPSS-based functional landscape resulting from (3) is highly multimodal which hinders its reliable optimization. Here, an exhaustive search oriented towards evaluation the objective function (3) for a set of designs $X = \{x_p\}_{1\leq p\leq P}$, followed by domination-based ranking of the obtained responses (so as to identify $X_{opt} \subset X$ that correspond to local minima) is performed. The final result $R_c^*$ is obtained as an average of solutions $R_c(x_\beta)$, $\beta = 1, 2, ...,$ and $x_\beta \in X_{opt}$, representing the $\sigma$-quantile of all locally optimal responses, where $\sigma = 0.1$.

## 3  Correction Results

The considered correction framework has been validated based on a total of four experiments conducted in a non-anechoic test site, i.e. an office room ($5.5 \times 4.5 \times 3.1$ m$^3$) that is considered not appropriate for far-field experiments (cf. Figure 1(a)). The AUT used in the experiments is a spline-based monopole depicted in Fig. 1(b), whereas the RA for the two-antenna system is a Vivaldi structure of Fig. 1(c). The test setup is configured with an angular resolution of 5°. The number of frequency points around $f_0$ is set to

$K = 201$ with bandwidth of $B = 3$ GHz. The AUT responses have been measured in the yz-plane at the specified frequencies of interest $f_0 \in \{4, 5, 6, 7\}$ GHz. For all of the considered experiments, the correction performance between the non-anechoic and AC-based measurements is quantified using the root-mean-square error, $e_R$ expressed in decibels [9].

The multitaper post-processing is independently applied at each $f_0$ and optimized according to the exhaustive search procedure outlined in Sect. 2.3. The correction process via DPSS taper functions is executed based on the set of test designs $X$ with the lower- and upper-bounds of $l_b = [101\ 0.1x_1]^T$ and $u_b = [901\ 0.9x_1]^T$, respectively. The $R_c(X)$ responses are ranked based on their corresponding $U(X)$ values and the final solution is determined as the average of $R_c(x_\beta)$ responses with the highest rank among the corrected measurements. The process involves the use of five designs, corresponding to the $\sigma$-quantile of the best locally optimal solutions found as a result of exhaustive search of Sect. 2.3 for $\sigma = 0.1$, and is repeated at all of the considered $f_0$ frequencies.

Figure 2 compares the radiation pattern characteristics of the antenna at the selected frequencies before and after multitaper-based post-processing. The obtained responses demonstrate a significant improvement of non-anechoic measurements fidelity due to correction. Additionally, the antenna performance characteristics before and after refinement, for all considered $f_0$ frequencies, are summarized in Table 1. The average improvement of the radiation pattern fidelity—expressed in terms of $\Delta = |e_R(R_c^*(f_0, \theta)) - e_R(R_u(f_0, \theta))|$—is nearly 13 dB. The maximum and minimum $\Delta$ changes are 15.4 dB at 4 GHz and 11 dB at 5 GHz, respectively.

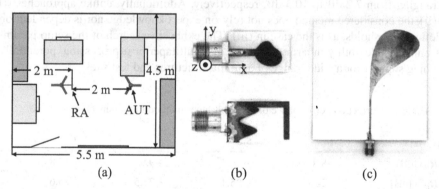

(a)                    (b)                    (c)

**Fig. 1.** A non-anechoic test site considered for experiments: (a) schematic view highlighting the location of rotary towers with tall (dark gray) and short (light gray) furniture, as well as photographs of (b) the spline-based monopole used as the AUT, and (c) the Vivaldi radiator used as the RA. Note that the considered antenna structures are not in scale.

# 4 Discussion and Comparisons

The discussed approach was compared against other state-of-the-art post-processing algorithms based on time-domain analysis. The selected benchmark methods—implemented for the same configuration parameters ($f_0$, $B$, $K$, $N$)—utilize: (i) manual estimation of RA-AUT distance (i.e., based on physical measurements of the distance) with

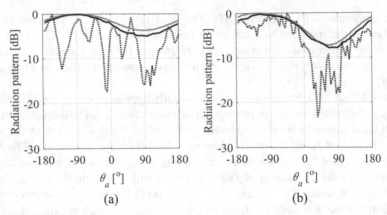

**Fig. 2.** Radiation patterns in yz-plane (cf. Fig. 1) obtained from AC (gray) and the non-anechoic measurements (black) before (···), and after correction (——) at: (a) 4 GHz and (b) 7 GHz frequencies.

a rectangular window function, (ii) cognition-based analysis of the impulse response with a Hann function, and (iii) modification of the interval identified based on relative thresholds using a composite window [8, 9, 15]. The correction performance, assessed by the averaged $e_R$ factor, for all considered methods is presented in Table 2. The considered multitaper approach consistently outperforms other methods (for the considered antenna and test frequencies), with improvements in the refined radiation characteristics ranging from 7.2 dB to 10.3 dB, respectively. Additionally, unlike approaches (i) and (ii), the considered method does not rely on expert knowledge nor is dependent on predefined thresholds, as is the case in (iii). The exhaustive search of the setup parameters enhances flexibility in terms of tuning the multitaper post-processing, potentially extending support for a wider range of antenna structures and test sites.

**Table 1.** Correction performance of the multitaper method in non-anechoic test site.

| $f_0$ [GHz] | 4 | 5 | 6 | 7 |
|---|---|---|---|---|
| $e_R(\boldsymbol{R}_u)$ [dB] | −8.8 | −14.1 | −9.9 | −12.8 |
| $e_R(\boldsymbol{R}_c{}^*)$ [dB] | −24.2 | −25.1 | −22.5 | −25.6 |
| $\Delta$ [dB] | 15.4 | 11.0 | 12.6 | 12.8 |

$\Delta = |e_R(\boldsymbol{R}_c{}^*(f_0, \boldsymbol{\theta})) - e_R(\boldsymbol{R}_u(f_0, \boldsymbol{\theta}))|$

**Table 2.** Correction performance of the chosen methods in non-anechoic test site.

| Root-mean-square error $e_R$, averaged over the frequencies of interest | | | |
|---|---|---|---|
| (i) | (ii) | (iii) | This work |
| $-14.0$ dB | $-15.9$ dB | $-17.1$ dB | $-24.3$ dB |

## 5 Conclusion

In this paper, we consider a multitaper post-processing framework for correcting antenna measurements performed in non-anechoic conditions. The method improves RA-AUT transmission obtained from one-shot experiments using mutually orthogonal DPSS taper functions that enable the extraction of the desirable part of the signal from interferences and noise. Demonstrated based on four measurements of a spline-based monopole structure in a non-anechoic test site, the method enhances the fidelity of the non-anechoic responses by an average of nearly 13 dB contrasted to uncorrected characteristics. It has been favorably compared to other techniques applicable to time-domain post-processing of far-field responses.

Future work will focus on enhancing the approach to enable automatic calibration of all relevant setup parameters and development of the ensemble method dedicated to further augment the information embedded within the one-shot measurements.

**Acknowledgments.** This work was supported in part by the National Science Centre of Poland Grant 2021/43/B/ST7/01856, National Centre for Research and Development Grant NOR/POLNOR/HAPADS/0049/2019-00, and Gdansk University of Technology (Excellence Initiative - Research University) Grant 16/2023/IDUB/IV.2/EUROPIUM.

**Disclosure of Interests.** The authors have no competing interests to declare that are relevant to the content of this article.

## References

1. Hemming, L.: Electromagnetic Anechoic Chambers: A fundamental Design and Specification Guide. IEEE Press, Piscataway (2002)
2. Kurokawa, S., Hirose, M., Komiyama, K.: Measurement and uncertainty analysis of free-space antenna factors of a log-periodic antenna using time-domain techniques. IEEE Trans. Instrum. Meas. **58**(4), 1120–1125 (2009)
3. Li, X., Chen, L., Wang, Z., Yang, K., Miao, J.: An ultra-wideband plane wave generator for 5G base station antenna measurement. Electronics **12**, 1824 (2022)
4. Zhang, F., Zhang, Y., Wang, Z., Fan, W.: Plane wave generator in non-anechoic radio environment. IEEE Antennas Wirel. Propag. Lett. **22**(12), 2896–2900 (2023)
5. Olencki, J., Waladi, V., Bekasiewicz, A., Leifsson, L.: A low-cost system for far-field non-anechoic measurements of antenna performance figures. IEEE Access. **11**, 39165–39175 (2023)

6. Piasecki, P., Strycharz, J.: Measurement of an omnidirectional antenna pattern in an anechoic chamber and an office room with and without time domain signal processing. In: Signal Processing Symposium, pp. 1–4, Debe, Poland (2015)
7. Dzwonkowski, M., Waladi, V., Bekasiewicz, A.: Multi-taper-based automatic correction of non-anechoic antenna measurements. Metrologia **61**(1), 1–9 (2024)
8. Loredo, S., Pino, M.R., Las-Heras, F., Sarkar, T.K.: Echo identification and cancellation techniques for antenna measurement in non-anechoic test sites. IEEE Antennas Propag. Mag. **46**(1), 100–107 (2004)
9. Soltane, A., Andrieu, G., Perrin, E., Decroze, C., Reineix, A.: Antenna radiation pattern measurement in a reverberating enclosure using the time-gating technique. IEEE Antennas Wirel. Propag. Lett. **19**(1), 183–187 (2020)
10. Fourestie, B., Altman, Z., Wiart, J., Azoulay, A.: On the use of the matrix-pencil method to correlate measurements at different test sites. IEEE Trans. Antennas Propag. **47**(10), 1569–1573 (1999)
11. Du, Z., Moon, J.I., Oh, S.-S., Koh, J., Sarkar, T.K.: Generation of free space radiation patterns from non-anechoic measurements using Chebyshev polynomials. IEEE Trans. Antennas Propag. **58**(8), 2785–2790 (2010)
12. Fiumara, V., Fusco, A., Iadarola, G., Matta, V., Pinto, I.M.: Free-space antenna pattern retrieval in nonideal reverberation chambers. IEEE Trans. Electromagn. Compat. **58**(3), 673–677 (2016)
13. Leon, G., Loredo, S., Zapatero, S., Las-Heras, F.: Radiation pattern retrieval in non-anechoic chambers using the matrix pencil algorithm. Prog. Electromagnet. Res. Lett. **9**, 119–127 (2009)
14. Fourestie, B., Altman, Z.: Gabor schemes for analyzing antenna measurements. IEEE Trans. Antennas Propag. **49**, 1245–1253 (2001)
15. Sao Jose, A.N., Deniau, V., Resende, U.C., Adriano, R.: Improving antenna gain estimations in non-ideal test sites with auto-tunable filters. Measurement **159**, 107720 (2020)
16. Koh, J., De, A., Sarkar, T.K., Moon, H., Zhao, W., Salazar-Palma, M.: Free space radiation pattern reconstruction from non-anechoic measurements using an impulse response of the environment. IEEE Trans. Antennas Propag. **60**(2), 821–831 (2012)
17. Froes, S.M., Corral, P., Novo, M.S., Aljaro, M., Lima, A.C.: Antenna radiation pattern measurement in a nonanechoic chamber. IEEE Antennas Wirel. Propag. Lett. **18**(2), 383–386 (2019)
18. Knapp, J., Kornprobst, J., Eibert, T.F.: Equivalent source and pattern reconstruction from oversampled measurements in highly reflective environments. IET Microwaves Antennas Propag. **13**(13), 2232–2241 (2019)
19. Woten, A., Lusth, J., El-Shenawee, M.: Interpreting artificial neural networks for microwave detection of breast cancer. IEEE Microwave Wirel. Compon. Lett. **17**(12), 825–827 (2007)
20. Lu, M., Xiao, X., Pang, Y., Liu, G., Lu, H.: Detection and localization of breast cancer using UWB microwave technology and CNN-LSTM framework. IEEE Trans. Microw. Theory Tech. **70**(11), 5085–5094 (2022)
21. Slepian, D.: Prolate spheroidal wave functions, Fourier analysis, and uncertainty – V: the discrete case. Bell Syst. Tech. J. **57**(5), 1371–1430 (1978)
22. Hristopulos, D.T.: Discrete Prolate Spheroidal Sequence. In: B.S. Daya Sagar et al. (eds.), Encyclopedia of Mathematical Geosciences. Encyclopedia of Earth Sciences Series, Springer, Cham (2021). https://doi.org/10.1007/978-3-030-26050-7_93-1

# Specification-Oriented Automatic Design of Topologically Agnostic Antenna Structure

Adrian Bekasiewicz[1]([✉]) [ID], Mariusz Dzwonkowski[1,2] [ID], Tom Dhaene[3] [ID],
and Ivo Couckuyt[3] [ID]

[1] Faculty of Electronics, Telecommunications and Informatics,
Gdansk University of Technology, Narutowicza 11/12, 80-233 Gdansk, Poland
bekasiewicz@ru.is
[2] Department of Radiology Informatics and Statistics, Faculty of Health Sciences,
Medical University of Gdansk, Tuwima 15, 80-210, Gdansk, Poland
[3] Department of Information Technology (INTEC), IDLab, Ghent University-imec, iGent,
Technologiepark-Zwijnaarde 126, 9052 Ghent, Belgium

**Abstract.** Design of antennas for modern applications is a challenging task that combines cognition-driven development of topology intertwined with tuning of its parameters using rigorous numerical optimization. However, the process can be streamlined by neglecting the engineering insight in favor of automatic determination of structure geometry. In this work, a specification-oriented design of topologically agnostic antenna is considered. The radiator is developed using a bi-stage algorithm that involves min-max classification of randomly-generated topologies followed by local tuning of the promising designs using a trust-region optimization applied to a feature-based representation of the structure frequency response. The automatically generated antenna is characterized by −10 dB reflection for over 600 MHz around the center frequency of 6.5 GHz and a dual-lobe radiation pattern. The obtained performance figures make the radiator of use for in-door positioning applications. The design method has been favorably compared against the frequency-based trust-region optimization.

**Keywords:** antenna design · numerical optimization · trust-region · topology-agnostic antenna · response features

## 1 Introduction

Design of modern antennas is an inherently cognitive process. It involves experience-driven development of topology followed by its tuning so as to fulfill the desired performance requirements [1–3]. When combined with robust optimization, this engineer-in-a-loop approach proved to be useful for the design of new (often unconventional) topologies [4, 5]. Although restricting the antenna development to a specific shape is considered pivotal for ensuring feasibility of the design process, it is also a subject to engineering bias and hence limits the potential in terms of achieving solutions characterized by unique (perhaps not expected) performance characteristics [4]. These might

L. Franco et al. (Eds.): ICCS 2024, LNCS 14834, pp. 11–18, 2024.
https://doi.org/10.1007/978-3-031-63759-9_2

include, for instance broadband/multiband behavior, but also improved radiation capabilities and/or small dimensions [3–5]. From this perspective, a streamlined design that rely only on numerical optimization methods represents an interesting alternative to the standard antenna development techniques [6].

Automatic antenna generation governed by numerical optimization is a challenging task. From the perspective of geometry, the design can be represented as a set of points (interconnected using, e.g., line-sections, or splines), or in the form of a binary matrix that defines the configuration of primitives (e.g., rectangles) constituting the antenna [6–11]. Point-based approaches are capable of supporting geometrically complex topologies. Owing to continuous and free-form nature of coordinates, "evolution" of geometry can be governed by standard numerical optimization algorithms [7, 9]. However, a large number of points and constraints on their distribution are required to ensure flexibility in terms of attainable topology and its consistency (here understood as lack of self-intersections of the coordinate-based curves) [7, 8]. Matrix-based methods naturally ensure consistency by representing topologies as compositions of partially overlapping primitives [6, 10]. Their main bottleneck is a large number of variables required even for relatively simple topologies. Besides activation/deactivation of primitives (based on contents of the matrix), their dimensions also need to be scaled which reveals a mixed-integer nature of the problem [10]. Finally, high dimensionality and the need to evaluate performance based on the expensive electromagnetic (EM) simulations make the discussed universal representations impractical for optimization when conventional algorithms—that require hundreds or even thousands of design evaluations to converge—are considered [4, 9, 10].

The problem related to unacceptable numerical cost can be mitigated using trust-region (TR) methods. TR is a class of techniques that enable iterative approximation of the desired solutions based on the optimization of data-efficient models [12, 13]. Regardless of proved usefulness, local character of the TR-based methods limit their applicability to improvement of already promising initial designs rather than exploration of the feasible space. Furthermore, due to complexity and multi-modal character of the problem, successful development of antennas using the outlined routines is subject to availability of appropriate response-processing mechanisms.

In this work, a framework for specification-oriented development of topologically agnostic antenna has been considered. The method involves classification of automatically generated topologies followed by TR-based tuning of the selected design in a setup where the frequency responses are represented in the form of carefully selected feature points so as to enable the algorithm convergence. The method has been used to design a 53-dimensional planar antenna represented using a set of points interconnected using line sections. The structure has been optimized so as to ensure reflection below $-10$ dB over a frequency range from 6.2 GHz to 6.8 GHz which corresponds to 5th channel of the ultra-wideband (UWB) spectrum. The antenna is also characterized by a dual-lobe radiation pattern and a relatively high gain of up to 7 dB. Owing to planar topology, the proposed radiator might be used as a component of in-door localization systems, or medical-imaging devices (e.g., for breast cancer detection). Comparison of the considered design approach against conventional, TR-based optimization over the frequency domain has also been considered.

## 2  Topologically Agnostic Antenna

Automatic specification-driven antenna development is subject to availability of a model that supports free-form adjustment of geometries. Consider a topology-agnostic planar antenna shown in Fig. 1. The structure is implemented on a substrate with permittivity and thickness of $\varepsilon_r = 2.55$ and $h = 1.52$ mm, respectively. It comprises a patch fed through a concentric probe. The radiator is represented using the interconnected points defined in a cylindrical coordinate system. To ensure feasibility of geometry each angular coordinate is set as relative to the previous one (the first is set to 0) and all angles are scaled w.r.t. their cumulative sum and normalized to the range between 0 and $2\pi$ (full circle). It should be noted that symmetry planes for the radiator are not defined so as to ensure its flexibility in terms of attainable performance specifications (even though this is achieved at the expense of increased computational cost associated with the size EM model discretization domain). Note that such a representation supports random generation of topologies. For the probe, an additional routine is used to ensure that it is enclosed within the generated shape.

The antenna is implemented in a CST Microwave Studio and evaluated using its time-domain simulator [14]. To ensure flexibility in terms of the number of design parameters, the structure EM model is generated dynamically and implements a safeguard mechanisms that verify consistency of the geometry and handle errors [14]. On average, the EM models are discretized using 100,000 tetrahedral mesh cells and the cost of single evaluation amounts to 60 s.

The antenna geometry is represented using the following vector of design parameters $x = [C\ \rho_f\ \varphi_f\ \rho\ \varphi]^T$, where $C$ is a scaling coefficient for the patch/probe coordinates, $\rho_f$ and $\varphi_f$ represent the radial and angular position of the feed w.r.t. the origin of the cylindrical system, whereas the vectors $\rho = [\rho_1\ ...\ \rho_l]^T$ and $\varphi = [\varphi_1\ ...\ \varphi_l]^T$ ($l = 1, ..., L$) comprise the points that represent the shape of the patch (cf. Figure 1). The parameters $o = 5$, $r_1 = 1.27$, and $r_2 = 2.84$ remain fixed, whereas $A = B = 2(C \cdot \max(\rho) + o)$. Note that the specific values of $r_1$ and $r_2$ ensure 50 $\Omega$ input impedance of the feed. All dimensions except for the fixed ones and $C$ (in mm) are unit-less. The antenna topology is considered feasible within the following lower and upper design bounds: $l = [25\ 0\ \varphi_{0.l}\ 0.1\mathbf{1}\ 0.01\mathbf{1}]^T$ and $u = [35\ \rho_{0.f}\ \varphi_{0.h}\ 0.9\mathbf{1}\ 0.8\mathbf{1}]$T, where $\mathbf{1}$ is an $L$-dimensional vector of ones; $\varphi_{0.l} = \varphi^{(0)} - \pi$, $\varphi_{0.h} = \varphi^{(0)} + \pi$, and $\rho_{0.f} = \max(\rho^{(0)})$ are the constraints determined based on the randomly generated design $x^{(0)}$ considered for numerical optimization based on the classification (cf. Sect. 3). Note that the vector $x$ comprises $D = 2L + 3$ design parameters.

## 3  Design Methodology

### 3.1  Problem Formulation

Let $R(x) = R(x, f)$ be the antenna reflection response obtained over the frequency sweep $f$ for the vector of input parameters $x$. The design problem can be formulated as the following non-linear minimization task:

$$x^* = \arg\min_{x \in X}(U(R(x)))\tag{1}$$

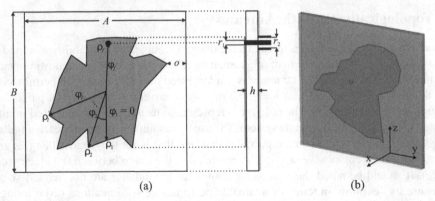

**Fig. 1.** The proposed topologically agnostic antenna: (a) geometry of the structure with highlight on the design parameters and (b) visualization of the structure optimized using the algorithm of Sect. 3. Note that $\rho_f' = C\rho_f$ and $\rho_l' = C\rho_l$ ($l = 1, ..., L$).

where $x^*$ is the optimal design to be found within the feasible region of the search space $X$ and $U$ is a scalar objective function. To enable unsupervised design of topology-agnostic antennas the problem (1) is solved using a bi-stage procedure where the promising candidates are first identified and then optimized using a gradient algorithm embedded within a TR framework.

The problem concerning automatic determination of the promising initial design is addressed through a random generation of the candidate solutions followed by their sorting based on a simple min-max classifier. Let $X_r = \{x_n\}$, $n = 1, ..., N$, ($X_r \subset X$) be an $N$-element set of random designs that represent the geometrically-flexible antenna and $\{R_n(x_n)\}$ be the set of their corresponding EM-based responses. Each of the obtained designs can be evaluated as:

$$E_n = \max(R(x_n))_{f_l \le f \le f_H} \tag{2}$$

where $f \in f$ represent the frequency points within the bandwidth from $f_l$ to $f_h$. Upon evaluation, the responses $E_1, ..., E_N$ are sorted according to their values and the design $x_k \in X_r$ that corresponds to the solution featuring the lowest value is selected as a starting point for local TR-based optimization.

## 3.2   Feature-Based Representation of Responses

The main challenge pertinent to specification-oriented design of topology-agnostic structures is that due to intricate relations between the input parameters, the resulting responses are non-linear functions of frequency. The problem can be mitigated by shifting the design problem into a feature-based domain where the key properties of the frequency response (from the optimization perspective) are represented using a set of carefully selected points [15].

Let $F(x) = P(R(x))$ be the antenna response expressed in terms of the feature points, where $P$ is the function used for their extraction. The feature-based response is defined as $F(x) = [\omega\ S]^T$, where $\omega = [\omega_1\ ...\ \omega_q]^T$ and $S = [S_1\ ...\ S_q]^T$ ($q = 1, ..., Q$) represent the specific points related to frequencies of interest and response levels (cf. Figure 2(a)). In other words, a pair of points $\omega_\gamma$, $S_\gamma$ ($\gamma \leq Q$) represents a specific frequency (e.g., pertinent to a local minimum of the response) and its corresponding reflection level, whereas $\omega_\kappa$, $S_\kappa$ ($\kappa \leq Q$; $\kappa \neq \gamma$) refers to the frequency point at a specific pre-determined reflection level (e.g., at the edge of the bandwidth). Note that although the pair of points is stored, one typically requires a specific component of the pair (e.g., pertinent to resonant frequency, or level at a local maximum). Compared to the original response, its feature-based representation is a much less-nonlinear function of input parameters and hence aids the design optimization process (especially when performed using linear approximation models). The outlined concept is illustrated in Fig. 2. For more comprehensive discussion on the feature-based representation of frequency responses, see [15, 16].

## 3.3 Optimization Algorithm

The antenna is optimized using a gradient-based algorithm embedded in a trust-region framework. The latter generates approximations, $i = 0, \ldots$, to the final design by solving [12]:

$$x^{(i+1)} = \arg\min_{\|x - x^{(i)}\| \leq \delta} \left( U\left( G_\varepsilon^{(i)}(x) \right) \right) \qquad (3)$$

The composite model $G_\varepsilon^{(i)} = [G_\omega^{(i)} \ G_S^{(i)}]^T$, where $G_\omega^{(i)} = \omega(x^{(i)}) + J_\omega(x^{(i)})(x - x^{(i)})$ and $G_S^{(i)} = S(x^{(i)}) + J_S(x^{(i)})(x - x^{(i)})$ represents the first-order surrogates generated around $x^{(i)}$ w.r.t. the frequency- and level-related features. The Jacobians are based on a large-step finite differences (FD) [15]:

$$J_\omega\left(x^{(i)}\right) = \left[ \cdots \left( \omega\left(x^{(i)} + p_d^{(i)}\right) - \omega(x^{(i)}) \right) \frac{1}{p_d^{(i)}} \cdots \right]^T$$
$$J_S\left(x^{(i)}\right) = \left[ \cdots \left( S\left(x^{(i)} + p_d^{(i)}\right) - S(x^{(i)}) \right) \frac{1}{p_d^{(i)}} \cdots \right]^T \qquad (4)$$

Note that the parameter $p_d^{(i)}$ ($d = 1, \ldots, D$) denotes the FD perturbation w.r.t. $d$th dimension of the (currently best) design $x^{(i)}$, whereas the components of vector $p_d^{(i)}$ are set to zero for all dimensions except $d$th which is equal to $p_d^{(i)}$. The FD steps are selected in proportion to the design $x^{(i)}$ and are updated after each successful iteration [17]. The radius $\delta$ is controlled based on the ratio $\rho = [U(F(x^{(i+1)})) - U(F(x^{(i)}))]/[U(G_\varepsilon^{(i)}(x^{(i+1)})) - U(G_\varepsilon^{(i)}(x^{(i)}))]$. The initial radius is set to $\delta = 1$. When $\rho < 0.25$ (poor prediction of $G_\varepsilon^{(i)}$) then $\delta = \delta/3$, whereas for $\rho > 0.75$, $\delta = 2\delta$ [15]. The gain coefficient $\rho$ is also used to accept or reject the candidate designs obtained from (3). The algorithm is terminated when $\delta^{(i+1)} < \varepsilon$, or $\|x^{(i+1)} - x^{(i)}\| < \varepsilon$ (here, $\varepsilon = 10^{-3}$). For more detailed discussion on TR-based optimization of problems represented in terms of response features, see [12, 13].

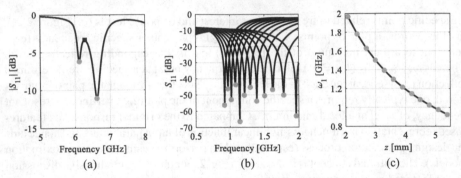

**Fig. 2.** Extraction of the response features: (a) level $S_1$ (■) at the local maximum and frequencies $\omega_1$ and $\omega_2$ (●) at a local minimum, (b) family of frequency responses obtained as a function of $z$, and (c) feature based response that corresponds to resonant frequencies along $z$. Note less non-linear changes of feature coordinates compared to frequency characteristics.

## 4  Numerical Results

The proposed topology-agnostic antenna is optimized using the method of Sect. 3. The initial design $x^{(0)} = [30\ 0.13\ 2.62\ 0.22\ 0.36\ 0.48\ 0.57\ 0.59\ 0.56\ 0.53\ 0.44\ 0.34\ 0.36\ 0.35$ $0.43\ 0.52\ 0.54\ 0.38\ 0.29\ 0.43\ 0.42\ 0.42\ 0.47\ 0.5\ 0.57\ 0.41\ 0.29\ 0.22\ 0\ 0.17\ 0.18\ 0.26\ 0.26$ $0.26\ 0.29\ 0.25\ 0.3\ 0.29\ 0.42\ 0.3\ 0.31\ 0.05\ 0.05\ 0.31\ 0.13\ 0.43\ 0.36\ 0.31\ 0.29\ 0.04\ 0.14$ $0.3\ 0.6]^T$ is selected from a set of 200 randomly generated candidate solutions evaluated using the min-max classifier (2). The objective function for TR-based optimization (note that FD perturbations are set as 2% of $x^{(i)}$) is given as:

$$U(x) = \sum \left( \max \left( S_{\{1,2\}}(x) - S_t, 0 \right)^2 \right) + \beta \left\| \begin{matrix} \omega_3(x) - \omega_{t.1} \\ \omega_4(x) - \omega_{t.2} \end{matrix} \right\| \tag{5}$$

The function (5) is calculated using four feature points: $S_1$ and $S_2$ refer to the response levels at local maxima within the bandwidth of interest, whereas $\omega_3$ and $\omega_4$ denote the local minima of the response that are close to the target frequencies $\omega_{t.1} = 6.2$ GHz and $\omega_{t.2} = 6.8$ GHz. The threshold on the reflection is $S_t = -10.2$ dB, whereas $\beta = 100$ (cf. Figure 2). The optimized design $x^* = [30.04\ 0.13\ 2.47\ 0.21\ 0.36\ 0.47\ 0.56\ 0.61$ $0.58\ 0.52\ 0.46\ 0.34\ 0.36\ 0.31\ 0.44\ 0.53\ 0.55\ 0.38\ 0.33\ 0.44\ 0.41\ 0.42\ 0.49\ 0.52\ 0.57$ $0.41\ 0.28\ 0.23\ 0.01\ 0.16\ 0.17\ 0.25\ 0.26\ 0.27\ 0.29\ 0.24\ 0.3\ 0.29\ 0.42\ 0.29\ 0.29\ 0.08$ $0.07\ 0.32\ 0.13\ 0.43\ 0.36\ 0.31\ 0.29\ 0.04\ 0.15\ 0.32\ 0.61]^T$ is found after 10 TR iterations (total cost: 221 EM simulations; see Fig. 1 for antenna visualization). It features −10 dB reflection within the operational range from 6.12 GHz to 6.84 GHz. The response at the initial and optimized designs, as well as the radiation pattern (in xz-plane) at 6.5 GHz— center frequency for 5th channel of UWB spectrum—are shown in Fig. 3. The antenna is characterized by a dual-lobe far-field response with local minimum and maximum at 0° and 35° angles, respectively. The realized gain for the latter is around 7 dB. It is worth noting that the obtained unorthodox radiation pattern is a by-product of the reflection-oriented optimization of the structure at hand.

The characteristics make the antenna suitable for installation in convex corners within the facilities dedicated for in-door monitoring using UWB-based systems [8]. On a

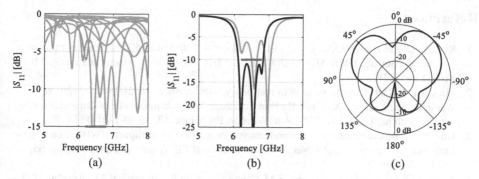

**Fig. 3.** Optimization of topology-agnostic antenna: (a) a few of randomly selected designs and bandwidth of interest for the min-max classifier (red line), (b) frequency responses at $x^{(0)}$ and $x^*$, as well as (c) xz-plane (cf. Figure 1) radiation pattern at 6.5 GHz.

conceptual level, application of the structure in a heterogeneous localization architecture (with antennas tailored to conditions) could reduce the overall cost of UWB system installation.

The antenna of Sect. 2 has also been optimized using a standard TR-based algorithm with min-max design objective given as $U(x) = \max(R(x))_{\omega t.1 \leq f \leq \omega t.2}$. The benchmark method has been terminated after 7 iterations due to lack of $U(x)$ improvement with the final solution being the same as an initial design $x^{(0)}$. The obtained results suggest that, for the considered design problem, feature-based optimization is an useful tool for determination of satisfactory antenna geometries.

## 5 Conclusion

In this work, a bi-stage specification-oriented design of topology-agnostic planar antenna has been considered. The method involves identification of the starting point from a set of randomly generated designs using a classifier based on a simple min-max metric followed by its local tuning in a feature-based trust-region framework. The 53-dimensional topology has been generated at a cost of 421 EM simulations. The final design is characterized by $-10$ dB reflection within 6.2 GHz to 6.8 GHz band and a dual-lobe radiation pattern with minima in the direction perpendicular to the radiator and a maxima at the $35°$ angle. The structure is applicable for in-door positioning in challenging propagation conditions. The considered feature-enhanced TR design has been favorably benchmarked against a standard TR optimization performed in the frequency-domain. Future work will focus on implementation of the radiator as a component of the localization system.

**Acknowledgments.** This work was supported in part by the National Science Centre of Poland Grant 2021/43/B/ST7/01856 and Gdansk University of Technology (Excellence Initiative - Research University) Grant 16/2023/IDUB/IV.2/EUROPIUM.

**Disclosure of Interests.** The authors declare no conflict of interests.

# References

1. Koziel, S., Bekasiewicz, A., Cheng, Q.S.: Conceptual design and automated optimization of a novel compact UWB MIMO slot antenna. IET Microwaves Antennas Propag. **11**(8), 1162–1168 (2017)
2. Mroczka, J.: The cognitive process in metrology. Measurement **46**, 2896–2907 (2013)
3. Wang, L., Xu, L., Chen, X., Yang, R., Han, L., Zhang, W.: A compact ultrawideband diversity antenna with high isolation. IEEE Ant. Wireless Prop. Lett. **13**, 35–38 (2014)
4. Koziel, S., Bekasiewicz, A.: Comprehensive comparison of compact UWB antenna performance by means of multiobjective optimization. IEEE Trans. Antennas Propag. **65**(7), 3427–3436 (2017)
5. Khan, M.S., Capobianco, A., Asif, S.M., Anagnostou, D.E., Shubair, R.M., Braaten, B.D.: A compact CSRR-enabled UWB diversity antenna. IEEE Antennas Wirel. Propag. Lett. **16**, 808–812 (2017)
6. Jacobs, J.P.: Accurate modeling by convolutional neural-network regression of resonant frequencies of dual-band pixelated microstrip antenna. IEEE Antennas Wirel. Propag. Lett. **20**(12), 2417–2421 (2021)
7. Ghassemi, M., Bakr, M., Sangary, N.: Antenna design exploiting adjoint sensitivity-based geometry evolution. IET Microwaves Antennas Propag. **7**, 268–276 (2013)
8. Bekasiewicz, A.: Optimization of the hardware layer for IoT systems using a trust-region method with adaptive forward finite differences. IEEE Internet Things J. **10**(11), 9498–9512 (2023)
9. Whiting, E.B., Campbell, S.D., Mackertich-Sengerdy, G., Werner, D.H.: Meta-atom library generation via an efficient multi-objective shape optimization method. IEEE Open J. Antennas Propag. **28**, 24229–24242 (2020)
10. Alnas, J., Giddings, G., Jeong, N.: Bandwidth improvement of an inverted-f antenna using dynamic hybrid binary particle swarm optimization. Appl. Sci. **11**, 2559 (2021)
11. Capek, M., Jelinek, L., Gustafsson, M.: Shape synthesis based on topology sensitivity. IEEE Trans. Antennas Propag. **67**(6), 3889–3901 (2019)
12. Conn, A., Gould, N.I.M., Toint, P.L.: Trust-region methods. MPS-SIAM Series on Optimization, Philadelphia (2000)
13. Koziel, S., Pietrenko-Dabrowska, A.: Performance-based nested surrogate modeling of antenna input characteristics. IEEE Trans. Antennas Propag. **67**(5), 2904–2912 (2019)
14. CST Microwave Studio, Dassault Systems, 10 rue Marcel Dassault, CS 40501, Vélizy-Villacoublay Cedex, France (2015)
15. Pietrenko-Dabrowska, A., Koziel, S.: Response Feature Technology for High-Frequency Electronics. Optimization, Modeling, and Design Automation. Springer, New York (2023). https://doi.org/10.1007/978-3-031-43845-5
16. Koziel, S., Bekasiewicz, A.: Expedited simulation-driven design optimization of UWB antennas by means of response features. Int. J. RF Microwave Comput. Aided Eng. **27**(6), 1–8 (2017)
17. Mathur, R.: An analytical approach to computing step sizes for finite-difference derivatives. PhD Thesis, University of Texas, Austin (2012)

# Accurate Post-processing of Spatially-Separated Antenna Measurements Realized in Non-Anechoic Environments

Adrian Bekasiewicz[1](✉) ⓘ, Vorya Waladi[1] ⓘ, Tom Dhaene[2] ⓘ,
and Bartosz Czaplewski[1] ⓘ

[1] Faculty of Electronics, Telecommunications and Informatics,
Gdansk University of Technology, Narutowicza 11/12, 80-233 Gdansk, Poland
bekasiewicz@ru.is
[2] Department of Information Technology (INTEC), IDLab, Ghent University-imec, iGent,
Technologiepark-Zwijnaarde 126, 9052 Ghent, Belgium

**Abstract.** Antenna far-field performance is normally evaluated in expensive laboratories that maintain strict control over the propagation environment. Alternatively, the responses can be measured in non-anechoic conditions and then refined to extract the information on the structure field-related behavior. Here, a framework for correction of antenna measurements performed in non-anechoic test site has been proposed. The method involves automatic synchronization (in time-domain) of spatially separated measurements followed by their combination so as to augment the fraction of the signal that represents the antenna performance while suppressing the interferences. The method has been demonstrated based on six experiments performed in an office room. The performance improvement due to proposed post-processing amounts to 9.4 dB, which is represents up to over 5 dB improvement compared to the state-of-the-art methods.

**Keywords:** Data post-processing · anechoic chamber · antenna measurements · non-anechoic experiments · time-domain analysis

## 1 Introduction

Experimental validation is an essential step in the development of antenna structures [1, 2]. Far-field responses are extracted from measurements of the two-port system that constitutes the antenna under test (AUT), reference antenna (RA) and the wireless medium (air) between them. The experiments are normally performed in laboratories such as anechoic chambers (ACs), or compact-range test sites that maintain strict control over propagation environment [3, 4]. Although capable of ensuring certification-grade accuracy, professional facilities are utterly expensive. Their high cost might not be justified for applications such as teaching, or budget-constrained research [5, 6]. Alternatively, tests can be performed in uncontrolled environments. Unfortunately, interferences and external electromagnetic (EM) noise make non-anechoic experiments useless for drawing conclusions on the AUT performance [7].

© The Author(s), under exclusive license to Springer Nature Switzerland AG 2024
L. Franco et al. (Eds.): ICCS 2024, LNCS 14834, pp. 19–27, 2024.
https://doi.org/10.1007/978-3-031-63759-9_3

The quality of far field responses obtained in uncontrolled environments can be improved using appropriate post-processing mechanisms. The available techniques fall into two main categories that include: (i) decomposition of noisy measurements and (ii) characterization of propagation environment [7–11]. The former involves extracting the fraction of the signal that corresponds to the line-of-sight (LoS) transmission within the RA-AUT system. This task can be realized in through conversion of the frequency measurements to time-domain followed by modification of the impulse response using appropriately tailored window functions [8, 9]. Alternatively, the frequency responses can be approximated using a composition of basis functions in the form of complex exponentials, or Chebyshev polynomials [11, 12]. The second class of methods involves comparative analyses of multiple measurements performed either in different locations within the test site, or characterized by temporal separation [9, 10]. Another approach includes characterization of the test site before and after introduction of obstacles using a suitable probe antenna [11].

The discussed post-processing methods are subject to challenges related to operational conditions, setup, and reliability. The algorithms are predominantly validated in idealized conditions that either include measurements in fully-featured ACs (yet with installed reflective surfaces), or in semi-anechoic chambers [8, 11, 12]. Contrary to non-anechoic test sites (e.g., office spaces, hallways)—not tailored to far-field experiments—the mentioned facilities are optimized for suppression of interferences and hence represent much less demanding propagation environments [7]. Furthermore, the performance of available signals correction routines is subject to algorithm-specific control parameters which are determined manually based on engineering-experience [13]. Mentioned simplifications and cognition-based adjustment procedures challenge the applicability of existing post-processing techniques to routine experiments.

In this work, a framework for correction of far-field measurements performed in challenging non-anechoic environments has been proposed. The method involves synchronization of spatially-separated responses in time-domain w.r.t. fraction of the signals that correspond to LoS transmission followed by their combination oriented towards augmenting the relevant part of the response while suppressing the noise and interferences. The main contribution of the work include automatic synchronization and correction of the non-anechoic data using the presented framework. The presented approach has been demonstrated based on six experiments (a total of 30 measurements) performed in a non-anechoic environment—in the form of a standard office room—using a geometrically small, spline-parameterized Vivaldi antenna [14]. A comparison of the proposed framework against the state-of-the-art methods from the literature has also been provided.

## 2  Post-processing Method

### 2.1  Problem Formulation

Let $R_p = R_p(\omega, \theta)$ be a frequency-domain matrix of uncorrected transmission responses obtained for the RA-AUT system within the non-anechoic site at $p$th test setup, $p = 1, 2, \ldots, P$, (i.e., the specific location of the antennas). The exact distance and/or position of the radiators are not important as they will be automatically extracted from the impulse responses in the course of post-processing. The vector $\boldsymbol{\omega} = [\omega_1 \ldots \omega_k]^T$ denotes the $k$-point sweep ($k = 1, \ldots, K$) defined over a bandwidth $B = \omega_K - \omega_1$ around the frequency of interest $f_0 = 0.5B$, whereas $\boldsymbol{\theta} = [\theta_1 \ldots \theta_a]^T$ ($a = 1, \ldots, A$) represents the angular positions of AUT w.r.t. RA. The goal of the method is to extract the information from all $\boldsymbol{R} = [\boldsymbol{R}_1 \ldots \boldsymbol{R}_p]^T$ non-anechoic measurements to obtain the corrected response $\boldsymbol{R}_c(f_0, \boldsymbol{\theta})$, which approximates the AC-based radiation pattern.

## 2.2 LoS Delay Extraction

Precise determination of LoS (i.e., the transmission delay on the shortest RA-AUT path) is crucial for time-domain-based synchronization of the $\boldsymbol{R}_p$ measurements. The latter is crucial to augment the useful part of the response while suppressing the interferences and noise from the external EM sources [14]. Note that the LoS delays have to be determined for all $\theta_a$ angles as the AUT rotation during measurements (required for extraction of radiation patterns) results in slight change of its distance from the RA and hence might affect the correction performance (cf. Fig. 1(a)). To enable post-processing, the LoS profiles have to be extracted for all $P$ measurements.

Let $\boldsymbol{T}_p = \boldsymbol{T}_p(t, \theta_a) = F^{-1}(\boldsymbol{R}_p(\omega, \theta_a), N)$ be the complex time-domain response extracted from the frequency measurements using Fourier transform ($F^{-1}(\cdot)$), where $t = [t_1, \ldots, t_N]^T = \partial t \cdot \boldsymbol{M}$ with $\partial t = 1/B$, $\boldsymbol{M} = [-N/2, \ldots, N/2-2, N/2-1]^T$, and $N = 2^{\lceil \log 2(K) \rceil + 3}$ ($\lceil \cdot \rceil$ is a round-up to the nearest integer) [15, 16]. The power response $\boldsymbol{P}_p = \boldsymbol{T}_p \circ \boldsymbol{T}_p^*$ ("$\circ$" and "$*$" denote the component-wise product and Hermitian transpose) is used to seek for the RA-AUT maxima that correspond to the LoS delay.

The profile extraction is performed iteratively. Let $\boldsymbol{d}_p{}^{(j)} = [d_{p.1}{}^{(j)} \ldots d_{p.a}{}^{(j)}]^T$ be a vector of delays at the $j$th step of the algorithm. For simplicity and compactness of notation, $\boldsymbol{d}_p{}^{(j)}$ and $d_{p.a}{}^{(j)}$ will be referred to as $\boldsymbol{d}^{(j)}$ and $d_a{}^{(j)}$, respectively. In the first step ($j = 1$), the vector $\boldsymbol{d}^{(j)}$ is obtained from:

$$d_a^{(j)} = \underset{d_a \in t : d_l^{(j)} \leq d_a \leq d_h^{(j)}}{\arg\max} \left( P_p(t, \theta_a) \right) \tag{1}$$

where $d_l{}^{(1)} = 0$ and $d_h{}^{(1)} = \partial t \cdot (N/2-1)$. Due to challenging conditions in non-anechoic environments, the delays obtained from (1) might be inaccurate (cf. Fig. 1(b)) [14]. To mitigate the problem, the LoS profile is re-set ($j = 2$) using (1), yet with bounds confined to an interval defined around $d_{\min} = \min(\boldsymbol{d}^{(1)})$ such that $d_l{}^{(2)} = d_{\min} - \gamma \cdot w_{\min}$ and $d_h{}^{(2)} = d_{\min} + \gamma \cdot w_{\min}$. The parameters $w_{\min}$ and $\gamma$ represent half-prominence of $\boldsymbol{P}_p$ (i.e., width of pulse at half of its height) at the RA-AUT angle that corresponds to $d_{\min}$ and the range factor (here, $\gamma = 3$) [14]. It should be noted that the profile $\boldsymbol{d}^{(2)}$ might still be a subject to distortions resulting from the external noise (more-or-less dependent on the specific $\theta_a$), as well as numerical errors resulting from the aliasing effects and zero-padding of the responses with $N > K$ (cf. Fig. 1(b)) [13, 15]. These are accounted for through modification of the LoS-profile using a combination of a median filter and a smoothing function in the form of a moving average ($j = 3$). The former is implemented as:

$$d_a^{(j)} = \begin{cases} \left(d_{a-1}^{(j-1)} + d_{a+1}^{(j-1)}\right)/2, & \text{when } d_a^{(j-1)} \leq M\left(d^{(j-1)}\right) - 2\sigma\left(d^{(j-1)}\right) \\ \left(d_{a-1}^{(j-1)} + d_{a+1}^{(j-1)}\right)/2, & \text{when } d_a^{(j-1)} \geq M\left(d^{(j-1)}\right) + 2\sigma\left(d^{(j-1)}\right) \\ d_a^{(j-1)}, & \text{otherwise} \end{cases} \quad (2)$$

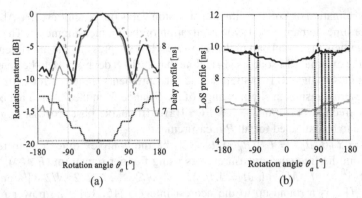

**Fig. 1.** LoS extraction: (a) AC-based (– –) vs. non-anechoic measurements (——) corrected using variable (black) and constant (gray) profiles (···), as well as (b) LoS profiles at $j = 1$ (···), $j = 2$ (– –) and $j = 3$ (——) for RA-AUT distance of around 1.5 m (gray) and 2.5 m (black). Note that a large peaks around $\theta_a = \pm 90°$ are due to incorrect identification of the LoS signals.

Note that $M(\cdot)$ and $\sigma(\cdot)$ denote a median and a standard deviation. The goal of the filter (2) is to alter the LoS profile at the angles that deviate from the expected (i.e., relatively small) changes of the delay. The threshold is set to two standard deviations according to the median of the $d^{(2)}$. Finally, a moving-average-based smoothing is performed to obtain the final profile $d_p{}^* = d^* = d^{(3)}$ which can be used for synchronization of the measurements performed in different setups.

### 2.3  Spatially-Enhanced Post-processing

The proposed spatially-enhanced post-processing algorithm has been conceptually explained in Fig. 2. The method involves synchronization of the individual $R_p$ measurements based on their corresponding $d_p{}^*$ LoS profiles followed by multiplication of the modified responses so as to augment the useful part of the transmitted signals while suppressing the noise and multi-path interferences.

Let $d_r$, $r \in \{1, 2,.., P\}$, be the LoS profile extracted from the measurements in the $R_r$ setup, i.e., where the RA-AUT delay is the shortest. Then, let $\delta_p = d_p{}^* - d_r$ be the vector that represents the shift between LoS transmission for spatially-separated setups (for $p = r$, $\delta_p = 0^T$, where $0$ is the $A$-dimensional vector of zeros). Now let $t_{p.a} = t - \delta_{p.a}$ represent the time-domain sweep at the $a$th angle shifted relative to $d_{r.a} \in d_r$ in order to synchronize the delays between measurements (note that $t_{r.a} = t$). The signal $T_c$ is obtained as a component-wise product of the synchronized responses:

$$T_c(t, \theta_a) = \prod_{p=1}^{P} T_p(t_{p.a}, \theta_a) \tag{3}$$

When compared to $T_p(t, \theta_a)$, the $T_p(t_{p.a}, \theta_a)$ represents the same—yet artificially shifted in time—response (cf. Fig. 2). The consequence of stacking the spatially-separated measurements using (3) is amplification of similar (albeit slightly distorted by external noise) fractions of the responses that coincide with LoS, while suppressing the weaker and non-synchronized (due to varying distances to reflective surfaces, and/or changing directions of noise sources) interferences. Note that multiplication of the spatially-separated measurements is performed using the complex impulse responses rather than the power in time-domain. The latter is used only for identification of the peaks, due their much less dynamic and "smoother" changes as a function of time, which aid accurate identification of the LoS profile [14, 16].

In the next step, the composite impulse response $T_c$ is converted to the frequency spectrum using an $N$-point Fourier transform $R_c(\Omega, \theta_a) = F(T_c(t, \theta_a), N)$. The vector of frequency points $\Omega = [\Omega_1 \, \Omega_2 \, ... \, \Omega_N]^T = \partial\omega \cdot M - B + f_0$ with $\partial\omega = 1/(t_N - t_1)$. The useful fraction of the corrected response is reconstructed as $R_c(\omega_c, \theta_a)$, where $\omega_c = [\Omega_1 \, \Omega_{P+1} \, \Omega_{2P+1} \, ... \, \Omega_{(K-1)P+1}]^T$. The procedure is repeated for all $A$ angles in order to extract $R_c(\omega_c, \theta)$. The corrected response based on the spatially-separated non-anechoic measurements at $f_0 \in \omega_c$ is represented as $R_c(f_0, \theta)$.

## 3 Results

### 3.1 Test Setup and Antenna Structure

The proposed post-processing procedure has been validated in a standard $6.7 \times 5.2 \times 3.1$ m$^3$ office room that has not been tailored to far-field measurements. The considered experiments involved measurements of the spline-parameterized antipodal Vivaldi antenna—used as both the AUT and RA—radiation patterns (in yz-plane) for five ($P = 5$) spatially-separated setups at a total of six frequencies of interest (cf. Fig. 3(a)). The obtained responses have been compared against the measurements performed in a professional anechoic chamber. The schematic view of the test system has been shown in Fig. 3(b). The angular resolution of the measurements has been set to 5°, whereas the bandwidth and the number of points around $f_0$ have been defined as $K = 201$ and $B = 1$ GHz, respectively [16].

### 3.2 Post-processing Results and Benchmark

The non-anechoic antenna measurements have been performed at the center frequencies of $f_0 \in \{3.5, 4, 5.5, 6.5, 7, 10.5\}$ GHz for a total of five spatially-separated setups concerning change of the RA positions relative to the AUT from 1 m to 3 m with a 0.5 m step (cf. Fig. 3(a)). The obtained results have been combined and corrected using the algorithm of Sect. 2. The procedure has been executed separately for each frequency and it involved identification of the LoS profiles (based on time-domain characteristics obtained from the frequency data) for each setup followed by synchronization of the responses, their stacking, conversion back to the frequency spectrum, and extraction of the useful part of the signals (i.e., the refined radiation pattern).

A comparison of the non-anechoic responses before (in a setup with 2 m distance) and after the correction is provided in Table 1. The performance of the responses is expressed—against the AC-based measurements—in terms of a decibel-based root-mean square error calculated for all $A$ angles of AUT rotation and then averaged [14]. The radiation patterns at the selected frequencies are shown in Fig. 4. The obtained results indicate that, for the considered Vivaldi antenna and the selected center frequencies, the average performance improvement of the measurements performed in the non-anechoic environment (averaged over all of the considered frequencies) due to post-processing amounts to 9.4 dB, i.e., from −15.8 dB for uncorrected to −25.2 dB for the corrected responses.

The proposed approach has been benchmarked against the existing time-domain-based methods in terms of the correction performance. The considered approaches involved the use of different windowing functions and selection of the intervals based on: (i) physical measurements of RA-AUT distance and the expected path for the shortest reflection (box window) [7], (ii) visual inspection of the impulse response (Hann window) [8], and (iii) automatic analysis of the impulse response with constant thresholds (composite window) [9, 15]. The RA-AUT distance for the experiments (i)-(ii) is set to 2 m (cf. Fig. 3). For all of the considered tests, the setup in terms of $B$ and $K$ remains the same (cf. Sect. 3.1). The results gathered in Table 1 indicate that the performance of the proposed method (averaged over all six of the considered frequencies) is 3 dB to over 6 dB better compared to the benchmark techniques [7–9]. The main advantage of the presented framework includes a streamlined data analysis that does not involve manual selection of the correction parameters. The mentioned feature makes the method suitable for supporting day-to-day measurements in challenging non-anechoic environments while ensuring acceptable accuracy of the results.

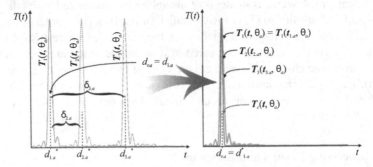

**Fig. 2.** A conceptual illustration of the proposed post-processing algorithm based on augmentation of the LoS transmission using a set of spatially-separated measurements.

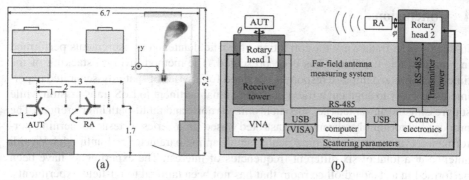

**Fig. 3.** Non-anechoic experiments: (a) the test site (dimensions in m) with highlight on RA locations (×) with a photograph of the Vivaldi antenna, as well as (b) a block diagram of the system for far-field antenna measurements [14].

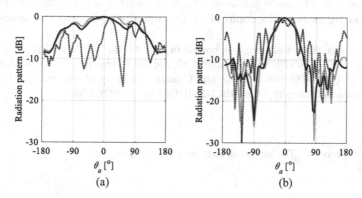

**Fig. 4.** AC-based (gray) and non-anechoic (black) measurements performed at the test site of Fig. 3(a) before (···) and after (—) correction at: (a) 3.5 GHz and (b) 10.5 GHz.

**Table 1.** Post-processing results and benchmark against the state-of-the-art methods.

| Results | | | | | | | Benchmark | | |
|---|---|---|---|---|---|---|---|---|---|
| Considered frequencies | | | | | | | Method | $E(e_R)^*$ [dB] | $\Delta(E(e_R))$ [dB] |
| $f_0$ [GHz] | 3.5 | 4 | 5.5 | 6.5 | 7 | 10.5 | This work | −25.2 | N/A |
| $e_R(\mathbf{R})^\$$ [dB] | −15.3 | −14.9 | −15.6 | −14.9 | −16.5 | −17.8 | (i) | −19.8 | 5.4 |
| $e_R(\mathbf{R}_c)$ [dB] | −25.4 | −28.5 | −23.1 | −25.1 | −24.1 | −25.1 | (ii) | −22.2 | 3 |
| $\Delta$ [dB]$^{\#}$ | −10.1 | −13.6 | −7.5 | −10.2 | −7.6 | −7.3 | (iii) | −19.1 | 6.1 |

$^{\#} \Delta = |e_R(\mathbf{R}_c(f_0, \boldsymbol{\theta})) - e_R(\mathbf{R}(f_0, \boldsymbol{\theta}))|$
$^{*}$ Averaged over all of the frequencies considered for the experiments
$^{\$}$ For one-shot measurement in a setup with the RA-AUT distance of 2 m (cf. Fig. 3)

# 4   Conclusion

In this work, a framework for correction of far-field antenna measurements performed in non-anechoic conditions has been presented. The method involves stacking of the time-synchronized impulse responses obtained for a series of spatially-separated experiments in order to augment a fraction of the signal pertinent to LoS transmission while suppressing the unwanted (i.e., unsynchronized) noise and multi-path interferences. The proposed algorithm has been demonstrated based on a series of tests concerning performance evaluation of a geometrically small spline-parameterized antipodal Vivaldi antenna at a total of six different frequencies of interest. The experiments have been performed in a standard office room that has not been tailored to far-field experiments. The obtained results have been compared against the AC-based radiation patterns. The benchmark of the method against the state-of-the-art algorithms from the literature has also been provided.

Future work will focus on demonstrating the performance of the method for setups with different distances and locations of the measurement system within the test site.

**Acknowledgments.** This work was supported in part by the National Science Centre of Poland Grant 2021/43/B/ST7/01856, National Centre for Research and Development Grant NOR/POLNOR/HAPADS/0049/2019-00, and Gdansk University of Technology (Excellence Initiative - Research University) Grant 16/2023/IDUB/IV.2/EUROPIUM.

**Disclosure of Interests.**   The authors declare no conflicts of interests.

# References

1. Koziel, S., Bekasiewicz, A., Cheng, Q.S.: Conceptual design and automated optimization of a novel compact UWB MIMO slot antenna. IET Microwaves Antennas Propag. **11**(8), 1162–1168 (2017)
2. Bekasiewicz, A., Koziel, S., Zieniutycz, W.: Design space reduction for expedited multi-objective design optimization of antennas in highly-dimensional spaces. In: Koziel, S., Leifsson, L., Yang, XS. (eds.) Solving Computationally Expensive Engineering Problems: Methods and Applications, pp. 113–147, Springer, Cham (2014). https://doi.org/10.1007/978-3-319-08985-0_5
3. Hemming, L.: Electromagnetic Anechoic Chambers: A fundamental Design and Specification Guide. IEEE Press, Piscataway (2002)
4. Kurokawa, S., Hirose, M., Komiyama, K.: Measurement and uncertainty analysis of free-space antenna factors of a log-periodic antenna using time-domain techniques. IEEE Trans. Instrum. Meas. **58**(4), 1120–1125 (2009)
5. Toh, B.Y., Cahill, R., Fusco, V.F.: Understanding and measuring circular polarization. IEEE Trans. Educ. **46**(3), 313–318 (2003)
6. Khan, M.S., Capobianco, A.-D., Asif, S.M., Anagnostou, D.E., Shubair, R.M., Braaten, B.D.: A compact CSRR-enabled UWB diversity antenna. IEEE Antennas Wirel. Propag. Lett. **16**, 808–812 (2017)

7. Piasecki P., Strycharz, J.: Measurement of an omnidirectional antenna pattern in an anechoic chamber and an office room with and without time domain signal processing. In: Signal Processing Symposium, pp. 1–4, Debe, Poland (2015)

8. Soltane, A., Andrieu, G., Perrin, E., Decroze, C., Reineix, A.: Antenna radiation pattern measurement in a reverberating enclosure using the time-gating technique. IEEE Antennas Wirel. Propag. Lett. **19**(1), 183–187 (2020)

9. de Sao Jose, A.N., Deniau, V., Resende, U.C., Adriano, R.: Improving antenna gain estimations in non-ideal test sites with auto-tunable filters. Measurement **159**, 107720 (2020)

10. Froes, S.M., Corral, P., Novo, M.S., Aljaro, M., Lima, A.C.C.: Antenna radiation pattern measurement in a nonanechoic chamber. IEEE Antennas Wirel. Propag. Lett. **18**(2), 383–386 (2019)

11. Sarkar, T.K., Salazar-Palma, M., Zhu, M.D., Chen, H.: Modern Characterization of Electromagnetic Systems and Its Associated Metrology. IEEE Press, Wiley, Hoboken (2021)

12. Du, Z., Moon, J.I., Oh, S.-S., Koh, J., Sarkar, T.K.: Generation of free space radiation patterns from non-anechoic measurements using Chebyshev polynomials. IEEE Trans. Antennas Propag. **58**(8), 2785–2790 (2010)

13. Mroczka, J.: The cognitive process in metrology. Measurement **46**, 2896–2907 (2013)

14. Bekasiewicz, A., Waladi, V.: Filter-Hilbert method for automatic correction of non-anechoic antenna measurements with embedded self-calibration mechanism. Measurement **222**, 113705 (2023)

15. Oppenheim, A.V., Schafer, R.W.: Discrete-Time Signal Processing. Prentice Hall, Upper Saddle River (2009)

16. Bekasiewicz, A., Koziel, S., Czyz, M.: Time-gating method with automatic calibration for accurate measurements of electrically small antenna radiation patterns in non-anechoic environments. Measurement **208**, 112477 (2023)

# TR-Based Antenna Design with Forward FD: The Effects of Step Size on the Optimization Performance

Adrian Bekasiewicz[1]($\boxtimes$) (ID), Slawomir Koziel[1,2] (ID), Tom Dhaene[3] (ID), and Marcin Narloch[1] (ID)

[1] Faculty of Electronics, Telecommunications and Informatics, Gdansk University of Technology, Narutowicza 11/12, 80-233 Gdansk, Poland
bekasiewicz@ru.is
[2] Department of Engineering, Reykjavik University, Menntavegur 1, 102, Reykjavík, Iceland
[3] Department of Information Technology (INTEC), IDLab, Ghent University-imec, iGent, Technologiepark-Zwijnaarde 126, 9052 Ghent, Belgium

**Abstract.** Numerical methods are important tools for design of modern antennas. Trust-region (TR) methods coupled with data-efficient surrogates based on finite differentiation (FD) represent a popular class of antenna design algorithms. However, TR performance is subject to FD setup, which is normally determined *a priori* based on rules-of-thumb. In this work, the effect of FD perturbations on the performance of TR-based design is evaluated on a case study basis concerning a total of 80 optimizations of a planar antenna structure. The obtained results demonstrate that, for the considered radiator, the performance of the final designs obtained using different FD setups may vary by as much as 18 dB (and by over 4 dB on average). At the same time, the *a priori* perturbations in a range between 1.5% and 3% (w.r.t. the initial design) seem to be suitable for maintaining (relatively) consistent and high-quality results.

**Keywords:** Trust-region methods · finite-differences · perturbation size · antenna design · numerical optimization

## 1 Introduction

Optimization methods are indispensable tools for the design of contemporary antennas. The main bottleneck of conventional algorithms is that they require dozens, or even hundreds of model evaluations to converge [1, 2]. At the same time, the simulation cost of antenna electromagnetic (EM) models is high—especially for modern multi-parameter topologies—which challenges the concept of their direct optimization, especially if the designer does not have access to high-performance computing clusters and a large number of software licenses [3, 4]. From this perspective, availability of data-efficient methods is of high importance for reliable, yet low-cost antenna design.

Data-efficient design of microwave and antenna circuits can be performed using surrogate-assisted methods (SAM). The goal of SAM is to embed the antenna design

L. Franco et al. (Eds.): ICCS 2024, LNCS 14834, pp. 28–36, 2024.
https://doi.org/10.1007/978-3-031-63759-9_4

into a meta-loop that involves optimization of a cheap auxiliary model (with limited accuracy) that is iteratively updated/re-constructed using only a handful of accurate (yet numerically expensive) data samples obtained from EM simulations [5, 6]. Trust-region (TR) methods belong to a popular class of SAM algorithms. They perform exploitation of the search space through optimization of the local, numerically cheap model that predicts the new design candidates. To alleviate the high cost of EM simulations, the model is often in the form of a first-order Taylor expansion constructed from the structure response and its derivatives [1, 7]. Although the latter can be obtained analytically—at a low computational overhead (w.r.t. zero-order simulation)—adjoint-capable simulations are available in only a handful of commercially available EM solvers [8, 9]. Alternatively, the sensitivity data can be approximated using finite-differences (FD), i.e., through evaluation of the structure response at the design of interest and series of small perturbations around it. In antenna engineering, linear models are often constructed using forward FDs as which require only one EM simulation per design parameter [2, 8].

Regardless of proved usefulness, the performance of FD-based TR optimization is subject to appropriate setup (perturbation-wise). Although algorithms dedicated to automatically determine suitable FD steps have been reported in the literature, they are prohibitively expensive when applied to EM-driven design problems [10, 11]. Instead, the perturbations are predominantly selected *a priori* based on rules-of-thumb. Popular methods include determination of FD steps as a fraction of the initial design, square-root of machine precision, or based on the experience-driven manual tuning of structure parameters [7, 10]. Despite explicit relation between the FD setup and TR optimization performance, the problem pertinent to determination of appropriate perturbations for EM-driven optimization of antennas remains open.

In this work, the effects of *a priori* selected FD steps on the performance of TR-based gradient optimization have been evaluated on a case study basis concerning the design of a planar quasi-patch antenna. A total of 80 numerical experiments—spanning across 10 different designs and 8 FD-setups each—have been performed. The results demonstrate that, for the considered antenna, the FD-induced performance discrepancy between the optimized designs is substantial and can vary by as much as 18 dB. At the same time, the average (i.e., calculated over all of the designs) best-to-worst-setup difference amounts to over 4 dB. The obtained data not only demonstrate that an appropriate FD setup is crucial to ensure high performance (and consistency of optimization) but also suggests that perturbations in a range between 1.5 to 3% of the initial design are suitable for mitigating the effects of numerical noise while supporting exploitation of the search space using the linear approximation models.

## 2 Optimization Algorithm

### 2.1 Problem Formulation

Let $\boldsymbol{R}(x)$ be the EM-simulation response of the structure obtained for the vector of input parameters $x$. The optimization problem is given as:

$$x^* = \arg \min_{x \in X} U(\boldsymbol{R}(x)) \qquad (1)$$

Here, the $x^* \in X$ represents the optimized design to be found within the feasible region of the search space $X$, which is defined by the lower/upper bounds $l/u$; $U$ is a scalar objective function. Due to high evaluation cost, direct optimization of $R$ is impractical. Instead, the task (1) can be embedded into a surrogate-assisted design framework to enable cost-efficient identification of $x^*$.

## 2.2  TR Optimization Framework

The goal of TR-based optimization is to generate a series of approximations, $i = 0, 1, 2, ...$, to the problem (1), as [6]:

$$x^{(i+1)} = \arg \min_{\delta^{(i)} \ge \|x - x^{(i)}\|} U\left(G^{(i)}(x)\right) \tag{2}$$

where $G^{(i)} = R(x^{(i)}) + J(x^{(i)})(x - x^{(i)})$ is a linear approximation model obtained at the $i$th step of the optimization process and $J$ represents its FD-based Jacobian [10]:

$$J\left(x^{(i)}\right) = \begin{bmatrix} 1/p_1 \left(R(x^{(i)} + p_1) - R(x^{(i)})\right) \\ \vdots \\ 1/p_D \left(R(x^{(i)} + p_D) - R(x^{(i)})\right) \end{bmatrix}^T \tag{3}$$

The problem (2) is solved by a gradient-based algorithm executed on the linear approximation models constructed from (3). It should be noted that the variable $p_d$ ($d = 1, ..., D$) represents the perturbation w.r.t. $d$th dimension of the design $x^{(i)}$, whereas $p_d = [0 ... p_d ... 0]^T$. The trust-region radius $\delta^{(i)}$ is adjusted based on the gain coefficient $\rho = [U(R(x^{(i+1)})) - U(R(x^{(i)}))]/[U(G^{(i)}(x^{(i+1)})) - U(G^{(i)}(x^{(i)}))]$ which expresses the ratio between the expected and the obtained objective function change [6]. The factor is also used to accept ($\rho > 0$), or reject ($\rho < 0$) the candidate designs generated by (2) as new approximations of the final solution. The initial radius is set to $\delta^{(0)} = 1$. For the remaining iterations it is decreased as $\delta^{(i+1)} = \alpha_1 \|x^{(i+1)} - x^{(i)}\|$ for $\rho < 0.05$ (poor performance of the local model), or increased as $\delta^{(i+1)} = \max(\alpha_2 \|x^{(i+1)} - x^{(i)}\|, \delta^{(i)})$ for $\rho > 0.9$ (acceptable improvement of the response). The scaling factors are set to $\alpha_1 = 0.25$, and $\alpha_2 = 2.5$ [6]. The algorithm is terminated when $\delta^{(i+1)} < \varepsilon$ or $\|x^{(i+1)} - x^{(i)}\| < \varepsilon$ (here, $\varepsilon = 10^{-2}$). For more detailed discussion of the method, see [2, 6, 8].

## 2.3  Perturbations in FD-Based Jacobians

On the conceptual level, the problem pertinent to selection of appropriate perturbations (also referred to as a step-size dilemma; cf. Figure 1(a)) involves balancing their size so as to minimize both the truncation (too large step) and round-off (too small step) errors [10]. Another challenge is that EM simulation models are inherently noisy which stems from inconsistency of their discretization for closely-located designs (as the ones required for construction of the linear models). As a consequence, for too small step the linear model might approximate noise, while the use of too large perturbations might result in poor representation of the objective function changes (see Fig. 1(b)) [12, 13]. Furthermore, the problem of FD-setup is both circuit- and design-dependent. Consequently, a series of model evaluations would be required for each candidate solution and w.r.t. each variable to find the optimal steps. Clearly, it is not practical when the EM-driven antenna development is considered [10, 11, 14].

The FD setup for antenna design problems is normally performed through *a priori* specification of perturbations that are (hopefully) small enough to capture local changes of the objective function but also large enough to mitigate the effects of numerical noise on the identified descent direction (see Fig. 1(b)). The popular rule-of-thumb methods include selection of steps that either represent a fraction of the initial design, or correspond to the machine precision (single for the EM-driven problems) [10]. Alternatively FD-steps can be obtained based on the experience-driven tuning that involves visual inspection of the objective function changes [7]. Notwithstanding, the question concerning practical usefulness of the mentioned concepts for EM-based problems, as well as their effects on the TR algorithm performance remains open.

## 3  Antenna Structure

Figure 2 illustrates the quasi-patch antenna considered for the experiments [9]. The structure is dedicated to ISM-band (industrial, scientific, medicine) applications. It is designed on a dielectric substrate with permittivity/thickness of 3.5/0.762 mm and comprises a bi-component radiator in the form of a deformed patch loaded by a monopole strip and fed through a 50 $\Omega$ microstrip line. Impedance matching of the patch and the feedline is maintained using two insets (notches). The ground plane length below the radiating component is parameterized so as to ensure its flexibility in terms of the operational bandwidth (w.r.t. conventional structures). The component is implemented in CST Microwave Studio, and evaluated using its time-domain solver [15]. On average, the EM model is discretized using 150,000 tetrahedral mesh cells (determined based on a series of numerical experiments oriented towards maintaining acceptable accuracy of the simulations), whereas its evaluation cost amounts to 120s. The model also implements the coaxial connector [16].

The design parameters are $x = [L\ l_2\ W\ w_2\ l_0\ o_0]^T$. The relative variables are $o = 0.22L$, and $l_s = 0.1L$, whereas dimensions $l_1 = 1.5$, $w_1 = 2.5$, $w_s = 0.5$, and $w_0 = 1.7$ remain constant during the optimization (all in mm). The feasible region of the design space $X$ is defined by the following lower and upper bounds: $l = [10\ 5\ 3.5\ 0.2\ 3\ 2]^T$ and $u = [25\ 25\ 10\ 3.2\ 15\ 10]^T$. The objective function $U(x) = U(R(x))$ is given as:

$$U(x) \;=\; \max\{R(x)\}_{f_L \le f \le f_H} \tag{4}$$

where $f_L = 5$ GHz and $f_H = 6$ GHz represent the corner frequencies for the antenna operational bandwidth, whereas $R(x) = R(x, f)$ represents its reflection response and $f \in f$ is the frequency sweep. Note that the function (4) defines a simple min-max problem, where the goal is to minimize a maximum value of the antenna response within the frequency range of interest (here, 5 GHz to 6 GHz).

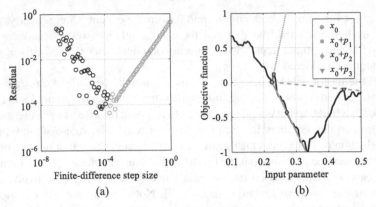

**Fig. 1.** FD steps: (a) a residual of analytical and FD-based derivatives for a function $\sin(\pi/4)$ with highlight on the round-off (black) and truncation errors (gray), (b) the quality of linear models for a noisy function when FD steps are too small ($\cdots$), too large ($--$), and correct ($—$).

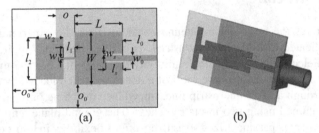

**Fig. 2.** Quasi-patch: (a) geometry with highlight on design variables and (b) visualization of the structure with coaxial connector [9]. Note that light gray represents the ground plane.

## 4  Numerical Results

The considered numerical experiments involve a series of TR-based optimizations of the antenna for ten—randomly selected within the $l/u$ bounds [9]—designs (for details, see Table 1), each using eight different FD setups. These included specification of steps as: a manually defined fraction of the initial design ($p$ from 0.5% to 3% with 0.5% step), or based on a single-point machine precision (double is not supported by the EM solver; $p = \varepsilon^{0.5} \approx 0.0032$; $\varepsilon = 10^{-7}$), as well as through manual tuning with $p = [p_1 \ldots p_D]$ = $0.001 \cdot [3\ 3\ 3\ 7\ 8\ 6]^T$ (here, $D = 6$; cf. Sections 2 and 3). Note that the computational cost of experience-driven tuning (pertinent to model evaluations necessary for visual inspection of function changes) corresponds to 10 EM simulations.

The results gathered in Table 2 indicate that the FD perturbations setup substantially affects the TR convergence and the performance of the obtained designs. For the considered problems and tests, the maximum discrepancy between the designs obtained using different settings amounts to 18 dB (per specific design – $x_7$) and 4.2 dB (on average for all of the designs). Furthermore, the results suggest that, although the best responses (collectively) are obtained using 3% steps, for design $x_9$ the performance of the optimized responses was almost 4 dB lower compared to the case with 2.5% perturbations. The result suggests that the local model slightly overshoot the optimum. On the other hand, the standard deviations $\sigma$ indicate that 3% step offers the most consistent responses among all of the experiments (the obtained value would be lower if not for the poor-quality design $x_{10}$). As it comes to the test cases that involved custom and precision-based perturbations, their performance is comparable to 0.5% and 1% steps (which is understandable given their ranges). Worsened results for small perturbations suggest that the driving factor behind premature convergence of TR is the numerical noise (which surpasses round-off errors). From this perspective, the use of steps in the range from 1.5% to 3% of $x$ seem to counteract the noise while (mostly) ensuring appropriate representation of the objective function changes.

The optimized designs (obtained for $p$ 3%) are collected in Table 3, whereas Fig. 3 shows the effects of selected FD steps on the TR performance for $x_1$ and $x_2$ designs. It should also be noted that the FD-induced change of computational cost is up to 22% for the considered tests, with the better performing setups being (understandably) more expensive, as more iterations are needed to accurately exploit the search space.

**Table 1.** Antenna designs used for TR algorithm benchmark.

| Designs used for benchmark | | | | | | | | | | |
|---|---|---|---|---|---|---|---|---|---|---|
|  | $x_1^{(0)}$ | $x_2^{(0)}$ | $x_3^{(0)}$ | $x_4^{(0)}$ | $x_5^{(0)}$ | $x_6^{(0)}$ | $x_7^{(0)}$ | $x_8^{(0)}$ | $x_9^{(0)}$ | $x_{10}^{(0)}$ |
| $U(x)$ | −2.03 | −4.50 | −0.34 | −3.53 | −2.97 | −0.87 | −2.04 | −2.45 | −6.23 | −1.23 |
| Parameters | 17.5 | 22.2 | 18.8 | 17.9 | 14.6 | 13.4 | 20.4 | 13.6 | 18.2 | 21.6 |
|  | 15.1 | 21.3 | 16.7 | 15.6 | 11.2 | 9.58 | 18.9 | 9.87 | 15.9 | 20.5 |
|  | 6.79 | 8.79 | 7.30 | 6.95 | 5.52 | 4.99 | 8.04 | 5.08 | 7.07 | 8.56 |
|  | 1.72 | 2.64 | 1.96 | 1.79 | 1.13 | 0.89 | 2.30 | 0.93 | 1.85 | 2.54 |
|  | 9.07 | 12.8 | 10.0 | 9.37 | 6.73 | 5.75 | 11.3 | 5.92 | 9.60 | 12.3 |
|  | 6.05 | 8.51 | 6.68 | 6.25 | 4.49 | 3.83 | 7.59 | 3.95 | 6.40 | 8.23 |

**Table 2.** TR-based optimization – benchmark results.

| Design | Perturbation size / type | | | | | | | | | | | | | $\varepsilon^{0.5}$ & | | Custom | |
| | $0.005x^{(0)}$ | | $0.01x^{(0)}$ | | $0.015x^{(0)}$ | | $0.02x^{(0)}$ | | $0.025x^{(0)}$ | | $0.03x^{(0)}$ | | | | | |
| | $U(x^*)$ | $R^\#$ | $U(x^*)$ | $R^\#$ | $U(x^*)$ | $R^\#$ | $U(x^*)$ | $R^\#$ | $U(x^*)$ | $R^\#$ | $U(x^*)$ | $R^\#$ | $U(x^*)$ | $R^\#$ | $U(x^*)$ | $R^{\#,!}$ |
|---|---|---|---|---|---|---|---|---|---|---|---|---|---|---|---|---|
| $x_1$ | −16.9 | 74 | −19.8 | 67 | −17.3 | 45 | −23.8 | 88 | −22.1 | 88 | −26.9 | 88 | −21.2 | 88 | −26.7 | 104 |
| $x_2$ | −27.2 | 35 | −31.1 | 58 | −24.1 | 39 | −29.6 | 43 | −30.5 | 43 | −31.2 | 43 | −28.0 | 45 | −24.3 | 61 |
| $x_3$ | −29.7 | 72 | −30.2 | 86 | −30.6 | 88 | −30.2 | 86 | −25.3 | 86 | −29.5 | 81 | −27.9 | 77 | −24.3 | 92 |
| $x_4$ | −30.7 | 52 | −30.8 | 52 | −28.1 | 52 | −27.5 | 36 | −30.9 | 36 | −30.9 | 85 | −30.2 | 81 | −27.6 | 61 |
| $x_5$ | −16.3 | 60 | −17.0 | 30 | −18.4 | 44 | −17.6 | 39 | −21.5 | 39 | −28.2 | 94 | −17.7 | 50 | −15.4 | 40 |
| $x_6$ | −27.1 | 88 | −26.3 | 72 | −31.0 | 63 | −30.2 | 65 | −28.6 | 65 | −30.6 | 94 | −27.0 | 81 | −29.6 | 95 |
| $x_7$ | −20.2 | 88 | −11.4 | 59 | −26.7 | 66 | −29.5 | 93 | −28.7 | 93 | −29.2 | 94 | −28.7 | 94 | −20.4 | 63 |
| $x_8$ | −27.0 | 72 | −26.8 | 38 | −29.7 | 58 | −23.3 | 37 | −29.4 | 37 | −30.6 | 72 | −27.9 | 36 | −27.3 | 76 |
| $x_9$ | −26.6 | 35 | −27.8 | 42 | −28.1 | 66 | −29.6 | 65 | −30.1 | 65 | −26.3 | 37 | −17.9 | 37 | −30.0 | 68 |
| $x_{10}$ | −16.1 | 80 | −16.4 | 81 | −16.3 | 74 | −16.4 | 81 | −16.0 | 81 | −16.3 | 73 | −17.0 | 64 | −16.3 | 84 |
| $E^\$$ | −23.8 | 65 | −23.8 | 59 | −25.0 | 60 | −25.8 | 63 | −26.3 | 63 | −28.0 | 76 | −24.3 | 65 | −24.2 | 74 |
| $\sigma^\$$ | 5.76 | 20 | 7.03 | 18 | 5.68 | 15 | 5.27 | 23 | 4.96 | 23 | 4.43 | 20 | 5.25 | 22 | 5.21 | 19 |

$ Average and standard deviation of the values obtained for the considered designs
# Computational cost of the optimization expressed in the number of EM model evaluations
& Perturbation size determined as square root of the machine precision $\varepsilon = 10^{-7}$
! Including the cost of manual adjustment of perturbations

**Table 3.** Designs optimized using FD setup with 3 percent perturbations.

| Optimized designs$^{\$}$ – $p = 0.003x$ | | | | | | | | | | |
|---|---|---|---|---|---|---|---|---|---|---|
| | $x_1^{*}$ | $x_2^{*}$ | $x_3^{*}$ | $x_4^{*}$ | $x_5^{*}$ | $x_6^{*}$ | $x_7^{*}$ | $x_1^{*}$ | $x_9^{*}$ | $x_{10}^{*}$ |
| Parameters | 20.2 | 20.7 | 20.5 | 20.6 | 20.4 | 20.5 | 20.5 | 20.6 | 19.9 | 19.0 |
| | 12.9 | 11.3 | 12.6 | 11.9 | 11.7 | 12.6 | 11.9 | 12.6 | 12.6 | 11.5 |
| | 4.10 | 3.56 | 3.92 | 3.72 | 3.84 | 3.84 | 3.73 | 3.85 | 4.32 | 5.22 |
| | 0.20 | 0.49 | 0.21 | 0.36 | 0.39 | 0.23 | 0.38 | 0.23 | 0.20 | 2.59 |
| | 10.1 | 11.1 | 10.4 | 10.8 | 10.4 | 10.8 | 11.0 | 10.7 | 9.70 | 3.00 |
| | 9.37 | 10.0 | 9.81 | 9.89 | 10.0 | 9.64 | 9.79 | 9.71 | 10.0 | 5.26 |

$^{\$}$For objective function values see Table 2

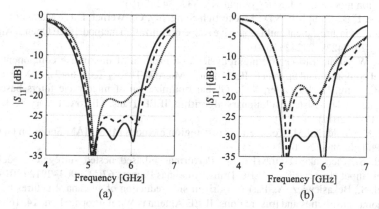

**Fig. 3.** Antenna design – comparison of the final designs obtained using TR-based optimization with forward FD steps of 1% ($\cdots$), 2% ($--$), and 3% (——) w.r.t. the center design: (a) design case $x_1$ and (b) design case $x_2$. Note that the selected FD perturbations have a notable effect on the quality of the optimized antenna response.

## 5  Conclusion

In this work, the effects of FD setup on performance of the TR-based antenna optimization have been demonstrated on a case study basis concerning a planar quasi-patch radiator. A total of 80 numerical experiments, spanned across 10 designs (each optimized using 8 different setups) have been performed. The numerical results demonstrate that the selected FD substantially affects the quality of the obtained designs (by as much as 18 dB per problem and around 4 dB on average), but also design cost (by up to 22%). Furthermore, the gathered—problem-specific data—suggest that determination of FD steps in a range between 1.5% and 3% of the initial design is suitable for ensuring high quality results for a range of test cases.

**Acknowledgments.** This work was supported in part by the National Science Centre of Poland Grants 2020/37/B/ST7/01448 and 2021/43/B/ST7/01856, National Centre for Research and Development Grant NOR/POLNOR/HAPADS/0049/2019-00, and Gdansk University of Technology (Excellence Initiative - Research University) Grant 16/2023/IDUB/IV.2/EUROPIUM.

**Disclosure of Interests.** The authors declare no conflicts of interest.

# References

1. Bekasiewicz, A., Koziel, S., Zieniutycz, W.: Design space reduction for expedited multi-objective design optimization of antennas in highly-dimensional spaces. In: Koziel, S., Leifsson, L., Yang, XS. (eds.) Solving Computationally Expensive Engineering Problems: Methods and Applications, vol. 97, pp. 113–147, Springer, Cham (2014). https://doi.org/10.1007/978-3-319-08985-0_5
2. Koziel, S., Bekasiewicz, A.: Multi-objective optimization of expensive electromagnetic simulation models. Appl. Soft Comput. **47**, 332–342 (2016)
3. Whiting, E.B., Campbell, S.D., Mackertich-Sengerdy, G., Werner, D.H.: Meta-atom library generation via an efficient multi-objective shape optimization method. IEEE Open J. Antennas Propag. **28**, 24229–24242 (2020)
4. Zhang, W., et al.: EM-centric multiphysics optimization of microwave components using parallel computational approach. IEEE Trans. Microw. Theory Tech. **68**(2), 479–489 (2020)
5. Liu, B., Yang, H., Lancaster, M.J.: Global optimization of microwave filters based on a surrogate model-assisted evolutionary algorithm. IEEE Trans. Microw. Theory Tech. **65**(6), 1976–1985 (2017)
6. Conn, A., Gould, N.I.M., Toint, P.L.: Trust-region methods MPS-SIAM Series on Optimization, Philadelphia (2000)
7. Koziel, S., Pietrenko-Dabrowska, A.: Performance-based nested surrogate modeling of antenna input characteristics. IEEE Trans. Antennas Propag. **67**(5), 2904–2912 (2019)
8. Koziel, S., Bekasiewicz, A.: Fast EM-driven size reduction of antenna structures by means of adjoint sensitivities and trust regions. IEEE Antennas Wirel. Propag. Lett. **14**, 1681–1684 (2015)
9. Bekasiewicz, A.: Optimization of the hardware layer for IoT systems using a trust-region method with adaptive forward finite differences. IEEE Internet Things J. **10**(11), 9498–9512 (2023)
10. Mathur, R.: An analytical approach to computing step sizes for finite-difference derivatives. PhD Thesis, University of Texas, Austin (2012)
11. Restrepo, R.L., Ocampo, C.: Automatic algorithm for accurate numerical gradient calculation in general and complex spacecraft trajectories. AAS/AIAA Space Flight Mechanics Meeting (2013)
12. Zhang, J.N., Feng, F., Jin, J., Zhang, Q.J.: Efficient yield estimation of microwave structures using mesh deformation-incorporated space mapping surrogates. IEEE Microwave Wirel. Compon. Lett. **30**(10), 937–940 (2020)
13. Lamecki, A.: A mesh deformation technique based on solid mechanics for parametric analysis of high-frequency devices with 3-D FEM. IEEE Trans. Microw. Theory Tech. **64**(11), 3400–3408 (2016)
14. Yang, W.Y., Cao, W., Chung, T.-S., Morris, J.: Applied Numerical Methods using MATLAB. Wiley (2005)
15. CST Microwave Studio, Dassault Systems, 10 rue Marcel Dassault, CS 40501, Vélizy-Villacoublay Cedex, France (2015)
16. Palaniswamy, S.K., et al.: 3D eight-port ultra-wideband (UWB) antenna array for diversity applications. IEEE Antennas Wirel. Propag. Lett. **16**, 569–572 (2017)

# Miniaturization-Oriented Design of Spline-Parameterized UWB Antenna for In-Door Positioning Applications

Adrian Bekasiewicz[1]([✉]) [iD], Tom Dhaene[2] [iD], Ivo Couckuyt[2] [iD], and Jacek Litka[1] [iD]

[1] Faculty of Electronics, Telecommunications and Informatics,
Gdansk University of Technology, Narutowicza 11/12, 80-233 Gdansk, Poland
bekasiewicz@ru.is
[2] Department of Information Technology (INTEC), IDLab, Ghent University-imec,
iGent, Technologiepark-Zwijnaarde 126, 9052 Ghent, Belgium

**Abstract.** Design of ultra-wideband antennas for in-door localization applications is a challenging task. It involves development of geometry that maintains appropriate balance between the size and performance. In this work, a topologically-flexible monopole has been generated using a stratified framework which embeds a gradient-based trust-region (TR) optimization algorithm in a meta-loop that gradually increases the structure dimensionality. The optimization has been performed using a composite objective function that maintains acceptable size/performance trade-off. The final design features a reflection below $-10$ dB within the UWB spectrum and a small footprint of only 182 mm$^2$. The considered method has been benchmarked against a standard TR-based routine executed directly on a multi-dimensional electromagnetic model of the antenna.

**Keywords:** Topologically-flexible antennas · trust-region methods · stratified optimization · antenna design · in-door localization

## 1 Introduction

Ultra-wideband (UWB) technology is a promising solution for the development of modern real-time localization services dedicated to in-door environments. Its advantage—compared to other radio-frequency methods—includes pulse operation over a broad frequency range which makes it less susceptible to interferences [1, 2]. Quality-of-service offered by positioning systems is a subject to availability of antennas which comply with the regulations that mandate the access to wireless medium while ensuring a high performance [3]. Apart from the electrical- and field-related requirements, applicability of UWB radiators in mobile terminals is also affected by their footprints [4]. Therefore, accounting for the size/performance trade-off is important when design of radiators for in-door positioning is considered [1, 4].

Conventional approaches to compact antennas development involve experience-driven determination of the topology followed by its trial-and-error modifications in

L. Franco et al. (Eds.): ICCS 2024, LNCS 14834, pp. 37–45, 2024.
https://doi.org/10.1007/978-3-031-63759-9_5

hope of achieving improved performance and/or area reduction [5, 6]. This inherently cognitive process is affected by the engineering bias as designers, understandably, lean towards the geometries and modifications they are familiar with [4, 7]. Although the structure development should be followed by rigorous optimization-based tuning, this step is often neglected in favor of parametric studies (aided by visual-inspection of performance changes) oriented towards achieving the satisfactory performance [8]. The main reasoning behind this laborious and prone-to-failure procedure is that accurate evaluation of complex antenna performance can only be performed using computationally expensive electromagnetic (EM) simulations. At the same time, a large number of EM simulations required to converge challenges the applicability of conventional algorithms. The undesirable consequence of the outlined scheme is that the antenna design is often governed by a mix of past experiences and a limited number of observations which, at best, lead to acceptable solutions rather than the ones that offers the best balance between the requirements [5, 7, 8].

The effects of engineering-bias on the antenna development process can be mitigated by shifting the design paradigm from the fixed-topology to flexible models where the final geometry is determined by numerical methods. This can be achieved using the so-called dummy EM models which represent the antenna as a set of interconnected points, or in the form of a binary matrix that govern the structure shape [9–11]. The bottleneck of generic models is that they require overwhelmingly large number of dimensions to support diverse geometries and hence the cost of their EM-driven optimization is numerically prohibitive. The challenges pertinent to unacceptable optimization cost can be mitigated using trust-region (TR) methods. TR algorithms exploit the promising regions of the search space based on evaluations of data-efficient models that approximate EM simulations [12, 13]. Despite being numerically-efficient, local nature of TR-based optimization might limit its applicability for solving complex and multi-modal problems such as automatic tuning of topology oriented towards ensuring a balance between the size and performance.

In this work, a framework for stratified design of topologically-flexible antenna has been considered. The method embeds the TR-based optimization into a meta-loop that enables a gradual increase of the structure dimensionality so as to first identify and then exploit the promising region of the search space. The design process is governed by a composite objective function that balances the trade-off between the antenna size and its performance. The method has been used for the development of a spline-parameterized UWB monopole dedicated to in-door positioning applications in wearable scenarios (i.e., for mobile tags) [1]. The optimized structure features a reflection at most $-10$ dB within 3.1 GHz to 10.6 GHz bandwidth and a footprint of only 182 mm$^2$. The method has been benchmarked against a standard TR-based optimization.

## 2  Spline-Parameterized Monopole

Optimization of the radiator in a dimensionally-flexible setup is subject to availability of a suitable EM simulation model. Here, the UWB antenna of Fig. 1 has been considered [14]. The structure is designed on a substrate with permittivity/thickness of 3.38/0.813 mm. It features a driven element in the form of a radiator fed through a microstrip line and a ground plane with an L-shaped extension. The antenna EM simulation model is implemented in CST Microwave Studio and evaluated using its time-domain solver [15]. The structure is discretized using 400,000 hexahedral mesh cells (on average), whereas its typical simulation time amounts to 160s.

In order to support dimensionality adjustment, the radiator and ground plane are represented in the form of spline-based curves defined using variable number of points (so-called knots) [11]. The vector of design parameters is $x = [x_c \ Y \cdot x_g \ S \cdot x_r]^T$, where $x_c = [X \ l_f \ l_1 \ l_{2r} \ w_1 \ o_r]^T$ represents topology-specific variables; $x_g = [x_{g.1} \ ... \ x_{g.l} \ ... \ x_{g.L}]^T$ and $x_r = [x_{r.1} \ ... \ x_{r.l} \ ... \ x_{r.L}]^T$ are the coordinates of the spline knots ($l = 1, ..., L$) spanned equidistantly between the antenna edges (for the ground plane) and along the $2\pi$ azimuth angles defined in the cylindrical system (for the radiator; cf. Figure 1). Maintaining equidistant distribution of points prevents self-intersections of the generated splines which is crucial to ensure feasible topologies. The parameters $l_2 = (X - w_1)l_{2r}$, $Y = l_1 + w_1$, $l_{fr} = \min(X, Y - l_f)/2$, $S = \min(X - o_r, Y - l_f)/2$, and $o = 0.5X + o_r$ are relative, whereas $w_f = 1.8$ is set to ensure 50 $\Omega$ input impedance. Note that $l_{2r}$ is dimensionless and all other variables are in mm. The antenna model is considered feasible within the following lower and upper bounds $l_b = [6 \ 4 \ 10 \ 0.05 \ 0.5 \ -1 \ 0.2\mathbf{1}^L \ 0.1\mathbf{1}^L]^T$ and $u_b = [30 \ 15 \ 30 \ 1 \ 2.5 \ 1 \ 0.8\mathbf{1}^L \ \mathbf{1}^L]^T$, where $\mathbf{1}^L$ is the $L$-dimensional vector of ones. The vector $x$ is represented using $2L + 6$ dimensions. For each EM simulation, the structure is generated dynamically using the specified number of points [15].

## 3 Design Methodology

### 3.1 Problem Formulation

Let $R(x) = R(x, f)$ be the response of the antenna obtained over a frequency sweep $f$ for the vector of input parameters $x$. The optimization task is given as:

$$x^* = \text{argmin}(U(R(x)))  \quad (1)$$

where $x^*$ is the optimal design to be found and $U(x) = U(R(x))$ is a scalar objective function. Direct solving of (1) is numerically impractical when multi-parameter structures are considered. Instead, the problem can be solved using a gradient-based algorithm embedded within the TR framework [12].

### 3.2 TR-Based Optimization Algorithm

The considered optimization engine generates a series of approximations, $i = 0, 1, 2, ...$, to the original problem by solving:

$$x^{(i+1)} = \arg \min_{\|x - x^{(i)}\| \leq \delta^{(i)}} \left( U\left( G^{(i)}(x) \right) \right)  \quad (2)$$

where $G^{(i)} = R(x^{(i)}) + J(x^{(i)})(x - x^{(i)})$ is a first-order Taylor expansion model, whereas $J$ is a Jacobian generated around the $x^{(i)}$ using a large-step finite differences [2, 12]. The TR radius, that determines the region of $G^{(i)}$ model validity (around $x^{(i)}$), is adjusted based on the gain coefficient $\rho = [U(\boldsymbol{R}(\boldsymbol{x}^{(i+1)})) - U(\boldsymbol{R}(\boldsymbol{x}^{(i)}))]/[U(\boldsymbol{G}^{(i)}(\boldsymbol{x}^{(i+1)})) - U(\boldsymbol{G}^{(i)}(\boldsymbol{x}^{(i)}))]$ which expresses the expected versus obtained change of the objective function. The radius is initialized as $\delta^{(0)} = 1$ and then adjusted according to the gain as $\delta^{(i+1)} = 2\delta^{(i)}$ when $\rho > 0.75$, and $\delta^{(i+1)} = \delta^{(i)}/3$ when $\rho < 0.25$. The algorithm is terminated when $\|\boldsymbol{x}^{(i+1)} - \boldsymbol{x}^{(i)}\| \le \varepsilon$, or $\delta^{(i+1)} \le \varepsilon$, where $\varepsilon = 10^{-2}$ ($\|\cdot\|$ represents the Euclidean norm). Note that the method is data-efficient as it requires only $D + 1$ EM simulations ($D$ is the problem dimensionality) for construction of $\boldsymbol{G}^{(i)}$ per successful iteration, i.e., when $\rho > 0$. Additional EM model evaluation is required for each unsuccessful step. For more comprehensive discussion on the TR-based optimization, see [2, 12, 14].

### 3.3 Stratified Design Framework

The TR algorithm is embedded within a stratified design framework which gradually increases the number of antenna parameters [14]. The goal of the process is to first identify (using a low number of parameters) and then exploit (by increasing problem dimensionality) the promising regions of the search space. The method involves a series of TR-based optimizations of the antenna of Sect. 2 being represented using a specified number of variables. Let $\boldsymbol{L} = [L^{(0)} \ldots L^{(j)} \ldots L^{(J)}]^T$ ($j = 0, 1, \ldots, J$) be a vector that defines the number of spline parameters used to construct the antenna ground plane and radiator in consecutive meta-iterations. The algorithm is initialized using $L^{(0)}$, whereas at the beginning of each consecutive iteration ($j > 0$) the vectors $\boldsymbol{x}_g^{(j+1)}$ and $\boldsymbol{x}_r^{(j+1)}$ required for construction of $\boldsymbol{x}_0^{(j+1)}$ are obtained though interpolation of their counterparts extracted from $\boldsymbol{x}_{\text{opt}}^{(j)}$ found as a result of previous meta-iteration. The considered stratified design framework can be summarized as follows:

1. Specify $L$, set $j = 0$ and define $\boldsymbol{x}_0^{(j)}$;
2. Optimize $\boldsymbol{x}_0^{(j)}$ using algorithm of Sect. 3.2 to obtain $\boldsymbol{x}_{\text{opt}}^{(j)} = \boldsymbol{x}^*$;
3. If $j = J$ or $U(\boldsymbol{x}_{\text{opt}}^{(j+1)}) \le U(\boldsymbol{x}_{\text{opt}}^{(j)})$, set $\boldsymbol{x}^* = \boldsymbol{x}_{\text{opt}}^{(j)}$ and END; otherwise obtain $\boldsymbol{x}_g^{(j+1)}$, $\boldsymbol{x}_r^{(j+1)}$ by $L^{(j+1)}$-point interpolation of $\boldsymbol{x}_g^{(j)}$, $\boldsymbol{x}_r^{(j)}$ extracted from $\boldsymbol{x}_{\text{opt}}^{(j)}$ and go to Step 4;

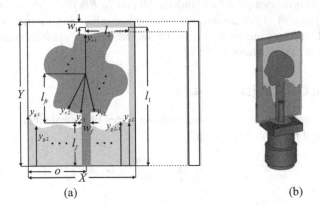

(a)                                            (b)

**Fig. 1.** A spline-based monopole: (a) geometry with highlight on the design parameters and (b) visualization of the structure. Note that $y_{g.l} = Y \cdot x_{g.l}$ and $y_{r.l} = S \cdot x_{r.l}$ ($l = 1, \ldots, L$; cf. Sect. 2).

4. Extract $x_c^{(j)}$ from $x_{opt}^{(j)}$ and define $x_0^{(j+1)} = [x_c^{(j)} \; x_g^{(j+1)} \; x_r^{(j+1)}]^T$, set $j = j + 1$ and go to Step 2.

It should be noted that the size of $L^{(j)}$ affects the optimization cost as the number of EM simulations required for construction of the linear model is proportional to the problem dimensionality (cf. Sect. 2.3). On the other hand, embedding the design task in the stratified framework mitigates the risk of getting stuck in a poor local optimum.

### 3.4 Objective Function

The objective function considered for optimization is given as (see Fig. 2 for conceptual illustration) [14]:

$$U(x, \alpha) = \begin{cases} U_1(x), & \text{when } \alpha = 0 \\ A(x) + \gamma_1 \max(U_1(x)/S_1, 0), & \text{when } \alpha = 1 \\ \max(S(x)) + \gamma_2 \max\left(\left(A(x) - A_1^{(i)}\right)/A_1^{(i)}, 0\right), & \text{when } \alpha = 2 \end{cases} \quad (3)$$

where $S(x) = |R(x)|_{f_L \leq f \leq f_H}$ represents the structure reflection response (in dB) within $f_L = 3.1$ GHz to $f_H = 10.6$ GHz, threshold $S_1 = -10$ dB, $A(x) = S \cdot Y$ denotes the antenna footprint (cf. Sect. 2), whereas $\gamma_1 = 1000$ and $\gamma_2 = 500$ are the scaling coefficients; $U_1(x) = \max(S(x)) - S_1$. The thresholds on the size $A_1^{(i)}$ are recorded when the function selector $\alpha$ (also referred to as a mode; see Fig. 2) is triggered.

The reasoning behind the use of composite objective (3) is to ensure the acceptable balance between antenna size and its electrical performance. When $\alpha = 0$ the optimization is oriented only towards minimization of the in-band reflection. Once attained (i.e., $U(x, \alpha = 0) < 0$), the algorithm records $A_1^{(i)}$ sets $\alpha = 1$ and performs explicit miniaturization of $A$ with penalty on $S_{11}$. When $\max(S(x)) > S_2$ (here, $S_2 = -9.5$ dB; cf. Figure 2), the mode is set to $\alpha = 2$, $A_1^{(i)}$ is stored, and the explicit minimization of reflection with penalty on size is performed until $\max(S(x)) < S_1$, when $\alpha = 1$ is selected again.

## 4 Numerical Results

The antenna is optimized using the framework of Sect. 3. The goal of the design process is to miniaturize the structure while maintaining its in-band reflection at the level of around $-10$ dB (controlled by $S_1$, $S_2$ thresholds). The vector of spline-knots for consecutive meta-iterations and the initial design are $L = [1 \; 8 \; 16 \; 24 \; 32]^T$, and $x_0^{(0)} = [10 \; 6 \; 16 \; 0.8 \; 1 \; 0 \; 0.35 \; 0.6]^T$, respectively. The final design $x^* = x_{opt}^{(2)} = [10.99 \; 4.87 \; 15.37 \; 1 \; 1.53 \; -0.31 \; 0.45 \; 0.49 \; 0.24 \; 0.2 \; 0.47 \; 0.2 \; 0.21 \; 0.35 \; 0.52 \; 0.49 \; 0.78 \; 0.39 \; 0.89 \; 0.68 \; 0.59 \; 0.6]^T$ is obtained after three meta-iterations. The algorithm termination is triggered at $j = 2$, due to lack of the objective function improvement.

The optimized design is characterized by reflection below $-10$ dB within the entire bandwidth and the dimensions of $10.99 \times 16.91$ mm. Although the final solution is larger by 18 mm$^2$ compared to the starting point (the footprint of 182.4 mm$^2$ at $x^*$ vs. 170 mm$^2$ at $x_0^{(0)}$), its in-band reflection is $-5.7$ dB lower—w.r.t. to maximum in-band reflection of $-4.2$ dB at $x_0^{(0)}$—which demonstrates flexibility of the discussed framework and composite objective function in terms of balancing the electrical performance and physical dimensions of the radiator. The convergence curves for the TR algorithm at the first two meta-steps, as well as the reflection responses at the selected designs are shown in Fig. 3. The geometry of the optimized structure and its radiation-pattern characteristics obtained at 4 GHz and 6.5 GHz (in xy-plane) are given in Fig. 4. Note that the considered test frequencies are centered w.r.t. channels 2 and 5 of the UWB band [3]. The radiation patterns are fairly omnidirectional. Their distortion around the $-90°$ direction is due to implementation of the L-shaped ground plane extension which supports miniaturization, but also acts as an obstacle for the radiated field (cf. Figure 1). Nonetheless, owing to small size, the radiator is suitable for wearable applications.

The considered framework has been compared against a state-of-the-art TR-based algorithm in terms of cost and performance [12]. The benchmark method has been used for direct optimization of the antenna represented by a 38-dimensional vector of parameters. The process was controlled only by the objective function (3) with $\alpha = 1$. For fair comparison, the same initial design $x_0^{(0)}$ has been set for both algorithms. The results gathered in Table 1 indicate that, although the benchmark algorithm generated smaller design at a lower cost, it failed at fulfilling the performance-related specifications. From this perspective, the considered stratified design approach offers improved balance between the antenna size and its electrical properties.

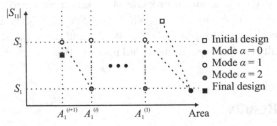

**Fig. 2.** TR-based optimization of the antenna using a composite objective function that balances the electrical- and size-related specifications.

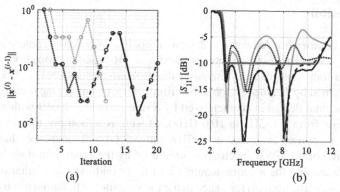

(a)                                             (b)

**Fig. 3.** A spline-based monopole: (a) convergence plots for the first 20 TR-based iterations at the meta-steps $j = 0$ (gray) and $j = 1$ with $\alpha = 0$ ($\cdots$), $\alpha = 1$ ($--$) and $\alpha = 2$ ($-$), as well as (b) responses at $x_0^{(0)}$ (gray) and $x_{opt}^{(0)}$ ($\cdots$), $x_{opt}^{(1)}$ ($—$), and $x_{opt}^{(2)}$ ($--$) designs. Note that the performance $x_{opt}^{(2)}$ is slightly worse w.r.t. $x_{opt}^{(1)}$ due to interpolation between meta-iterations.

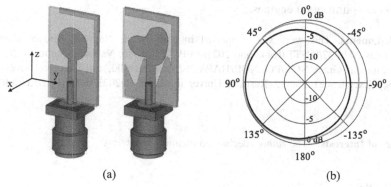

(a)                                             (b)

**Fig. 4.** A spline-based antenna: (a) in-scale comparison of geometries at $x_0^{(0)}$ (left) and $x^*$ (right) designs, as well as (b) xy-plane radiation patterns at 4 GHz (gray) and 6.5 GHz (black).

**Table 1.** Benchmark of the stratified design optimization framework.

| Method | Meta-step – cost [$R$] | | | Total cost [$R$] | Size [mm$^2$] | max($S(x)$) [dB] |
|---|---|---|---|---|---|---|
| | 1 | 2 | 3 | | | |
| TR with $\alpha = 1$ | 241 | – | – | 241 | 171.6 | $-8.5^*$ |
| This work | 76 | 491 | 39 | 606 | 182.4 | $-10$ |

* The optimized design violates the performance specifications resulting from $S_1$

## 5  Conclusion

In this work, a framework dedicated to design of topologically flexible antenna structures has been discussed. The method involves a series of TR-based optimizations embedded within a meta-loop that gradually increases the problem dimensionality. The TR-based design is governed by a composite objective function that balances the size/performance trade-off. The final design is characterized by a reflection of $-10$ dB within the UWB bandwidth (i.e., from 3.1 GHz to 10.6 GHz), as well as a footprint of only 182 mm$^2$. The far-field responses evaluated at 4 GHz and 6.5 GHz frequencies (that represent channel 2 and 5 of the UWB spectrum – often used by the existing in-door localization systems) indicate that the antenna features a fairly omnidirectional radiation patterns. Small dimensions and high performance make the structure of potential use for wearable in-door positioning devices. The benchmark of the considered framework against a standard TR-based optimization demonstrates that, despite higher computational cost, the method offers improved balance between the performance figures and dimensions. Future work will focus on application of the antenna prototypes as components of a real-world in-door positioning system, as well as comparison of the considered method against other optimization engines.

**Acknowledgments.** This work was supported in part by the National Science Centre of Poland Grants 2020/37/B/ST7/01448 and 2021/43/B/ST7/01856, National Centre for Research and Development Grant NOR/POLNOR/HAPADS/0049/2019-00, and Gdansk University of Technology (Excellence Initiative - Research University) Grant 16/2023/IDUB/IV.2/EUROPIUM.

**Disclosure of Interests.**  The authors declare no conflicts of interest.

## References

1. Zafari, F., Gkelias, A., Leung, K.K.: A survey of indoor localization systems and technologies. IEEE Commun. Surv. Tutorials. **21**(3), 2568–2599 (2019)
2. Koziel, S., Bekasiewicz, A., Cheng, Q.S.: Conceptual design and automated optimization of a novel compact UWB MIMO slot antenna. IET Microwaves Antennas Propag. **11**(8), 1162–1168 (2017)
3. Coppens, D., Shahid, A., Lemey, S., Van Herbruggen, B., Marshall C., De Poorter, E.: An Overview of UWB standards and organizations (IEEE 802.15.4, FiRa, Apple): interoperability aspects and future research directions. IEEE Access. **10**, 70219–70241 (2022)
4. Bekasiewicz, A., Koziel, S., Zieniutycz, W.: Design space reduction for expedited multi-objective design optimization of antennas in highly-dimensional spaces. In Solving Computationally Expensive Engineering Problems: Methods and Applications, pp. 113–147, Springer, Heidelberg (2014). https://doi.org/10.1007/978-3-319-08985-0_5
5. Al-Bawri, S.S., et al.: Compact ultra-wideband monopole antenna loaded with metamaterial. Sensors **20**, 796 (2020)
6. Khan, M.S., Capobianco, A., Asif, S.M., Anagnostou, D.E., Shubair, R.M., Braaten, B.D.: A compact CSRR-enabled UWB diversity antenna. IEEE Antennas Wirel. Propag. Lett. **16**, 808–812 (2017)

7. Mroczka, J.: The cognitive process in metrology. Measurement **46**, 2896–2907 (2013)
8. Koziel, S., Bekasiewicz, A.: Comprehensive comparison of compact UWB antenna performance by means of multiobjective optimization. IEEE Trans. Antennas Propag. **65**(7), 3427–3436 (2017)
9. Alnas, J., Giddings, G., Jeong, N.: Bandwidth improvement of an inverted-f antenna using dynamic hybrid binary particle swarm optimization. Appl. Sci. **11**, 2559 (2021)
10. Dong, B., Yang, J., Dahlstrom, J., Flygare, J., Pantaleev, M., Billade, B.: Optimization and realization of quadruple-ridge flared horn with new spline-defined profiles as a high-efficiency feed from 4.6 GHz to 24 GHz. IEEE Trans. Antennas Propag. **67**(1), 585–590 (2019)
11. Ghassemi, M., Bakr, M., Sangary, N.: Antenna design exploiting adjoint sensitivity-based geometry evolution. IET Microwaves Antennas Propag. **7**, 268–276 (2013)
12. Conn, A., Gould, N.I.M., Toint, P.L.: Trust-region Methods. MPS-SIAM Series on Optimization, Philadelphia (2000)
13. Koziel, S., Pietrenko-Dabrowska, A.: Performance-based nested surrogate modeling of antenna input characteristics. IEEE Trans. Antennas Propag. **67**(5), 2904–2912 (2019)
14. Bekasiewicz, A., Kurgan, P., Koziel, S.: Numerically efficient miniaturization-oriented optimization of an ultra-wideband spline-parameterized antenna. IEEE Access. **10**, 21608–21618 (2022)
15. CST Microwave Studio, Dassault Systems, 10 rue Marcel Dassault, CS 40501, Vélizy-Villacoublay Cedex, France (2015)

# Beneath the Facade of IP Leasing: Graph-Based Approach for Identifying Malicious IP Blocks

Zhenni Liu[1,2], Yong Sun[1,2(✉)], Zhao Li[1,2], Jiangyi Yin[1,2], and Qingyun Liu[1,2]

[1] Institute of Information Engineering, Chinese Academy of Sciences, Beijing, China
{liuzhenni,sunyong,lizhao,yinjiangyi,liuqingyun}@iie.ac.cn
[2] School of Cyber Security,
University of Chinese Academy of Sciences, Beijing, China

**Abstract.** With the depletion of IPv4 address resources, the prevalence of IPv4 address leasing services by hosting providers has surged. These services allow users to rent IP blocks, offering an affordable and flexible solution compared to traditional IP address allocation. Unfortunately, this convenience has led to an increase in abuse, with illegal users renting IP blocks to host malicious content such as phishing sites and spam services. To mitigate the issue of IP abuse, some research focuses on individual IP identification for point-wise blacklisting. However, this approach leads to a game of whack-a-mole, where blacklisted IPs become transient due to content migration within the IP block. Other studies take a block perspective, recognizing and classifying IP blocks. This enables the discovery of potentially malicious IPs within the block, effectively countering service migration issues. However, existing IP block identification methods face challenges as they rely on specific WHOIS fields, which are sometimes not updated in real-time, leading to inaccuracies. In terms of classification, methods rely on limited statistical features, overlooking vital relationships between IP blocks, making them susceptible to evasion. To address these challenges, we propose BlockFinder, a two-stage framework. The first stage leverages the temporal and spatial stability of services to identify blocks of varying sizes. In the second stage, we introduce an innovative IP block classification model that integrates global node and local subgraph representations to comprehensively learn the graph structure, thereby enhancing evasion difficulty. Experimental results show that our approach achieves state-of-the-art performance.

**Keywords:** IP blocks detection · Graph representation learning

## 1 Introduction

In recent years, IPv4 address leasing services offered by hosting providers have become increasingly prevalent. These services often involve the rental of consecutive IP addresses, known as "IP Blocks," for hosting Internet services. However, recent research [1] has highlighted a challenge where illegal users exploit these IP

L. Franco et al. (Eds.): ICCS 2024, LNCS 14834, pp. 46–53, 2024.
https://doi.org/10.1007/978-3-031-63759-9_6

blocks, frequently migrating malicious services within them, making it difficult for network authorities to efficiently identify malicious IPs.

To tackle this issue, we propose BlockFinder, an automated framework designed for the identification of malicious IP blocks. BlockFinder focuses on identifying malicious IPs at the block level, effectively combating the issue of malicious service migration within IP blocks. The framework consists of two stages:

**IP Block Identification.** According to the previous work [6], the IP blocks leased from hosting providers may not be meticulously recorded by RIRs as with IP allocations. Relying on public data, such as IP WHOIS records, to obtain IP blocks is not practical. Therefore, we propose a novel method based on the service stability of IP blocks. This approach ensures that all IPs within each IP block belong to the same entity, rather than roughly treating the entire Autonomous System (AS) IP address space as a single IP block.

**IP Block Classification.** Current IP block classification methods [1] often rely on passive flow statistics, which are susceptible to evasion. We propose a model that utilizes a comprehensive analysis of statistical features and graph-based behaviors. For graph-based detection, we leverage the observation that services migrate between different blocks, and IP block subgraphs reveal communication patterns. Specifically, we integrate a novel combination of node and subgraph representation, enhancing identification effectiveness even in scenarios with isolated nodes.

## 2   Related Work

Existing methods for identifying malicious IPs can be categorized into those that recognize from the perspective of individual IPs and those that identify from the perspective of IP blocks.

Individual IP perspective methods, like those by Alvarez et al. [2] and Coskun et al. [3], cluster IPs based on communication destination to find similarities with blacklisted IPs. However, these methods may not promptly detect the migration of malicious content within blocks.

IP block perspective methods include AS or hosting provider reputation-based approaches and IP block reputation-based methods. The former calculates a maliciousness score using meta-information from AS or hosting provider IPs [5, 8], but may miss smaller abused IP blocks. IP block reputation methods evaluate services' maliciousness within smaller IP blocks. For instance, Alrwais et al. [1] identified sub-allocated IP blocks from reputable hosting providers exploited for hosting malicious content, using IP WHOIS and PDNS. However, challenges arise due to outdated WHOIS records [6].

The rise in hosting malicious content on IP blocks from reputable providers underscores the need for effective identification methods. Yet, existing WHOIS-based methods face limitations due to delayed updates [6]. Meanwhile, relying solely on PDNS for IP block classification lacks data diversity and is vulnerable to evasion. Thus, more effective identification methods are necessary (Fig. 1).

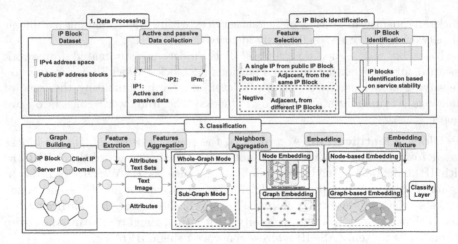

**Fig. 1.** The architecture of BlockFinder.

# 3   Approach

## 3.1   IP Block Identification

We propose a technical approach to achieve fine-grained IP block identification based on the service stability of IP blocks. Through extensive observation of both active probing and passive traffic data, we have identified that the stability of services within the IP block is inherently maintained, demonstrating in two aspects:

- **Temporal stability.** Significant changes in services within the block are uncommon over short periods. As shown in the Fig. 2, on IP Block1 consisting of two consecutive IP addresses 108.\*.\*.249[1] and 108.\*.\*.249, the service set {zimm.\*.com, harri.\*.com} is observed at T1, and {harri.\*.com, zimm.\*.com} is observed at T2. Despite dynamic services changes on individual IPs, the overall service set of the entire IP block remains the same, indicating high stability.
- **Spatial stability.** The service set distribution within an IP block remains stable. Distinct service sets are observed on different IP blocks. As depicted in Fig. 2, the service set {zimm.\*.com, harri.\*.com} consistently resides on IP Block1, while IP Block2 hosts the stable service set {luc.\*.com, weiss.\*.com}.

Based on the description, we define the stability contribution resulting from dividing two IPs into the same IP block as $W(IP_p, IP_q)$, shown in Eq. 1. Here, $Sim(IP_p, IP_q)$ represents the similarity of their service sets, and $I(IP)$ represents the information amount of the service set. $S(i,j)$ denotes the sum of stability contributions brought by each IP in the IP block. The objective of IP block identification is to determine division locations in the continuous IP address space to maximize the overall stability value of identified IP blocks.

---

[1] In order to protect user privacy, we use "\*" to represent key locations.

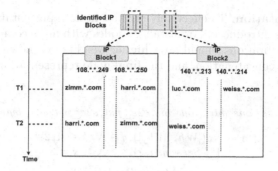

**Fig. 2.** Service stability.

$$W(IP_p, IP_q) = \begin{cases} Sim(IP_p, IP_q) & I(IP_p) > \alpha_i, I(IP_q) > \alpha_i, \\ & Sim > \alpha_s \\ 0 & I(IP_p) < \alpha_i \, or \, I(IP_q) < \alpha_i \\ punish(punish < 0) & I(IP_p) > \alpha_i, I(IP_q) > \alpha_i, \\ & Sim < \alpha_s \end{cases} \quad (1)$$

$$S(i, j) = \frac{\frac{1}{2} \sum_{p \in [i,j]} \sum_{q \in [i,j]} W(IP_p, IP_q)}{j - i + 1}, (p \neq q) \quad (2)$$

### 3.2  IP Block Classification

Based on the obtained IP block in the previous section, the constructed heterogeneous graph is depicted in Fig. 3. Building upon this graph, we introduce the HGNT-Net algorithm (Heterogeneous Graph Node and Topology representation learning Network). This algorithm advocates for incorporating both the node representation of IP blocks and the heterogeneous subgraph topology centered around IP blocks during the learning phase.

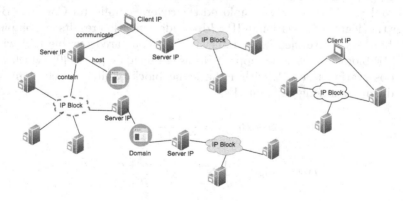

**Fig. 3.** Example of IP Block Graph.

**Node Representation.** To effectively extract node representations, an attention mechanism is introduced. For instance, nodes with more communication sessions and longer durations should have higher weights, as shown in Eq. 4. Here, || denotes feature concatenation, and the final node representation is determined as shown in Eq. 5.

$$Session(i,j) = [sessions, duration\_ms, c2s\_pkt, s2c\_pkt, c2s\_byte, s2c\_byte] \quad (3)$$

$$e_{ij}^{lk} = \begin{cases} a([Wh_i \| Wh_j]), j \in \mathcal{N}_i \ k \in [1,2] \\ Session(i,j), j \in \mathcal{N}_i \quad k = 3 \end{cases} \quad (4)$$

$$h_i^{l+1} = \sigma(\|_{k=1}^{K} \sum_{j \in \mathcal{N}_i} \frac{exp(LeakyReLU(e_{ij}^{lk}))}{\sum_{q \in \mathcal{N}_i} exp(LeakyReLU(e_{iq}^{lk}))} W_k^l h_j^l + W_0^l h_i^l) \quad (5)$$

**Subgraph Representation.** The heterogeneous subgraph representation learning begins by extracting the heterogeneous subgraph centered on each IP block. This subgraph exhibits unique topology and content characteristics. Content representation $h_{gc}$ focuses on the content of all relevant nodes in the IP block subgraph. The topology representation $h_{gt}$ primarily depicts the network structure of the IP block subgraph, including its degree, network density, diameter, and clustering coefficient. Finally, the node representation and subgraph representation are merged and fed into the classification layer, as shown in the Eq. 6.

$$h = \beta h_n \| \gamma(h_{gc} \| h_{gt}) \quad (6)$$

## 4   Experimental Evaluation

### 4.1   IP Block Identification

We collected a total of 999 IP blocks from FireHOL[2] IP List within the 122 /24 IP prefixes. Since IP block identification results are represented as ranges and cannot be evaluated using traditional accuracy metrics, we utilized modified evaluation indicators commonly used in clustering scenarios: the IP block outline coefficient (S-Block) and Davies-Bouldin coefficients (DB coefficients), as shown in Eq. 7 and 8. Additionally, we employed the coverage indicator Coverage-Block to assess the degree of deviation in IP address identification results by comparing test IP blocks with verified block intervals. We compared our method with a classical IP block identification approach based on signal pulses [9], which relies on the observation that IPs within the same block exhibit closely connected feature distances among adjacent IPs.

$$S - Block = \frac{b - a}{\max(a,b)} \quad (7)$$

$$DB - Block = \frac{1}{n} \sum_{i=1}^{n} \max(\frac{\sigma_i + \sigma_j}{dist(c_i, c_j)}) \quad (8)$$

---

[2] FireHOL. https://firehol.org/.

Our method identified a total of 968 IP blocks within the same 122 /24 IP prefixes. The S-Block value for our result is 0.55, and the DB-Block is 8.7. In contrast, the S-Block for the signal pulse method is 0.31, with a DB-Block of 11.9. This demonstrates that our method achieves better cohesion within identified IP blocks and better separation between blocks. The coverage results are provided in Table 1. Here, ValBlock represents the collected IP blocks, TestBlock represents the IP blocks identified by our method, and ValBlock rate represents the proportion of total ValBlocks. It's observed that nearly 90% of the ValBlocks are entirely encompassed within the TestBlock. This is primarily because ValBlock is manually collected and may not provide comprehensive coverage. Our method offers broader observation capabilities, hence including more IPs.

**Table 1.** Coverage-Block results.

| Coverage Type | Coverage-Block | ValBlock Rate |
|---|---|---|
| $ValBlock \subseteq TestBlock$ | 100% | 86.99% |
| $TestBlock = ValBlock$ | 100% | 1.9% |
| $TestBlock \subseteq ValBlock$ | 59.09% | 2.4% |
| $ValBlock \cap TestBlock \subseteq ValBlock$ | 61.5% | 10.91% |

### 4.2 IP Block Classification

The labeled IP block dataset is derived from the 968 IP blocks identified in Sect. 4.1. Among these, 300 IP blocks were labeled as malicious by FireHOL, while 100 blocks were labeled as benign. Table 2 shows the number of nodes and edges associated with the aforementioned 968 IP blocks. Evaluation indicators are presented in Eqs. 9, 10 and 11.

**Table 2.** Heterogeneous graph nodes.

| Node Type | Node Count | Edge Type | Edge Count |
|---|---|---|---|
| ip_block | 968 | IP_block-server_ip | 9399 |
| server_ip | 9399 | server_ip-client_ip | 14350 |
| client_ip | 14350 | server_ip-domain | 6485 |
| domain | 5251 | – | – |

$$Specificity = Recall@neg = \frac{TN}{TN + FP} \tag{9}$$

$$Precision@neg = \frac{TN}{TN + FN} \tag{10}$$

$$F1@neg = \frac{2 \cdot Precision@neg \cdot Recall@neg}{Precision@neg + Recall@neg} \tag{11}$$

To comprehensively verify the effectiveness of HGNT-Net, we conducted two sets of experiments. The first set compares our approach with classic heterogeneous graph node classification algorithms: RGCN [7] and HGT [4]. RGCN is a foundational work on basic heterogeneous graph node classification, while HGT introduces an attention mechanism. Our algorithm, HGNT-Net, integrates both node and subgraph representation. The second set comprises ablation experiments, comparing: 1) using only node features without considering relationships between nodes; 2) considering relationships between nodes without incorporating subgraph topology; and 3) integrating both relationships between nodes and subgraph topology.

Experimental results are summarized in Table 3. Compared to RGCN, HGT introduces an attention mechanism allocating different attention scores to contributions of different paths. HGNT-Net not only integrates the attention mechanism for nodes but also incorporates the topological characteristics of subgraphs to further enhance model performance. Experiments demonstrate that our method outperforms others across various indicators.

Table 3. The experimental results of comparison algorithm.

| Methods | Accuracy | Precision | Specificity | F1@neg | AUC |
|---|---|---|---|---|---|
| RGCN | 0.88 | 0.76 | 0.76 | 0.82 | 0.85 |
| HGT | 0.83 | 0.81 | 0.8 | 0.74 | 0.9 |
| Without using graph | 0.77 | 0.75 | 0.75 | 0.77 | 0.82 |
| Only node representation | 0.85 | 0.83 | 0.83 | 0.73 | 0.86 |
| HGNT-Net | **0.91** | **0.87** | **0.87** | **0.86** | **0.96** |

### 4.3   Measurement

Using BlockFinder, we conducted measurements on the top 5 IP address spaces of hosting provider M247. Initially, we identified IP blocks within the target IP address spaces using our method based on service stability. Subsequently, we updated the heterogeneous graph based on the identified IP blocks. Finally, we utilized HGNT-Net to identify malicious IP blocks within these spaces. In total, we identified 4,672 IP blocks out of the 5 IP address spaces, comprising a total of 300,000 IPs, with 1,101 of them classified as malicious. We systematically measured the size and utilization rate of these malicious blocks. Our findings revealed that the average size of malicious IP blocks is approximately 32, ensuring swift evasion without arousing suspicion due to excessively large malicious scope. The utilization rate of nearly half of the IP blocks is only 40%, indicating underutilization of IPs within the blocks. In the event of an IP being blacklisted, services can be migrated to other IPs within the same block, thus prolonging the service's survival time. The measurements were conducted on a CPU AMD Ryzen 5 5600H with Radeon Graphics @ 3.30 GHz. Since IP block leases from cloud hosting providers typically have minimum durations, such as one month,

IP blocks usually do not change frequently. Additionally, the number of IP block nodes is typically around $1/100$ of individual IP nodes. Due to the relatively infrequent updates of IP blocks and the significant reduction in scale compared to individual IPs, efficient periodic recognition can be achieved.

## 5 Conclusion

In this paper, we introduce BlockFinder, a novel framework for identifying malicious IP blocks. Initially, we employ an IP block identification method based on service stability to identify blocks of varying sizes. Subsequently, our classification model, HGNT-Net, enhances performance in scenarios with isolated nodes and limited information dissemination by integrating node and subgraph representations centered on IP blocks. The algorithm's effectiveness is demonstrated through ablation experiments and comparisons with classic algorithms. Our framework effectively addresses the challenge of malicious services migrating within IP blocks by detecting malicious IPs from a block perspective.

**Acknowledgments.** This work is supported by National Key R&D Program of China (Grant No.2021YFB3101001) and the Scaling Program of Institute of Information Engineering, CAS (Grant No.E3Z0191101).

## References

1. Alrwais, S., et al.: Under the shadow of sunshine: understanding and detecting bulletproof hosting on legitimate service provider networks. In: 2017 IEEE Symposium on Security and Privacy (SP), pp. 805–823. IEEE (2017)
2. Cid-Fuentes, J.Á., Szabo, C., Falkner, K.: An adaptive framework for the detection of novel botnets. Comput. Secur. **79**, 148–161 (2018)
3. Coskun, B.: (un) Wisdom of crowds: accurately spotting malicious IP clusters using not-so-accurate IP blacklists. IEEE Trans. Inf. Forensics Secur. **12**(6), 1406–1417 (2017)
4. Hu, Z., Dong, Y., Wang, K., Sun, Y.: Heterogeneous graph transformer. In: Proceedings of the Web Conference 2020, pp. 2704–2710 (2020)
5. Konte, M., Perdisci, R., Feamster, N.: Aswatch: an as reputation system to expose bulletproof hosting ASEs. In: Proceedings of the 2015 ACM Conference on Special Interest Group on Data Communication, pp. 625–638 (2015)
6. Noroozian, A., et al.: Platforms in everything: analyzing {Ground-Truth} data on the anatomy and economics of {Bullet-Proof} hosting. In: 28th USENIX Security Symposium (USENIX Security 19), pp. 1341–1356 (2019)
7. Schlichtkrull, M., Kipf, T.N., Bloem, P., van den Berg, R., Titov, I., Welling, M.: Modeling relational data with graph convolutional networks. In: Gangemi, A., et al. (eds.) ESWC 2018. LNCS, vol. 10843, pp. 593–607. Springer, Cham (2018). https://doi.org/10.1007/978-3-319-93417-4_38
8. Shue, C.A., Kalafut, A.J., Gupta, M.: Abnormally malicious autonomous systems and their internet connectivity. IEEE/ACM Trans. Network. **20**(1), 220–230 (2011)
9. Xie, Y., Yu, F., Achan, K., Gillum, E., Goldszmidt, M., Wobber, T.: How dynamic are IP addresses? In: Proceedings of the 2007 Conference on Applications, Technologies, and Protocols for Computer Communications, pp. 301–312 (2007)

# A New Highly Efficient Preprocessing Algorithm for Convex Hull, Maximum Distance and Minimal Bounding Circle in E²: Efficiency Analysis

Vaclav Skala[✉][iD]

Faculty of Applied Sciences, Department of Computer Science and Engineering,
University of West Bohemia, Pilsen, CZ 301 00, Czech Republic
skala@kiv.zcu.cz

**Abstract.** This contribution describes an efficient and simple preprocessing algorithm for finding a convex hull, maximum distance of points or convex hull diameter, and the smallest enclosing circle in $E^2$. The proposed algorithm is convenient for large data sets with unknown intervals and ranges of the data sets. It is based on efficient preprocessing, which significantly reduces points used in final processing by standard available algorithms.

**Keywords:** Preprocessing · maximum distance · smallest enclosing circle · smallest enclosing ball · algorithm complexity · preprocessing · convex hull · convex hull diameter

## 1 Introduction

Many sophisticated algorithms are solving geometrical or computational problems, mostly evaluated according to their computational (asymptotic) complexity expecting the number of processed elements $N \mapsto \infty$.

Algorithms like maximum distance of points, i.e. convex hull diameter, convex hull, and minimal enclosing circle in $E^2$ are typical examples with known computational complexities with many modifications claiming better asymptotic computational complexity and faster run-time. Usually, a small attention is given to possible preprocessing strategies, which can significantly improve the total run-time and memory needed in some cases.

Let's consider an elementary problem: Find the maximum distance of two points in $E^2$ for the given a set $\Omega$ of points $\mathbf{x}_i$, $i = 1, \dots, N$ and $N$ is "reasonably" high. There are several strategies:

1. Simple algorithm based on mutual finding $d_{max} = \max\limits_{i,j, \ \& \ i<j} \|\mathbf{x}_i - \mathbf{x}_j\|_2$

   It leads to $O(N^2)$ the computational complexity (the algorithm requires $N(N-1)/2$ steps), which is prohibitive even for a small $N$.

This work was supported by the MEYS CZ, Institutional Support for LCDRO.

L. Franco et al. (Eds.): ICCS 2024, LNCS 14834, pp. 54–61, 2024.
https://doi.org/10.1007/978-3-031-63759-9_7

2. Convex Hull (CH) computation, e.g. using the Kirkpatrick-Seidel algorithm [1] with $O(N \log h)$, followed by finding a maximum distance of the remaining $h \ll N$ convex hull points using the algorithm with computational complexity $O(h^2)$. It should be noted, that of the given set $\Omega$ of unordered points form a circle, the number of processed points is $h = N$.

Using the Convex Hull algorithm, can be also used to accelerate the minimum enclosing circle algorithm [2,3,9,11–14]. A simple and efficient algorithm for finding the minimum distance of points was introduced in [5–8], see Fig. 2.

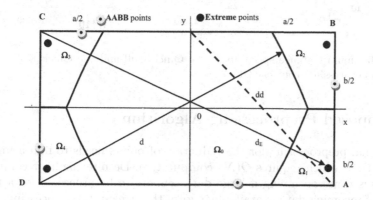

**Fig. 1.** Data domain subdivision; courtesy [10].

The algorithm uses a preprocessing with computational complexity $O(N)$ based on simple steps:

1. Find the Axis Aligned Bounding Box (AABB) of points in the given data set $\Omega$.
2. Find the maximum mutual distance $d = \max\{dist(AC), dist(BD)\}$.
3. Split points into data subsets $\Omega_0, \ldots, \Omega_4$, see Fig. 1, subset are defined by arcs given by the radius $d$ and by the corners of the AABB.
   Points inside the subsets $\Omega_0$ can be removed from the further processing directly.
4. Find the maximum mutual distance of points in pairs of subsets: $(\Omega_1, \Omega_3)$, $(\Omega_2, \Omega_4)$, $(\Omega_1, \Omega_2)$, $(\Omega_2, \Omega_4)$, $(\Omega_3, \Omega_4)$, $(\Omega_4, \Omega_1)$.

In the case of the uniform point's distribution, this approach leads to a maximum distance algorithm with computational complexity $O_{expected}(N)$, see Fig. 2. Detailed descriptions can be found in [5,10].

Several modifications of the that based on orthogonal and polar space subdivision [6,9], extension used in 3D convex hull algorithm [8], etc. However, by a deeper analysis of the preprocessing step, additional significant improvements can be made.

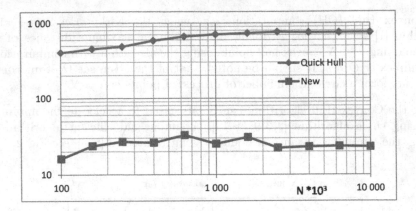

**Fig. 2.** Maximum distance speed-up of the Quick Hull and the Quick Hull with the original preprocessing; courtesy [10].

## 2   Proposed Preprocessing Algorithm

The original preprocessing for the reduction of points needs to find a min-max box (AABB), which requires $O(N)$ computation. Despite the liner complexity, this step is time-consuming if large data sets are to be processed, or a limited storage of incoming data is available, e.g. in the stream data processing.

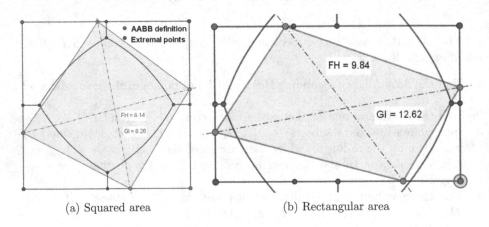

(a) Squared area                    (b) Rectangular area

**Fig. 3.** Squared and rectangular areas and the selection function influence

### 2.1   Basic Idea

Let us consider situations in Fig. 3. In the case of the square data domain, Fig. 3a, the tests based on a circular segment containment or a half-space test seem to be more or less equivalent. In the case of the rectangular AABB, Fig. 3b, the circular segment test is more efficient. However, the areas $\Omega_i$, $i = 1, \ldots, 4$ are too large.

It can be seen that the area of points which can be directly excluded is significantly larger. If the AABBox is known, the closest points to the AABBox corners can be found, see Fig. 4, with the computational complexity $O(N)$.

If the AABBox is known, the nearest points to the AABBox vertices can be found with computational complexity $O(N)$. Then the area $\Omega_0$ is defined by a convex polygon and all points inside to the area $\Omega_0$ can be removed from further processing. In the following step, the remaining points will be split into the other areas $\Omega_i$. However, this step requires additional large memory allocation.

However, the removal steps of finding AABBox and finding the closest points to the AABBox vertices lead to a more efficient algorithm.

## 2.2   Increasing Efficiency

Let us assume, that from a small sample of points, the AABBox and the closest points to the AABBox vertices are obtained. Then the extreme points on the AABBox edges form separating half-planes defined by points $A, C$ and $B, D$, see Fig. 4.

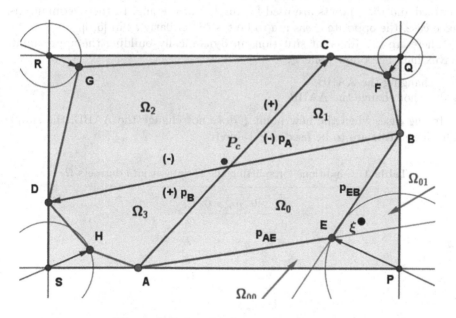

**Fig. 4.** Splitting the dataset to subsets $\Omega_i$

It means, that for any point determining to which set of points it belongs to, is computationally simple. Even more, each basic area is split to areas $\Omega_i$, $\Omega_{i0}$ and $\Omega_{i1}$ which are formed by the points $E, F, G, H$, i.e. the points closest the

AABBox vertices found so far. The point-in-area test is computationally simple as only tests for two half-planes given by points $A, E$ and $E, B$ are needed in the case of $\Omega_i$, $\Omega_{i0}$ and $\Omega_{i1}$, similarly for other areas.

Then, for all points $\xi$ in the given data set $\Omega$ the following steps are made.

1. will fall into the convex hull of those points or one of the side areas outside; the point is excluded from further processing, or
2. change of the position of some points of the convex hull, i.e. new closest point to an AABBox vertex found, then the half-planes given by points $A, E$ and $E, B$ has to be recomputed, or
3. change of the position of extreme points forming AABBox, e.g. position of the point $A$, then the half-planes $A, E$ and $H, A$ have to be recomputed and check if the vertices $H$ and $E$ remained convex; in the concave case, the relevant vertex has to be replaced by a virtual one laying the line $AB$ or $DA$, see Fig. 4.

Note, that the AABBox and eight-point convex hull are changing dynamically as points are processed with the complete computational complexity $O(N)$.

However, the areas $\Omega_{i0}$ and $\Omega_{i1}$ contain also invalid points due to their incremental construction, and have to be rechecked and non-relevant points have to be removed.

After those two steps above, eight subsets $\Omega_{i0}$ and $\Omega_{i1}$, $i = 0,\ldots,3$ are obtained and their points are used for further processing. In the maximum distance case, the opposite areas are to be tested similarly as in [5,6].

There are two different situations in dynamically building the approximate convex hull, i.e. a new point $\xi$:

1. is changing the AABB
2. does not change the AABB

In the case, when the new point $\xi$ does not change the AABB, the simple half-planes tests are to be used, see Table 1:

**Table 1.** Conditions for splitting the $\Omega$ dataset into datasets $\Omega_i$

| $p_A > 0$ | $p_B > 0$ | $\Omega_i$ |
|---|---|---|
| + | + | $\Omega_0$ |
| + | − | $\Omega_1$ |
| − | + | $\Omega_3$ |
| − | − | $\Omega_2$ |

Let us consider a new point $\xi$ in the $\Omega_0$ area. There are the following possible cases:

– the point $\xi$ is closer to the AABB corner $P$ than the CH corner $E$, then the point $\xi$ replaces the CH point $E$, Fig. 5, and the line $p_P$ has to be recomputed, i.e. the areas $\Omega_{00}$ and $\Omega_{01}$ are changed.

- the position of the point $\xi$ can be in an area $\Omega_{00}$ or $\Omega_{01}$. If the point $\xi$ is in the area $\Omega_{00}$, resp. $\Omega_{01}$, the point is stored in the relevant list.
- the point $\xi$ is not in area $\Omega_{00}$ nor $\Omega_{01}$, the point is removed from the future processing.

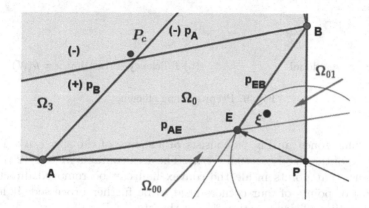

**Fig. 5.** Corner detail with $\Omega_{00}$ and $\Omega_{01}$ subareas specification

Now, an approximate convex hull (A-CH) has been constructed. It is defined by eight points, i.e. $A, E, B, F, C, G, D, H, A$. It means, that after this preprocessing step of all points with $O(N)$ complexity, points are:

- some points are stored in the lists $\Omega_{i0}$, resp. $\Omega_{i1}$, $i = 0, \ldots, 3$,
- some points have been removed directly during this preprocessing step.

Now, points in $\Omega_{i0}$ and $\Omega_{i1}$ are re-checked and points inside the A-CH are to be removed. It means, that points inside the $\Omega_{i0}$, resp. $\Omega_{i1}$, $i = 0, \ldots, 3$ areas form only a very small fraction of the original data set.

## 2.3   Preprocessing Efficiency

The preprocessing algorithm described above is of the $O(N)$ computational complexity and computational requirements are very low, actually half-plane tests only[1]. As the preprocessing algorithm is intended for algorithms with a higher computational complexity, the efficiency of the preprocessing algorithm is an important issue.

For the sake of simplicity, let us consider data from the domain $[0, 1] \times [0, 1]$ with $N = n^2$ points, and points forming the AABBox are in the middle of the AABB edges. Figure 6a presents a detail of the AABB corner areas, formed by a square and two triangles.

---

[1] Half-plane test is implemented as the dot-product, and a separating plane line is determined by the outer-product (cross product in this case) [4].

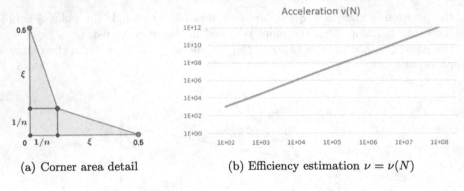

(a) Corner area detail          (b) Efficiency estimation $\nu = \nu(N)$

**Fig. 6.** Preprocessing efficiency

Then the "blue zones" in Fig. 6a consists of a square of the size $1/n \times 1/n$ and two triangles of the size $1/n \times (0.5 - 1/n)$. The area contains $M = 1/(2n)$ points.

It means, that points inside the convex hull can be removed directly and only $4/(2n)$ of points of four corners need to be further processed. It leads to the preprocessing efficiency estimation of this step $\nu$, Fig. 6b:

$$\nu = \frac{N}{4M} = \frac{n^2}{4\frac{1}{2n}} = \frac{n^3}{2} \quad , \quad \nu = O(n^3) = O(N\sqrt{N}) \tag{1}$$

The efficiency of preprocessing grows $O(n^3)$, resp. $O(N\sqrt{N})$, where $N$ is the number of points in the given data set.

## 3  Conclusion

The proposed preprocessing algorithm with the $O(N)$ computational complexity has been introduced. It can be used directly with an advantage for the solution of many geometrical problems with higher computational complexity the $O(N)$, e.g. for a convex hull, a diameter of a convex hull, smallest enclosing circle computations, etc. An extension for the $E^3$ case is expected.

**Acknowledgment.** The author would like to thank colleagues from Zhejiang University (Hangzhou), Shandong University in Jinan (China), colleagues and students at the University of West Bohemia in Pilsen (CZ) for the suggestions and fruitful discussions, and special thanks to Alexander Drdak for experimental counter implementation. Thanks also go to the anonymous reviewers for their critical comments, advice provided, and unknown relevant references.

# References

1. Kirkpatrick, D.G., Seidel, R.: Ultimate planar convex hull algorithm? SIAM J. Comput. **15**(1), 287–299 (1986). https://doi.org/10.1137/0215021
2. Matoušek, J., Sharir, M., Welzl, E.: A subexponential bound for linear programming. Algorithmica (New York) **16**(4-5), 498–516 (1996). https://doi.org/10.1007/bf01940877, the code available at https://news.ycombinator.com/item?id=14475832
3. Shen, K.W., Wang, X.K., Wang, J.Q.: Multi-criteria decision-making method based on smallest enclosing circle in incompletely reliable information environment. Comput. Ind. Eng. **130**, 1–13 (2019). https://doi.org/10.1016/j.cie.2019.02.011
4. Skala, V.: Barycentric coordinates computation in homogeneous coordinates. Comput. Graph. (Pergamon) **32**(1), 120–127 (2008). https://doi.org/10.1016/j.cag.2007.09.007
5. Skala, V.: Fast $o_{expected}(n)$ algorithm for finding exact maximum distance in E2 instead of $o(n^2)$ or $o(n \; lgn)$. AIP Conf. Proc. **1558**, 2496–2499 (2013). https://doi.org/10.1063/1.4826047
6. Skala, V.: Diameter and convex hull of points using space subdivision in $E^2$ and $E^3$. In: Gervasi, O., et al. (eds.) ICCSA 2020. LNCS, vol. 12249, pp. 286–295. Springer, Cham (2020). https://doi.org/10.1007/978-3-030-58799-4_21
7. Skala, V., Majdisova, Z.: Fast algorithm for finding maximum distance with space subdivision in $E^2$. In: Zhang, Y.-J. (ed.) ICIG 2015. LNCS, vol. 9218, pp. 261–274. Springer, Cham (2015). https://doi.org/10.1007/978-3-319-21963-9_24
8. Skala, V., Majdisova, Z., Smolik, M.: Space subdivision to speed-up convex hull construction in E3. Adv. Eng. Softw. **91**, 12–22 (2016). https://doi.org/10.1016/j.advengsoft.2015.09.002
9. Skala, V., Smolik, M., Majdisova, Z.: Reducing the number of points on the convex hull calculation using the polar space subdivision in E2. SIBGRAPI **2016**, 40–47 (2017). https://doi.org/10.1109/SIBGRAPI.2016.015
10. Skala, V.: Fast $o_{expected}(n)$ algorithm for finding exact maximum distance in $e^2$ instead of $o(n^2)$ or $o(n \lg n)$. In: AIP Conference Proceedings. AIP (2013). https://doi.org/10.1063/1.4826047
11. Skala, V., Cerny, M., Saleh, J.Y.: Simple and efficient acceleration of the smallest enclosing ball for large data sets in $E^2$: analysis and comparative results. In: Groen, D., de Mulatier, C., Paszynski, M., Krzhizhanovskaya, V.V., Dongarra, J.J., Sloot, P.M.A. (eds.) ICCS 2022. LNCS, vol. 13350, pp. 720–733. Springer, Cham (2022). https://doi.org/10.1007/978-3-031-08751-6_52
12. Smolik, M., Skala, V.: Efficient speed-up of the smallest enclosing circle algorithm. Informatica (Netherlands) **33**(3), 623—633 (2022). https://doi.org/10.15388/22-INFOR477
13. Welzl, E.: Smallest enclosing disks (balls and ellipsoids). In: Maurer, H. (ed.) New Results and New Trends in Computer Science. LNCS, vol. 555, pp. 359–370. Springer, Heidelberg (1991). https://doi.org/10.1007/BFb0038202
14. Welzl, E.: The smallest enclosing circle - a contribution to democracy from switzerland? Algorithms Unplugged, pp. 357–360 (2011). https://doi.org/10.1007/978-3-642-15328-0_36

# Strategic Promotional Campaigns for Sustainable Behaviors: Maximizing Influence in Competitive Complex Contagions

Arkadiusz Lipiecki[(✉)] [iD]

Wrocław University of Science and Technology, Wrocław, Poland
arkadiusz.lipiecki@pwr.edu.pl

**Abstract.** We address the research gap in evaluating the effectiveness of network seeding strategies in maximizing the spread of beliefs within non-progressive competing complex contagions. Our study focuses on management perspective of devising promotional campaigns for sustainable and health behaviors. We conduct an extensive computational analysis on two empirical datasets, comparing four established strategies in two different scenarios. Our results show that it is possible to achieve widespread adoption of beliefs, even under very limited network information. However, this success requires a strategic approach that includes additional efforts to prevent the targeted influencers from abandoning these attitudes in the future.

**Keywords:** Influence Maximization · Complex Contagion · Networks · Seeding Strategies · Belief Dynamics · Adoption of Sustainable Behaviors

## 1 Introduction

In 2021 the International Energy Agency published the world's first comprehensive report studying how to achieve net-zero carbon dioxide emissions. *Net Zero by 2050* highlights behavioral change as a crucial factor in decarbonization. This raises a question of what factors can drive the collective adoption of sustainable behaviors and how to facilitate it. Kowalska-Pyzalska [7] notes that environmental behaviors are strongly connected to, i.a., environmental beliefs, norms and social influence. Similarly, social norms among close ties has been identified as strong predictors of vaccine intentions [11]. In view of this, our attention centers on harnessing the power of social contagion, with particular focus on belief spreading. Within the joint realms of complex systems, network science and social simulation, considerable attention has been focused on network seeding strategies – methods for optimal selection of early adopters that maximizes the spread of their influence. From the perspective of promotional campaigns, seeding strategies attempt to answer the question of who to target with the

This research was funded by the National Science Center (NCN, Poland) through grant no. 2019/35/B/HS6/02530.

campaign to maximize its outreach potential. For example, one-hop strategy, which exploits the friendship paradox to find seed candidates, was shown to yield higher product adoption and health knowledge dissemination than targeting highly connected individuals [6]. Recently, in the face of empirical findings suggesting that social spreading is not a simple, but a complex contagion, i.e. it requires the influence of a group rather than a single individual, Guilbeault and Centola [4] introduced a new measure for identifying central nodes and shown its efficiency in maximizing influence across various progressive models. However, progressive models do not allow the individuals to revert to their previous states. While this limitation is justified in many cases, our beliefs and behaviors can be highly variable, and efforts to promote socially responsible attitudes are vulnerable to being undermined by uncertainty and competition between conflicting views. The need for influence maximization under reversible opinions has already been addressed for simple competing contagions within the voter model [3,12]. Yet despite the great interest in spreading processes based on complex contagion, there is little to no research devoted to evaluating the effectiveness of network seeding strategies in the non-progressive case of competitive complex contagions [14]. In order to address this gap we perform a simulation study on two empirical social networks, providing a comparison of selected seeding strategies in facilitating collective change of beliefs within such a framework. The code is open-source and available on GitHub: github.com/lipiecki/qvoter-seeding.

## 2   Methods

**Model.** To model the non-progressive competing complex contagions we implement the $q$-Voter Model ($q$VM) [2] with $n$ agents placed in nodes of a social network, where neighborhood $N(i)$ of node $i$ corresponds to the set of $k_i$ nodes directly linked to $i$. The $q$VM is a generalization of the voter model, in which an agent can change its state only when influenced by a unanimous group of $q$ randomly selected neighbors. In our model each agent can be in one of two states: *adopter* or *rejecter*. We use the term *adopter* to refer to an agent in the state that we aim to promote, and *rejecter* for an agent in the opposite state. A single elementary update of the model consists of the following steps:

1. Choose a random agent $i$.
2. Form the $q$-panel – choose at random $q$ neighbors of $i$.
3. If all agents in the $q$-panel are adopters – $i$ becomes an adopter.
   Otherwise, if all agents in the $q$-panel are rejecters – $i$ becomes a rejecter.

The general model of competing complex contagion studied in [14] corresponds to the $q$VM in which the $q$-panel is drawn with replacement. However, drawing with replacement allows an agent to change its state under the influence of less than $q$ neighbors due to multiple selections of the same neighbor. In order to ensure that simple contagion dynamics do not affect the results of simulations, we adopt the approach of drawing without replacement. Additionally, we examine two settings of the model with respect to seed behavior: flexible seeds, which undergo an updating procedure and differ from non-seeds solely in their initial state, and inflexible seeds (zealots) [12], which are indefinitely fixed as adopters and do not change their state.

**Seeding Strategies.** To provide a concise but comprehensive picture of how susceptible the $q$VM is to various seeding methods, we have selected four distinct strategies, differing in computational complexity and information requirement. We will refer to the number of nodes that are seeded within a given experiment as the seeding budget.

**High Degree (HD)** – seeding nodes with the highest degree, which is one of the most straightforward node centrality measures. The number of direct neighbors is a simple proxy of node's importance and does not require full information of the network structure. Seeding high degree nodes was shown to be an optimal strategy for the simple voter model [3].

**PageRank (PR)** – seeding nodes with the highest PageRank centrality, which was initially introduced for identifying the most important web pages. Since then PageRank has gained an incredible amount of attention, also in the context of social influence. It is a promising candidate for evaluating the importance of nodes in the $q$VM. To understand why, consider the influence probability $P_{ij}$, i.e. the probability that node $i$ selects $j \in N(i)$ as part of the $q$-panel during an elementary update. Since $q$ sources are drawn, $P_{ij} = q/k_i$. Notice that $P/q$ is a stochastic matrix of the random walk on the same network. PageRank centrality measure is closely tied to the stationary probability distribution of such a random walk.

**Complex Centrality (CC)** – seeding nodes with the highest complex centrality was proposed as an effective strategy in maximizing the spread of complex contagions [4]. Complex centrality of a node is defined as an average length of complex paths extending from that node, where the length of a complex path between nodes $i$ and $j$ is equal to the shortest path length between them in the subgraph induced by the spread of progressive complex contagion from the neighborhood of $i$ to node $j$ [4]. Complex centrality depends on the underlying dynamics of the contagion through thresholds $T_i$, corresponding to the number of adopters in $N(i)$ required to activate $i$ within a progressive complex contagion. Hence, to adapt this method to the non-progressive case, we set $T_i$ for each individual $i$ as a minimal number of adopters in its neighborhood for which the probability of gathering a unanimous $q$-panel of adopters is greater or equal to the probability of gathering a unanimous $q$-panel of rejecters: $T_i = \max\{q, \lceil k_i/2 \rceil\}$.

**One-Hop** – this strategy identifies seed nodes through a simple two-step process, in which we first select a random set of individuals with size equal to the seeding budget and then query them to nominate one of their neighbors (in simulations selected at random) as the seed [10]. If the number of nominated unique seeds is smaller than the budget, we randomly draw nodes from the initially selected random set. One-hop strategy works under very limited network information, as it does not require any prior knowledge about the network structure, and the number of links discovered during the querying is no greater than the seeding budget. Moreover, due to simplicity of the process, it can be readily implemented in field experiments, which is a significant advantage from the perspective of promotional strategies. Finally, it is worth noting that the process of selecting seeds within the one-hop strategy corresponds to performing single-step random walks starting from random initial nodes, which relates one-hop to PR centrality and to the influence probability within the $q$VM.

**Simulations.** To evaluate the effectiveness of seeding strategies in maximizing the spread of beliefs, we perform simulation experiments on two empirical networks – Facebook dataset from the Stanford Network Analysis Project (SNAP) [9] and Facebook dataset of verified pages of politicians (Politicians) [13]. According to the model specification, in the process of updating a state of a node, its $q$ random neighbors are selected. However, it may occur that the degree of a node is smaller than $q$. Then we can either omit such nodes and do not update their state, or assume a different size of the $q$-panel, equal to the degree of a node. We adopt the latter approach, but this raises another issue. In the empirical networks we study there exist nodes with degree equal to one, which means that updating such nodes would correspond to a simple contagion dynamics. To avoid this, we conduct preprocessing of the network data to ensure that every node has at least two neighbors. For each node with a single neighbor we perform a *triad formation* step of the Holme-Kim algorithm [5], where we add a link between the single-neighbor node and one of its randomly selected next-nearest neighbors. Since this preprocessing introduces randomness to the networks, we conduct simulations on an ensemble of $10^2$ networks generated this way. The agent systems are evolved for $10^5$ Monte Carlo steps (each consisting of $n$ elementary updates) or until the absorbing state (full adoption or full rejection) is reached.

# 3  Results

The first question that ought to be answered in the context of maximizing influence within the $q$VM is whether there exists a substantial difference between the spreading behavior in the case of a simple voter model ($q = 1$) and the nonlinear $q$VM with complex contagion ($q > 1$). Figure 1 presents the comparison of the percentage of adopters for varying fraction of random seed nodes for $q = 1$ and $q = 2$. In the flexible seeds scenario, the simple voter model always evolves towards either full adoption or total absence thereof. The probability of adoption increases linearly with the number of seed nodes. The adoption within the 2VM is qualitatively different, the average number of adopters follows a sharp S-shape transition pattern with an inflection point at 50%.

Belief dynamics with zealots highlight that collective change is significantly more difficult if the voter-like dynamics follow complex contagion. For $q = 1$ agent systems reach full adoption for any non-zero seeding budget, while for $q = 2$ it is reached only after exceeding a sufficient fraction of seeds. These results show evident qualitative differences between the susceptibility to collective change of

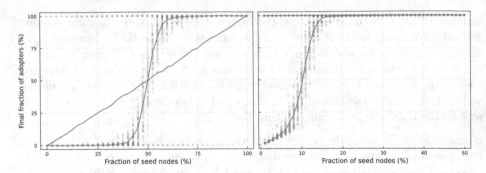

**Fig. 1.** The final fraction of adopters with respect to the fraction of randomly seeded nodes on the SNAP network within the $q$VM for $q = 1$ (blue) and $q = 2$ (orange). Colored lines represent the ensemble averages ($10^3$ networks), while colored circles mark the individual simulation outcomes. Left panel shows the results for flexible seeds, and the right panel – for zealots. (Color figure online)

the simple voter system and the non-linear $q$VM, which motivates us to examine influence maximization strategies within complex contagions, i.e. $q > 1$.

**Table 1.** Network statistics (size $n$, average node degree $\langle k \rangle$, average clustering coefficient $\rho$, average shortest path length $d$) and a minimal seeding budget for which the ensemble median reached a specified adoption level $\tau$ within the 2VM.

| Network | $n$ | $\langle k \rangle$ | $\delta$ | $d$ | $\tau$ | Flexible | | | | Zealots | | | |
|---|---|---|---|---|---|---|---|---|---|---|---|---|---|
| | | | | | | HD | PR | CC | One-hop | HD | PR | CC | One-hop |
| SNAP | 4039 | 43.7 | 0.62 | 3.7 | 80% | 40% | 25.0% | 35.0% | 47.5% | 17.5% | 5.0% | 17.5% | 10.0% |
| | | | | | 90% | 50.0% | 27.5% | 37.5% | 47.5% | 20.0% | 7.5% | 20.0% | 10.0% |
| | | | | | 95% | 50.0% | 32.5% | 40.0% | 50.0% | 30.0% | 15.0% | 27.5% | 12.5% |
| Politicians | 5908 | 14.3 | 0.49 | 4.6 | 80% | 25.0% | 20.0% | 27.5% | 37.5% | 7.5% | 5.0% | 10.0% | 5.0% |
| | | | | | 90% | 47.5% | 25.0% | 47.5% | 42.5% | 10.0% | 7.5% | 15.0% | 7.5% |
| | | | | | 95% | 67.5% | 35.0% | 67.5% | 47.5% | 60.0% | 20.0% | 60.0% | 10.0% |

We therefore proceed to analyze the seeding experiments for $q = 2$. Statistics of the network ensemble and the seeding budgets required for reaching specified adoption levels are presented in Table 1, while the entire adoption curves are shown in Fig. 2. Firstly, in the setting of flexible seeds, the strategy based on PR can be considered as the most efficient among the examined contenders. It significantly outperforms one-hop, and although HD and CC lead to higher fractions of adopters for low seeding budgets, they are more costly than PR in reaching the adoption levels close to 50% and higher (see left panels of Fig. 2). However, simulations with zealots paint a different picture, with a surprising performance of the one-hop strategy. While PR is the most efficient at reaching 80% and 90% adoption levels (see Table 1) and leads to highest adoption at small fractions of seed nodes (see right panels of Fig. 2), one-hop strategy outperforms PR in achieving nearly full adoption. For zealots, seeding with HD and CC performs

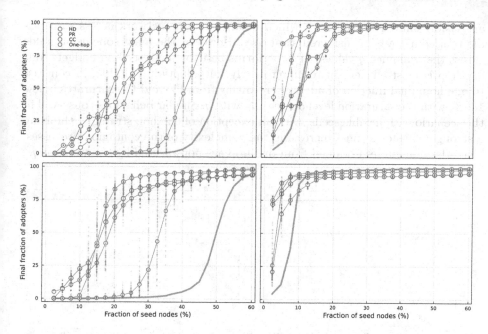

**Fig. 2.** The final fraction of adopters with respect to the fraction of seed nodes for selected seeding strategies within the 2VM. Empty colored circles correspond to ensemble averages ($10^2$ networks), while full colored markers correspond to individual simulation outcomes. Grey lines correspond to ensemble averages for random seeding. Top panels correspond to simulations on the SNAP network, while bottom panels – on the Politicians network. Left panel shows the results for flexible seeds, and the right panel – for zealots. (Color figure online)

rather poorly when we aim for widespread adoption, requiring c.a. twice larger budgets to reach target levels of 80% and above in the SNAP network, and up to six times larger budget to reach 95% adoption in the Politicians network. Moreover, one-hop is the only strategy that is consistently, i.e. across all fractions of seed nodes, no worse that random seeding in the case of zealots. One possible explanation for this result is the fact that random and one-hop strategies seed nodes in various parts of the network, while highly central nodes can cluster at the core of the network, thus leading to higher difficulty in affecting network's periphery. In general, given the same seeding budget, adoption levels are significantly higher when seed nodes behave as zealots, which is to be expected given that they serve as constant sources of pro-adoption influence.

Finally, let us examine how the susceptibility of the social network to collective adoption varies for different sizes of influence group. Since the size of the $q$-panel affects the spreading of both adoption and rejection, it is difficult to form expectations of how it affects the efficiency of seeding strategies. Figure 3 shows the comparison between the efficiency of selected strategies in seeding the Facebook SNAP network for $q = 2, 3$ and $4$. For zealot seeds, all the adoption curves are visibly shifted to the right, which means that the network is less susceptible to seeding by all considered strategies. However, the main conclusion holds across the examined $q$ values – PR performs best for small seeding budgets, but one-hop is most efficient in reaching nearly full adoption. Notably, in contrast to one-hop, final fraction of adopters resulting from PR and CC saturates below 100%, with the saturation level decreasing with $q$. Similar behavior is observed in the scenario with flexible seeds, with an exception of one-hop strategy exhibiting a strong resilience against increasing the complexity of the contagion processes. The adoption curves obtained from one-hop are not affected by the value of $q$.

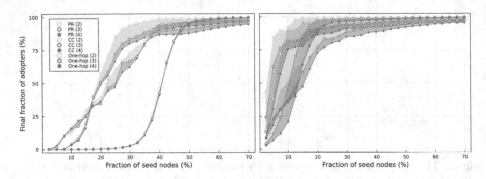

**Fig. 3.** Comparison of the final fraction of adopters with respect to the fraction of seed nodes on the SNAP network for selected seeding strategies within the $q$VM for $q = 2, 3$ and 4. Values presented on the plots are ensemble averages ($10^2$ networks). Left panel shows the results for flexible seeds, and the right panel – for zealots.

## 4   Discussion

In the face of current problems that require widespread collective action, such as reaching herd immunity or climate change mitigation, computational social science can provide valuable insights to support strategic promotion of environmental and health behaviors [7,8]. Therefore, in this study we addressed the research gap in evaluating the effectiveness of network seeding strategies within the framework of competitive complex contagions, which accounts for both group influence and variability in opinions. Notably, many problems remain for future research. For example, considering the spreading behavior with the presence of zealots that oppose the promoted attitude [12] or examining the effects of network coevolution on the widespread belief adoption [1] can provide insights on how to facilitate collective change in the presence of high hesitancy and polarization. From the managerial perspective, an important next step is to evaluate more sophisticated strategies that work under limited network information [10].

Our results indicate that increasing the complexity of competing contagions decreases the susceptibility of social networks to seeding strategies. Yet influence maximization methods that do not account for contagion complexity emerged as promising solutions. PageRank centrality outperformed all other strategies in reaching adoption majority in the setting of flexible seeds. In the scenario of pro-adoption zealots, PageRank was most effective for small seeding budgets, but one-hop turned out as the best strategy for achieving nearly full adoption. Since one-hop requires very little information about the network and can be implemented by conducting a single round of surveys, our findings suggest that facilitating collective adoption of beliefs and behaviors through social influence is achievable in real-life scenarios. Moreover, zealot seeds required drastically smaller budgets than flexible ones. Therefore, the key recommendation for decision-makers managing promotional campaigns is that dedicating resources for maintaining consistency in the attitudes of targeted influencers can be more beneficial than acquiring additional network information.

# References

1. Borges, H.M., Vasconcelos, V.V., Pinheiro, F.L.: How social rewiring preferences bridge polarized communities. Chaos Solitons Fract. **180**, 114594 (2024)
2. Castellano, C., Muñoz, M.A., Pastor-Satorras, R.: Nonlinear $q$-voter model. Phys. Rev. E **80**, 041129 (2009)
3. Even-Dar, E., Shapira, A.: A note on maximizing the spread of influence in social networks. Inf. Process. Lett. **111**(4), 184–187 (2011)
4. Guilbeault, D., Centola, D.: Topological measures for identifying and predicting the spread of complex contagions. Nat. Commun. **12**(1), 4430 (2021)
5. Holme, P., Kim, B.J.: Growing scale-free networks with tunable clustering. Phys. Rev. E **65**, 026107 (2002)
6. Kim, D.A., et al.: Social network targeting to maximise population behaviour change: a cluster randomised controlled trial. Lancet **386**(9989), 145–153 (2015)
7. Kowalska-Pyzalska, A.: What makes consumers adopt to innovative energy services in the energy market? A review of incentives and barriers. Renew. Sustain. Energy Rev. **82**, 3570–3581 (2018)
8. Latkin, C.A., Knowlton, A.R.: Social network assessments and interventions for health behavior change: a critical review. Behav. Med. **41**(3), 90–97 (2015)
9. McAuley, J., Leskovec, J.: Learning to discover social circles in ego networks. In: NIPS (2012)
10. Ou, J., Buskens, V., van de Rijt, A., Panja, D.: Influence maximization under limited network information: seeding high-degree neighbors. J. Phys. Complex. **3**(4), 045004 (2022)
11. Rabb, N., Bowers, J., Glick, D., Wilson, K.H., Yokum, D.: The influence of social norms varies with "others" groups: evidence from COVID-19 vaccination intentions. Proc. Natl. Acad. Sci. **119**(29), e2118770119 (2022)
12. Romero Moreno, G., Manino, E., Tran-Thanh, L., Brede, M.: Zealotry and influence maximization in the voter model: when to target partial zealots? In: Barbosa, H., Gomez-Gardenes, J., Gonçalves, B., Mangioni, G., Menezes, R., Oliveira, M. (eds.) Complex Networks XI. SPC, pp. 107–118. Springer, Cham (2020). https://doi.org/10.1007/978-3-030-40943-2_10

13. Rozemberczki, B., Davies, R., Sarkar, R., Sutton, C.: GEMSEC: graph embedding with self clustering. In: ASONAM '19, pp. 65–72. ACM (2019)
14. Vasconcelos, V.V., Levin, S.A., Pinheiro, F.L.: Consensus and polarization in competing complex contagion processes. J. Roy. Soc. Interface **16**(155), 20190196 (2019)

# Inference Algorithm for Knowledge Bases with Rule Cluster Structure

Agnieszka Nowak-Brzezińska[ID] and Igor Gaibei[✉][ID]

Faculty of Science and Technology, Institute of Computer Science,
University of Silesia, Bankowa 12, 40-007 Katowice, Poland
{agnieszka.nowak-brzezinska,igor.gaibei}@us.edu.pl

**Abstract.** This paper presents an inference algorithm for knowledge bases with a rule cluster structure. The research includes the study of the efficiency of inference, measured by the number of cases in which the inference was successful. Finding a rule whose premises are true and activating it leads to extracting new knowledge and adding it as a fact to the knowledge base. We aim to check which clustering and inference parameters influence the inference efficiency. We used four various real datasets in our experimental stage. Overall, we proceeded with almost twenty thousand experiments. The results prove that the clustering algorithm, the amount of input data, the method of cluster representation, and the subject of clustering significantly impact the inference efficiency.

**Keywords:** inference algorithm · rule-based knowledge bases · clustering algorithms

## 1 Introduction

Expert systems are a significant artificial intelligence (AI) branch that has been developed for several decades. Recently, we have seen a lot of movement in this field. Many applications are being developed to have the role of an assistant with access to knowledge and share this knowledge with the user. Such an assistant is often simply an expert system with built-in expert knowledge and implemented one of two inference algorithms: forward and backward. Expert knowledge is often stored in the form of IF-THEN rules, and classical inference requires analysis of each rule, one by one, to assess which one can activate since all premises are true. When there are a lot of such rules, a reasoning process take a very long time, which may not be acceptable to the user who needs information without a delay. Our concept is based on the idea that we will cluster similar rules into groups and assign representatives to these rule clusters. These representations will later be reviewed in the inference process. The paper presents the following topics: rule clustering algorithms, comparing classical inference with inference which operates on rule cluster representations, studying the efficiency of inference (measured by the number of cases with successful inference), assessing the impact of various parameters on this efficiency.

© The Author(s), under exclusive license to Springer Nature Switzerland AG 2024
L. Franco et al. (Eds.): ICCS 2024, LNCS 14834, pp. 71–78, 2024.
https://doi.org/10.1007/978-3-031-63759-9_9

## 1.1    Literature Review

In the literature, one can find many articles on the analysis and application of inference algorithms, analyzing expert systems based on the knowledge representations, the comparison of the $K$ - $Means$ and the $AHC$ algorithm, the use of different distance measures, methods of combining clusters or methods of analyzing the quality of clustering. In [1], the authors present a rule-based forward chaining system that decides the measurement category. In [2], the authors propose an expert system that helps cyclists decide whether to fix the issue themselves or search for an expert. The paper [3] presents an expert system built using the forward chaining method for rules $IF\text{-}THEN$, but not for rule group representations. The paper [4] presents the design of an expert system based on a forward inference for plant disease identification. [5] presents a project designed for medical diagnosis based on the provided symptoms. The expert system performs inference using $IF\text{-}THEN$ production rules. In [6], the authors evaluate the effectiveness of various clustering methods. In [7], authors compare and assess five clustering algorithms, however, not for rule-based knowledge representation. The authors of [8] compare $K\text{-}Means$ and $AHC$ algorithms in terms of the number of clusters, the number of objects in clusters, the number of iterations, and clustering time for small and large data. In [9], the authors compare clustering algorithms and methods of cluster quality assessment (F-measure, Entropy) for different values of the number of clusters. In our previous work [10], we examined the $K\text{-}Means$ and the $AHC$ clustering algorithm in the context of rule-based knowledge representation. Although, it is impossible to find papers that would combine these issues in one study. We wanted to investigate which algorithm ($AHC$ or $K\text{-}Means$ algorithm) and method of creating a cluster representative (mean or median) is more effective in terms of the efficiency of inference on clusters.

## 1.2    Article Structure

Section 2 describes the clustering algorithms used in this research. The inference algorithm is presented in Sect. 3. The procedure for creating rule clusters and representatives of groups is presented in Sect. 4. Then, in Sect. 5, the description of the experiments was included with the results and their analysis. Section 6 contains the summary of the research.

## 2    Clustering Algorithms

This part presents a concise description of clustering algorithms applied to the rules in the knowledge base. Among the available clustering techniques, *non-hierarchical* and *hierarchical* methods can be used. When we have a lot of rules in the knowledge base, then, unfortunately, the efficiency of an inference decreases because the inference time increases and the system user has problems interpreting too many newly generated facts. Rules can be grouped, and we only need to search the representatives of rule clusters and select the cluster that best matches the given facts. Two clustering algorithms were used in our work: the non-hierarchical algorithm ($K$-$Means$) with the hierarchical algorithm $AHC$. Both are presented in the following subsections.

## 2.1 Partitional $K$-$Means$ Clustering Algorithm

$K$-$Means$ clustering algorithm tries to group similar rules in the form of $K$ clusters. Each cluster is represented by its center (arithmetic mean of data points). In each iteration we try to divide $N$ original rules into $K$ rule clusters so well that each rule belongs to the cluster to which it is most similar. The main idea of the algorithm is as follows:

1. Select the number of rule clusters ($K$) and assign $K$ centers.
2. For each rule, the nearest cluster center is determined.
3. The rule cluster representative is created - an average value for each attribute value (conditional and decisional) or a mode value for qualitative attributes.
4. New cluster center is formed.
5. The 3rd and 4th steps repeat iteratively.
6. The algorithm ends when no rule cluster changes occur at some iteration.

## 2.2 Hierarchical $AHC$ Clustering Algorithm

$AHC$ hierarchical clustering algorithm for rules creates a tree of rule clusters. The $AHC$ algorithm works as follows:

1. Each rule forms a separate cluster. We must calculate the distance between each pair of rule clusters.
2. Find and join the two most similar rule clusters.
3. Repeat the second step until obtaining the declared final number of rule clusters ($K$) or combining all rules into one big cluster.

Each algorithm will create an entirely different structure of the focus of rules.

# 3    Forward Chaining Inference

There are two inference algorithms: forward chaining (from premises to conclusions) and backward chaining (from hypothesis/conclusion to premises). In this work, we have only dealt with the first method. Each rule is analyzed to determine whether all its' premisses are satisfied. The rule is activated, and its conclusion is added to the set of facts. The algorithm stops where no more rules can be activated or when the starting hypothesis is added to the set of facts. The more rules, the longer the inference time. Therefore, we aim to cluster similar rules, hoping we reduce the inference time significantly when we divide such a large rule set into clusters of similar rules. This is possible because we only need to search the representatives of rule clusters and select the cluster that best matches the given facts. The rule cluster representatives are achieved as follows: at the cluster level, we calculate the similarity of the representative of each cluster to the fact vector, using Euclidean distance to calculate the similarity. We select the most promising cluster. Within the selected cluster of rules, we calculate the similarity of facts to each rule. We then apply the forward inference algorithm only on the selected cluster. The next step is to look within the loop for rules to fire. If more than one rule could be fired at a time, we chose one of them using the appropriate rule selection strategy. In our experiments we fire the first rule in the list of all rules to be fired (sequential strategy). Firing (activation) of a rule leads to adding the conclusion of such rule to the fact base and blocking this rule from being activated again. The inference ends when there are no more rules to be fired or the set of facts (after the last activation) includes the inference target set at the beginning.

### 3.1   Group Representative: Methods of Creating a Group Representative

In our experiments, we used two methods for creating a representative of a group of rules - the mean and the median. Our goal is to verify which method provides better inference efficiency. Figure 1 shows an example of a hierarchical algorithm where the choice of a rule cluster depends on the similarity of the fact representative to the cluster representative. As we can see, in some cases, the mean or median method can lead to entirely different branches of the binary tree. Therefore, it is crucial to study the influence of the cluster representative on the efficiency of inference.

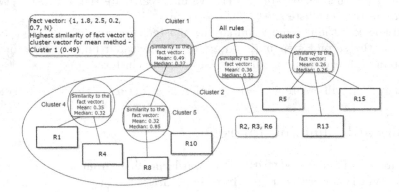

**Fig. 1.** Binary tree of rule clusters

   To find the median value for a rule cluster representative with an odd number of numbers, one would find the number in the middle with an equal number of numbers on either side of the median. To find the median, we should first arrange the numbers, usually from lowest to highest. Mean is the average of all the numbers within each group of rules. Therefore, to represent the rules in vector format, we should use the values of the attributes of each rule and zero out the attributes that have no value and are not considered in the rule. This is an essential operation for the correct calculation of vector similarity. It may turn out that the use of each of these representation methods can finally lead us to different results. We want to examine it in this work. We will perform the same experiments for both methods and compare the results in the experimental part.

## 4 Methodology

The experiments proceeded as follows. The source dataset was loaded into the *RSES* tool, where decision rules were generated using the *LEM 2* algorithm. The rules are then loaded into the Python environment in the customized software and were grouped using two clustering algorithms: *K-Means* and *AHC*. We tested different distance measures, different clustering methods (for the *AHC* algorithm), and different values for the parameter representing the number of rule groups created, different numbers of generated facts (inputs), and different methods for creating a representative. We studied the clustering time and cluster quality indices, the *Dunn* and the *Davies-Bouldin* indexes. The next step was to calculate the vector of initial facts. In our experiments, we generate the initial facts that constitute 5%, 25%, and 50% of all unique descriptors in the premises for each algorithm. We used two methods for creating a representative of a group of rules - the mean and the median. The next step was to calculate the representations of all rule clusters. Let us assume that the example rule cluster contains the following rules: $R1$ : IF $a1 = 1$ THEN $dec = T$, $R4$ : IF $a1 = 2$ THEN $dec = T$, $R8$ : IF $a2 = 2$ THEN $dec = T$ and $R10$ : IF $a3 = 3$ THEN $dec = T$. If the number of attributes equals 5 rule vectors are as follows: $[1, 0, 0, 0, 0, T]$, $[2, 0, 0, 0, 0, T]$, $[0, 2, 0, 0, 0, T]$, $[0, 0, 3, 0, 0, T]$ and the rule cluster representative is the following *Representative mean* = {1.5 2 3 0 0 T} or *Representative median*= {1 1 0 0 0 T}.

## 5 Results of Experiments

This section presents the course of experiments and the analysis of the results achieved. In the experiments, we included real knowledge bases with different structures. There were the following datasets: $kb_1$ with 4435 instances, 37 attributes and 937 rules [11], $kb_2$ with 7027 instances, 65 attributes and 4125 rules [12], $kb_3$ with 527 instances, 38 attributes and 123 rules [13] and $kb_4$ with 17898 instances, 9 attributes, 6432 rules [14]. The runtime for the experiments had the following configuration: Spyder compiler with *Python* version 3.9 from the Anaconda platform. The computer parameters on which all experiments were carried out are as follows: Intel Core i5-7500K, 16 Gb RAM. To run the experiments, we used self-written software in the Python programming language. The following libraries were used: *Pandas* for data processing and analysis and *NumPy* for basic operations on n-arrays and matrices. Finally, we used the

$RSES$ system and the $LEM2$ algorithm to generate rules, although we also checked the *exhaustive* algorithm.

## 5.1 Experiments Procedure

We perform clustering sequentially for each algorithm ($K$-*Means*, $AHC$) for $K = 2, 3, \ldots, 22$, using one of the three distance measures (Euclidean, Chebyshev, Manhattan). For the $AHC$, we repeated the algorithm for single, complete, and average linkages. We repeat each algorithm for three different inputs: the conditions alone, the conclusions alone, and the conditions and conclusions of the rules together. For each algorithm, we generate the initial facts constituting 5%, 25%, and 50% of all unique descriptors in the premises. We used two methods for creating a representative of a group of rules - the mean and the median. For four knowledge bases, this gives a total of $18,144$ experiments.

## 5.2 Results

This section presents the results of selected experiments. We decided to examine whether the following parameters affect more or less inference efficiency, and thus the frequency of successful inference process: selected clustering algorithm ($K$-*Means* or $AHC$), number of input facts (5%, 25%, and 50%), what was clustered (premises, conclusions, or both), rule cluster representative method (mean, median). Table 1 presents the inference efficiency achieved for two clustering algorithms. You can see much greater effectiveness of the inference process when using the $K$-*Means* method. This may be surprising at first glance. Generally, the $AHC$ method clusters the objects naturally by connecting the most similar pair of rules (or rule clusters) in each algorithm iteration. But it is a structure that is further searched as a binary tree, selecting only one child node from two given nodes at every level. It can cause cases when we wrongly choose the node for further searching. In the case of the $K$-*Means* algorithm, we achieve a flat rule cluster structure, which means that we search every rule in a selected cluster. Therefore, the chance that we will wrongly select a proper rule cluster (and then a proper rule) is much smaller than in the case of the $AHC$ algorithm. In the case of the $AHC$ algorithm, the average time is $O(log2N)$, while in the case of the $K$-*Means*, it is linear $O(K) + O(N/K)$[1], which generally lasts longer than the $AHC$ searching time. The results presented in Table 2 show the effectiveness of the proposed idea, except that it depends on how the rules were grouped (what algorithm); however, how many input facts were used is crucial. It can be seen that as the number of input facts increases, the effectiveness of inference increases. In the $K$-*Means* algorithm, with 50% of facts, the effectiveness of inference is about 95%, while for the same number of facts, the $AHC$ algorithm is only 70%.

Table 3 showed that the grouping algorithm and the selected method of creating a representative (out of two proposed methods of concentration representation: mean and median) significantly impact the inference's effectiveness. We

---

[1] $N$ is the rule number and $K$ is the number of rule clusters.

**Table 1.** Inference efficiency vs. clustering algorithms

|           | Succeed         | Failed         |
|-----------|-----------------|----------------|
| K − Means | 3618 (79.76%)   | 918 (20.24%)   |
| AHC       | 7291 (53.58%)   | 6317 (46.42%)  |
| sum       | 10909 (60.12%)  | 7235 (39.88%)  |

**Table 2.** Inference efficiency vs. clustering algorithms and the amount of input facts

|           | % of facts | Succeed | Failed  | clustering object | Succeed | Failed  |
|-----------|------------|---------|---------|-------------------|---------|---------|
| K − Means | 5%         | 73.61%  | 26.39%  | cond              | 72.29%  | 27.71%  |
|           | 25%        | 71.63%  | 28.37%  | dec               | 90.15%  | 9.85%   |
|           | 50%        | 94.05%  | 5.95%   | cond + dec        | 76.85%  | 23.15%  |
| AHC       | 5%         | 35.45%  | 64.55%  | cond              | 70.26%  | 29.74%  |
|           | 25%        | 54.94%  | 45.06%  | dec               | 16.40%  | 83.60%  |
|           | 50%        | 70.35%  | 29.65%  | cond + dec        | 74.07%  | 25.93%  |

**Table 3.** Inference efficiency vs. clustering algorithms and representative method

|           | representative method | Succeed        | Failed         |
|-----------|-----------------------|----------------|----------------|
| K − Means | mean                  | 1849 (81.53%)  | 419 (18.47%)   |
|           | median                | 1769 (78.00%)  | 499 (22.00%)   |
| AHC       | mean                  | 3672 (53.97%)  | 3132 (46.03%)  |
|           | median                | 3619 (53.19%)  | 3185 (46.81%)  |

know that using the *K-Means* algorithm is successful in about 80% of cases. Still, when we consider the representation method, you can see that it will be more effective than the median when the representative is created using the mean method. The *K-Means* algorithm will behave entirely differently than *AHC*. When we use the *K-Means* algorithm and we only cluster by rule decisions (so large groups of rules are created), the inference is just over 90% effectiveness; when we group by the rule conditions, this efficiency is over 72%, and when we cluster. For *AHC*, it is entirely different. When we only group under decisions, the effectiveness of inference is only 16.4%. At the same time, clustering by rule conditions results in achieving efficiency at over 70%, and clustering by both conditions and decisions brings the best results, 74%, respectively.

# 6   Summary

This paper presents an inference algorithm for knowledge bases with a rule cluster structure. The research includes the study of the efficiency of inference, which will be measured by the number of cases in which the inference was successful.

We aim to check which clustering and inference parameters influence the inference efficiency. We proceeded with almost twenty thousand experiments. The results prove that the clustering algorithm, the amount of input data, the method of cluster representation, and the subject of clustering significantly impact the inference efficiency. We observed a much greater effectiveness of the inference process when using the *K-Means* method compared to *AHC*. As the number of input facts increases, the effectiveness of inference increases.

# References

1. Wang, D., Bai, Y.: Data processing and information retrieval of atmospheric measurements. In: CIVEMSA, Germany, pp. 1–5. IEEE (2022). https://doi.org/10.1109/CIVEMSA53371.2022.9853650
2. Hernández, M.P., et. (eds) Advances in Computational Intelligence – MICAI 2023. LNCS vol. 14502. Springer, Cham (2023). https://doi.org/10.1007/978-3-031-51940-6_8
3. Mustafidah, H., Alfiansyah, B.R., Suwarsito, P., Hidayat, N.: Expert system using forward chaining to determine freshwater fish types based on water quality and area conditions. In: ICIC, Indonesia, pp. 1–5 (2023). https://doi.org/10.1109/ICIC60109.2023.10382066
4. Sitanggang, D., et al.: Application of forward chaining method to diagnosis of onion plant diseases. J. Phys.: Conf. Ser. **1007**, 012048 (2018). https://doi.org/10.1088/1742-6596/1007/1/012048
5. Jabeen, S.H., Zhai, G.: A prototype design for medical diagnosis by an expert system. In: WCSE 2017, pp. 1413–1417 (2017)
6. MVNM, R.K., Sharma, V., Gupta, K., Jain, A., Priya, B., Prasad, M.S.R.: Performance evaluation and comparison of clustering algorithms for social network dataset. In: IC3I, India, pp. 111–117 (2023). https://doi.org/10.1109/IC3I59117.2023.10397806
7. Teslenko, D., Sorokina, A., Smelyakov, K., Filipov, O.: Comparative analysis of the applicability of five clustering algorithms for market segmentatio. In: IEEE Open Conference of Electrical, Electronic and Information Sciences, Lithuania, pp. 1–6 (2023). https://doi.org/10.1109/eStream59056.2023.10134796
8. Saleena, T.S., Sathish, A.J.: Comparison of K-means algorithm and hierarchical algorithm using Weka tool. Int. J. Adv. Res. Comput. Commun. Eng. **7**, 74–79 (2018)
9. Steinbach, M.S., Karypis, G., Kumar, V.: A Comparison of Document Clustering Techniques. Department of Computer Science and Engineering, Computer Science (2000)
10. Nowak-Brzezińska, A., Gaibei, I.: Decision rule clustering-comparison of the algorithms. In: Campagner, A., Urs Lenz, O., Xia, S., Ślęzak, D., Wąs, J., Yao, J. (eds.) IJCRS 2023. LNCS, vol. 14481, pp. 387–401. Springer, Cham (2023). https://doi.org/10.1007/978-3-031-50959-9_27
11. https://archive-beta.ics.uci.edu/dataset/146/statlog+landsat+satellite . Accessed Dec 2023
12. https://archive-beta.ics.uci.edu/dataset/365/polish+companies+bankruptcy+data. Accessed Dec 2023
13. https://archive-beta.ics.uci.edu/dataset/106/water+treatment+plant . Accessed Dec 2023
14. https://archive-beta.ics.uci.edu/dataset/372/htru2 . Accessed Dec 2023

# Towards a Framework for Multimodal Creativity States Detection from Emotion, Arousal, and Valence

Sepideh Kalateh[1,2]([✉]), Sanaz Nikghadam Hojjati[1,2], and Jose Barata[1,2]

[1] Centre of Technology and Systems (CTS-UNINOVA) and Associated Lab on
Intelligent Systems (LASI), 2829-516 Caparica, Portugal
{sepideh.kalateh,sanaznik,jab}@uninova.pt
[2] NOVA University of Lisbon, 2829-516 Caparica, Portugal

**Abstract.** In the multi-disciplinary context of computational creativity
and affective human-machine interaction, understanding and detecting
creative processes accurately is advantage. This paper introduces a novel
computational framework for creatively state detection, employing a multimodal approach that integrates emotions, arousal, and valence. The
framework utilizes multimodal inputs to capture the creativity states,
with emotion detection forming a foundational element. By fusioning
emotions and emotional dimension, arousal, and valence. This paper outlines the theoretical foundations, key components, and integration principles of the proposed framework, paving the way for future advancements
in computational creativity and affective computing.

**Keywords:** Computational Creativity · Affective Human Machine
interaction · Multimodal · Emotion detection · Creativity state
detection

## 1 Introduction

As technology becomes increasingly important in our daily lives, its impact on
societal aspects also becomes a critical consideration. Technology can take care
of the routine and repetitive tasks and provides time for humans to feel more
and be creative more [1]. Technology provides the essential elements for human-
centric purpose-driven and creative lives which leads to society 0.5 [2,3]. It can
simultaneously facilitate and hinder creativity, as creativity itself possesses both
beneficial and destructive aspects.

Computational creativity [4] a growing technological aspect, involves harnessing computational algorithms and methodologies to generate valuable ideas,
solutions, or expressions. For instance, generative models empower machines to
produce innovative outputs by amalgamating new permutations or adaptations
of existing data [5]. Notably, natural language processing techniques further
augment machine creativity by facilitating the generation of coherent and contextually relevant textual content [6]. Despite remarkable strides in simulating

creativity, the detection of creative states in humans remains a nuanced pursuit. Traditionally, methods such as eye tracking [7], physiological signals [8], and textual or large language analyses [9] have been employed across various contexts to detect creativity.

Creativity, essential for human advancement, fuels innovation across diverse fields, from ancient discoveries to contemporary societal progressions [10]. Emotion serves as a crucial factor, both inspiring and impeding creativity, shaping motivation, idea generation, and perseverance in overcoming obstacles [11,12]. Understanding creative states empowers individuals to navigate life's challenges, nurturing adaptability and resilience while fostering collaboration and problem-solving [13]. In light of this, our research focuses on assessing creative states through emotions, with the aim of enhancing human-machine interactions and fostering innovative solutions to societal challenges.

*How can a multimodal framework be formulated to assess creativity states, considering the factors of emotions, arousal, and valence as the inputs?*

Various applications of creativity-aware systems include personalized educational tools, entertainment content, therapeutic interventions, marketing strategies, human-robot collaboration, and assistive technologies, all tailored to individuals' emotional and creative states for enhanced experiences and outcomes [14,15]. The remainder of this paper is organized to go through the fundamental concepts and then introduce a framework in an attempt provide an answer for the asked question.

## 2   Background: Creativity and Emotion

**Creativity** is a multifaceted cognitive process characterized by the generation of novel and valuable ideas, solutions, or expressions that diverge from conventional thinking. It involves the ability to connect seemingly unrelated concepts, think outside established norms, and produce outcomes that exhibit originality, relevance, and often, a degree of surprise [16]. It spans various domains, including the arts, sciences, and everyday problem-solving, reflecting an individual's capacity to navigate ambiguity, embrace curiosity, and engage in the synthesis of disparate elements [17]. The conventional understanding of creativity, as outlined by Runco and Jaeger [18], comprises two elements: **originality** and **effectiveness**, which is also known as appropriateness, usefulness, or meaningfulness. These benchmarks are applicable to the evaluation of creative thinking. Creative thinking is commonly assessed through the lens of generating ideas, particularly emphasizing divergent thinking. **Creative state** is a mental or cognitive condition characterized by enhanced creativity and the ability to generate novel and valuable ideas, solutions, or expressions. Individuals may experience different cognitive and psychological states when engaging in creative tasks. Several drives can contribute to the emergence of a creative state [19]. **Creative Drives** refer to the underlying motivations, forces, or impulses that propel individuals toward creative expression and problem-solving. These drives are the internal or external factors that stimulate and energize the creative process [20]. Internal drives play a crucial role in fostering creativity, as they emanate from personal passions, curiosities, and the intrinsic motivation to explore, innovate, or express oneself creatively, also they are easier and more accessible to control or be acquire the skills to navigate them [21]. Internal creative drives can encompass a range of psychological factors, such as the emotions, arousal, valence, or the need for self-expression.

**Emotions** are complex psychological and physiological states that are triggered by various stimuli or situations. They involve a range of subjective experiences, such as feelings, moods, and affective states, that can influence a person's thoughts, behavior, and physiological responses [22]. Research on emotion suggests a concept of emotional abilities, which involves the capacity to think and reason about emotions, engage in problem-solving related to emotions, and utilize emotions for cognitive processes [23,24]. This perspective is rooted in the idea that emotions convey valuable information that can guide thinking and problem-solving, while also recognizing the intrinsic aspect of emotional experiences that involves emotion regulation [25]. Dimensional theories propose that every emotional experience can be characterized along a minimum of two dimensions, arousal and valence [26,27]. **Arousal** considered as an emotional dimension. It signifies the degree of physiological and psychological activation within an individual and is linked to the intensity of emotional encounters. It encompasses heightened states of responsiveness across various dimensions, encompassing both bodily reactions and mental engagement [28]. **Valence** denotes the inherent pleasantness or unpleasantness associated with an emotional encounter. It serves as a measure of the subjective appeal or aversion individuals attribute to a specific emotional experience. Valence encapsulates the positive or negative evaluative aspect of emotions, offering insights into the affective quality and the overall emotional tone of a given experience [28].

**Circumplex model of Emotion**, together, arousal and valence create a two-dimensional space commonly known as the "affective space" or "emotion space." Different emotions can be plotted within this space based on their levels of arousal and valence in the Circumplex model. The model of emotion was developed by James Russell [29]. It organizes emotions into four quadrants within a circular arrangement based on their valence (positive or negative) and arousal (intensity or activation level). Below is an overview of each quadrant along with dominant emotions (see Fig. 1).

**Circumplex model and Creativity mapping**, creativity is enhanced in the presence of positive emotions such as happiness, cheerfulness, and even anger, as well as during states of boredom [11,30]. In a series of studies, Isen and her colleagues [31] investigated the impact of emotion on creativity. They revealed that inducing a positive emotions consistently enhanced creative thinking. The hypothesis behind this observation stems from the idea that positive information in memory tends to be more intricately interconnected than negative information for the majority of individuals. Consequently, positive emotional states are posited to facilitate spreading activation, leading to a broader and more diverse range of cognitive materials being cued, thereby forming a complex cognitive context [32].

The presence of both positive and negative activated emotional states was linked to increased concurrent creative involvement and in this regard most research resulted that positive emotions and positive activation are in more favoured of creative states [11, 20]. On the other hand, positive and negative deactivating emotional states were connected to decreased levels of creative engagement. In recent research, researchers introduce a holistic model that maps creativity characteristic onto a circular model, illustrating how each quadrant of the model, shaped by the presented emotions, arousal, and valence, can impact creativity as a creativity drives [33].

The model illustrated in Fig. 1 represents a mapping to all four quadrants of the Circumplex model to creativity state. The primary aim is to establish a metric that facilitates the assessment of creativity states for machines. *High Arousal, Positive Valence (Excited-Happy Quadrant)* is often conducive to creative thinking and production. The high arousal fuels enthusiasm and energy, while the positive valence supports a mindset open to exploration and generating innovative ideas. It's a state where individuals feel motivated and inspired to bring their creative visions to life. *High Arousal, Negative Valence (Anxious-Angry Quadrant)*, while high arousal in this quadrant signal intensity and focus, the negative valence introduces challenges. Anxiety and frustration hinder the free flow of creative thinking, but controlled stress still drive attention to detail and meticulous creative refinement. *Low Arousal, Negative Valence (Depressed-Sad Quadrant)* presents challenges for creativity. It is associated with feelings of sadness, potentially hindering the motivation and energy required for creative thinking and production. However, periods of introspection and reflection can still be valuable for ideation. *Low Arousal, Positive Valence (Calm-Content Quadrant)* can be beneficial for certain aspects of creativity. A state of calmness and contentment foster a focused and tranquil environment for creative thinking. However, excessive relaxation leads to complacency, so it's essential to strike a balance between tranquility and the necessary arousal for creative energy.

**Emotion recognition** is the process of identifying and interpreting the emotional state of an individual based on various cues and signals, such as facial expressions, speech, physiological responses, body language, and text. It involves understanding and categorizing the specific emotional experiences that a person is undergoing at a given moment [34]. **Unimodal Emotion Recognition**, early efforts in human emotion recognition primarily focused on single modes of expression. Unimodal involves identifying emotions using data from a single source, such as facial expressions or speech signals. It focuses on analyzing one type of modality to infer emotional states [35]. **Multimodal Emotion Recognition**, human tends to show its emotion through several ways. Combining information from various sources or modalities is highly advantageous for improving the accuracy of recognizing emotions [36]. So, essentially, recognizing emotions is best approached as a challenge of understanding emotions through multiple sources or multimodal emotion recognition [37]. Unlike recog-

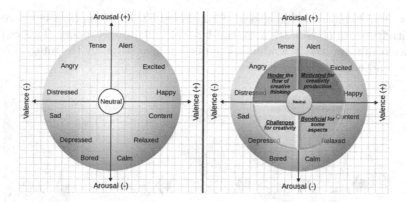

**Fig. 1.** *Left:* The Circumplex model of emotion. *Right:* Mapping of the creativity states to Circumplex model of emotion.

nizing emotions from just one source, multimodal emotion recognition looks at many sources of information all at once. However, it can be tricky because even among these methods, some are better at predicting emotions than others [38].

## 3   Multimodal Creativity State Detection Framework

Equipped with the necessary conceptual foundations and a metric for quantifying creativity states in machines, this section addresses the **MQR** question by formulating a Framework for Multimodal Creativity State Detection. The framework includes Theoretical and Technical components which has been formulated and shown in Fig. 2. The presented framework provides a conceptual overview of multimodal creativity state detection.

*Theoretical Approaches* is integration of explained psychological and neuroscience theories concerning creativity and its dynamic interaction with emotion, specifically addressing emotional dimensions such as arousal and valence, offers a foundation for establishing a measurable metric for creative state detection. **Technical Approaches** includes, *Input Modality* involves determining the modalities of input data for the framework. Modalities may include video, audio, text, and physiological signals. Understanding the nature of the input is crucial for subsequent stages of the framework. *Feature extraction* is capturing key characteristics or patterns from the raw data that are indicative of emotional and cognitive states. *Fusion and classification* In this step, features extracted from different modalities are integrated or combined. This fusion process aims to create a representation of the input data. *Emotion, Arousal, and Valence classification* involves the detection of emotion, arousal, and valence from the input modalities. *Creativity State classification*, using the detected emotional components, the framework proceeds to classify the current state of creativity to the measurable metrics. Creativity state classification involves categorizing the individual's creative mindset or expression based on the identified emotional and emotional dimension. ***Creativity State Detection*** is the result of creativity state classification by the help of Circumplex model. This output provides result into the individual's current creative state, by the integration of emotion and emotional dimensions.

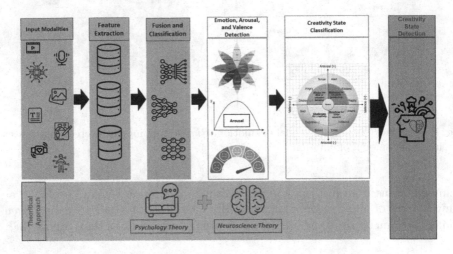

**Fig. 2.** Multimodal Creativity State Detection Framework

## 4   Conclusion and Outlooks

This paper introduces a novel metric for assessing creative state using Emotion, Valence, and Arousal, along with a novel multimodal creativity recognition framework within a concrete background review. Creativity is increasingly pivotal in human-machine and human-human interactions. This framework introduces a pathway for fostering higher cognitive collaboration between them. Considering the complexity and evolving nature of the field both theoretical and technological, there are several avenues for future work in the development and enhancements. *Select the appropriate modalities*, identifying suitable modalities and determining effective fusion methods stand out as prominent challenges. Taking into account both theoretical foundations—acknowledging that humans express emotions unevenly across various modalities [39]—and technological preparedness. *Experiments and Validation in real world setting*, and *Consider other creative drives as input modality*, Various creative drivers impact the creative state, and attention stands out as a particularly influential drives [40].

**Acknowledgement.** This work was funded by Fundação para a Ciência e Tecnologia through the program UIDB/00066/2020 and Center of Technology and Systems (CTS).

# References

1. Rust, R.T., Huang, M.-H.: The Feeling Economy. Springer, Cham (2021). https://doi.org/10.1007/978-3-030-52977-2
2. Deguchi, A., et al.: What is society 5.0. Society **5**, 1–24 (2020)
3. Nikghadam-Hojjati, S., Barata, J.: Computational creativity to design cyber-physical systems in Industry 4.0. In: Camarinha-Matos, L.M., Afsarmanesh, H., Antonelli, D. (eds.) PRO-VE 2019. IAICT, vol. 568, pp. 29–40. Springer, Cham (2019). https://doi.org/10.1007/978-3-030-28464-0_4
4. Llano, M.T., et al.: Explainable computational creativity. arXiv preprint arXiv:2205.05682 (2022)
5. Doshi, A.R., Hauser, O.: Generative artificial intelligence enhances creativity. Available at SSRN (2023)
6. Yalçın, Ö.N., Abukhodair, N., DiPaola, S.: Empathic AI painter: a computational creativity system with embodied conversational interaction. In: NeurIPS 2019 Competition and Demonstration Track, pp. 131–141. PMLR (2020)
7. Schindler, M., Lilienthal, A.J., Chadalavada, R., Ögren, M.: Creativity in the eye of the student. Refining investigations of mathematical creativity using eye-tracking goggles. In: Proceedings of the 40th Conference of the International Group for the Psychology of Mathematics Education (PME), vol. 4, pp. 163–170 (2016)
8. Muldner, K., Burleson, W.: Utilizing sensor data to model students' creativity in a digital environment. Comput. Hum. Behav. **42**, 127–137 (2015)
9. Zhao, Y., et al.: Assessing and understanding creativity in large language models. arXiv preprint arXiv:2401.12491 (2024)
10. Moran, S.: The Roles of Creativity in Society. The Cambridge Handbook of Creativity, pp. 74–90 (2010)
11. Baas, M., De Dreu, C.K., Nijstad, B.A.: A meta-analysis of 25 years of mood-creativity research: hedonic tone, activation, or regulatory focus? Psychol. Bull. **134**(6), 779 (2008)
12. De Dreu, C.K., Baas, M., Nijstad, B.A.: Hedonic tone and activation level in the mood-creativity link: toward a dual pathway to creativity model. J. Pers. Soc. Psychol. **94**(5), 739 (2008)
13. Sawyer, R.K., Henriksen, D.: Explaining Creativity: The Science of Human Innovation. Oxford University Press, Oxford (2024)
14. Kalateh, S., Estrada-Jimenez, L.A., Pulikottil, T., Hojjati, S.N., Barata, J.: The human role in human-centric industry. In: IECON 2022–48th Annual Conference of the IEEE Industrial Electronics Society, pp. 1–6. IEEE (2022)
15. Zhang, T., Liu, M., Yuan, T., Al-Nabhan, N.: Emotion-aware and intelligent internet of medical things toward emotion recognition during Covid-19 pandemic. IEEE Internet Things J. **8**(21), 16002–16013 (2020)
16. Young, J.G.: What is creativity? J. Creative Behav. **19**(2), 77–87 (1985)
17. Gaut, B.: The philosophy of creativity. Philos. Compass **5**(12), 1034–1046 (2010)
18. Runco, M.A., Jaeger, G.J.: The standard definition of creativity. Creat. Res. J. **24**(1), 92–96 (2012)
19. Kaufman, J.C., Sternberg, R.J.: The International Handbook of Creativity. Cambridge University Press, Cambridge (2006)
20. Khalil, R., Godde, B., Karim, A.A.: The link between creativity, cognition, and creative drives and underlying neural mechanisms. Front. Neural Circ. **13**, 18 (2019)
21. Henriksen, D., Richardson, C., Shack, K.: Mindfulness and creativity: implications for thinking and learning. Thinking Skills Creativity **37**, 100689 (2020)

22. Scarantino, A., De Sousa, R.: Emotion. The Metaphysics Research Lab (2018)
23. Mayer, J.D., Salovey, P.: Emotional intelligence and the construction and regulation of feelings. Appl. Prev. Psychol. 4(3), 197–208 (1995)
24. Mayer, J.D., Caruso, D.R., Salovey, P.: The ability model of emotional intelligence: principles and updates. Emot. Rev. 8(4), 290–300 (2016)
25. Cheng, R., Lu, K., Hao, N.: The effect of anger on malevolent creativity and strategies for its emotion regulation. Acta Psychol. Sin. 53(8), 847 (2021)
26. Barrett, L.F.: Discrete emotions or dimensions? the role of valence focus and arousal focus. Cogn. Emotion 12(4), 579–599 (1998)
27. Bliss-Moreau, E., Williams, L.A., Santistevan, A.C.: The immutability of valence and arousal in the foundation of emotion. Emotion 20(6), 993 (2020)
28. Bestelmeyer, P.E., Kotz, S.A., Belin, P.: Effects of emotional valence and arousal on the voice perception network. Soc. Cogn. Affect. Neurosci. 12(8), 1351–1358 (2017)
29. Russell, J.A.: A circumplex model of affect. J. Pers. Soc. Psychol. 39(6), 1161 (1980)
30. Averill, J.R., Thomas-Knowles, C.: Emotional Creativity. Handbook of Positive Psychology, pp. 172–185 (2005)
31. Isen, A.M.: An influence of positive affect on decision making in complex situations: theoretical issues with practical implications. J. Consum. Psychol. 11(2), 75–85 (2001)
32. Kaufmann, G.: The Effect of Mood on Creativity in the Innovative Process. The International Handbook on Innovation, pp. 191–203 (2003)
33. Gu, S., Gao, M., Yan, Y., Wang, F., Tang, Y.Y., Huang, J.H.: The neural mechanism underlying cognitive and emotional processes in creativity. Front. Psychol. 9, 1924 (2018)
34. Jackson, J.C., et al.: Emotion semantics show both cultural variation and universal structure. Science 366(6472), 1517–1522 (2019)
35. Akputu, O.K., Seng, K.P., Lee, Y.: Affect recognition for web 2.0 intelligent e-tutoring systems: exploration of students' emotional feedback. In: E-Learning 2.0 Technologies and Web Applications in Higher Education (2013)
36. Keltner, D., Sauter, D., Tracy, J., Cowen, A.: Emotional expression: advances in basic emotion theory. J. Nonverbal Behav. 43, 133–160 (2019)
37. Baltrušaitis, T., Ahuja, C., Morency, L.P.: Multimodal machine learning: a survey and taxonomy. IEEE Trans. Pattern Anal. Mach. Intell. 41(2), 423–443 (2018)
38. Cimtay, Y., Ekmekcioglu, E., Caglar-Ozhan, S.: Cross-subject multimodal emotion recognition based on hybrid fusion. IEEE Access 8, 168865–168878 (2020)
39. Mehrabian, A.: Communication without words. In: Communication Theory, pp. 193–200. Routledge (2017)
40. Carruthers, L., MacLean, R., Willis, A.: The relationship between creativity and attention in adults. Creat. Res. J. 30(4), 370–379 (2018)

# Coupling PIES and PINN for Solving Two-Dimensional Boundary Value Problems via Domain Decomposition

Krzysztof Szerszeń[(✉)] [iD] and Eugeniusz Zieniuk [iD]

Faculty of Computer Science, University of Bialystok, Konstantego Ciołkowskiego 1M, 15-245 Białystok, Poland
{k.szerszen,e.zieniuk}@uwb.edu.pl

**Abstract.** The paper proposes coupling Parametric Integral Equation System (PIES) and Physics-Informed Neural Network (PINN) for solving two-dimensional potential boundary value problems defined by the Laplace equation. As a result, the computational domain can be decomposed into subdomains, where solutions are obtained independently using PIES and PINN while simultaneously satisfying interface connection conditions. The efficacy of this approach is validated through a numerical example.

**Keywords:** Parametric Integral Equation System (PIES) · Physics-Informed Neural Network (PINN) · PIES-PINN coupling · 2D boundary value problems

## 1 Introduction

In recent years, the field of machine learning has attracted considerable attention within the scientific community, also due to its ability to solve problems formulated by partial differential equations (PDEs). Notably, the application of Physics-Informed Neural Networks (PINN) [1] has proven to be successful in addressing both forward and inverse problems [2]. PINN transforms a boundary value problem defined by PDE into an optimization problem, where the objective function can be directly defined by the PDE through automatic differentiation. The exponential increase in recent publications and the diverse range of applications position PINN as a viable alternative to established computational methods such as FEM, FDM, and meshless methods. While PINN shows promising potential, it faces several challenges, including the computational overhead linked with training neural networks, scalability concerns for complex geometries and the tendency of networks to learn functions with higher variability at a slower pace than those with simpler distributions. Enhancing the efficiency of PINN can be achieved by decomposing the computational domain into subdomains and employing a distinct neural network in each. Different PINN variants, such as cPINN [3], xPINN [4] as well as [5], have been introduced using this approach. Moreover, the possibility of accelerating computations through the utilization of multiple graphics processing units for network training in subdomains has been demonstrated [6].

© The Author(s), under exclusive license to Springer Nature Switzerland AG 2024
L. Franco et al. (Eds.): ICCS 2024, LNCS 14834, pp. 87–94, 2024.
https://doi.org/10.1007/978-3-031-63759-9_11

This paper proposes coupling PINN with Parametric Integral Equation System (PIES) for solving two-dimensional boundary value problems. PIES enables a mathematical reduction of the dimension of the given boundary value problem by one. Consequently, the process of obtaining solutions within the computational domain relies on analyzing the solution of the problem at its boundary. In the proposed approach, the computational domain is decomposed into subdomains, where solutions are independently obtained using PIES and PINN while simultaneously satisfying compatibility conditions at the interfaces of these subdomains. This approach is facilitated by the fact that PIES does not require discretization of the domain, as is the case with the FEM, or just the boundary, as in the BEM. The boundary between PIES and PINN subdomains can be depicted using parametric curves, while the Chebyshev series can approximate the field and flux density functions.

The proposed hybrid approach combines the advantages of domain-based methods, such as PINN, with methods based on the analysis of solutions at the boundary, such as PIES. PINN solves for unknowns within the domain, whereas PIES only deals with unknowns at the boundaries. Moreover, PIES is efficient and relatively straightforward to use in treating bounded or unbounded domains with linear material behavior. On the other hand, PINN is better suited for domains with inhomogeneities and nonlinearities. The experimental section provides a preliminary accuracy analysis of the proposed approach using a two-dimensional linear potential problem defined by Laplace's equation. To the best of the authors' knowledge, this represents the first known attempt to integrate PINN with existing computational methods for solving PDEs.

## 2   Problem Statement

We consider a boundary value problem defined in the domain $\Omega$ bounded by the boundary $\Gamma$. As shown in Fig. 1, this domain can be partitioned into subdomains: $\Omega_A$ where the solution is obtained using PINN and $\Omega_B$, where the solution is determined using PIES.

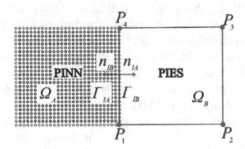

**Fig. 1.** Declaration of the subdomain $\Omega_A$ in PINN, the boundary of the subdomain $\Omega_B$ in PIES, with the normal vectors $n_{IA}$ and $n_{IB}$ to the boundary at the interface of the subdomains.

In this section, we present a concise overview of these methods and their integration at the subdomain interface. The practical aspects of the presented approach are illustrated through the analysis of a stationary temperature field problem, mathematically described by the Laplace equation.

## 2.1 Parametric Integral Equation System

Parametric Integral Equation Systems (PIES) is a computational method designed for solving boundary value problems. It eliminates the need for discretization of both the domain and boundary into elements. The PIES formulation for the 2D problem considered in this work, described by Laplace's equation, is as follows [7]:

$$0.5u_l(\bar{s}) = \sum_{j=1}^{n} \int_{s_{j-1}}^{s_j} \left\{ \overline{U}_{lj}^*(\bar{s}, s)p_j(s) - \overline{P}_{lj}^*(\bar{s}, s)u_j(s) \right\} J_j(s)ds,$$

$$l = 1, 2, \ldots, n, \; s_{l-1} \leq \bar{s} \leq s_l, \; s_{j-1} \leq s \leq s_j. \tag{1}$$

As depicted in Fig. 1, to obtain solutions, it is necessary to specify solely the boundary of the domain $\Omega$, whose shape is analytically embedded in the subintegral functions $\overline{U}_{lj}^*(\bar{s}, s)$ and $\overline{P}_{lj}^*(\bar{s}, s)$ defined as follows:

$$\overline{U}_{lj}^*(\bar{s}, s) = ln\frac{1}{\left(\eta_1^2 + \eta_2^2\right)^{0.5}}, \overline{P}_{lj}^*(\bar{s}, s) = \frac{\eta_1 n_j^{(1)}(s) + \eta_2 n_j^{(2)}(s)}{\eta_1^2 + \eta_2^2}, \tag{2a,b}$$

$$\eta_1 = \Gamma_l^{(1)}(\bar{s}) - \Gamma_j^{(1)}(s), \eta_2 = \Gamma_l^{(2)}(\bar{s}) - \Gamma_j^{(2)}(s), \tag{3}$$

where $\Gamma_l^{(1)}$ $\Gamma_l^{(2)}$ are components of Bézier curves that we use to define the boundary, $n_j^{(1)}$, $n_j^{(2)}$ denote the the the normals to the boundary. Solving the boundary value problem in PIES entails determining the field and flux distribution on the boundary, expressed in formula (1) through functions $u_j(s)$ and $p_j(s)$ defined on the $j$-th Bézier curve. Based on these functions in the second stage of the analysis, the solution can be derived at any point within the domian. The methodology for obtaining the solution to the boundary value problem in PIES, both on the boundary and within the domain have been presented in several earlier works, including [7]. In this study, $u_j(s)$ and $p_j(s)$ are defined through Chebyshev series in the following manner:

$$u_j(s) = \sum_{k=0}^{K-1} u_j^{(k)} T_j^{(k)}(s), p_j(s) = \sum_{k=0}^{K-1} p_j^{(k)} T_j^{(k)}(s), \tag{4a,b}$$

where $u_j^{(k)}$, $p_j^{(k)}$ are the coefficients of the series associated with the basis functions $T_j^{(k)}(s)$, representing Chebyshev polynomials of the first kind, and $K$ is the number of terms in these series. The functions (4a,b) will be used to integrate with PINN, as described in Sect. 2.3.

## 2.2 Physics-Informed Neural Network

PINN is a machine learning technique that utilizes neural networks to obtain approximate solutions to PDEs. The fundamental idea of PINN is to determine the parameters of the neural network, denoted as $\boldsymbol{\theta}$, in such a way that it can approximate the solution $u(\boldsymbol{x}) = \mathcal{N}(\boldsymbol{x}; \boldsymbol{\theta})$, satisfying the given PDE for points $\boldsymbol{x} \in \Omega$ and the prescribed boundary conditions for points $\boldsymbol{x} \in \Gamma$. This can be expressed as [1]:

$$D[u(\boldsymbol{x})] = f(\boldsymbol{x}), \boldsymbol{x} \in \Omega, \quad B[u(\boldsymbol{x})] = g(\boldsymbol{x}), \boldsymbol{x} \in \Gamma, \tag{5a,b}$$

where $D[\cdot]$ is the differential operator acting on the function $u(x)$, and $B[\cdot]$ is the boundary operator. The terms $f(x)$ and $g(x)$ specify forcing and boundary conditions, respectively. In the context of the Laplace's equation examined in this study, the operators take the following form: $D[\cdot] = \frac{\partial^2}{\partial^2 x}$, $B[\cdot] = \frac{\partial}{\partial x}$, with $f(x)$ being set to zero. Finding neural network parameters $\theta$ requires minimizing the loss functions linked to the partial differential equation and boundary conditions, formulated as:

$$\mathcal{L} = \mathcal{L}_{PDE} + \mathcal{L}_{BC}$$
$$= \frac{1}{N_{PDE}} \sum_{i=1}^{N_{PDE}} |D[\mathcal{N}(x_i; \theta)] - f(x_i)|^2 + \frac{1}{N_{BC}} \sum_{i=1}^{N_{BC}} |B[\mathcal{N}(x_i; \theta)] - g(x_i)|^2. \quad (6)$$

In practical implementation, this is achieved through the iterative training of the network using a set of $N_{PDE}$ points within the domain and $N_{BC}$ on the boundary of the problem, also referred to as collocation points.

### 2.3 Coupling PIES and PINN

To couple PIES and PINN, it is necessary to take into account compatibility conditions at the interface of the $\Gamma_{IA}$ and $\Gamma_{IB}$ subdomains, as illustrated in Fig. 1. These conditions can be formulated as follows:

$$u(x)|_{\Gamma_{IA}} = u(x)|_{\Gamma_{IB}}, \quad \nabla u(x)|_{\Gamma_{IA}} n_{IA} = -\nabla u(x)|_{\Gamma_{IB}} n_{IB}, \quad (7a,b)$$

where $n_{IA}$ and $n_{IB}$ are normal vectors to the boundary at the interface. It is worth noting the distinct representation and obtaining of $u(x)$ and $\nabla u(x)$ for PIES and PINN. In the case of PIES, they are defined as $u(x)|_{\Gamma_{IB}} = u_{j=IB}(s)$ and $\nabla u(x)|_{\Gamma_{IB}} n_{IB} = p_{j=IB}(s)$ using Chebyshev series (4a) and (4b), and their values depend on the coefficients $u_{j=IB}^{(k)}$, $p_{j=IB}^{(k)}$, where $IB$ is the index of the Bézier curve defining the boundary in PIES at the interface with the PINN subdomain. On the other hand, in PINN, an iterative procedure is required to determine $u(x)|_{\Gamma_{IA}} = \mathcal{N}(x; \theta)$ and $\nabla u(x)|_{\Gamma_{IA}} n_{IA} = B[\mathcal{N}(x; \theta)]$ for $x \in \Gamma_{IA}$. As a result, solving the boundary problem in PIES and PINN will be carried out through an iterative procedure using a modified form of the objective function compared to (6), given by:

$$\mathcal{L} = \mathcal{L}_{PDE} + \mathcal{L}_{BC} + \mathcal{L}_I, \quad (8)$$

$$\mathcal{L}_I = \frac{1}{N_I} \sum_{i=1}^{N_I} |B[\mathcal{N}(x_i; \theta)] - p_{j=IB}(s_i)|^2. \quad (9)$$

The last term $p_{j=IB}(s_i)$ in formula (9) represents the Chebyshev series (4b) approximating the value of the flux on the boundary $\Gamma_{IB}$, where $s_i$ denotes the parameter values at points corresponding to $x \in \Gamma_{IA}$ on the common interface. To determine the coefficients $p_{j=IB}^{(k)}$ of this series, we need to solve PIES, employing the collocation method as described in the context of PIES in [7, 8]. By expressing formula (1) at collocation points identified with $\bar{s}$, we obtain a system of algebraic equations that approximate PIES, presented in the following matrix formula:

$$[H]\left\{\begin{matrix} p \\ p_{IB} \end{matrix}\right\} = [G]\left\{\begin{matrix} u \\ u_{IB} \end{matrix}\right\}. \quad (10)$$

The generation of the $[H]$ and $[G]$ matrix elements involves calculating the regular and singular integrals, with detailed information provided in [8]. In turn, $u$ represents the set of coefficients of the Chebyshev series in the function $u_j(s)$ (4a) on the outer boundary of the problem, while $u_{IB}$ denotes the set of coefficients on the interface $\Gamma_{IB}$. Similarly, $p$ and $p_{IB}$ are sets of coefficients of Chebyshev series approximating the flux $p_j(s)$ (4b) on the outer boundary and the interface $\Gamma_{IB}$, respectively. The coefficients, denoted by $u$ in (10), are determined using the least squares method for applied boundary conditions specifed on the outer boundary. The coefficients $u_{IB}$ are calculated utilizing the same least squares method, but this time relying on the current solution $u(x)|_{\Gamma_A}$ obtained from the PINN in each iteration. In the following step, we can derive $p_{IB}$ from (10). It should be emphasized that both the matrices $[H]$ and $[G]$, as well as $u$, are determined only once at the beginning of the iterative procedure. The schematic of the algorithm cooupling PIES and PINN is presented in Fig. 2.

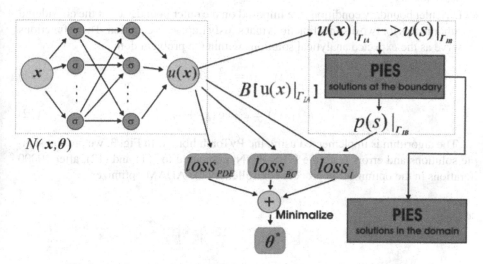

**Fig. 2.** A schematic diagram illustrating the coupling of PIES and PINN.

After the iterative procedure, we obtain solutions within both the domain and at the boundary in PINN, as well as solutions at the boundary in the case of PIES. The final step involves determining the solution within the domain $\Omega_B$ using the integral identity in PIES, as detailed in [9].

## 3 Numerical Example

Below, we present the preliminary studies on the PIES-PINN coupling. We examine a two-dimensional boundary value problem defined by Laplace's equation within subdomains $\Omega_A$ and $\Omega_B$, as illustrated in Fig. 1. The following assumptions are considered:

- The square boundary of subdomain $\Omega_B$ in PIES is defined using 4 first-degree Bézier segments determined by 4 corner points $P_1, P_2, P_3, P_4$, with one of these segments used to define the interface $\Gamma_{IB}$;
- In the square subdomain $\Omega_A$, $100 \times 100$ collocation points are uniformly declared, with 100 points on $x \in \Gamma_{IA}$;
- The functions $u(x)|_{\Gamma_{IB}}$ and $\nabla u(x)|_{\Gamma_{IB}} n_{IB}$ are approximated using the first 5 terms of the Chebyshev series (4a) and (4b), respectively;
- The initial analysis presented here utilizes a fully connected neural network with 7 hidden layers, each comprising 100 neurons and employing Gaussian error linear unit (GELU) activations. Future research is planned to explore how different neural network architectures affect calculation accuracy. The network takes a two-element vector $x = \{x_1, x_2\}$ as the input and produces a scalar value $u(x)$ at the output, representing the approximated pointwise field distribution within the domain and on the boundary;
- Dirichlet boundary conditions are imposed on the outer boundaries of the considered subdomains for two distinct functions that satisfy Laplace's equation. These functions serve as the expected analytical solutions within the problem domain:

$$u(x_1, x_2) = x_1^2 - x_2^2, \tag{11}$$

$$u(x_1, x_2) = e^{x_1} \cos x_2 + x_1. \tag{12}$$

The algorithm is implemented using the PyTorch library. In Fig. 3, we present sample solutions and errors from the PIES-PINN, compared to (11) and (12), after 10000 iterations in the optimization process controlled by the ADAM optimizer.

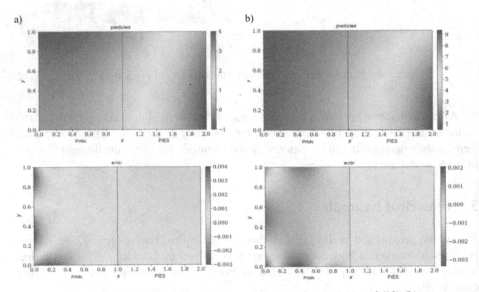

**Fig. 3.** PIES-PINN solutions and errors compared to (11) (a) and (12) (b).

As shown in Fig. 3, the approximation error of PINN reaches the value of $1-3e$ within the subdomain $\Omega_A$. Meanwhile, the error for PIES is even lower, reaching the value of $1-5e$ within the subdomain $\Omega_B$. It should be noted that the solution in PIES near the boundary is inherently subject to an error due to the singular nature of the integral singularity used for this purpose, known as the boundary layer effect. To eliminate this error, the algorithm presented in [9] is applied.

Furthermore, Fig. 4 shows the the individual components of the loss function (8) with iterations.

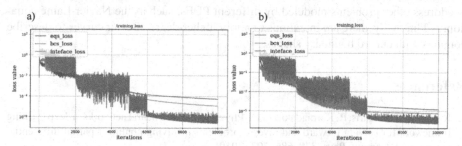

**Fig. 4.** Evolution of losses $\mathcal{L}_{PDE}$, $\mathcal{L}_{BC}$, $\mathcal{L}_I$ with iterations for (11) (a) and (12) (b).

We can observe fluctuations in the interface loss $\mathcal{L}_I$ (9) during iterations for both (11) and (12) boundary conditions. However, as depicted in Fig. 5, the approximated values of $u(x)|_{\Gamma_{IB}}$ and $\nabla u(x)|_{\Gamma_{IB}} n_{IB}$ at the interface $\Gamma_{IB}$ closely match the analytical solutions, which is crucial for the accuracy of coupling PINN and PIES.

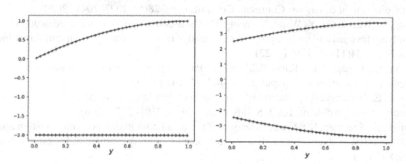

**Fig. 5.** The exact (black solid line) and predicted $u(x)|_{\Gamma_{IB}}$ (red dashed line), $\nabla u(x)|_{\Gamma_{IB}} n_{IB}$ (blue dashed line) solutions comparison at the interface $\Gamma_{IB}$ in PIES for (11) (a) and (12) (b).

Here, $u(x)|_{\Gamma_{IB}}$ and $\nabla u(x)|_{\Gamma_{IB}} n_{IB}$ are approximated using the first 5 terms of the Chebyshev series. To enhance the accuracy of this approximation, we can increase the number of terms $K$ in (4a,b). This is the subject of further research.

## 4  Conclusions

The paper proposes the coupling PIES and PINN to solve two-dimensional boundary value problems, using Laplace's equation as an example. This is advantageous because in PIES, it is only necessary to define the boundary of the domain, while concurrently separating the representation of such a shape into Bézier curves from the approximation of solutions on the boundary in the form of Chebyshev series. In the upcoming research stage, we aim to assess the proposed initial concept with more complex shapes consisting of numerous sub-domains. Additionally, there are plans to broaden the concept to address other problems modeled by different PDEs, such as the Navier-Lamé equation. Additionally, there are intentions to apply higher-degree Bézier curves to describe domains with curved boundary shapes.

## References

1. Raissi, M., Perdikaris, P., Karniadakis, G.E.: Physics-informed neural networks: a deep learning framework for solving forward and inverse problems involving nonlinear partial differential equations. J. Comput. Phys. **378**, 686–707 (2019)
2. Cuomo, S., Di Cola, V.S., Giampaolo, F., Rozza, G., Raissi, M., Piccialli, F.: Scientific machine learning through physics–informed neural networks: where we are and what's next. J. Sci. Comput. **92**(3), 88 (2022)
3. Jagtap, A.D., Kharazmi, E., Karniadakis, G.E.: Conservative physics-informed neural networks on discrete domains for conservation laws: applications to forward and inverse problems. Comput. Methods Appl. Mech. Eng. **365**, 113028 (2020)
4. Jagtap, A.D., Karniadakis, G.E.: Extended physics-informed neural networks (XPINNs): a generalized space-time domain decomposition based deep learning framework for nonlinear partial differential equations. Commun. Comput. Phys. **28**(5), 2002–2041 (2020)
5. Zhang, B., Wu, G., Gu, Y., Wang, X., Wang, F.: Multi-domain physics-informed neural network for solving forward and inverse problems of steady-state heat conduction in multilayer media. Phys. Fluids **34**(11), 116116 (2022)
6. Shukla, K., Jagtap, A.D., Karniadakis, G.E.: Parallel physics-informed neural networks via domain decomposition. J. Comput. Phys. **447**, 110683 (2021)
7. Zieniuk, E.: Bézier curves in the modification of boundary integral equations (BIE) for potential boundary-values problems. Int. J. Solids Struct. **40**(9), 2301–2320 (2003)
8. Zieniuk, E., Szerszeń, K.: A regularization of the parametric integral equation system applied to 2D boundary problems for Laplace's equation with stability evaluation. J. Comput. Sci. **61**, 101658 (2022)
9. Zieniuk, E., Szerszeń, K., Bołtuć, A.: A novel strategy for eliminating the boundary layer effect in the regularized integral identity in PIES for 2D potential problem. Int. J. Comput. Methods **20**(03), 2250053 (2023)

# Energy Efficiency of Multithreaded WZ Factorization with the Use of OpenMP and OpenACC on CPU and GPU

Beata Bylina [ID] and Jarosław Bylina[(✉)] [ID]

Institute of Computer Science, Marie Curie-Skłodowska University, Lublin, Poland
{beata.bylina,jaroslaw.bylina}@umcs.pl

**Abstract.** Energy efficiency research aims to optimize the use of computing resources by minimizing energy consumption and increasing computational efficiency. This article explores the effect of the directive-based parallel programming model on energy efficiency for the multithreaded WZ factorization on multi-core central processing units (CPUs) and multi-core Graphics Processing Units (GPUs). Implementations of the multithreaded WZ factorization (both the basic and its variant optimized by strip-mining vectorization) are based on OpenMP and OpenACC. Strip-mining gave clear enhancement in comparison to the basic version. The energy efficiency is much better on GPU than on CPU; and on CPU, the use of OpenMP is more energy-efficient; however, for GPU, OpenACC gives better results.

**Keywords:** WZ factorization · OpenACC · OpenMP · energy efficiency

## 1 Introduction

Research focused on enhancing the energy efficiency of algorithms through various programming techniques plays a crucial role in the broader context of sustainable development. In this paper, we will examine energy efficiency using only the ratio of the total number of floating-point operations to the total energy consumption (Flop/J) metric, as presented in paper [11].

The development of multithreaded parallel applications on multi-core architectures can be based on various parallel programming frameworks, including OpenMP (Open Multi-Processing) [9], OpenCL (Open Computing Language) [5], and OpenACC (Open ACCelerators) [8]. Choosing the one that is appropriate for the target context is not easy. OpenMP and OpenACC are two of the most widely used directive-based parallel programming models for parallelization.

The WZ matrix factorization was introduced in 1979 by D.J. Evans and M. Hatzopoulos [2]. It is also known as Quadrant Interlocking Factorization (QIF). The aim of their work was to design a factorization for greater parallelization. The WZ factorization is investigated by various researchers, as [1,12].

© The Author(s), under exclusive license to Springer Nature Switzerland AG 2024
L. Franco et al. (Eds.): ICCS 2024, LNCS 14834, pp. 95–102, 2024.
https://doi.org/10.1007/978-3-031-63759-9_12

The WZ factorization was chosen because it contains a lot of arithmetic operations that could be performed in parallel and distributed among many cores on both the CPU and GPU. factorization were selected for further research described in this article, namely, the basic one and its variant optimized by strip-mining vectorization, which is the best in terms of performance and energy consumption. In this paper, the focus is on describing the experience of moving ideas and concepts from the OpenMP programming model on multi-core architecture CPU to OpenACC both on multi-core architecture CPU and GPU. In particular, energy efficiency is analyzed and compared for two directive-based programming models on CPU, namely OpenMP and OpenACC, on two platforms (GPU and CPU). Experimental evaluations prove the expected improvement in energy efficiency.

The rest of the paper is organized as follows. In Sect. 2, contemporary research on parallel programming and its impact on energy consumption was presented. Section 3 details multithreaded implementations of the WZ factorization algorithm, utilizing OpenMP and OpenACC. In Sect. 4, experimental evaluations of energy efficiency on multi-core CPUs and many-core GPUs are presented. Lastly, Sect. 5 concludes the paper and outlines future research directions.

## 2    Related Work

In high-performance computing, energy efficiency via directive-based parallel programming is crucial. Articles [7,10] explore OpenMP's impact on energy efficiency, optimizing execution time and energy consumption. [10] delves into OpenMP's runtime power level adjustments, while [7] studies performance and energy usage in linear algebra kernels. Other works [3,6] compare parallel programming models on multi-core CPUs and GPUs, including OpenMP and OpenACC. The paper [3] assesses CUDA, OpenMP, and OpenACC on Nvidia Tesla V100 GPU for matrix multiplication, highlighting data size's impact on performance. None of them compare OpenMP with OpenACC in dense linear algebra algorithms regarding energy-performance trade-offs.

## 3    Multithreaded Implementations of the WZ Factorization

The WZ factorization [12] consists in transforming a nonsingular square matrix **A** into a product of two matrices, namely **W** and **Z**. The form of the matrix **W** is a butterfly and the for of the matrix **Z**—of an hourglass (details can be seen in [1], for example).

The easiest way was to transform OpenMP to OpenACC is to replace `#pragma omp parallel for` enforcing the parallelism of the loops. In OpenACC, parallel directives `#pragma acc parallel loop gang` will be used by the compiler for different architectures (both CPU and GPU). The `loop` directive uses the `gang` clause to instruct the compiler about the level of parallelism.

There is no direct equivalent of #pragma omp simd in OpenACC. However, a certain substitute of that pragma is #pragma acc loop vector. Additionally, we put independent which is to notify the compiler that the iterations of the loop are data independent. In the GPU version only, we need to provide the compiler with additional information on how to manage the transfer between the device (GPU) and the host (CPU) and therefore we added #pragma acc data copy(a) copy(w). Lastly, in the OpenMP version on GPU we have to put #pragma omp target device(0)—to enforce computations offloading to GPU.

Figure 1 shows the pseudocode (only the inner loop fragment) of the basic version of WZ factorization with the use of OpenMP (blue pragmas) and OpenACC (green pragmas). Purple pragmas are added to make the OpenMP code run on a GPU device.

Therefore, the strip-mining technique is manually applied to the algorithm as shown in Fig. 2. A loop in the process of strip-mining is divided into two loops, where the inner one has BLOCK_SIZE iterations and the outer one has n/BLOCK_SIZE iterations (n is the number of iterations in the original loop). The strip-mining alone can have some positive impact on the performance (by easing the automatic vectorization process).

In such a process we improve the temporal and spatial locality of the data. By dividing the data into pieces of BLOCK_SIZE, we cause them to fit in cache memory and stay there as long as needed to conduct current computations. This minimizes the frequency of cache memory swaps. The theoretical value of BLOCK_SIZE can be estimated from: $\left(\text{BLOCK\_SIZE}^2 + 4 \cdot \text{BLOCK\_SIZE}\right) \cdot s \leq M$

```
#pragma omp target device(0)
#pragma acc data copy(a[0:n*n]) copy(w[0:n*n])
for(k = 0; k < n/2-1; k++) {
  p = n-k-1; akk = a[k][k]; akp = a[k][p];
  apk = a[p][k]; app = a[p][p]; detinv = 1 / (apk*akp - akk*app);
#pragma omp parallel for
#pragma acc parallel
  {
#pragma acc loop gang
    for(i = k+1; i < p; i++) {
      w[i][k] = (apk*a[i][p] - app*a[i][k]) * detinv;
      w[i][p] = (akp*a[i][k] - akk*a[i][p]) * detinv;
// INNER LOOP
#pragma omp simd
#pragma acc loop vector independent
      for(j = k+1; j < p; j++)
        a[i][j] = a[i][j] - w[i][k]*a[k][j] - w[i][p]*a[p][j]; } } }
```

**Fig. 1.** The version of the basic algorithm with the use of OpenMP and OpenACC—pseudocode. (Color figure online)

where $M$ is the size (in bytes) of the cache memory considered and $s$ is the size (also in bytes) of one floating-point number. Thus we can see that it should be satisfied (for the L2 cache size $M = 1024$ kB and the size of the double precision $s = 8$ B) if BLOCK_SIZE $\leq 256$ (we use only divisors of the size of the matrix).

## 4  Numerical Experiments

We tested two types of the WZ factorization algorithm: basic, which is the algorithm presented in Fig. 1, and sm-$b$, which refers to the strip-mining algorithms depicted in Fig. 2.

All the implementations were tested on randomly generated dense matrices which had the WZ factorization with no pivoting needed. The sizes of the matrices were: 8192, 16384, 32768 (as in [7]). The experiment was conducted on a CPU Intel(R) Xeon(R) Gold 5218R with codename Cascade Lake-SP, featuring 40 cores (20 per socket), 80 threads (2 per core), and a SIMD register size of 512 bits. During the tests, Intel ICC (version 2021.5.0) was used, with the following compiler options: -O3 -qopenmp -xHOST -ipo -no-prec-div -fp-model fast=2 -qopt-zmm-usage=high. The -xHOST option generates optimized code based on the Intel(R) Xeon(R) Gold processor capabilities, but does not utilize 512-bit ZMM registers. To utilize AVX2 instructions on ZMM registers, the -qopt-zmm-usage=high option is added. To analyze the impact of all versions of the algorithm on energy consumption on CPU, we used measurements from the RAPL (Running Average Power Limit) interface [4].

The experiment was also conducted on a Tesla V100S-PCIE-32GB GPU. It features a 1370 MHz core clock, 32 GB memory, and 5120 CUDA cores. The compiler used during testing was nvc. GPU power sensors were monitored using the NVIDIA System Management Interface (nvidia-smi), based on the NVIDIA Management Library (NVML).

In Fig. 3, we have graphs showing the energy efficiency of the sm-$b$ version for block size $b = 64, 128, 256$ on both CPU and GPU, using both OpenMP and OpenACC (NB: different scales were used for GPU and CPU).

```
// INNER LOOP
      start = RDTTNM(k+1, BLOCK_SIZE);
      for(jj = start; jj < p; jj += BLOCK_SIZE) {
         __assume(jj % BLOCK_SIZE == 0);
#pragma omp simd
#pragma acc loop vector independent
         for(j = jj; j < jj+BLOCK_SIZE; ++j)
            a[i][j] = a[i][j] - w[i][k]*a[k][j]- w[i][p]*a[p][j]; } } }
```

Fig. 2. Strip-mining in the basic algorithm with the use of OpenMP and OpenACC—pseudocode (the inner loop only).

Analyzing these data, we can observe that there is no single block size that consistently yields optimal results across all dataset sizes. The experimental results indicate that the energy efficiency of both OpenMP and OpenACC may deteriorate with larger input data sizes. Thus, comprehensive studies are necessary for all dataset sizes.

Contrary to common belief, the best runtime performance does not always correspond to the best energy consumption. Tables 1, 2, and 3 present the time, performance, energy consumption, and energy efficiency of the algorithm versions that performed best during the experiments across all dataset sizes. Analyzing data from these tables, we observe that the GPU version using OpenACC (values in bold) outperforms other versions of the algorithm. Conversely, the worst-performing version is on CPU with OpenACC. Additionally, applying strip-mining to the algorithm results in improvements across time, performance, and energy consumption. For instance, in terms of energy efficiency, the sm-64 variant on GPU exhibits about a 61% improvement compared to the second-best variant, namely basic on GPU for a size of 8192.

Analyzing the collected data, reveals that as matrix size increases, time, performance, energy consumption, and energy efficiency all decline. Similar performance trends are noted in [3].

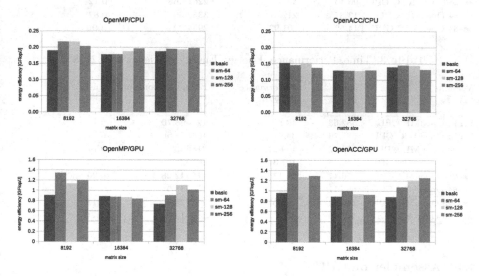

**Fig. 3.** Energy efficiency for sm-b for $b = 64, 128, 256$ on CPU (top) using OpenMP and OpenACC and GPU (bottom) using OpenMP and OpenACC

**Table 1.** Time, Performance, Energy, Energy efficiency for 8192

| Versions | Time [s] | Performance [Gflops] | Energy consumption [J] | Energy efficiency [Gflop/J] |
|---|---|---|---|---|
| basic OpenMP CPU | 9.64 | 38.01 | 1932.87 | 0.18 |
| basic OpenACC CPU | 9.99 | 36.68 | 2414.23 | 0.15 |
| basic OpenMP GPU | 2.58 | 142,06 | 405.55 | 0.90 |
| basic OpenACC GPU | 2.53 | 144.86 | 383.62 | 0.95 |
| sm-64 OpenMP CPU | 8.67 | 42.27 | 1687.53 | 0.21 |
| sm-128 OpenACC CPU | 10.40 | 35.22 | 2419.90 | 0.15 |
| sm-64 OpenMP GPU | 2.43 | 150.82 | 272.77 | 1.34 |
| sm-64 OpenACC GPU | 2.33 | 157.29 | 238.00 | **1.53** |

**Table 2.** Time, Performance, Energy, Energy efficiency for 16384

| Versions | Time [s] | Performance [Gflops] | Energy consumption [J] | Energy efficiency [Gflop/J] |
|---|---|---|---|---|
| basic OpenMP CPU | 75.96 | 38.59 | 16512.92 | 0.17 |
| basic OpenACC CPU | 92.03 | 31.86 | 22845.20 | 0.12 |
| basic OpenMP GPU | 19.17 | 152.95 | 3312.43 | 0.88 |
| basic OpenACC GPU | 17.72 | 165.46 | 3306.49 | 0.88 |
| sm-256 OpenMP CPU | 68.54 | 42.77 | 14925.40 | 0.19 |
| sm-256 OpenACC CPU | 90.13 | 32.52 | 22618.68 | 0.12 |
| sm-64 OpenMP GPU | 13.81 | 212.31 | 3352.50 | 0.87 |
| sm-64 OpenACC GPU | 14.52 | 201.93 | 2934.79 | **0.99** |

**Table 3.** Time, Performance, Energy, Energy efficiency for 32768

| Versions | Time [s] | Performance [Gflops] | Energy consumption [J] | Energy efficiency [Gflop/J] |
|---|---|---|---|---|
| basic OpenMP CPU | 582.63 | 40.26 | 125098.70 | 0.19 |
| basic OpenACC CPU | 681.70 | 34.41 | 168578.50 | 0.14 |
| basic OpenMP GPU | 138.62 | 169.21 | 31869.74 | 0.74 |
| basic OpenACC GPU | 140.18 | 167.33 | 26657.43 | 0.88 |
| sm-256 OpenMP CPU | 556.17 | 42.17 | 118442.28 | 0.20 |
| sm-64 OpenACC CPU | 685.11 | 34.24 | 162233.58 | 0.14 |
| sm-128 OpenMP GPU | 140.82 | 166.57 | 21172.60 | 1.12 |
| sm-256 OpenACC GPU | 117.10 | 200.31 | 18661.66 | **1.26** |

## 4.1   Assembler on CPU

The compiler's task is complex, translating high-level source code into machine code while optimizing processor resources. Assembly code varies across compilers, affecting execution times and energy efficiency. In [6], execution times differ significantly between OpenMP and OpenACC implementations for the same application, likely due to compiler maturity with directives. Our analysis reveals energy consumption and efficiency also depend on compiler maturity. Examining assembler code on CPU with ICC (OpenMP) and nvc (OpenACC), we observe differences in register usage, impacting code vectorization and thus performance and energy efficiency. Longer registers enable better vectorization, but memory bandwidth limitations may mitigate their impact on performance (Fig. 4).

```
/ICC compiler,  OpenMP, CPU without  -qopt-zmm-usage=high option:
  vmovupd    8(%r15,%rsi,8), %ymm1                          #26.68
  vbroadcastsd (%rdi,%r11,8), %ymm2                         #26.51
  vbroadcastsd (%rdi,%r10,8), %ymm3                         #26.85
  vfnmadd213pd 8(%r12,%rsi,8), %ymm1, %ymm2                 #26.68
  vfnmadd132pd 8(%r14,%rsi,8), %ymm2, %ymm3                 #26.103
********************************************
/ICC compiler,  OpenMP, CPU with  -qopt-zmm-usage=high option:
  vmovups    8(%r15,%r13,8), %zmm3                          #26.68
  vbroadcastsd (%rdi,%r12,8), %zmm4                         #26.51
  vbroadcastsd (%rdi,%r8,8), %zmm5                          #26.85
  vfnmadd213pd 8(%rsi,%r13,8), %zmm3, %zmm4                 #26.68
  vfnmadd132pd 8(%r9,%r13,8), %zmm4, %zmm5                  #26.103
********************************************
/NVC compiler, OpenACC, CPU:
  vmovsd        -8(%r14,%r15,8), %xmm1        # xmm1 = mem[0],zero
  vmovsd        (%r11), %xmm2                 # xmm2 = mem[0],zero
  vfnmadd213sd -8(%r14,%rsi,8), %xmm1, %xmm2
                                    # xmm2 = -(xmm1 * xmm2) + mem
  vmovsd        -8(%r14,%r10,8), %xmm1        # xmm1 = mem[0],zero
  vfnmadd132sd (%r8), %xmm2, %xmm1            # xmm1 = -(xmm1 * mem) + xmm2
```

**Fig. 4.** The snippet of assembly code for element calculations a[i][j] with the use of OpenMP (ICC compiler—on the top) and OpenACC (NVC—on the bottom) on CPU.

## 5   Conclusion

This paper develops a multithreaded implementation of the WZ factorization using OpenACC, comparing energy efficiency between OpenMP for multi-core CPUs and OpenACC for both CPUs and GPUs. Strip-mining enhances performance and energy efficiency. Choosing between OpenMP and OpenACC affects time, performance, energy consumption, and efficiency. Porting OpenMP to OpenACC on CPU doesn't improve performance or efficiency, but on GPU, results are better. In general, OpenMP and OpenACC may increase energy efficiency.

The energy efficiency findings in multithreaded WZ factorization can be extended to similar numerical algorithms in computational linear algebra (especially ones, which employ quite regular nested loops), providing valuable insights for developers and researchers.

Future work will focus on developing mathematical models for energy efficiency with OpenMP and OpenACC, and software for optimizing energy efficiency on CPU and GPU automatically.

# References

1. Bylina, B., Bylina, J., Piekarz, M.: Influence of loop transformations on performance and energy consumption of the multithreded WZ factorization. In: Ganzha, M., Maciaszek, L.A., Paprzycki, M., Slezak, D. (eds.) Proceedings of the 17th Conference on Computer Science and Intelligence Systems, FedCSIS 2022, Sofia, Bulgaria, 4–7 September 2022. Annals of Computer Science and Information Systems, vol. 30, pp. 479–488 (2022). https://doi.org/10.15439/2022F251
2. Evans, D.J., Hatzopoulos, M.: A parallel linear system solver. Int. J. Comput. Math. **7**(3), 227–238 (1979). https://doi.org/10.1080/00207167908803174
3. Khalilov, M., Timofeev, A.: Performance analysis of CUDA, OpenACC and OpenMP programming models on TESLA V100 GPU. J. Phys. Conf. Ser. **1740**, 012056 (2021). https://doi.org/10.1088/1742-6596/1740/1/012056
4. Khan, K.N., Hirki, M., Niemi, T., Nurminen, J.K., Ou, Z.: RAPL in action: experiences in using RAPL for power measurements. ACM Trans. Model. Perform. Eval. Comput. Syst. **3**(2) (2018). https://doi.org/10.1145/3177754
5. Khronos Group: OpenCL overview. https://www.khronos.org/opencl/. Accessed Apr 2024
6. Larrea, V., Budiardja, R., Gayatri, R., Daley, C., Hernandez, O., Joubert, W.: Experiences in porting mini-applications to OpenACC and OpenMP on heterogeneous systems. Concurrency Comput. Pract. Experience **32** (2020). https://doi.org/10.1002/cpe.5780
7. Lima, J.V.F., Raïs, I., Lefèvre, L., Gautier, T.: Performance and energy analysis of OpenMP runtime systems with dense linear algebra algorithms. Int. J. High Performance Comput. Appl. **33**(3), 431–443 (2019). https://doi.org/10.1177/1094342018792079
8. OpenACC Organization: Homepage | OpenACC. https://www.openacc.org/. Accessed Apr 2024
9. OpenMP ARB: Home—OpenMP. https://www.openmp.org/. Accessed Apr 2024
10. Shahneous Bari, M.A., Malik, A.M., Qawasmeh, A., Chapman, B.: Performance and energy impact of OpenMP runtime configurations on power constrained systems. Sustain. Comput. Inf. Syst. **23**, 1–12 (2019). https://doi.org/10.1016/j.suscom.2019.04.002
11. Szustak, L., Wyrzykowski, R., Olas, T., Mele, V.: Correlation of performance optimizations and energy consumption for stencil-based application on Intel Xeon scalable processors. IEEE Trans. Parallel Distrib. Syst. **31**, 2582–2593 (2020). https://doi.org/10.1109/TPDS.2020.2996314
12. Yalamov, P., Evans, D.: The WZ matrix factorisation method. Parallel Comput. **21**(7), 1111–1120 (1995). https://doi.org/10.1016/0167-8191(94)00088-R. https://www.sciencedirect.com/science/article/pii/016781919400088R

# PGAS Data Structure for Unbalanced Tree-Based Algorithms at Scale

Guillaume Helbecque[1,2]([✉]) [ID], Tiago Carneiro[3], Nouredine Melab[2], Jan Gmys[2], and Pascal Bouvry[1]

[1] Université du Luxembourg, DCS-FSTM/SnT , Esch-sur-Alzette, Luxembourg
**guillaume.helbecque@uni.lu**
[2] Université de Lille, CNRS/CRIStAL, Centre Inria de l'Université de Lille, Lille, France
[3] Interuniversity Microelectronics Centre, Leuven, Belgium

**Abstract.** The design and implementation of algorithms for increasingly large and complex modern supercomputers requires the definition of data structures and workload distribution mechanisms in a productive and scalable way. In this paper, we propose a PGAS data structure along with a Work-Stealing mechanism for the class of parallel tree-based algorithms that explore unbalanced trees using the depth-first search strategy. The contribution has been implemented and packaged as an open-source module in the Chapel PGAS language. The experimentation of the contribution in a single-node setting using backtracking applied to fine-grained Unbalanced Tree-Search (UTS) benchmark shows that 68% of the linear speed-up can be achieved. In addition, the scalability of the contribution has been evaluated using the Branch-and-Bound algorithm to solve big instances of the Flowshop Scheduling problem on a large cluster. The reported results reveal that 50% of strong scaling efficiency is achieved using 400 computer nodes (51,200 processing cores).

**Keywords:** Depth-first Tree Search · PGAS · Scalable Data structures · Work Stealing

## 1 Introduction

In the landscape of modern programming environments, the definition of efficient and versatile data structures is a fundamental requirement. This need becomes even more pronounced in Partitioned Global Address Space (PGAS) environments that are inherently tailored for distributed computing, where the ability to effectively manage data structures across clusters is pivotal. In this work, the focus is on tree-based algorithms that explore unbalanced solution spaces. This class of algorithms has garnered significant attention due to their capacity to offer viable solutions to problems in different areas, such as Operations Research, Artificial Intelligence, Bio-informatics, and Machine Learning [11,19].

These algorithms, such as backtracking and Branch-and-Bound (B&B), are able to efficiently explore solution spaces. However, they exhibit large irregular

L. Franco et al. (Eds.): ICCS 2024, LNCS 14834, pp. 103–111, 2024.
https://doi.org/10.1007/978-3-031-63759-9_13

trees making their design in a parallel distributed context raising multiple challenges. The major of them is the design of efficient and scalable data structures and dynamic adaptive load balancing mechanisms.

To raise that challenge, we introduced a PGAS data structure, called DistBag_DFS, along with a Work Stealing (WS) mechanism for the design and implementation of unbalanced Depth-First (DFS) tree-search at scale [13]. The implementation is based on the Chapel programming language [3] and packaged in the open-source DistributedBag_DFS module [12].

In this paper, we extend this previous work addressing its limitations. First, we provide a comprehensive description of DistBag_DFS and associated WS, along with a performance evaluation of this latter mechanism on the backtracking fine-grained Unbalanced Tree-Search (UTS) benchmark. Then, we investigate its performance in a large-scale distributed-memory setting using B&B applied to the Permutation Flowshop Scheduling problem. Up to 400 computer nodes (51,200 processing cores) are used to evaluate the scalability of the WS mechanism in solving large instances. The results demonstrate the high performance of the PGAS data structure and WS, even in highly-demanding scenarios.

In the following, we first give a short background and some related works. Then, we present DistBag_DFS and associated WS. After a performance evaluation, we outline some conclusions and future directions.

## 2   Background and Related Work

Tree-based search algorithms are powerful techniques that have the ability to efficiently explore solution spaces. The exploration is often guided by the principles of backtracking and B&B involving a systematic search through the decision tree, incrementally building and evaluating potential solutions. Those algorithms generally involve highly irregular and unpredictable search trees, and explore the tree in a DFS manner. Implemented using a last-in, first-out (LIFO) stack to store generated but not yet visited nodes, DFS is favored in combinatorial algorithms due to its memory efficiency, especially when compared to memory-intensive strategies such as breadth-first.

Tree-search algorithms are inherently recursive, making them well-suited for parallelization. The most general and frequently used approach is the parallel tree exploration, which consists in exploring several disjoint subspaces in parallel [7]. In asynchronous mode, adopted in this paper, the search processes communicate in an unpredictable way making non trivial the sharing of knowledge among workers. Therefore, defining an efficient data structure, to store the workload, and its associated management policy is highly crucial for performance.

In this work, the multi-pool strategy is used. Each worker manages its own pool of generated, but not yet evaluated, nodes and maintaining them in a DFS order. This strategy requires a sophisticated communication model since it raises the issue of balancing the workload between multiple pools. A popular and provably efficient dynamic load balancing approach is the WS paradigm [1]. Under WS, each process usually maintains a double-ended queue (deque) of nodes. Each worker processes nodes from the tail of the deque and steals work items from the head of another deque when its pool is empty.

In this context, some data structures based on traditional programming environments have been proposed [2,9]. The latter allow high-performance and often benefit from problem-specific optimizations. In contrast, several papers discuss scalable dynamic load balancing techniques for unbalanced tree-search using PGAS-based environments [4,6,15,17]. Similarly, we introduced in [13] the DistBag_DFS distributed data structure designed for unbalanced tree-search at scale. The latter is implemented using Chapel and has been exploited for a generic parallel distributed tree-search. We demonstrated the competitiveness of our approach compared to OpenMP and MPI+X baselines, considering both intra- and inter-node performance, and productivity-awareness. This work extends [13] with a comprehensive description of the data structure and its WS mechanism while investigating its performance at scale.

## 3   The DistBag_DFS Data Structure and Work Stealing

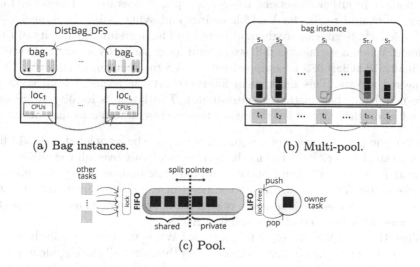

(a) Bag instances.          (b) Multi-pool.

(c) Pool.

**Fig. 1.** Illustration of the DistBag_DFS components: (a) bag instances, (b) multi-pool, and (c) pool based on non-blocking split deque.

**Design.** Figure 1 illustrates the hierarchical structure of the data structure. First of all, DistBag_DFS maintains one *bag* instance (multi-pool) per locale, as shown in Fig. 1a. In Chapel, a *locale* is a subset of the target architecture that can be used to control and reason about affinity for the sake of performance and scalability. For most target architectures, a locale is equivalent to a computer node. In that sense, this component of the data structure handles the inter-node level of parallelism. While the data structure is safe to use in a distributed manner, it provides a mean to obtain a privatized instance of it for maximized performance, and each locale always operates on its privatized instance.

As shown in Fig. 1b, each bag instance contains multiple pools, called *segments*. More precisely, one segment is maintained per parallel task ($T$ in the figure). Each task has a unique identifier $1, \ldots, T$, used to map it to a segment. The latter is used in the `DistBag_DFS`'s insertion (resp. retrieval) procedure to specify the segment into (resp. from) which an element node gets inserted (resp. retrieved). This is required by the DFS, because when a node is evaluated, the entire subtree below it must be explored before another sibling node is processed. However, when children nodes are inserted into a different segment than the one from which the parent was taken, that necessary condition cannot be ensured. In addition, it is worth to mention that while each segment locally guarantees a DFS order, the multi-pool do not.

The implementation of segments is based on non-blocking split deques [5]. Under this scheme, each segment is logically split into a "shared" and "private" region using an atomic *split pointer*, as shown in Fig. 1c. This scheme allows lock-free local access to the private portion of the deque and copy-free transfer of work between the shared and private portions. Work transfer is done by moving the split pointer in either directions using appropriate operators. The other tasks access the shared region for load balancing, and synchronize themselves using an atomic lock. Segments are dynamic-sized and have an initial capacity of 1,024 elements. When a segment is full, we extend its capacity by a power of two.

Finally, `DistBag_DFS` is equipped with a WS mechanism, as shown in Fig. 1 by the red arrows. This mechanism intervenes at the levels of both segments (intra-node) and bag instances (inter-node). The latter is locality-aware and, depending on the state of the data structure, different scenarios may occur:

- When the private region of a segment is empty, the associated task will first try to steal a work item locally. It iterates randomly over all the eligible segments from the same bag instance, and only one node is stolen, if applicable. Indeed, the WS is performed at the head of the deque and thanks to DFS the stolen node is the shallowest one in the search tree. Thus, it is expected to generate a large number of children nodes.
- When the local WS attempts fail, a global WS is triggered. Similarly, it iterates over all the eligible bag instances, and then over all the eligible segments on it. Since remote accesses generate high overheads, multiple nodes are stolen at once. In addition, only one global steal attempt is allowed per bag instance, meaning that when a task is performing a global WS, the other tasks can not.
- When all the local and global WS attempts fail, nothing is returned and one can be sure that the whole `DistBag_DFS` is either empty or few elements remain inside.

**Implementation Aspects.** `DistBag_DFS` is designed to be as simple as possible for the user. It implements a multi-pool and encapsulates a load balancing mechanism transparently to the user, thanks to the PGAS paradigm. Indeed, the latter includes a unified global address space, implicit communications, better data locality, and expressive memory models. Furthermore, the data structure is generic and can contain any types, even user-defined or external ones.

Regarding the user's interface, the data structure is composed of a set of two initialization variables (`eltType` and `targetLocales`) and seven methods. The operators `add`, `addBulk`, and `remove` allow the user to insert an element, insert elements in bulk, and remove an element from a given segment, respectively. Each of these procedures applies to the bag instance of the locale it is called from. In addition, the data structure contains four global methods that apply to the whole `DistBag_DFS`. To avoid holding onto locks, we take a snapshot approach, increasing memory consumption but also increasing parallelism. This allows other concurrent, even mutating, operations while iterating, but opens the possibility to iterating over duplicated or missing elements from concurrent operations. These methods are `clear`, `these`, `contains`, and `getSize`, and allow the user to clear `DistBag_DFS`, iterate over it, search for a specific element in it, and get its global size, respectively. Finally, the data structure owns some configuration parameters that can be used to fine-tune its capacity and WS. The latter allow the user to set the initial and maximum capacities of the segments and also the minimum number of elements a segment must have to become eligible to be stolen from. This may be useful if some segments contain less elements than others and should not be stolen from.

One aspect that requires further investigation is the required "task id" in the insertion/retrieval operations of the data structure. In the current version of `DistBag_DFS`, it is required to ensure the local DFS ordering. We could implement an automated way to deal with this index, but *a priori* Chapel's design intentionally avoids supporting a standard language-level way to query a task's id. One can still exploit the internal `chpl_task_ID_t` opaque type, that refers to the task id that the runtime uses, but this could raise portability issues since Chapel includes different runtime tasking options, and the support is not guaranteed to continue across future versions of the language.

## 4    Performance Evaluation

**Experimental Protocol.** We first evaluate the performance of the WS mechanism in a shared-memory setting on the UTS benchmark [16]. The latter is widely used to evaluate dynamic load balancing of fine-grained applications and is solved using backtracking. Moreover, we assess the scalability of our implementation in a large scale distributed-memory setting. As test-cases, large Permutation Flowshop Scheduling Problem (PFSP) instances proposed by E. Taillard in [18] are solved using the B&B technique. They consist in finding an optimal processing order for $n$ jobs on $m$ machines, such that the completion time of the last job on the last machine (makespan) is minimized. The so-called two-machine bound [14] and the dynamic *minBranch* branching technique [10] are used.

The Luxembourg national petascale MeluXina - Cluster module is used for the experiments. Each computer node has 2 AMD EPYC Rome 7H12 64 cores @ 2.6 GHz CPUs and 512 GB of RAM. In addition, the nodes are interconnected *via* the InfiniBand HDR high-speed fabric and operate under Rocky Linux 8.7. Chapel 1.31.0 is used in a fine-tuned configuration environment, along with the gcc 11.3.0 back-end compiler.

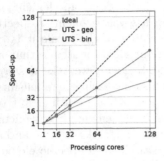

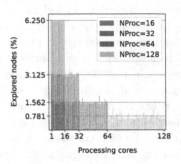

**Fig. 2.** Speed-up achieved solving geometrical and binomial synthetic UTS trees, compared to a sequential version.

**Fig. 3.** Percentage of explored nodes per processing cores solving the UTS-bin instance. The grey lines represent the ideal percentage, i.e., $100/NProc$.

**Table 1.** Summary of the instances used, along with some execution statistics.

| Inst. | Nb. of nodes ($10^6$) | Time (s) | nodes/s ($10^3$) | WS attempts (% success) |
|-------|------------------------|----------|-------------------|--------------------------|
| UTS-geo | 171.1 | 37.38 | 4,577 | 48,433 (99.0%) |
| UTS-bin | 131.7 | 37.11 | 3,548 | 1,473,048 (96.8%) |

**Evaluation of Work-Stealing.** In this section, two synthetic UTS trees with different types, binomial and geometric, are solved. A binomial tree is an optimal adversary for load balancing strategies, since there is no advantage to be gained by choosing to move one node over another for WS: the expected work at all nodes is identical (i.e., at most two children nodes). In contrast, in a geometric tree the expected size of the subtree rooted at a node increases with proximity to the root. One can see on Fig. 2 that for the best results, 68% of the ideal speed-up is achieved using 128 processing cores and solving the UTS-geo instance. This represents 40% more than the UTS-bin instance, which is directly related to the branching factor, as explained above.

Table 1 provides some execution statistics of the solved instances. In order to allow a fair comparison between instances, we make sure that the sequential time are approximately the same. One can see that for each instance, the percentage of WS attempts failed is less than 9%, which demonstrate the relevance of the WS. Moreover, Fig. 3 shows the percentage of explored nodes per processing cores, solving the `UTS-bin` instance. One can see that for each experiment the total workload is almost evenly balanced among all the processing units, meaning that all the allocated resources are fully exploited.

These experiments show that our WS mechanism is able to achieve good performance, as well as a good workload distribution between all the allocated resources, even at low-granularity. The latter takes advantage of the fact that the shallowest nodes are stolen first, and that those nodes generally have a higher branching factor than the others. Nevertheless, it was observed that the performance may be impacted when it is not the case, like solving `UTS-bin`.

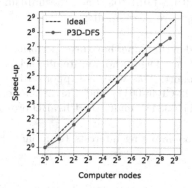
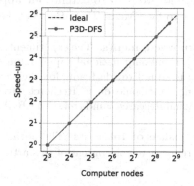

**Fig. 4.** Speed-up achieved solving `ta056`, compared to a multi-core version.

**Fig. 5.** Relative speed-up achieved solving `ta057`.

**Performance at Scale.** Figure 4 shows the speed-up reached solving the `ta056` PFSP instance up to 400 nodes, compared to a multi-core version. The latter exhibits $173 \times 10^9$ nodes and requires 1.26 node-hour. The experimental results revealed that up to 70% of the ideal speed-up can be achieved using up to 128 computer nodes, and around 50% using 400 nodes. In the latter experiment, 51,200 processing cores are used, and therefore as many segments are maintained in parallel. This leads to a large number of potentially remote communications, which can explain this limit in performance.

As a preliminary experiment towards the resolution of open instances, `ta057` is also solved. The latter exhibits a tree composed of $28,340 \times 10^9$ nodes, requires on average 220 node-hour, and was first solved to the optimality in 2022 [8], exploiting on average 384 GPUs during 1 h11. For time limitation reasons, we consider the processing time on 8 nodes as the reference time for the speed-up computation. Figure 5 shows a high relative scalability, with 98% of the ideal speed-up reached using 400 nodes.

# 5  Conclusions and Future Works

We investigated a PGAS data structure and WS for the class of unbalanced tree-based algorithms, focusing on DFS. According to the experimental results, it is shown that the data structure and WS allow to achieve 68% of the linear speed-up on a fine-grain backtracking application in single-node setting. Furthermore, large scale experiments revealed that 50% of strong scaling efficiency is achieved using 400 computer nodes (51,200 processing cores) solving large PFSP instances using the B&B technique.

This work opens the road toward the resolution of open COPs instances. To that end, we plan to extend our approach with a fault-tolerance mechanism in order to face Mean-Time-Between-Failure that are ever smaller.

**Acknowledgments.** The experiments presented in this paper have been performed on the Luxembourg national supercomputer MeluXina. The authors gratefully acknowledge the LuxProvide teams for their expert support. This work is supported by the Agence Nationale de la Recherche (ref. ANR-22-CE46-0011) and the Luxembourg National Research Fund (ref. INTER/ANR/22/17133848), under the UltraBO project.

**Disclosure of Interests.** The authors have no competing interests to declare that are relevant to the content of this article.

# References

1. Blumofe, R.D., Leiserson, C.E.: Scheduling multithreaded computations by work stealing. J. ACM **46**(5), 720–748 (1999)
2. Carneiro Pessoa, T., Gmys, J., de Carvalho Júnior, F.H., et al.: GPU-accelerated backtracking using CUDA Dynamic parallelism. Concurrency Comput. Pract. Experience **30**(9), e4374 (2018)
3. Chamberlain, B.L., Ronaghan, E., Albrecht, B., et al.: Chapel Comes of Age: Making Scalable Programming Productive (2018)
4. Cong, G., Kodali, S., Krishnamoorthy, S., et al.: Solving large, irregular graph problems using adaptive work-stealing. In: 37th International Conference on Parallel Processing, pp. 536–545 (2008)
5. van Dijk, T., van de Pol, J.C.: Lace: non-blocking split deque for work-stealing. In: Lopes, L., et al. (eds.) Euro-Par 2014. LNCS, vol. 8806, pp. 206–217. Springer, Cham (2014). https://doi.org/10.1007/978-3-319-14313-2_18
6. Dinan, J., Larkins, D.B., Sadayappan, P., et al.: Scalable work stealing. In: Proceedings of the Conference on High Performance Computing Networking, Storage and Analysis. Association for Computing Machinery (2009)
7. Gendron, B., Crainic, T.G.: Parallel Branch-and-Branch algorithms: survey and synthesis. Oper. Res. **42**(6), 1042–1066 (1994)
8. Gmys, J.: Exactly solving hard permutation flowshop scheduling problems on peta-scale GPU-accelerated supercomputers. INFORMS J. Comput. **34**(5), 2502–2522 (2022)
9. Gmys, J., Leroy, R., Mezmaz, M., et al.: Work stealing with private integer-vector-matrix data structure for multi-core Branch-and-Bound algorithms. Concurrency Comput. Pract. Experience **28**(18), 4463–4484 (2016)

10. Gmys, J., Mezmaz, M., Melab, N., et al.: A computationally efficient Branch-and-Bound algorithm for the permutation flow-shop scheduling problem. Eur. J. Oper. Res. **284**(3), 814–833 (2020)
11. Grama, A., Kumar, V.: Parallel search algorithms for discrete optimization problems. ORSA J. Comput. **7**(4), 365–385 (1995)
12. Helbecque, G., Gmys, J., Carneiro, T., et al.: Productivity- and performance-aware parallel distributed depth-first search, October 2023. https://github.com/Guillaume-Helbecque/P3D-DFS
13. Helbecque, G., Gmys, J., Melab, N., et al.: Parallel distributed productivity-aware tree-search using Chapel. Concurrency Comput. Pract. Experience **35**(27), e7874 (2023)
14. Lageweg, B.J., Lenstra, J.K., Rinnooy Kan, A.H.G.: A general bounding scheme for the permutation flow-shop problem. Oper. Res. **26**(1), 53–67 (1978)
15. Machado, R., Lojewski, C., Abreu, S., et al.: Unbalanced tree search on a manycore system using the GPI programming model. Comput. Sci. Res. Dev. **26**(3), 229–236 (2011)
16. Olivier, S., et al.: UTS: an unbalanced tree search benchmark. In: Almási, G., Caşcaval, C., Wu, P. (eds.) LCPC 2006. LNCS, vol. 4382, pp. 235–250. Springer, Heidelberg (2007). https://doi.org/10.1007/978-3-540-72521-3_18
17. Olivier, S., Prins, J.: Scalable dynamic load balancing using UPC. In: 37th International Conference on Parallel Processing, pp. 123–131 (2008)
18. Taillard, E.: Benchmarks for basic scheduling problems. Eur. J. Oper. Res. **64**(2), 278–285 (1993)
19. Zhang, W.: Branch-and-Bound search algorithms and their computational complexity. Technical report, Defence Technical Information Center (1996)

# Assessing the Stability of Text-to-Text Models for Keyword Generation Tasks

Tomasz Walkowiak[(✉)] [iD]

Wroclaw University of Science and Technology, Wroclaw, Poland
`tomasz.walkowiak@pwr.edu.pl`

**Abstract.** The paper investigates the stability of text-to-text T5 models in keyword generation tasks, highlighting the sensitivity of their results to subtle experimental variations such as the seed used to shuffle fine-tuning data. The authors advocate for incorporating error bars and standard deviations when reporting results to account for this variability, which is common practice in other domains of data science, but not common in keyphrase generation. Through experiments with T5 models, they demonstrate how small changes in experimental conditions can lead to significant variations in model performance, particularly with larger model sizes. Furthermore, they analyze the coherence within a family of models and propose novel approaches to assess the stability of the model. In general, the findings underscore the importance of considering experimental variability when evaluating and comparing text-to-text models for keyword generation tasks.

**Keywords:** keywords extraction · T5 · model stability

## 1 Introduction

Keywords play a crucial role in a summary and retrieval of documents, providing the reader with a brief overview of the content of the document. Maintaining high-quality keywords is vital to improve content visibility and allow users to easily find relevant information. Various techniques are available to automate keyword extraction [14], including unsupervised and supervised methods. One of the most promising approaches is the text-to-text generation technique, which generates tags dynamically based on the input text. This method allows us to capture not only the extracted keywords but also relevant terms that may not be explicitly mentioned in the text. Techniques in this field employ transformer-based neural networks such as T5 [13] and BART [6] to achieve cutting-edge results [9–11].

In many fields of data science, it is considered best practice to incorporate error bars when presenting findings, accounting for variations stemming from different random seeds across multiple experiments. This methodology is frequently endorsed by various conferences, as evidenced by the guidelines outlined

© The Author(s), under exclusive license to Springer Nature Switzerland AG 2024
L. Franco et al. (Eds.): ICCS 2024, LNCS 14834, pp. 112–119, 2024.
https://doi.org/10.1007/978-3-031-63759-9_14

in the NeurIPS checklist[1]. However, in the field of keyphrase generation, it is not customary to repeat experiments with different seeds and subsequently report standard deviations [11].

In this study, our objective is to examine this conventional practice. Our inquiry is prompted by the experiments described in [12], where two models exhibit disparities in their results, even starting from the same base model and probably trained on the same dataset. We seek to investigate the degree of sensitivity of the T5 models to subtle alterations under experimental conditions.

## 2    Related Work

There are various techniques to automate keyword extraction [14], including unsupervised [9,17] and supervised methods. The text-to-text approach to keyword extraction utilized in this paper falls within the supervised domain, which can be further classified into two groups: (i) classification with a closed set of labels [8,12], and (ii) classification with an open set of labels [7,18], both of which have been applied to this task. Recently, there has been increasing attention among researchers towards the use of T5 for keyword extraction [9–11]. The sensitivity of text-to-text models to subtle experimental conditions, to the best of the authors' knowledge, has not been thoroughly explored in the literature. However, the impact of minor experimental variations has been studied in other domains of machine learning [4], including natural language processing [1], image recognition [2], and out-of-distribution detection [15].

## 3    Method

The T5 [13] models are trained for the keyword extraction task using transfer learning. We start with a pre-trained base model and fine-tune it to generate a sequence of keywords from the original text, which is prefixed with a prompt. Our chosen prompt is "generate keywords:". During experiments, we used base models for English [13] and Polish (pltT5) [3]. Evaluation of fine-tuned models requires metrics that compare the results obtained with those of the target ones. We utilize three metrics: F1 score, a custom metric based on set similarity and semantic similarity. In this paper, we use the "micro" F1 score, which is calculated globally by summing the total true positives, false negatives, and false positives. One of the drawbacks of the F1 metric is its lack of symmetry, which means that its value can change when we replace targets with the results obtained. To address this, we propose a modification of the F1 metric, denoted here as *keysim*. The concept originates from the idea of similarity between sets (the target and the obtained keywords). We propose to calculate the similarity between sets $a$ and $b$ as the number of common elements divided by the number of elements in the first set ($a$). Since this metric is asymmetric, we symmetrize

---

[1] https://neurips.cc/Conferences/2022/PaperInformation/PaperChecklist.

it using the harmonic mean (similar to the F1 metric). Finally, averaging these similarities, among the $n$ examples, we can derive the desired metric:

$$keysim = \frac{1}{n} \sum_{i=1}^{n} \frac{2|a_i \cap b_i|}{|a_i| + |b_i|}, \tag{1}$$

where $a_i$, $b_i$ represent the targets and results.

The F1 and *keysim* evaluate keywords as discrete symbols, without considering the existence of synonyms in the language and the subjective nature of keywords. Therefore, it is valuable to evaluate keyword generators by incorporating semantic similarity between them. Once again, we will average the similarities for each sample. The semantic similarity between two words is commonly calculated using word embeddings [5]. We will follow this approach by initially defining the similarity between a keyword $x$ and a set of keywords $a$ as the maximum cosine similarity between an embedding of the word ($word2vec(x)$) and all keyword embeddings from the set ($word2vec(a_i)$), i.e.:

$$sim(x, a) = \max_{i=1...|a|} cos\left(word2vec(x), word2vec(a_i)\right). \tag{2}$$

In the experiments reported, we generated embeddings for keywords using fastText models [5] for English and Polish. With the word-to-set similarity established, we can define the similarity between two sets as the average of similarities calculated using Eq. 2. As the compared sets of keywords may have different numbers of elements, we must average the similarities between set A and set B and between set B and set A. Finally, by averaging over all samples (pairs of keyword sets), we obtain the semantic similarity metric, i.e.:

$$semsim = \frac{1}{2n} \sum_{i=1}^{n} \left( \frac{1}{|a_i|} \sum_{j=1}^{|a_i|} sim(a_{i,j}, b_i) + \frac{1}{|b_i|} \sum_{j=1}^{|b_i|} sim(b_{i,j}, a_i) \right). \tag{3}$$

## 4    Experiments and Results

### 4.1    Datasets

The data used to train and test the T5 models were corpora developed within the CURLICAT project [16], exactly the Polish Open Science Metadata Corpus (POSMAC) [9,10] which is part of CURLICAT. POSMAC contains English and Polish text annotated with keywords. For Polish texts, two versions were used: a smaller subset S-PL (accessible at[2]) and an almost three times larger one L-PL from [10].

---

[2] https://clip.ipipan.waw.pl/POSMAC.

## 4.2  Stability for Different Family of Models

T5 models are available in various sizes, typically classified as small, base, and large, and for different languages [13]. The fine-tuned T5 model retains the structure of the base model, but its weights are adjusted during fine-tuning. For clarity, in the following analysis, we will refer to a group of models derived from the same base model and tuned on the same training set as a "family of models." The train data set is typically randomly shuffled. So, if we keep the learning rate scheduler method unchanged, it seems as the only source of randomness that can affect the final model and hence its quality metrics.

To test this hypothesis, we performed a series of experiments. We used three data sets: two for Polish with different sizes (details are given in Sect. 4.1) and one for English. We train each each model on the same data set five times and show the results not as a single value, but as the mean and standard deviation. The results of these experiments are shown in Table 1. It is evident that the metrics vary slightly across each run.

A striking observation is the poor performance of large models, which exhibit a wide range of results. We conducted a deeper investigation of this phenomenon. They showed that for three runs of the large model, we observe notably poor results. However, for the other two runs, the results are significantly better compared to those of small models. Further examination revealed that these under-performing models could be identified during the training phase, as they exhibited significantly higher losses (even after the first epoch) than the runs, yielding better results on the training sets. Similar patterns were observed with the L-PL and EN datasets. This suggests that large models tend to struggle during training, but such behavior can be detected early. Therefore, it is recommended to identify such runs (based on random seed values) during training and exclude such models from further analysis.

## 4.3  Coherence of the Model Family

The findings of earlier sections reveal that simply changing the order of the training data can lead to different models. Having metrics to gauge the similarities between these models would be advantageous. The results presented thus far offer some insight by revealing the similarity between the model output and the target, making the standard deviation a potential indicator of stability, especially when it exhibits low values. However, the standard deviation of quality metrics might be overlooked because it primarily estimates the distances to the target values rather than assessing the similarities between models within the same family. Given that the *keysim* and *semsim* metrics are symmetric, we can utilize them to measure the similarities between two models. By treating one model as the target and then averaging over all pairs of models within a family, we can obtain an indicator of the family's coherence (with larger values indicating greater coherence).

The results depicting the similarity between models, obtained for the test datasets, are shown in Table 2. For a family of models comprising five fine-tuned

**Table 1.** Stability analysis of T5 models of varying sizes and fine-tuned on diverse datasets. Each row corresponds to a model family. The results are the averages of the quality metrics computed from five runs with different shuffle seeds. Standard deviations and ranges (computed as the difference between the maximum and minimum values) are also provided. Low values of standard deviation and range indicate higher stability of the model family.

| dataset | model | $keysim$ [%] | | | F1 [%] | | | $semsim$ [%] | | |
|---------|-------|---------|------|-------|---------|------|-------|---------|------|-------|
| | | average | std | range | average | std | range | average | std | range |
| S-PL | small | 14.34 | 0.84 | 2.39 | 17.76 | 1.03 | 3.03 | 60.56 | 0.59 | 1.65 |
| | base | 20.28 | 1.10 | 2.96 | 24.34 | 0.64 | 1.80 | 63.69 | 0.11 | 0.31 |
| | large | 15.28 | 6.87 | 16.54 | 20.66 | 6.28 | 15.90 | 61.93 | 2.82 | 6.65 |
| L-PL | small | 15.50 | 0.27 | 0.72 | 19.05 | 0.26 | 0.72 | 61.83 | 0.12 | 0.36 |
| | base | 21.42 | 1.04 | 2.96 | 25.72 | 0.78 | 2.13 | 64.87 | 0.31 | 0.84 |
| | large | 15.07 | 5.29 | 15.98 | 21.17 | 5.32 | 16.20 | 61.30 | 2.92 | 8.91 |
| EN | small | 11.25 | 0.76 | 2.03 | 14.53 | 0.76 | 2.02 | 57.89 | 0.47 | 1.26 |
| | base | 16.67 | 0.46 | 1.25 | 21.02 | 0.40 | 0.99 | 60.88 | 0.23 | 0.65 |
| | large | 13.68 | 3.40 | 8.84 | 17.69 | 3.30 | 8.65 | 60.17 | 1.62 | 4.18 |

models (derived from five runs with different seeds), we have a total of ten pairs. Therefore, we show the average, standard deviation, and range for each metric. To maintain clarity, we omit the results for large models, as discussed in the preceding section. Higher metrics values (with a maximum of 100%) indicate greater coherence of a model family. We observe that the metrics values are higher than those in Table 1, suggesting that the models are more similar to each other than to the training targets. However, the values are still relatively small, indicating significant differences between the models obtained. For the smaller datasets (S-PL and EN), we notice that the base models are more stable (with higher values of metrics) than the small models. However, for the L-PL model, the trend is the opposite.

## 4.4 Coherence of the Family of Models for Texts Further Away from the Original Training Set

A crucial feature of a keyword generator by text-to-text models is its ability to generate keywords for text that span various domains [9,11]. Since the training set typically covers a limited domain, researchers evaluate keyword generators in various topical domains and text genres that differ from those used during training [9,10], or even for languages not used during training [11]. That is why it is worth examining the coherence of each family of models against data sets that are increasingly distant from the original training set in the sense of topic area and languages. The benefit of the approach outlined in the preceding section lies in its independence from the need for targets to evaluate the coherence of the model family on a given dataset. We evaluated various model families using

**Table 2.** The mean value, standard deviation, and range of the *keysim* and *semsim* metrics calculated for all pairs of models within the same family. We suggest utilizing these metrics as indicators of family model coherence, with higher average values suggesting greater coherence.

| dataset | model | keysim [%] | | | semsim [%] | | |
|---------|-------|------------|------|-------|------------|------|-------|
| | | average | std | range | average | std | range |
| S-PL | small | 60.08 | 10.88 | 24.51 | 85.88 | 3.99 | 9.04 |
| | base | 66.41 | 6.03 | 15.77 | 87.86 | 1.52 | 4.18 |
| L-PL | small | 72.77 | 1.10 | 3.92 | 90.95 | 0.38 | 1.40 |
| | base | 68.20 | 4.19 | 11.91 | 88.72 | 1.25 | 3.88 |
| EN | small | 66.99 | 5.25 | 14.07 | 87.16 | 2.11 | 5.64 |
| | base | 69.65 | 2.29 | 7.56 | 87.88 | 0.83 | 2.93 |

**Table 3.** The mean value and standard deviation of the *keysim* metric calculated for all pairs of models within the same family, considering test datasets and texts originating farther from the original training data. The results indicate a consistent decrease in the mean values, suggesting that models within the same family tend to diverge increasingly as texts are drawn from further domains.

| model family | | evaluation datasets | | | |
|--------------|-------|---------------------|------------|------------|------------------|
| data | model | test | novels | French | Lorem Ipsum |
| S-PL | small | 60.08±10.88 | 55.71±9.17 | 45.41±11.62 | 42.46±13.19 |
| | base | 66.41±6.03 | 65.67±3.60 | 57.72±3.00 | 24.41±17.70 |
| L-PL | small | 72.77±1.10 | 66.03±1.21 | 55.99±1.17 | 44.49±4.49 |
| | base | 68.20±4.19 | 60.21±2.69 | 55.59±2.04 | 16.44±15.32 |
| EN | small | 73.16±1.34 | 64.23±1.88 | 65.89±0.96 | 36.85±3.16 |
| | base | 72.99±0.66 | 67.23±0.73 | 62.71±1.01 | 29.01±4.72 |

their respective testing sets (so texts within an identical domain). Next, we used texts from novels (in the same language as the training data). So texts from other domain. It is followed by texts in French and finally random texts (Lorem Ipsum). The results are shown in Table 3. It is noticeable that the coherence metric, defined as the average *keysim* score between all pairs of models from each family, declines as texts originate from increasingly distant domains. This suggests a degradation in the coherence of model outputs, indicating a trend toward randomness as the input data strays further from the training set.

# 5   Conclusion

We have demonstrated that text-to-text models fine-tuned for key generation downstream tasks are sensitive to experimental factors, such as the order of training data. These nuances influence the achieved metric values. Therefore, we propose that the research community conducting key generation tasks should replicate experiments using different seeds and subsequently report the results as mean values and standard deviations. Although this approach is a de facto standard in many areas of AI, it is not yet adopted in the domain of key generation [11].

We conducted an in-depth analysis of the coherence within a family of models. We analyzed models derived from the same training dataset and base model, differing only in the order of their training sets used in fine-tuning, to examine how their results vary. To assess this difference, we used the standard deviation of quality metrics. Additionally, we proposed an original approach that analyzes output between models without requiring a target dataset to evaluate model coherence. The results illustrate how model coherence decreases across datasets from increasingly diverse domains. This demonstrates how the responses of models within the same family, expected to be highly similar, become increasingly distinct for texts that are farther removed from the training set.

**Acknowledgments.** The work was financed as part of the investment: "CLARIN ERIC - European Research Infrastructure Consortium: Common Language Resources and Technology Infrastructure" (period: 2024–2026) funded by the Polish Ministry of Science and Higher Education (Programme: "Support for the participation of Polish scientific teams in international research infrastructure projects"), agreement number 2024/WK/01.

# References

1. Belz, A., Agarwal, S., Shimorina, A., Reiter, E.: A systematic review of reproducibility research in natural language processing. In: Proceedings of the 16th Conference of the European Chapter of the Association for Computational Linguistics: Main Volume, pp. 381–393 (2021)
2. Bouthillier, X., Laurent, C., Vincent, P.: Unreproducible research is reproducible. In: International Conference on Machine Learning, pp. 725–734. PMLR (2019)
3. Chrabrowa, A., et al.: Evaluation of transfer learning for Polish with a text-to-text model. In: Proceedings of the Thirteenth Language Resources and Evaluation Conference, pp. 4374–4394. European Language Resources Association, Marseille (2022). https://aclanthology.org/2022.lrec-1.466
4. Gundersen, O.E., Coakley, K., Kirkpatrick, C.: Sources of irreproducibility in machine learning: a review. arXiv preprint arXiv:2204.07610 (2022)
5. Joulin, A., Grave, E., Bojanowski, P., Mikolov, T.: Bag of tricks for efficient text classification. In: Proceedings of the 15th Conference of the European Chapter of the Association for Computational Linguistics: Volume 2, Short Papers, pp. 427–431. Association for Computational Linguistics, Valencia (2017). https://aclanthology.org/E17-2068
6. Lewis, M., et al.: BART: denoising sequence-to-sequence pre-training for natural language generation, translation, and comprehension. In: Proceedings of the 58th Annual Meeting of the Association for Computational Linguistics, pp. 7871–7880. Association for Computational Linguistics (2020). https://doi.org/10.18653/v1/2020.acl-main.703

7. Martinc, M., Škrlj, B., Pollak, S.: TNT-KID: transformer-based neural tagger for keyword identification. Nat. Lang. Eng. **28**(4), 409–448 (2022). https://doi.org/10.1017/S1351324921000127

8. Morales-Hernández, R.C., Juagüey, J.G., Becerra-Alonso, D.: A comparison of multi-label text classification models in research articles labeled with sustainable development goals. IEEE Access **10**, 123534–123548 (2022)

9. Pezik, P., Mikolajczyk, A., Wawrzynski, A., Niton, B., Ogrodniczuk, M.: Keyword extraction from short texts with a text-to-text transfer transformer. In: ACIIDS 2022. Communications in Computer and Information Science, vol. 1716, pp. 530–542. Springer, Cham (2022). https://doi.org/10.1007/978-981-19-8234-7_41

10. Pezik, P., et al.: Transferable keyword extraction and generation with text-to-text language models. In: Mikyska, J., et al. (eds.) ICCS 2023. LNCS, vol. 14072, pp. 398–405. Springer, Cham (2023). https://doi.org/10.1007/978-3-031-36021-3_42

11. Piedboeuf, F., Langlais, P.: A new dataset for multilingual keyphrase generation. In: Advances in Neural Information Processing Systems, vol. 35, pp. 38046–38059. Curran Associates, Inc. (2022)

12. Pogoda, M., Oleksy, M., Wojtasik, K., Walkowiak, T., Bojanowski, B.: Open versus closed: a comparative empirical assessment of automated news article tagging strategies. In: Knowledge-Based and Intelligent Information and Engineering Systems: Proceedings of the 27th International Conference KES-2023, Athens, 6–8 September 2023. Procedia Computer Science, vol. 225, pp. 3203–3212. Elsevier (2023). https://doi.org/10.1016/J.PROCS.2023.10.314

13. Raffel, C., et al.: Exploring the limits of transfer learning with a unified text-to-text transformer. J. Mach. Learn. Res. **21**(140), 1–67 (2020). http://jmlr.org/papers/v21/20-074.html

14. Song, M., Feng, Y., Jing, L.: A survey on recent advances in keyphrase extraction from pre-trained language models. In: Vlachos, A., Augenstein, I. (eds.) Findings of the Association for Computational Linguistics: EACL 2023, pp. 2153–2164. Association for Computational Linguistics, Dubrovnik (2023). https://doi.org/10.18653/v1/2023.findings-eacl.161

15. Szyc, K., Walkowiak, T., Maciejewski, H.: Why out-of-distribution detection experiments are not reliable - subtle experimental details muddle the OOD detector rankings. In: Uncertainty in Artificial Intelligence, UAI 2023, 31 July–4 August 2023, Pittsburgh. Proceedings of Machine Learning Research, vol. 216, pp. 2078–2088. PMLR (2023)

16. Váradi, T., et al.: Introducing the CURLICAT corpora: seven-language domain specific annotated corpora from curated sources. In: Proceedings of the Thirteenth Language Resources and Evaluation Conference, pp. 100–108. European Language Resources Association, Marseille (2022)

17. Wan, X., Xiao, J.: Single document keyphrase extraction using neighborhood knowledge. In: Proceedings of the 23rd National Conference on Artificial Intelligence (AAAI 2008), vol. 2, pp. 855–860. AAAI Press (2008)

18. Ye, J., Gui, T., Luo, Y., Xu, Y., Zhang, Q.: One2Set: generating diverse keyphrases as a set. In: Proceedings of the 59th Annual Meeting of the Association for Computational Linguistics and the 11th International Joint Conference on Natural Language Processing (Volume 1: Long Papers), pp. 4598–4608. Association for Computational Linguistics (2021).https://doi.org/10.18653/v1/2021.acl-long.354

# A Dictionary-Based with Stacked Ensemble Learning to Time Series Classification

Rauzan Sumara[1]($\boxtimes$) [iD], Wladyslaw Homenda[1,2] [iD], Witold Pedrycz[3] [iD], and Fusheng Yu[4] [iD]

[1] The Faculty of Mathematics and Information Science, Warsaw University of Technology, Warsaw, Poland
{rauzan.sumara.dokt,wladyslaw.homenda}@pw.edu.pl
[2] The Faculty of Applied Information Technology, University of Information Technology and Management, Rzeszow, Poland
[3] The Department of Electrical and Computer Engineering, University of Alberta, Edmonton, Canada
wpedrycz@ualberta.ca
[4] The School of Mathematical Sciences, Beijing Normal University, Beijing, China
yufusheng@bnu.edu.cn

**Abstract.** Dictionary-based methods are one of the strategies that have grown in the realm of time series classification. Particularly, these methods are effective for time series data that have different lengths. Our contribution involves introducing the integration of a dictionary-based technique with stacked ensemble learning. This study is unique since it combines the symbolic aggregate approximation (SAX) with stacking gated recurrent units (GRU) and a convolutional neural network (CNN), referred to as SGCNN, which has not been previously investigated in time series classification. Our approach uses the SAX technique to transform unprocessed numerical data into a symbolic representation. Next, the classification process is done using the SGCNN classifier. Empirical experiments demonstrate that our approach performs admirably across various datasets. In particular, our method achieves the second position among current advanced dictionary-based methods.

**Keywords:** Dictionary-based method · Stacked ensemble learning · Time series · Classification

## 1 Introduction

The field of time series classification has garnered significant interest over the past decade. Within this domain, the prominence of dictionary-based classifiers has grown notably, finding application in diverse contexts. Recent investigations

This research was supported by the Warsaw University of Technology within the Excellence Initiative: Research University (IDUB) programme.

into neural networks in the field of natural language processing (NLP) served as the inspiration for this research.

We introduce a novel time series classification framework, commencing with transforming time series into a sequence of words using the SAX and employing stacked ensemble learning for classification. Our objective is to explore the efficacy of a stacking the GRU and CNN, called SGCNN, as a classifier for time series classification. A distinctive aspect of this study is the original integration of SAX and SGCNN, a combination previously unexplored in time series classification. The rationale behind employing ensemble learning lies in its recognized capability to enhance predictive accuracy when contrasted with individual models.

We consider transforming numeric data into a symbolic representation and generating sequences of words. This new form of the converted time series is then used to train the SGCNN model. The empirical experiment was conducted on 30 benchmark datasets from www.timeseriesclassification.com. Comparative analysis contrasts the classification results obtained by our method with those achieved by state-of-the-art approaches. We also provide the code to be publicly available through the link[1].

## 2   Literature Review

Various groups of algorithms are dedicated to the task of time series classification, with dictionary-based methods being particularly relevant to the scope of this study. In this context, a dictionary contains segments of time series represented as symbols, where each segment is construed as a word. Several dictionary-based algorithms have been introduced. For instance, the Bag of Patterns (BOP) algorithm stands out as a famous dictionary-based approach [1]. This algorithm employs time series data through a histogram representation, illustrating similarities to the well-known bag of words technique utilized in NLP. An extension of the BOP method is the SAX and Vector Space Model (SAXVSM) [2]. This method involves the generation of random subsequences from a time series, followed by the extraction of features for each subsequence. Another example from this category is dynamic time warping features (DTW-F) [3].

One widely used technique that relies on a dictionary is the bag of symbolic Fourier approximation symbols (BOSS) method [4]. The method employs symbolic Fourier approximation (SFA), which contributes to its notable resistance to noise. Another variation of this methodology, called contract BOSS (cBOSS) [5], has also been presented. The alteration that has been implemented pertains to the selection of parameters, which is an inherent part of the cBOSS algorithm and enhances the speed of the procedure. An additional classifier within the domain of dictionary-based methods is word extraction for time series classification with dilation (WEASEL-Dilation) [6]. Representing a novel iteration of a prior version, WEASEL, WEASEL-Dilation seeks to mitigate the issue of the considerable memory footprint associated with WEASEL. This is achieved by

---

[1] github.com/rauzansumara/dictionary-based-with-stacked-ensemble-learning.

managing the search space through a randomly parameterized SFA transformation.

## 3    The Method

The proposed methodology can be deconstructed into four distinct stages. The initial stage involves standardizing the time series. The second step focuses on transforming the time series into a symbolic representation. Afterward, the third step involves creating word sequences and adding them to a corpus. The final step utilizes the generated corpus to train the SGCNN model.

**Standardizing the Time Series.** The initial phase of the process involves preprocessing the data through z-standardization. This standardization process is accomplished by applying the z-score formulation $z_t = \frac{x_t - \mu_x}{\sqrt{\sigma_x^2}}$, with $t = 1, 2, \ldots, n$, where $z_t$ denotes the standardized time series for the $t$-th element, $x_t$ represents the observed value of the time series at the $t$-th time point, $\mu_x$ is the mean of the time series, and $\sigma_x$ denotes the standard deviation of the time series.

**Converting Numeric Time Series into Symbolic Time Series.** In the subsequent step of the technique, we used the simple SAX procedure as a potential approach for our study, where a time series will have the same length as the sequence of symbols. The size of the alphabets or symbols is represented by $a$.

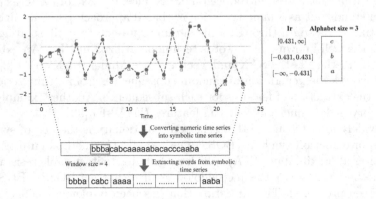

**Fig. 1.** The concept of transforming numerical data into symbols and generating a sequence of words.

The discretization process involves equally dividing the area under the Gaussian curve into distinct intervals, denoted as $Ir = \{Ir_0, Ir_1, Ir_2, \ldots, Ir_{a-1}, Ir_a\}$. The first interval, $Ir_0$, represents negative infinity, while the last interval, $Ir_a$, represents positive infinity. It is important to note that each interval, $Ir_j$, is smaller than the subsequent interval, $Ir_{j+1}$, where $j$ is less than $a$. If the time point falls inside a specified interval, the numerical time point is substituted with the designated alphabetical symbol. Using three alphabets, Fig. 1 demonstrates the example process of transforming numerical time series into symbols.

**Generating Words from Symbolic Time Series.** The window size $(w)$ must be selected to extract words. The selection of the word length range is at the researchers' discretion. Therefore, we chose the condition $2 \le w \le 8$ because the average English word often consists of two to eight characters. If $n$ represents the length of the symbolic time series, then a word list consisting of $n/w$ words is generated for each symbolic time series. The method of extracting words is illustrated in Fig. 1 with a window size of 4.

**Training Stacked Ensemble Learning Model.** In this paper, we introduced the SGCNN model, which is a combination of GRU positioned at the top and followed by CNN layers. It is a modified architecture from stacked ensemble learning introduced in the NLP paper [7]. Modification of this architecture, presented in Fig. 2, revolved around removing several layers and trying with different building blocks to make sure that it is not prone to either overfitting or underfitting.

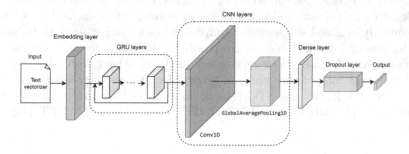

**Fig. 2.** Proposed stacked deep learning architecture.

Firstly, a 4-dim embedding layer is acquired through training in order to align with the surrounding context of each word. Embeddings are then passed through GRU layers. In the GRU, the update gate $(u_t)$ enables the model to determine the information from the previous hidden states $(H_{t-1})$ used to describe the future. Moreover, the reset gate $(r_t)$ works to determine the information of $H_{t-1}$ to be forgotten, and it is used to construct the current memory $(c_t)$ for capturing the relevant information in $H_t$, where $u_t = \sigma(W_u z_t + U_u H_{t-1})$, $r_t = \sigma(W_r z_t + U_r H_{t-1})$, $c_t = tanh(W_c z_t + r_t \odot U_c H_{t-1})$, and $H_t = (1 - z_t) \odot H_{t-1} + z_t \odot c_t$. These components rely on learning parameter matrices $W$, $U$, and a sigmoid activation function $\sigma(\dots)$.

Next, 16 units of the GRU layer process this input to extract features for subsequent layers followed by a CNN layer that incorporates the output generated by the GRU model. A 1×1 kernel window size and 16 filters are utilized in the CNN layer. Another important type of layer is a global average pooling (GAP). Following that, there is a dense layer with 32 neurons, and a dropout layer with a rate of 0.1 is utilized to decrease the complexity of SGCNN. Finally, this architecture uses a softmax activation function for the final layer.

## 4    Results

The study conducted experiments using a set of time series datasets that are consisted of thirty univariate time series, as shown in Table 1. All required calculations were performed in Python programming language. We fitted the models with different hyperparameters, such as using alphabet size $a = \{5, 7, 8, 9, 11, 12, 13, 15, 17, 18, 19, 20\}$ and words of length $w = \{2, 3, 5, 7, 8\}$. We constructed the model using an Adam optimizer with a loss function of a categorical cross-entropy, 100 epochs, 0.001 learning rate and 16-batch size. During the evaluation, the compared outcomes correspond only to the test sets. Last but not least, we did a post-hoc test using the Wilcoxon-Holm with a significance level $(\alpha)$ of 5%.

Heatmap plots for a few chosen datasets are displayed in Fig. 3, allowing us to examine the effects of alphabet size and word length on classification accuracy. The figures illustrate the variations in accuracy that occur when we alter the word length $w$ and alphabet size $a$. To achieve optimal outcomes, various datasets are frequently attained while utilizing alphabet sizes ranging from 11 to 20 and word lengths of 7 to 8. Table 1 corresponds to the best accuracy achieved by our approach and other dictionary-based methods. Apart from this, we also consider several competitors under a group of non-dictionary-based methods, such as TSF [8], RISE, and STC [9], proximityforest [10], InceptionTime [11], Catch22 [12], TS-CHIEF [13], HIVE COTE or HC1 [14], ROCKET [15], and RSTSF [16].

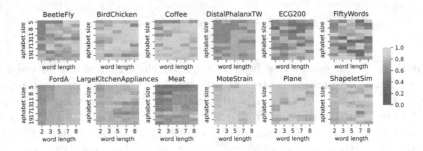

**Fig. 3.** Test accuracy for selected datasets based on alphabet size and word length.

Overall, the proposed method performs very competitively compared to other classification methods. The fact that our model is parameterized contributes to

**Table 1.** Our method compared to two groups of methods in terms of accuracy (Acc in %)

| Datasets | Dictionary-based | | Non dictionary-based | | Proposed | | |
|---|---|---|---|---|---|---|---|
| | Acc | Algoritm | Acc | Algoritm | Acc | $a$ | $\omega$ |
| BeetleFly | 95.00 | BOSS | **100.00** | TS_CHIEF | **100.00** | 11 | 7 |
| BirdChicken | **100.00** | BOSS | 95.00 | InceptionTime | **100.00** | 9 | 8 |
| Coffee | **100.00** | BOSS | **100.00** | TSF | **100.00** | 12 | 3 |
| CricketZ | 80.00 | WEASEL-D | **85.90** | ROCKET | 78.70 | 11 | 7 |
| DistalPhalanxOutlnCor | 78.62 | WEASEL-D | **80.43** | ProximityForest | 77.90 | 11 | 7 |
| DistalPhalanxTW | 70.50 | WEASEL-D | 70.50 | HC1 | **77.54** | 11 | 8 |
| Earthquakes | 74.82 | SAXVSM | 75.54 | ProximityForest | **83.69** | 11 | 8 |
| ECG200 | 89.00 | WEASEL-D | **92.00** | ROCKET | 76.81 | 8 | 8 |
| FaceFour | **100.00** | BOSS | **100.00** | TS_CHIEF | 95.45 | 15 | 3 |
| FiftyWords | 83.30 | WEASEL-D | 84.84 | TS_CHIEF | **98.90** | 11 | 2 |
| FordA | 95.61 | WEASEL-D | **98.03** | RSTSF | 89.74 | 18 | 8 |
| GunPoint | **100.00** | BOSS | **100.00** | STC | **100.00** | 8 | 8 |
| Herring | 65.63 | WEASEL-D | 70.31 | InceptionTime | **70.94** | 17 | 8 |
| InlineSkate | 47.45 | cBOSS | **67.45** | RSTSF | 42.02 | 15 | 7 |
| ItalyPowerDemand | 96.02 | DTW_F | **97.08** | RSTSF | 74.67 | 9 | 7 |
| LargeKitchenAppliances | 87.73 | SAXVSM | **90.40** | InceptionTime | 84.27 | 9 | 7 |
| Lightning7 | 76.71 | WEASEL-D | **84.93** | ProximityForest | 73.97 | 13 | 8 |
| Meat | 91.67 | WEASEL-D | **96.67** | RSTSF | 92.67 | 18 | 8 |
| MedicalImages | 78.42 | WEASEL-D | **82.63** | RSTSF | 72.76 | 20 | 8 |
| MiddlePhalanxOutlnCor | **84.54** | WEASEL-D | 84.19 | HC1 | 75.53 | 5 | 5 |
| MiddlePhalanxTW | 55.84 | cBOSS | 59.74 | RISE | **62.02** | 15 | 7 |
| MoteStrain | 95.21 | WEASEL-D | **95.85** | HC1 | 92.33 | 19 | 5 |
| OliveOil | **96.67** | WEASEL-D | 93.33 | RISE | 91.93 | 17 | 5 |
| Plane | **100.00** | BOSS | **100.00** | TSF | **100.00** | 20 | 8 |
| ProximalPhalanxOutlnCor | 91.07 | WEASEL-D | **93.13** | InceptionTime | 83.85 | 12 | 8 |
| ProximalPhalanxTW | 81.95 | BOSS | **82.44** | TS_CHIEF | 72.60 | 9 | 7 |
| ShapeletSim | **100.00** | BOSS | **100.00** | STC | 98.89 | 11 | 8 |
| SyntheticControl | 99.67 | WEASEL-D | **100.00** | ROCKET | 62.33 | 5 | 5 |
| Trace | **100.00** | BOSS | **100.00** | ProximityForest | 88.00 | 11 | 7 |
| Worms | 68.83 | WEASEL-D | **81.82** | TS_CHIEF | 62.34 | 19 | 5 |

this outcome, where we evaluate and select the optimal model for a specific dataset using the best hyperparameter configurations. As a result, our approach surpasses most dictionary-based methods in terms of accuracy except WEASEL-Dilation. Our method also performs at comparable levels or even better than several non-dictionary-based methods.

A critical difference diagram is shown in Fig. 4, which illustrates the average ranks of each method on 30 datasets. A lower mean rank implies the better

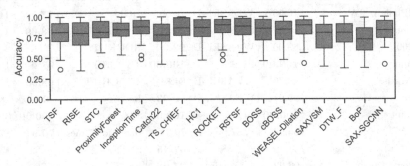

**Fig. 4.** Plot of critical difference based on average ranks from 30 datasets on test accuracy in comparison with (a) the dictionary-based and (b) the non-dictionary-based techniques.

**Fig. 5.** A box plot representing the accuracy of test set from 30 datasets.

accuracy of a particular method, while a solid horizontal bar implies statistical insignificance between two or more methods. According to average ranks, our approach, SAX-SGCNN, is slightly better than BOSS and cBOSS, even though it is significantly insignificant. The most accurate dictionary method is WEASEL-Dilation, which is in the same group as other dictionary-based methods. However, the horizontal bar connects to our approach, which means they are statistically indifferent. The similarity in behavior observed among SAX-SGCNN, BOSS, cBOSS, and WEASEL-Dilation can be attributed to the similarity in their respective approaches. All of them convert the time series into a symbolic representation and then analyze the sequence of words. Moreover, the most accurate non-dictionary-based methods such as HC1, ROCKET, TS-CHIEF, InceptionTime, RSTSF, and ProximityForest are statistically insignificant in comparison to our approach. The box plots in Fig. 5 show how close the accuracy of our approach to the current state-of-the-art methods. Top-ranking approaches have a low interquartile range (IQR), few outliers, and a median accuracy above 80%. Based on the results obtained, the proposed approach performs relatively better than the others in many cases.

## 5    Conclusion

The time series classification method that we describe in this paper works well with various datasets. Among dictionary-based methods, our approach, SAX-

SGCNN, outperforms all other methods except WEASEL-Dilation. One challenge of the presented method is regarding hyperparameter tunings, such as the number of alphabets and the word length, which must be tuned for each dataset separately to obtain the best results. Future work related to this research can be focused on extending this approach to classify the multivariate time series data, and choosing a better way of arranging hyperparameters to eliminate the problem will also be the subject of further modifications.

# References

1. Hatami, N., Gavet, Y., Debayle, J.: Bag of recurrence patterns representation for time-series classification. Pattern. Anal. Applic. **22**, 877–887 (2019)
2. Senin, P., Malinchik, S.: SAX-VSM: interpretable time series classification using SAX and vector space model. In: 2013 IEEE 13th International Conference on Data Mining, pp. 1175–1180. IEEE, Dallas (2013)
3. Franses, P.H., Wiemann, T.: Intertemporal similarity of economic time series: an application of dynamic time warping. Comput. Econ. **56**, 59–75 (2020)
4. Schäfer, P.: The BOSS is concerned with time series classification in the presence of noise. Data Min. Knowl. Disc. **29**, 1505–1530 (2015)
5. Middlehurst, M., Vickers, W., Bagnall, A.: Scalable dictionary classifiers for time series classification. In: Yin, H., Camacho, D., Tino, P., Tallón-Ballesteros, A.J., Menezes, R., Allmendinger, R. (eds.) IDEAL 2019. LNCS, vol. 11871, pp. 11–19. Springer, Cham (2019). https://doi.org/10.1007/978-3-030-33607-3_2
6. Schäfer, P., Leser, U.: WEASEL 2.0 – A Random Dilated Dictionary Transform for Fast, Accurate and Memory Constrained Time Series Classification (2023)
7. Lee, E., Rustam, F., Washington, P.B., Barakaz, F.E., Aljedaani, W., Ashraf, I.: Racism detection by analyzing differential opinions through sentiment analysis of tweets using stacked ensemble GCR-NN model. IEEE Access **10**, 9717–9728 (2022)
8. Tan, C.W., Bergmeir, C., Petitjean, F., Webb, G.I.: Time series extrinsic regression: predicting numeric values from time series data. Data Min. Knowl. Disc. **35**, 1032–1060 (2021)
9. Shu, W., Yao, Y., Lyu, S., Li, J., Chen, H.: Short isometric shapelet transform for binary time series classification. Knowl. Inf. Syst. **63**, 2023–2051 (2021)
10. Lucas, B., et al.: Proximity forest: an effective and scalable distance-based classifier for time series. Data Min. Knowl. Disc. **33**, 607–635 (2019)
11. Ismail Fawaz, H., et al.: InceptionTime: finding AlexNet for time series classification. Data Min. Knowl. Disc. **34**, 1936–1962 (2020)
12. Lubba, C.H., Sethi, S.S., Knaute, P., Schultz, S.R., Fulcher, B.D., Jones, N.S.: catch22: CAnonical Time-series CHaracteristics: selected through highly comparative time-series analysis. Data Min. Knowl. Disc. **33**, 1821–1852 (2019)
13. Shifaz, A., Pelletier, C., Petitjean, F., Webb, G.I.: TS-CHIEF: a scalable and accurate forest algorithm for time series classification. Data Min. Knowl. Disc. **34**, 742–775 (2020)
14. Lines, J., Taylor, S., Bagnall, A.: HIVE-COTE: the Hierarchical Vote Collective of Transformation-Based Ensembles for Time Series Classification. In: 2016 IEEE 16th International Conference on Data Mining (ICDM), pp. 1041–1046. IEEE, Barcelona (2016)

15. Dempster, A., Petitjean, F., Webb, G.I.: ROCKET: exceptionally fast and accurate time series classification using random convolutional kernels. Data Min. Knowl. Disc. **34**, 1454–1495 (2020)
16. Cabello, N., Naghizade, E., Qi, J., Kulik, L.: Fast, Accurate and Interpretable Time Series Classification Through Randomization (2021)

# Agent Based Simulation as an Efficient Way for HPC I/O System Tuning

Diego Encinas[1,2]([✉]) [iD], Marcelo Naiouf[1] [iD], Armando De Giusti[1] [iD],
Román Bond[2] [iD], Sandra Mendez[3] [iD], Dolores Rexachs[3] [iD], and Emilio Luque[3] [iD]

[1] Informatics Research Institute LIDI, CIC's Associated Research Center, National
University of La Plata 50 y 120, La Plata 1900, Argentina
dencinas@lidi.info.unlp.edu.ar
[2] SimHPC-TICAPPS, Universidad Nacional Arturo Jauretche, Florencio Varela
1888, Argentina
dencinas@unaj.edu.ar
[3] Computer Architecture and Operating Systems Department, Escola d'Enginyeria.
Universitat Autònoma de Barcelona, Edifici Q Campus UAB, Bellaterra 08193, Spain

**Abstract.** Generally, evaluating the performance offered by an HPC
I/O system with different configurations and the same application allows
selecting the best settings. This paper proposes to use agent-based modeling and simulation (ABMS) to evaluate the performance of the I/O
software stack to allow researchers to choose the best possible configuration without testing the real system. Testing configurations in a simulated environment minimizes the risk of disrupting real systems. In
particular, this paper analyzes the communication layer of an HPC I/O
system, more specifically the communication layer of the parallel file system (PVFS2). ABMS has been selected because it enables a rapid change
in the level of analysis for modeling and implementing a simulator. It can
focus on both macro- and micro-levels (how the aggregate behavior of
system agents is born and analyzing their individual behavior).

**Keywords:** Agent-Based Modeling and Simulation (ABMS) · HPC
I/O System · Parallel File System

## 1 Introduction

Improving the processing and storage of large amounts of data has become a challenge in parallel and distributed systems. Generally, evaluating the performance
offered by an I/O system with different configurations and the same application
allows selecting the best settings. However, to make changes in the configuration,
analyzing the performance that the applications will be able to achieve before
tuning the system (hardware and software) could be advantageous. Since at the
hardware level a correct configuration encompasses the arrangement of computing and input/output nodes, and at the software level, it includes selecting the
configurable parameters for the type of application used (intensive writing or
reading, method for performing I/O, and so forth). On top of this there can be

© The Author(s), under exclusive license to Springer Nature Switzerland AG 2024
L. Franco et al. (Eds.): ICCS 2024, LNCS 14834, pp. 129–136, 2024.
https://doi.org/10.1007/978-3-031-63759-9_16

a user level both for hardware and software such as an administrator, to generate the appropriate configurations. In other words, both users and researchers oftentimes want to make changes to these shares systems to analyze how this affects their applications, but they are unable to do so because these systems often run 24/7 or administration permissions are required. For this reason, one of the methods used to predict different configurations and the same application behavior in a computer system is using modeling and simulation techniques [1].

In this paper, we propose using Agent-Based Modeling and Simulation (ABMS) to develop a modeling of the Parallel Virtual File System 2 (PVFS2) communications layer called BMI (Buffered Message Interface) that will allow evaluating much of the performance of the I/O software stack. ABMS allows an easy change at the analysis level: it can focus on both macro- and micro-levels (how the aggregate behavior of system agents is born and, also, analyzing their individual behavior). The agent paradigm is used in various scientific fields and is of special interest in Artificial Intelligence (AI); it allows successfully solving complex problems compared with other classic techniques [2] like event-based simulation paradigm (Discrete Event Simulation, DES).

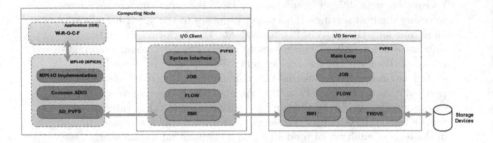

**Fig. 1.** At the left box, the layers in computer node and the right box represents the layers in the I/O server.

Generally, both paradigms operate in discrete time, but DES is used for low to medium abstraction levels. In ABMS, system behavior is defined at an individual level, and global behavior appears when the communication and interaction activities among the agents in an environment start. In fact, ABMS is easier to modify, since model debugging is usually done locally rather than globally. With the idea of using bottom-up modeling [3] and thus working in parts of the system, the agent paradigm was chosen.

To model the system and to have data to calibrate and validate, the test platforms developed have been previously explained [4] and add the appropriate monitoring tasks to the I/O software stack layers (Fig. 1), as well as the different functional and temporal models achieved [5] and their implementation to check the different stages of the model. Furthermore, storage nodes can be classified into data servers and metadata servers. The file system divides the files to be stored into fragments or stripes that are distributed in the data servers. The fragments are stored in the aforementioned buffers, both in client nodes and in server nodes.

In a nutshell, the simulator developed allows configuring the selected scenario (computing nodes, data servers and metadata servers), the file size for read and write operations, the access pattern (shared file or one file per process), and the number of iterations for each experiment. The output will allow quantifying I/O operations (immediate and non-immediate), bandwidth, operation time for the analyzed layer in the software stack. At a functional level, node-layer interactions can be viewed on the dashboard. To achieve a better control the test platform, each client node was configured to run a single application process.

## 2    Functional Analysis

The difficulty in the analysis of the I/O system lies in hardware heterogeneity and software stack complexity. In particular, work has been going on analyzing the PVFS file system as well as its OrangeFS equivalent, monitoring the different functions that make up each of the software layers. In the last publication [6] both client and server system interface and main loop layers, respectively, have been modeled in detail, and functional and temporal modeling and simulation were achieved. In the Modeling section below, the development achieved is explained.

To obtain data sets that allow carrying out a spatial modeling of the I/O system, monitoring tools were used on the Job, Flow and BMI layers of both the client and the server. These tools were used in previous works [6] but only focusing on obtaining temporary metrics. To further describe spatial behavior, the functions on the BMI layer were taken into account. BMI is a high-performance communications library used in PVFS2 that facilitates data transfer between the nodes of a computing system transparently. BMI will affect the model achieved through an agent that represents it and is explained in the Modeling section below.

Both `BMI_tcp_post_send_list()` and `BMI_tcp_post_recv_list()` are functions used to send and receive data asynchronously. Both functions are called by the PVFS2 client process to send a list of data buffers (BMI buffer) over the network to the dataservers.

By default, the size of the buffers at this level is 256 KiB. As regards the number of elements in the buffer, this strictly depends on the size of the selected stripe defined by the file system in the system interface layer. Generally:

$$stripesPerBuffer = \frac{BMIBufferSize}{StripeSize} \tag{1}$$

The number of stripes used by each client can be calculated as the size divided by the size of stripe. To know the number of stripes handled by each client, divide the previous result by the number of clients. Therefore:

$$StripesPerClient = \frac{FileSize \div StripeSize}{amountOfClients} = \frac{workload_i}{StripeSize} \tag{2}$$

where $workload_i$ represents a portion of the total file size. Thus, the sum of all workload_i would be the size of the entire file. As mentioned in the previous section, to achieve greater control of the test platform, it is configured so that each client node executes a single application process. Thus, the size of the file is divided equally for each client node.

Another issue to consider is the number of BMI_tcp_post_send_list operations that are called to send a certain workload in bytes. Generally:

$$BMIClientOps = \frac{bytesPerClient}{BMIBufferSize} = \frac{FileSize \div amountOfClients}{BMIBufferSize} \quad (3)$$

If the number of dataservers is odd, the BMIclientOps relation must be modified by adding the file size as a new term.

All of this regarding `tcp_post_send/recv_generic` and `work_on_recv/send_op` applies to both client and server. The server differs from the client in terms of the function responsible for sending data. On the client side, the function `BMI_tcp_post_send/recv_lists` used, while the equivalent server-side function is `BMI_tcp_post_send/recv`. It should be noted that, in the context of the server, a write operation corresponds to the reception of data and a read operation corresponds to the sending of data.

For operations in each dataserver, this involves taking into account that the total number of client operations must reach all dataservers. That is, the number of BMI client operations are sent to the different dataservers equally, using a round robin method [7].

# 3 Modeling

In previous works, the functional and temporal analysis has been carried out by monitoring the interactions of the I/O system at the software layer level in the I/O nodes as well as between the functions that make up each of these layers. Also, the use of state machines and sequence diagrams allowed obtaining a more detailed model to describe the interactions that are triggered throughout the system when write, read or close files operations are performed.

In the area of modeling, the concepts of macro- and micro-modeling have been introduced to describe other systems [8]. It could be said that the modeling carried out in the parallel I/O system is a macro-modeling, since the interactions found are due to the communications necessary to represent the behavior of the system without individualizing spatial parameters such as I/O operations, stripes and number of bytes per second. A micro-modeling would imply modeling the behavior of the fragments to be distributed in each of the I/O nodes.

## 3.1 Macro-Modeling

In a previous work has been [6] detailed client-server interactions for the PVFS2 file system, the interactions between their system interface and main loop layers, to analyze the spatial parameters for the BMI layers. In the BMI layer, immediate message quantification had to be added to the model; these messages are sent immediately to the receiver with the certainty that they will be received [9]. Next, the sequences of events that occur to send messages between client and server through the BMI layer are explained.

The `BMI_tcp_post_recv()` function is called to receive a 256-KiB buffer. Then, the `tcp_post_recv_generic()` function is started, but it fails to receive the entire buffer. At this point, the server receives only a few bytes or nothing at all. If some bytes of data are received, it is considered as an immediate operation (with the amount of bytes received). Therefore, `BMI_tcp_post_recv()` ends with errors. Next, the function `work_on_recv_op()` is started, which is responsible for receiving the remaining bytes that `tcp_post_recv_generic()` could not receive immediately. Several operations are necessary to receive the data buffer. Finally, `work_on_recv_op()` manages to receive the entire buffer and the `BMI_tcp_testcontext()` function is called to confirm the correct reception of the data. By means of the above-mentioned functions, the state machine corresponding to the BMI layer has been generated.

## 3.2   Micro-modeling

To find spatial parameters data sets, the elements that make up the data buffers, the operations and the times in which they are transmitted have been monitored, but with certain configuration conditions in the real system. The deployed test scenarios used PVFS2 as parallel file system and MPICH distribution as middleware. File and transfer sizes were 1, 2, 3, and 4 GiB using the IOR (Interleaved-or-Random) benchmark (application layer) [10] on a 6-node physical cluster. Each cluster node has a specific and dedicated role on different physical machines: 2 client nodes, 3 dataserver nodes and 1 metadataserver node.

An important feature of PVFS2 is that it has an event logging system called Gossip that generates a text file with all the events ordered chronologically with relevant information such as timestamp for each operation and bytes involved. In this stage, the number of total operations, IOPS and Bps are obtained.

Finally, the data sets obtained are analyzed to find the algorithms that are shown below and model the behavior of the spatial parameters, and implement them in the simulator. That is, the verification and calibration stages of the simulation are carried out considering state machines, state variables and agents as in previous works [5,6].

It is observed that, to send data, the client uses immediate and non-immediate operations. In the case of reception by the client, only non-immediate operations are observed. Table 1 summarizes client node send operations.

With this information, equations are proposed to model total times, immediate times, number of total operations, number of immediate operations and bytes. The development of the modeling equations was carried out using the regression technique. Based on the above, the file size is used as the independent variable; in the following equations it is represented by the parameter $x$.

On the client side, for data send operations:

$$- \ inmmediateOperations = 104.553571x^2 + 259.8107142x - 30.19285714$$

**Table 1.** Results for sending data by client.

| BMI_tcp_post_send_list() | File Size (GiB) | | | |
|---|---|---|---|---|
| | 1 | 2 | 3 | 4 |
| BMI Client Ops | 2049 | 4098 | 6147 | 8196 |
| Immediate operations | 235 | 1023 | 1635 | 2690 |
| Non-immediate operations | 1814 | 3075 | 4512 | 5506 |
| Total time; immediate ops. (s) | 0.015 | 0.063 | 0.1 | 0.16 |
| Total time; immediate and non-immediate ops. (s) | 130.54 | 254.39 | 378.03 | 493.21 |
| bps (MB/seg) | 3.92 | 4.02 | 4.06 | 4.15 |
| BMI Client Total IOPS | 15.69 | 16.11 | 16.26 | 16.62 |
| Immediate IOPS | 15187.25 | 16095.15 | 16190.76 | 16284.94 |

- $totalTime = 123.39x + 4.46$
- $immediateOpsTotalTime = 0.00336x^4 - 0.00274x^2 + 0.07493x^2 - 0.0354x - (1.365E - 16)$
- $totalOperations = \frac{x \div amountOfClients}{BMIBufferSize} - 1 + 2x$
- $bytes = \frac{x}{amountOfClients}$

On the client side, for data receive operations:

- $totalTime = 131.61x + 0.097$
- $bytes$ and $totaloperations$ same as send operations.

As regards the implementation of the simulator, it was developed using the NetLogo framework. This tool is used especially in the context of complex system simulation. Once the equations explained in the previous section have been found, they are implemented in the simulator for their corresponding tuning and validation.

## 4    Results and Discussion

The scenario generated for the micro-modeling is configured in the simulator, and the following metrics are obtained to represent the spatial parameters related to operations and bytes over time. Besides, simulation total times are minimal.

Table 2 shows the difference between the values obtained from the real system and the values generated by the simulation model of the system. Table 3 show predictions/estimates for 5 and 6 GiB for client.

Based on the network equipment and the interfaces used, the amount of main memory available to the nodes that make up the cluster, the type of protocol and the defined file size, it was expected to find a low percentage of operations completed immediately and a high percentage of non-immediate operations.

As regards buffers, it was observed on the server side that many small operations are needed to fill a 262,144 KiB buffer. Data reception on the server refers to write operations, so this behavior is reasonable/expected. On the client side,

**Table 2.** Simulated vs. physical with absolute error.

| BMI_tcp_post_recv() | Executed | | | | Simulated | | | | Absolute error | | | |
|---|---|---|---|---|---|---|---|---|---|---|---|---|
| | File Size (GiB) | | | | File Size (GiB) | | | | File Size (GiB) | | | |
| | 1 | 2 | 3 | 4 | 1 | 2 | 3 | 4 | 1 | 2 | 3 | 4 |
| BMI Server Ops | 1366 | 2732 | 4098 | 5464 | 1366 | 2732 | 4098 | 5464 | 0 | 0 | 0 | 0 |
| Bytes (GiB) | 0.33 | 0.66 | 1 | 1.33 | 0.33 | 0.66 | 1 | 1.33 | 0 | 0 | 0 | 0 |
| Total time (s) | 87.14 | 169.87 | 252.39 | 329.35 | 86.22 | 168.88 | 250.95 | 333.62 | 0.9133 | 0.9985 | 1.4472 | 4.2624 |
| BMI Server Total IOPS | 15.67 | 16.08 | 16.24 | 16.59 | 15.84 | 16.18 | 16.33 | 16.38 | 0.166 | 0.095 | 0.0936 | 0.2119 |
| Bps (MB/s) | 3.88 | 3.98 | 4.06 | 4.13 | 3.92 | 4 | 4.08 | 4.08 | 0.0411 | 0.0235 | 0.0234 | 0.0528 |

**Table 3.** Prediction of 5 and 6 GiB for sending data to the client.

| BMI_tcp_post_send_list() | File Size (GiB) | |
|---|---|---|
| | 5 | 6 |
| BMI Client Ops | 10249 | 12299 |
| Immediate operations | 3888 | 5299 |
| Non-immediate operations | 6331 | 7000 |
| Total time; immediate ops. (s) | 0.37 | 0.93 |
| Total time; immediate and non-immediate ops. (s) | 622.022 | 745.61 |
| Bps (MB/s) | 4.11 | 4.12 |
| BMI Client Total IOPS | 16.47 | 16.49 |
| Immediate IOPS | 10384.98 | 5715.14 |

a similar behavior is observed when receiving (reading) data; to fill reception buffers, a considerable number of operations are necessary.

The times in the different functions scale reasonably as file size increases. As indicated by the predictions, it is expected that the times will skyrocket exponentially if the physical performance of the cluster does not change. Some of the improvements in physical features may include solid state drives in data servers or a change in the speed of the local network (to 1Gbps, for example). With the proposed improvements, the number of immediate operations is expected to increase considerably, consequently increasing IOPS.

# 5   Conclusions

This article presents an analysis and conceptual modeling of the PVFS2 communications layer, using the agent paradigm with the objective of better tuning a costly part of HPC such as communications I/O. When creating the model, a distinction is made to achieve a macro- and micro-model.

For macro-modeling, the interactions between client and server of PVFS2 are taken into account. Finite state machines allow modeling the exchanges and defining agent implementation in the simulator. Additionally, the relations that represent spatial parameters functionality are obtained.

For micro-modeling, the mathematical equations are obtained to quantify the different parameters, such as number of operations, bytes, bandwidth and IOPS. Time is the only temporal parameter used. Consequently, the agents are BMI_client and BMI_server along with functions that help to the communication process between layers.

Finally, with the purpose of making system predictions, the simulator is run with input parameters of 5 and 6 GiB. As regards future work, different analysis, monitoring and tuning strategies will continue to be evaluated, such as different input and output stack elements in the system; and communications in metadata servers or at the metadata level. Another possibility is to include several processes on different nodes.

**Acknowledgements.** This research has been supported by the Agencia Estatal de Investigacion (AEI), Spain and the Fondo Europeo de Desarrollo Regional (FEDER) UE, under contract PID2020-112496GB-I00 and partially funded by the Fundacion Escuelas Universitarias Gimbernat (EUG).

# References

1. Boito, F.Z., et al.: A checkpoint of research on parallel i/o for high-performance computing. ACM Comput. Surv. **51**(2), 1–35 (2018)
2. Paradiso, M., et al.: An approach to parallelization of respiratory disease spread simulations in emergency rooms. In: 2022 International Conference on Computational Science and Computational Intelligence (CSCI), Las Vegas, pp. 1359–1365 (2022). https://doi.org/10.1109/CSCI58124.2022.00244
3. Siebers, P., Macal, C., Garnett, J., et al.: Discrete-event simulation is dead, long live agent-based simulation! J. Simulat. **4**, 204–210 (2010). https://doi.org/10.1057/jos.2010.14
4. Gomez-Sanchez, P., et al.: Using AWS EC2 as test-bed infrastructure in theI/O system configuration for HPC applications. J. Comput. Sci. Technol. **16**(2), 65–75 (2016)
5. Encinas, D., et al.: On the calibration, verification and validation of an agent-based model of the HPC input/output system. In: Proceedings from The Eleventh International Conference on Advances in System Simulation (SIMUL 2019) (2019)
6. Encinas, D., et al.: An agent-based model for analyzing the HPC input/output system. Int. J. Adv. Syst. Measur. **13**(3), 192–202 (2020). ISSN: 1942-261x
7. Koziol, Q. (ed.): High Performance Parallel I/O. CRC Press (2014)
8. Helbing, D., et al.: Micro-and macro-simulation of freeway traffic. Math. Comput. Model. **35**(5–6), 517–547 (2002)
9. Amerson, G., Apon, A.: Implementation and design analysis of a network messaging module using virtual interface architecture. In: 2004 IEEE International Conference on Cluster Computing (IEEE Cat. No. 04EX935). IEEE (2004)
10. Shan, H., Shalf, J.: Using IOR to Analyze the I/O Performance for HPC Platforms. No. LBNL-62647. Lawrence Berkeley National Lab (LBNL), Berkeley (2007)

# A Virtual Clinical Trial for Evaluation of Intelligent Monitoring of Exacerbation Level for COPD Patients

Mohsen Hallaj Asghar[1]([✉])[ID], Alvaro Wong[1][ID], Francisco Epelde[3][ID],
Manel Taboada[2][ID], Dolores Isabel Rexachs del Rosario[1][ID],
and Emilio Luque[1][ID]

[1] Autonomous University of Barcelona, 08193 Barcelona, Spain
Mohsen.HallajAsghar@autonoma.cat, Alvaro.Wong@uab.cat,
{dolores.rexachs,emilio.luque}@uab.es
[2] Escuelas Universitarias Gimbernat, Computer Science School, Bellaterra, 08193
Barcelona, Spain
manel.taboada@eug.es
[3] Consultant Internal Medicine, Hospital Universitari Parc Tauli, Autonomous
University of Barcelona, Sabadell, Barcelona, Spain
fepelde@gmail.com

**Abstract.** Chronic Obstructive Pulmonary Disease (COPD) is a progressive respiratory condition, ranking as the third leading cause of global morbidity and mortality. In this project, we simulate Real Clinical Trials using Virtual Clinical Trials (VCT) for COPD patients, offering possibilities not feasible in traditional trials, such as exploring treatment adherence levels and creating virtual cohorts with specific characteristics. We propose a cohort-based management strategy leveraging data analytics to identify patterns within the COPD patient population and advocate for employing a finite-state machine (FSM) approach to model COPD exacerbations. Further research and validation are crucial to refine and scale this integrated model.

**Keywords:** virtual clinical trial · Agent based modeling · Finite-state machine (FSM) · Self-management COPD patient

## 1 Introduction

Simulating Real Clinical Trials via Virtual Clinical Trials (VCT) offers unique advantages unmatched by traditional trials. Once validated with real trial results, VCTs allow us to: 1. Explore scenarios and outcomes ethically restricted in real trials, such as varying patient adherence levels; 2. Create virtual cohorts with diverse characteristics, enhancing; intervention evaluation; 3. Control and manipulate environmental conditions for testing, including healthcare service occupancy rates and prevalent health conditions. Given these benefits, We are introducing a 'Virtual Clinical Trial' focusing on monitoring and follow-up to evaluate its impact on the care of COPD patients. COPD is projected to become the third leading cause of mortality by 2030 [1], imposing a significant global

L. Franco et al. (Eds.): ICCS 2024, LNCS 14834, pp. 137–144, 2024.
https://doi.org/10.1007/978-3-031-63759-9_17

socioeconomic burden [2]. It is a persistent ailment initiated by factors like smoking or environmental pollutants, with manifestations including coughing, chest tightness, and breathlessness. Enhanced exacerbation management is crucial due to its correlation with decreased lung function and increased healthcare expenditures [3]. Traditional COPD management, reliant on clinic-based care, often overlooks the disease's dynamic nature. In response, we propose integrating virtual clinical trials with tailored self-management programs for COPD patients. This innovative model bridges clinical research and patient-centered care, empowering individuals to actively participate in their care and improve their well-being [4]. Additionally, cohort-based strategies leverage population-level insights to inform targeted interventions, employing data analytics to identify patterns within the COPD patient population. The synergy between self-management and cohort-based strategies aims to create a holistic COPD management framework [5]. Exacerbations occur due to various scenarios, requiring systematic modeling using finite-state machine (FSM) techniques to identify and predict exacerbation episodes accurately.

## 2   Methods

This section outlines the process, commencing with the FSM model for identifying exacerbation episodes from COPD patients' symptoms. Following this, it details the model of state patient care for extracting variables representing "stable" and "exacerbation" periods to generate training data for an algorithm based on robust features.

### 2.1   Finite-State Machine (FSM) for COPD Exacerbation

The Finite State Machine (FSM) is a valuable approach for understanding and categorizing different states of COPD exacerbations [6]. By representing patients' states and transitions based on symptoms and medication usage, it provides a structured framework for analyzing exacerbation events. In this model, patients can be in one of three states: normal, transitional, or exacerbation. Transitions between these states depend on inputs that encode changes in symptoms and medication usage. The inclusion of a transitional state accounts for instances where patients deviate from their normal state but do not meet the criteria for a full exacerbation. The inputs, representing medication usage and self-reported symptoms, allow for the dynamic characterization of patients' conditions. This enables healthcare professionals to track the progression of COPD and intervene appropriately based on the observed transitions. The definitions of exacerbation and symptom improvement provided by clinical experts offer clear criteria for categorizing events and determining when patients should transition between states. By aligning with clinical expertise and extensive literature review, this model enhances the accuracy and reliability of exacerbation identification. Overall, the FSM model provides a structured and pragmatic approach to understanding COPD exacerbations, addressing some of the challenges associated with existing definitions. Its incorporation of clinical input and clear criteria for state transitions makes it a valuable tool for both research and clinical practice in managing COPD.

## 2.2 Model of State Patient Care

The patient care process is modeled using an FSM, depicted in Fig. 1.

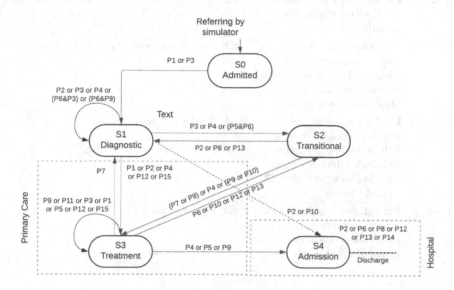

**Fig. 1.** A finite state machine (FSM) models a patient treatment process, where events and the state transitions they initiate are depicted as labels.

The initial state begins with the patient's admission (S0) to a medical ward. Subsequently, the physician examines the patient to gather diagnostic information (S1) and plan further transitions (S2). This diagnostic phase may involve single or multiple examinations and tests. Based on the assessment, a treatment plan is prescribed (S3) in primary care. The patient undergoes comprehensive examination and evaluation, leading to either discharge or hospitalization based on their condition. Throughout treatment, additional examinations may necessitate therapy adjustments or repeat diagnostics. Upon satisfactory conclusion of therapy, the patient is discharged (S4) from the ward. This example identifies five states: Q = (S0, S1, S2, S3, S4), and outlines eight process activities triggering state transitions from admission to discharge states (S0 to S4). The details are presented in Table 1.

In our formal framework, an activity denoted as E is characterized by the pair $< S, PE >$ where:

$$S \subseteq Q \tag{1}$$

**Table 1.** Defined activities and events within the patient treatment process.

| State at: | Process event (outcome) | Activated |
|---|---|---|
| S0, S1, S2 | P1: Confirms the declared symptoms; P2: Without problem; P3: New symptom emerge; P4: Medical test required | E1: Physical examination in Primary care |
| S1, S2, S3 | P4: Medical test required; P5: Positive result; P6: Negative result | E2: Medical laboratory test in primary care |
| S1, S2, S3 | P7: Diagnosed confirm and treatment assigned; P8: Diagnosed not confirm and patient discharge; P4: Medical test required | E3: Case assessment in primary care |
| S2, S3, S4 | P3: New symptom emerge; P9: Symptom Increasing; P10: Symptoms decreasing; P11: Side effect emerged; P12: Stable situation; P13: End of therapy | E4: Therapy in hospital |
| S3, S4 | P2: without problem; P14: End of recovery therapy; P15: New medical or personal evidence received | E5: Evidence received |

- Signifies the set of states from which this activity can be invoked (though not necessarily required).

$$PE \subseteq \sum \qquad (2)$$

- PE subset of sigma represents the set of events resulting from the activity's execution, with the potential to initiate a state transition. For each activity E, the state transitions that may occur upon its completion can be determined as follows:

$$\delta A : S * PE = P(S) \qquad (3)$$

Consider activity E2 (Medical laboratory test) with states S: S1, S2, S3 and events E: P5, P6, P4. Certain events, like P10 - condition improved, can be contextual and independent of specific activities. To predict exacerbation episodes, it's vital to identify intervals of stable to unstable conditions and deterioration trends preceding exacerbations. The FSM portrays states like mild, moderate, severe, and very severe exacerbation, linked by transitions reflecting health condition shifts. The table-set complements the FSM, detailing parameters like respiratory rate, blood pressure, and BMI influencing state transitions (Table 2).

This organized overview enhances understanding of COPD patient health progression during a flu episode.

## 3   Proof of Concept

In this project, data processing revolves around time-specific captured information, managing baseline data, and environmental variables. The raw data is grouped into categories, each indicating specific situations through numeric and string variables [7]. Table 2 presents a sample of ten virtual patients from the cohort, each identified by a unique ID number for easy data access. Patient characteristics include gender, where women may have a higher predisposition to COPD. Body Mass Index (BMI) plays a crucial role, with a lower BMI potentially increasing COPD risk. COPD is influenced by factors like smoking status, categorized into current smokers, ex-smokers, and never-smokers. The mMRC scale assesses dyspnea, while respiratory rate, heart rate, and sputum are evaluated based on various criteria. Oxygen saturation (SpO2) is categorized into four groups, and FEV1 measures lung function [8].

**Table 2.** Defined 10 sample datasets for COPD patients in the cohort

| ID | Age | Sex | BMI | mMRC | Smoke | RR | HR | Spo2 | Sputum | EFV1 |
|---|---|---|---|---|---|---|---|---|---|---|
| 1 | 47 | Male | $21 < BMI < 25$ | 0 | 3 | 13 | Higher | %97 | Normal | 80–above |
| 2 | 60 | Female | $< 21\,kg/m2$ | 3 | 1 | 18 | Higher | %95 | Purulent | 30–49 |
| 3 | 55 | Female | $30 > BMI < 35$ | 4 | 1 | 24 | Normal | %89 | Purulent | 29 or less |
| 4 | 78 | Male | $25 > BMI < 30$ | 1 | 2 | 15 | Higher | %98 | Normal | 50–79 |
| 5 | 71 | Male | $30 > BMI < 35$ | 4 | 1 | 22 | Lower | %88 | Purulent | 29–less |
| 6 | 82 | Female | $> 35\,kg/m2$ | 3 | 1 | 20 | Lower | %94 | Purulent | 30–49 |
| 7 | 79 | Male | $21 < BMI < 25$ | 2 | 1 | 18 | Higher | %95 | Normal | 50–79 |
| 8 | 69 | Female | $21 < BMI < 25$ | 0 | 3 | 13 | Higher | %96 | Normal | 80–above |
| 9 | 45 | Male | $21 < BMI < 25$ | 1 | 2 | 15 | Higher | %97 | Normal | 50–79 |
| 10 | 88 | Male | $< 21\,kg/m2$ | 3 | 1 | 21 | Lower | %93 | Purulent | 30–49 |

### 3.1   Specify the Outcome Variable

The study aims to create a fuzzy logic-driven medical decision system for diagnosing COPD using symptoms and test results. Employing fuzzy sets to quantify symptoms and results, it seeks to provide precise COPD severity measures and guide treatment.

Fuzzy logic-based medical decision framework involves: 1. Formulate fuzzy rules for COPD diagnosis: A. IF high respiratory rate OR high heart rate OR low oxygen saturation OR exposure to occupational dust and chemicals AND current smoker, THEN severe/very severe COPD; B. IF normal respiratory rate OR high heart rate OR purulent sputum OR exposure to home and office environment AND current ex-smoker, THEN moderate COPD; C. IF young age OR normal respiratory rate OR normal body temperature OR current never-smoker, THEN mild COPD [9]; 2. Specify fuzzy logic operations: A. AND: minimum operator; B. OR: maximum operator; C. NOT: complement operator; 3. Assess input variables using fuzzy sets and logic operations; 4. Combine fuzzy output values with the weighted average aggregation method; 5. Transform fuzzy output value into definite value using the centroid defuzzification method; 6. Present diagnostic outcome based on determined crisp output value; 7. Validate and assess the system using the dataset of confirmed COPD patients, comparing diagnostic accuracy to conventional methods.

### 3.2  Performance

The dataset includes 10 virtual patients as a sample for developing a comprehensive model. Table 2 presents a subset of these patients. The fuzzy logic algorithm was implemented by loading the dataset into a data frame. The algorithm's time complexity is:

$$O(n^2) \tag{4}$$

With a dataset size of 10 entries and 10 input variables, the time complexity is $O(10^2 * 10) = O(1000)$. The execution time was 0.47 ms. Overall, the fuzzy logic algorithm showed low memory usage, no redundancy, and reasonable execution speed for this dataset size. The output classifies patients into four categories: control, direct assessment, modifying medicine, and hospital referral (Table 3). Figures 2 illustrate the characteristics of these classes.

**Table 3.** Defined 10 sample datasets for COPD patients in the cohort

| ID | Age | Sex | Smoke | RR | HR | Spo2 | Sputum | Categories |
|----|-----|-----|-------|----|----|------|--------|------------|
| 1 | 47 | Male | 3 | 13 | Higher | %97 | Normal | **Control** |
| 2 | 60 | Female | 1 | 18 | Higher | %95 | Purulent | **Ref-Primary care** |
| 3 | 55 | Female | 1 | 24 | Normal | %89 | Purulent | **Ref-Hospital** |
| 4 | 78 | Male | 2 | 15 | Higher | %98 | Normal | **D-Assessment or M-Medicine** |
| 5 | 71 | Male | 1 | 22 | Lower | %88 | Purulent | **Ref-Hospital** |
| 6 | 82 | Female | 1 | 20 | Lower | %94 | Purulent | **Ref-Hospital** |
| 7 | 79 | Male | 1 | 18 | Higher | %95 | Normal | **D-Assessment or M-Medicine** |
| 8 | 69 | Female | 3 | 13 | Higher | %96 | Normal | **Control** |
| 9 | 45 | Male | 2 | 15 | Higher | %97 | Normal | **D-Assessment or M-Medicine** |
| 10 | 88 | Male | 1 | 21 | Lower | %93 | Purulent | **Ref-Primary care** |

This study quantifies the predictive efficacy of three key physiological parameters (Age, SpO2, and RR) and emphasizes their reliability in home monitoring with the mMRC scale. Incorporating the mMRC scale in future remote monitoring studies is advisable due to challenges in logging symptom-worsening episodes among COPD patients. Allowing patients to self-report with the mMRC scale at home provides real-time insight, aiding timely intervention. Figure 2 illustrates the output of the simulator, with two patients categorized into the Not-monitored (control) group and eight patients in the monitored group. Within the monitored group, three patients are identified by the simulator for hospital referral based on their variables.

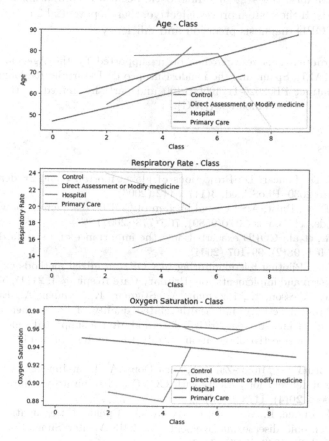

**Fig. 2.** The outcome of age, respiratory rate, and oxygen saturation was comprehensively assessed within the monitored group, highlighting the rates of each variable

Simultaneously, three patients are placed into the direct assessment and modification group, managed by a doctor or authorized personnel. Additionally, two patients are referred to primary care. Furthermore, Fig. 2 demonstrates the combination of patients in both groups, showcasing the variables of each class for the specific patient groups.

# 4   Conclusion

Previous research hasn't proposed FSM or similar models for COPD cohort identification, continuous patient monitoring, and addressing external variables like pollution and daily activity on vital signs. Our findings shed light on vital sign distribution during stable and exacerbation periods, based on data from 10 COPD patients. This method can deepen our understanding of symptom deterioration and medication impact. Future research will involve using real patient data. We'll evaluate the system's diagnostic accuracy, comparing it to conventional methods. If the system proves effective, there's potential for clinical testing to improve COPD diagnosis precision and efficiency

**Acknowledgment.** This research has been supported by the Agencia Estatal de Investigacion (AEI), Spain, and the Fondo Europeo de Desarrollo Regional (FEDER) UE, under contract PID2020-112496GB-I00 and partially funded by the Fundacion Escuelas Universitarias Gimbernat (EUG).

# References

1. Mathers, C.D., Loncar, D.: Projections of global mortality and burden of disease from 2002 to 2030. PLoS Med. **3**(11), e442 (2006)
2. Mannino, D.M., Buist, A.S.: Global burden of COPD: risk factors, prevalence, and future trends. The Lancet **370**(9589), 765–773 (2007)
3. Pauwels, R., et al.: COPD exacerbations: the importance of a standard definition. Respirat. Med. **98**(2), 99–107 (2004)
4. Ure, J., et al.: Piloting tele-monitoring in COPD: a mixed methods exploration of issues in design and implementation. Primary Care Respirat. J. **21**(1), 57–64 (2012)
5. Sturesdotter Åkesson, K., Beckman, A., Stigmar, K., Sundén, A., Ekvall Hansson, E.: Physical activity and health-related quality of life in men and women with hip and/or knee osteoarthritis before and after a supported self-management programme–a prospective observational study. Disabil. Rehabilit. **44**(16), 4275–4283 (2022)
6. Sanchez-Morillo, D., Muñoz-Zara, P., Lara-Doña, A., Leon-Jimenez, A.: Automated home oxygen delivery for patients with COPD and respiratory failure: a new approach. Sensors **20**(4), 1178 (2020)
7. Shojaei, E., Luque, E., Rexachs, D., Wong, A., Epelde, F.: Evaluation of lifestyle effects on chronic disease management. In: 2018 Winter Simulation Conference (WSC), pp. 1037–1048. IEEE (2018)
8. Brandsma, C.-A., Van den Berge, M., Hackett, T.-L., Brusselle, G., Timens, W.: Recent advances in chronic obstructive pulmonary disease pathogenesis: from disease mechanisms to precision medicine. J. Pathol. **250**(5), 624–635 (2020)
9. Jang, S., Kim, Y., Cho, W.-K.: A systematic review and meta-analysis of telemonitoring interventions on severe COPD exacerbations. Int. J. Environ. Res. Public Health **18**(13), 6757 (2021)

# Computational Modelling of Complex Multiphase Behavior of Environmentally-Friendly Materials for Sustainable Technological Solutions

Akshayveer[1], Federico C. Buroni[2], Roderick Melnik[1(✉)],
Luis Rodriguez-Tembleque[3], and Andres Saez[3]

[1] MS2Discovery Interdisciplinary Research Institute, Wilfrid Laurier University,
Waterloo, ON N2L3C5, Canada
{aakshayveer,rmelnik}@wlu.ca

[2] Department of Mechanical Engineering and Manufacturing, Universidad de Sevilla,
Camino de los Descubrimientos s/n, Seville 41092, Spain
fburoni@us.es

[3] Department of Continuum Mechanics and Structural Analysis, Universidad de
Sevilla, Camino de los Descubrimientos s/n, Seville 41092, Spain
{luisroteso,andres}@us.es

**Abstract.** This study presents a computational framework to investigate and predict the complicated multiphase properties of eco-friendly lead-free piezoelectric materials, which are crucial for sustainable technological progress. Although their electromechanical properties vary by phase, lead-free piezoelectric materials show a considerable thermo-electromechanical response. Lead-free materials such as $Bi_{0.5}Na_{0.5}TiO_3$ (BNT) and other BNT-type piezoelectric materials transition to rhombohedral (R3c), orthorhombic (Pnma), tetragonal (P4bm), and cubic (Cc) phases with temperature variation. These phases are determined by the symmetry and alignment of the ferroelectric domains. Multiple phases can occur simultaneously under specific thermal, electrical, and mechanical conditions, leading in complex multiphase behaviour. These materials' performance must be assessed by studying such behaviour. This study uses Landau-Ginzburg-Devonshire theory to simulate material micro-domain phase transitions. The computational model for BNT-type piezoelectric material covers temperature-induced ferroelectric domain switching and phase transitions. Therefore, the developed computational approach will assist us in better understanding the influence of these materials' complex multiphase behaviour on creating sustainable solutions with green technologies.

**Keywords:** Phase-field modelling · Multiphase co-existence · Complex materials and systems · Remnant polarization · Lead-free haptic devices · Human-computer interfaces · Sustainable technologies

© The Author(s), under exclusive license to Springer Nature Switzerland AG 2024
L. Franco et al. (Eds.): ICCS 2024, LNCS 14834, pp. 145–153, 2024.
https://doi.org/10.1007/978-3-031-63759-9_18

# 1    Introduction

Piezoelectric materials transform electrical energy into mechanical energy, essential in electronic devices like sensors, actuators, transducers, and energy harvesters [1]. They also have features like energy storage, field-induced strain, pyroelectricity, and polarization switching, leading to their use in dielectric capacitors, ferroelectric memory, and infrared detectors [2]. Lead-based ceramics (such as $Pb(Zr,Ti)O_3$ (PZT)) are commonly used due to their exceptional piezoelectric characteristics (piezoelectric coefficient $d_{33}$ of 600–700 $pC/N$, high Curie temperature $T_c$ of around 200 °C, and high strain responsiveness of 0.8%) [3]. However, lead oxide is hazardous, leading to the need for lead-free alternatives such as barium titanate ($BaTiO_3$, BT), potassium sodium niobite ($K_{0.5}Na_{0.5}NbO_3$, KNN), and bismuth sodium titanate ($Bi_{0.5}Na_{0.5}TiO_3$, BNT) [4].

Moreover, KNN and BT-based piezoelectric materials show improved response ($d_{33} \sim 500 pc/N$) and ($d_{33} \sim 445 \pm 20 pc/N$), however; the low Curie temperature ($T_c$) for KNN-based composites ($T_c \sim 200\,°C$) and BT-based composites ($T_c < 100\,°C$) restricts their use for high temperature applications [4]. Pure BNT exhibited moderate piezoelectric characteristics ($d_{33} < 100 pc/N$) and a strong coercive electric field ($E_c \sim 70\,kV/cm$) [5]. BNT's high $T_c$ (320 °C) [6] improves its suitability for high-temperature haptic applications. Additionally, BNT-based composites show higher strain than KNN and BT-based composites at temperatures above 200 °C [7], making them ideal for high-temperature actuators.

BNT has complex phase structure. The structure can be R3c, monoclinic (Cc), or a mix of both below depolarization temperature ($T_d \sim$200 °C) [8,9], depending on thermal, electrical, and mechanical treatments. BNT, with its nonergodic relaxor (NR) and ferroelectric (FE) properties, exhibits a square polarization-electric field loop, high remnant polarization $P_r$, and distinct macropiezoelectricity $d_{33}$ from room temperature to dipolar freezing temperature $T_f$ (190 °C) [10], making it suitable for high-temperature haptic applications. Higher temperatures than $T_f$ cause the BNT material to transition from antiferroelectric (AFE) to ferroelectric (FE) phase, reducing its piezoelectric properties. Beyond $T_d$, the material exhibits ergodic relaxor (ER) behaviour and becomes fully AFE. Several studies [11] identified an intermediate Pnma phase with AFE properties between 200–320 °C. However, most researchers now consider it a non-polar or weakly-polar phase [8], making it more suitable for actuator applications due to its high starin response. At temperatures above 320 °C, BNT phase transition becomes more complicated [9]. BNT enters P4bm, a tetragonal symmetric paraelectric phase, at 320 °C. The P4bm symmetry shows octahedral tilting and anti-parallel displacement of $Na^+/Bi^{3+}$ and $Ti^{4+}$ cations along $[001]P_C$, with just 0.2 percent cubic structural deformation [9]. At temperatures over 520 °C, the crystal structure becomes cubic Pm3m. A transition between cubic and tetragonal ferroelectric phases or between cubic and super-paraelectric phases may occur around 520 °C. The BNT exhibits non-negligible hysteresis and $P_r$ at high temperatures (320–520 °C) [12], making it suitable for sensor and haptic applications. Low $P_r$ values in the BNT's paraelectric phase over 520 °C restrict sensing and haptic applications.

The forgoing research shows that BNT exhibits difficult multi-phase co-existence over different temperature regimes, affecting strain response, polarization, and energy storage density. BNT's elastic, piezoelectric, and energy storage characteristics make it useful in various regimes. Novel uses of this material need a careful investigation of its complicated phase change behaviour at different temperatures. Under varied thermal, mechanical, and electrical boundary conditions, we studied BNT material phase transition changes using an experimentally validated computational model. Considering Landau-Ginzburg-Devonshire free energy in the computational model allows BNT material micro-domain switching during phase development and transitions at different temperatures. Although BNT-type piezoelectric materials have a convoluted phase structure, our computational method records correctly micro-domain switching and phase transitions under various heat settings. Thus, this research will help us better understand complicated materials' behaviours, leading to sustainable and eco-friendly sensors, actuators, haptic applications, and human-computer interfaces.

## 2    Methodology and Model Development

A two-dimensional piezoelectric composite design with micro-scale piezoelectric inclusions (BNT) is examined for transient thermo-electromechanical behavior with complicated phase shift and domain switching. The following sub-sections will discuss the composite architecture, the coupled thermo-piezoelectric model used to study the composite's behavior, the materials models that govern its dielectric and mechanical properties, and the boundary conditions used to compute specific effective electro-elastic coefficients of interest.

### 2.1    Geometrical Description and Boundary Conditions

In Fig. 1, the piezoelectric ceramic is shown as a two-dimensional RVE in the $x_1 - x_3$ plane. A 20 $\mu m$ square BNT composite ceramic is used. We investigate the composite material's P-E hysteresis curve to assess its piezoelectric and ferroelectric phase transition capabilities. Two boundary conditions, BC1 and BC2, are needed to derive these curves, as shown in Fig. 1(a) and (b). The reference temperature $\theta_R$ is the ambient temperature (27 °C), whereas the thermal boundary conditions at the right wall can reach up to 520 °C. The electric potential in BC2 follows a sine function of time (t) to provide P-E curves under certain thermal boundary conditions for polarized and depolarized BNT composites. Initial conditions are critical for transient phase-field thermo-electromechanical modelling of BNT-based piezoelectric composite, together with boundary conditions. Initial conditions are $u_i(t = 0) = 0$, $i = 1, 3$; $V(t = 0) = 0$, $\theta(t = 0) = \theta_R$.

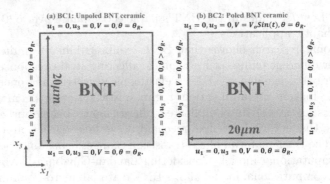

**Fig. 1.** Schematic showing the geometric description and boundary conditions of BNT ceramic.

## 2.2    Mathematical Model

We introduce the thermo-electromechanical model used to examine the BNT ceramic design described in Sect. 2.1 and shown in Fig. 1. The model examines phase transition and ferroelectric domain transitions by incorporating a Landau-Ginzburg-Devonshire free energy (e.g. in [13] and Table 1) in the Helmholtz free energy of the system. The constants are denoted as $C$ for elastic, $\epsilon$ for dielectric, $e$ for piezoelectric, $\mu$ for flexoelectric, $\beta$ for thermal expansion coefficient, and $\eta$ for thermoelectric coefficient. Additionally, $\overrightarrow{E}$, $\varepsilon(\overrightarrow{u})$, $\varepsilon^{el}$, $\varepsilon^{t}(\overrightarrow{p})$, and $\theta$ represent the electric field intensity vector, total mechanical strain tensor, elastic strain tensor, the transformation strain tensor, and temperature respectively. In Table 1, we have collected all governing equations and constitutive relationships for the computational domain and boundary conditions, as shown in Fig. 1.

## 2.3    Material Properties and Computational Implementation

Microscale BNT piezoelectric ceramic was chosen for its ability to regulate grain size and create excellent polycrystallinity and piezoelectric response. The temperature dependent material characteristics of BNT ceramic at the specified reference temperature $(\theta_R)$ are adopted from Hiruma et al. [14].

All phenomenological relations are discretized for the computational domain and related to the governing equations. The computational region was partitioned into a reasonable number of varied triangular mesh components using mesh convergence analysis. The residual polarization $P_r$ was analyzed at 75 °C for various grid counts. Results indicate that 15625 grids result in little change (less than 0.5%) in $P_r$ values, indicating that this is the optimal number. Grid layouts with mesh sizes between $0.16\mu m$ and $10nm$ are employed in this study. The discretized equations are solved using finite elements in the computational domain with BNT material boundary conditions. Modelling phase transitions at specific temperatures is challenging due to varying polarization vector orienta-

**Table 1.** Equations for computational model.

---

**Total free energy of the system:**

$$\phi(\vec{E}, \varepsilon(\vec{u}), \nabla\varepsilon, \vec{p}, \nabla\vec{p}, \theta) = \tfrac{1}{2}\lambda|\nabla\vec{p}|^2 + W(\theta, \vec{p}) + \tfrac{1}{2}\varepsilon^{el}C\varepsilon^{el} - \mu\vec{E}\nabla\varepsilon^{el} - \tfrac{1}{2}\epsilon\vec{E}^2$$
$$- (\theta - \theta_R)\beta\varepsilon(\vec{u}) - e\vec{E}\varepsilon^{el} - \vec{p}\vec{E} - \eta(\theta - \theta_R)\vec{E},$$

**Strain tensors:**

$$\varepsilon(\vec{u}) = \tfrac{1}{2}(\nabla\vec{u} + \nabla\vec{u}^T) = \varepsilon^{el} + \varepsilon^t(\vec{p}),\ \varepsilon^t(\vec{p}) = \gamma|\vec{p}|(\vec{n}\otimes\vec{n} - \tfrac{1}{3}I),\ \vec{n} := \frac{\vec{p}}{|\vec{p}|},$$

**Landau-Ginzburg-Devonshire free energy function:**

$$W(\theta, \vec{p}) = \alpha_1\frac{\theta_c - \theta}{\theta_c}(p_1^2 + p_2^2 + p_3^2) + \alpha_{11}(p_1^4 + p_2^4 + p_3^4) + \alpha_{12}\frac{\theta_c - \theta}{\theta_c}(p_1^2 p_2^2 + p_2^2 p_3^2 + p_3^2 p_1^2)$$
$$+ \alpha_{111}(p_1^6 + p_2^6 + p_3^6).$$

**Strategies of minimizing $W(\theta, \vec{p})$ for zero electric field at $\theta < \theta_0$:**

Cubic phase: $p_1 = p_2 = p_3 = 0$,

Tetragonal phase: $p_1 = p_2 = 0,\ \alpha_1\frac{\theta_c - \theta}{\theta_c} + 2\alpha_{11}p_3^2 + 3\alpha_{111}p_3^4 = 0$,

Orthorombic phase: $p_1 = 0, p_2 = p_3 \quad \alpha_1\frac{\theta_c - \theta}{\theta_c} + (2\alpha_{11} + \alpha_{12}\frac{\theta_c - \theta}{\theta_c})p_3^2 + 3\alpha_{111}p_3^4 = 0$,

Rhombohedral phase: $p_1 = p_2 = p_3 \quad \alpha_1\frac{\theta_c - \theta}{\theta_c} + 2(\alpha_{11} + \alpha_{12}\frac{\theta_c - \theta}{\theta_c})p_3^2 + 3\alpha_{111}p_3^4 = 0$.

**$W(\theta, \vec{p})$ for different phases:**

Cubic phase: $W(\theta, \vec{p}) = 0$, Tetragonal phase: $W(\theta, \vec{p}) = \alpha_1\frac{\theta_c - \theta}{\theta_c}p_3^2 + \alpha_{11}p_3^4 + \alpha_{111}p_3^6$,

Orthorombic phase: $W(\theta, \vec{p}) = 2\alpha_1\frac{\theta_c - \theta}{\theta_c}p_3^2 + (2\alpha_{11} + \alpha_{12}\frac{\theta_c - \theta}{\theta_c})p_3^4 + 2\alpha_{111}p_3^6$,

Rhombohedral phase: $W(\theta, \vec{p}) = 3\alpha_1\frac{\theta_c - \theta}{\theta_c}p_3^2 + 3(\alpha_{11} + \alpha_{12}\frac{\theta_c - \theta}{\theta_c})p_3^4 + 3\alpha_{111}p_3^6$,

**$W(\theta, \vec{p})$ in proximity of Curie temperature $\theta_c$:**

$$W(\theta, \vec{p}) = \alpha_{1c}p_{3c}^2 + \alpha_{11}p_{3c}^4 + \alpha_{111}p_{3c}^6 = 0$$

**The values of $\alpha_1$, $\alpha_{11}$, $\alpha_{12}$, and $\alpha_{111}$:**

$$2\alpha_1 = \frac{1}{\chi} = \frac{(\theta - \theta_0)}{(\theta_c - \theta_0)}\alpha_{1c},\ \alpha_{11} = \frac{-2\alpha_{1c}}{p_{3c}^2},\ \alpha_{12} = -a\alpha11,\ \alpha_{111} = \frac{\alpha_{1c}}{p_{3c}^4}.$$

**Constitutive equations:**

$$\phi_\varepsilon = \sigma = C(\varepsilon(\vec{u}) - \varepsilon^t(\vec{p})) - e\vec{E} - \beta(\theta - \theta_R),\ \phi_{\nabla\varepsilon} = \hat{\sigma} = \mu\vec{E},$$
$$-\phi_{\vec{E}} = \vec{D} = \epsilon\vec{E} + e(\varepsilon(\vec{u}) - \varepsilon^t(\vec{p})) + \eta(\theta - \theta_R) + \mu(\nabla\varepsilon(\vec{u}) - \nabla\varepsilon^t(\vec{p})) + \vec{p},$$
$$-\phi_\theta = S = \beta\varepsilon(\vec{u}) - \eta\vec{E} + W_\theta(\theta, \vec{p}),\ \phi_{\vec{p}} = W_{\theta,\vec{p}}(\vec{p}),\ \vec{q} = -k\nabla\theta,\ \vec{E} = -\nabla V, k > 0.$$

**Governing equations:**

$$\rho\ddot{\vec{u}} = \nabla\cdot(\sigma - \hat{\sigma}) + \vec{F},\tau\dot{\vec{p}} = \nabla\cdot(\lambda\nabla\vec{p}) - \phi_{\vec{p}}, -\theta\dot{\phi}_\theta = \tau\dot{\vec{p}}^2 + \nabla\cdot(k\nabla\theta), \nabla\cdot\vec{D} = \rho_e,$$

$\rho$: mass density,$\rho_e$: electric charge density,$k$: thermal conductivity,$\vec{F}$: external force, $\chi$: dielectric susceptibility,$\tau$: inverse mobility coefficient,$\lambda$: Interface energy coefficient.

---

tions (see Table 1). The complicated coupling of phase transitions was addressed by selective scale switching of polarization vector coefficients.

# 3   Results and Discussions

Lead-free piezoelectric materials like BNT are versatile for ecologically friendly sustainable technology due to their complex multiphase properties. At various temperatures, BNT exhibits complex ferroelectric domain switching, resulting in multiple phases and complex multiphase states. BNT material's polarization values and strain response vary with temperature, making it suited for sensor, actuator, and haptic device application in computer-human interactions across multiple temperatures. We provide some outstanding scenarios of BNT's P-E characteristics curve analysis to establish its potential as a multifunctional eco-friendly material. Prominent P-E features imply BNT's sensor and haptic compatibility.

## 3.1 Experimental Validation of the Developed Computational Model

To validate our computational models, we reproduced the polarization hysteresis loops at 1 Hz with an external electric field of 160 $kV/cm$ from the experimental study in [15]. Our computational model replicates the experimental results in Fig. 2(a) at 75 °C(R3c phase). The numerical model's P-E curve closely matches the experimental curve presented in [15]. Furthermore, the computational model's ability to closely approximate residual polarization and coercive electric field values, differing by less than 1% (as it is not possible to replicate experimental circumstances fully), underscores its potential for studying BNT's phase transition performance and its implications on various applications under diverse physical and environmental conditions. The computational model established here can be used for future investigations on the complicated behaviour of BNT and BNT-based composites.

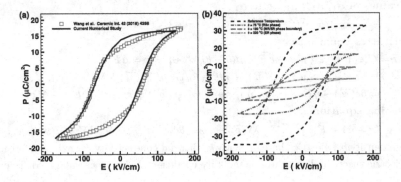

**Fig. 2.** (a) Comparison of P-E hysteresis curve of computational model to experimental work [15], and (b) P-E hysteresis curves for BNT ceramics at different temperatures and phase regimes.

## 3.2 Complex Nonlinear P-E Characteristics Curves in Different Phase Regimes

BNT ceramics, a lead-free and eco-friendly material, display varying polarization properties at different temperatures. In particular, BNT possesses an R3c phase structure before depolarization, although remnant polarization should diminish with temperature. The investigation was conducted at 1 Hz, 160 kV/cm, and different temperatures from ambient to 220 °C in the orthorombhic ER phase with some octahedral tilting. The P-E curve area peaks at room temperature and decreases with temperature due to reduced residual polarization and coercive electric field. The hysteresis curve intersects the electric field axis at the coercive electric field and cuts on the residual polarization at polarization axis. Refer to Fig. 2 (b) for the P-E characteristics curve at various temperatures and phases. At the reference temperature, the curve in R3c phase shows maximum residual polarization and P-E hysteresis region, suggesting good piezoelectric behaviour. The remnant polarization reduces to 18 $\mu C/m^2$ at 75 °C, which was 30 $\mu C/m^2$ at reference temperature. When the temperature is raised to 75 °C from reference temperature, the coercive field drops from 85 to 70 $kV/cm$. Increasing temperature is expected to decrease the size of the hysteresis curve. At $T_f$, 190 °C, the NR/ER phase boundary exists. BNT has strong residual polarization and high coercive electric fields, making it ideal for piezoelectric and high-temperature sensor applications. After reaching this temperature, the NR/ER phase border emerges, causing the domain to transition to the ER phase relaxor, reducing the P-E hysteresis area and residual polarization. Negligible hysteresis area and residual polarization enable great strain response in the BNT, which will be studied further. The ER phase's hysteresis loop makes it ideal for high-temperature actuators, lead-free haptic devices, and human-computer interfaces. The development of P-E characteristics with temperature allows us to examine BNT's versatility and promise in sustainable technology applications. BNT's complicated microdynamics enable the creation of several eco-friendly, sustainable technical solutions for various operational situations.

## 4  Conclusions

Experimentally verified phase-field thermo-electromechanical computational model explored BNT ceramics' phase change and micro-domain switching behaviour. The model predicts BNT ceramics' complicated multi-phase behaviour. Some related study findings are as follows: (a) The maximal P-E curve area occurs at reference temperature with strong residual polarization and coercive electric field, but decreases with temperature progression, (b) BNT ceramics have significant piezoelectric response up to the NR/ER phase boundary, making them an eco-friendly option for high-temperature haptics and sensors, and (c) Since the ER phase has low residual polarization at this temperature range, the hysteresis curve area diminishes, giving it a viable candidate for sustainable high-temperature actuator applications.

The present study fully captures phase change and micro-domain switching. The development of P-E characteristics with temperature allows us to examine BNT's piezoelectric behaviour at various temperatures, which helps us better comprehend its use in high-temperature sensors, actuators, and haptic device applications. Computational modelling, which is cheaper than experiments, supports these conclusions. This study will contribute to establishing advanced computational frameworks for predicting complicated material behaviours and providing innovative, environmentally friendly solutions in sustainable technologies.

**Acknowledgments.** The authors are grateful to the NSERC and the CRC Program (Canada) for their support. This publication is part of the $R^+D^+i$ project, PID2022-137903OB-I00, funded by MICIU/AEI/10.13039/ 501100011033/ and by FEDER, EU. This research was enabled in part by support provided by SHARCNET (www.sharcnet. ca) and Digital Research Alliance of Canada (www.alliancecan.ca).

**Disclosure of interests.** The authors have no conflicts of interest to declare that are relevant to the content of this article.

# References

1. Hans, J.: Piezoelectric ceramics. J. Am. Ceram. Soc. **41**(11), 494–498 (1958)
2. Maurya, D., et al.: Lead-free piezoelectric materials and composites for high power density energy harvesting. J. Mater. Res. **33**(16), 2235–2263 (2018)
3. Dutta, I., Singh, R.: Dynamic in situ X-ray diffraction study of antiferroelectric-ferroelectric phase transition in strontium-modified lead zirconate titanate ceramics. Integr. Ferroelectr. **131**(1), 153–172 (2011)
4. Wang, H., Yuan, H., Hu, Q., Wu, K., Zheng, Q., Lin, D.: Practical high-performance lead-free piezoelectrics: structural flexibility beyond utilizing multiphase coexistence. J. Alloy. Compd. **853**, 157167 (2021)
5. Hao, J., Li, W., Zhai, J., Chen, H.: Progress in high-strain perovskite piezoelectric ceramics. Mater. Sci. Eng. R. Rep. **135**, 1–57 (2019)
6. Kumari, M., Chahar, M., Shankar, S., Thakur, O.: Temperature dependent dielectric, ferroelectric and energy storage properties in $Bi_{0.5}Na_{0.5}TiO_3$ (BNT) nanoparticles. Mater. Today: Proc. **67**, 688–693 (2022)
7. Zhou, X., Xue, G., Luo, H., Bowen, R., Zhang, D.: Phase structure and properties of sodium bismuth titanate lead-free piezoelectric ceramics. Prog. Mater. Sci. **122**, 100836 (20221)
8. Rao, B., et al.: Local structural disorder and its influence on the average global structure and polar properties in $Na_{0.5}Bi_{0.5}TiO_3$. Phys. Rev. B: Condens. Matter **88**(22), 224103 (2013)
9. Jones, G., Thomas, P.: Investigation of the structure and phase transitions in the novel A-site substituted distorted perovskite compound $Na_{0.5}Bi_{0.5}TiO_3$. Acta Crystallogr. Sect. B: Struct. Sci. **58**(2), 168–178 (2002)
10. Kreisel, J., Dkhil, B., Bouvier, P., Kiat, J.: Effect of high pressure on relaxor ferroelectrics. Phys. Rev. B: Condens. Matter **65**(17), 172101 (2002)
11. Dorcet, V., Trolliard, G., Boullay, P.: The structural origin of the antiferroelectric properties and relaxor behavior of $Na_{0.5}Bi_{0.5}TiO_3$. J. Magnetism Magnet. Mater. **321**(11), 1758–1761 (2009)
12. Li, M., et al.: Constructing layered structures to enhance the breakdown strength and energy density of $Na_{0.5}Bi_{0.5}TiO_3$-based lead-free dielectric ceramics. J. Mater. Chem. C **7**(48), 15292–15300 (2019)
13. Ahluwalia, R., Tagantsev, A.K., Yudin, P., Setter, N., Ng, N., Srolovitz, D.J.: Flexoelectricity in solids: role of flexoelectricity in multidomain. Ferroelectrics **C8**, 285–310 (2016)

14. Hiruma, Y., Nagata, H., Takenaka, T.: Thermal depoling process and piezoelectric properties of bismuth sodium titanate ceramics. J. Appl. Phys. **105**(8), 084112 (2009)
15. Wang, C., Zhao, L., Liu, Y., Withers, R., Zhang, S., Wang, Q.: The temperature-dependent piezoelectric and electromechanical properties of cobalt-modified sodium bismuth titanate. Ceram. Int. **42**(3), 4268–4273 (2016)

# Sustainability in the Digital Age: Assessing the Carbon Footprint of E-commerce Platforms

Adam Wasilewski[1]([✉])[iD] and Grzegorz Kołaczek[2][iD]

[1] Faculty of Management, Wroclaw University of Science and Technology,
Wroclaw 50-370, Poland
adam.wasilewski@pwr.edu.pl
[2] Faculty of Information and Communication Technology,
Wroclaw University of Science and Technology, Wroclaw 50-370, Poland
grzegorz.kolaczek@pwr.edu.pl

**Abstract.** Sustainability is one of the development trends of various businesses, including those focused on digital channels. One example of the practical engagement is the care taken to minimize emissions of the various greenhouse gases, e.g. carbon dioxide. While e-commerce does not directly affect emissions, it does consume electricity, the generation of which increases the amount of $CO_2$ in the atmosphere.

This paper focuses on analyzing the carbon footprint of the top 100 most popular Polish e-shops in order to verify their commitment to sustainability. The research uses a measurement method used in online carbon footprint calculators, which, despite significant simplifications, allows a rough estimate of a website's impact on carbon dioxide emissions. Nevertheless, the perceived limitations of the algorithm used made it possible to suggest directions for its development, which could significantly affect the accuracy of the calculations.

**Keywords:** E-commerce · Sustainability · User interface · Carbon footprint

## 1 Introduction

In an era of digital transformation, e-commerce has become an integral part of everyday life, revolutionizing the way people shop and do business. The convenience and accessibility offered by online storefronts have contributed significantly to the growth of online retail, and e-commerce is projected to take over 41% of global retail sales by 2027 [2]. However, the rapid expansion of the digital market comes with a number of environmental challenges, particularly in terms of carbon emissions. The widespread digitization of life today also has an impact on the environment, both positive and negative. Notable among these is the increase in global greenhouse gas emissions attributable to digitization, which is estimated to be around 4% [17]. One reason is the need to power increasingly large data centers, which are energy-intensive and account for about 1% of global electricity consumption [14].

L. Franco et al. (Eds.): ICCS 2024, LNCS 14834, pp. 154–161, 2024.
https://doi.org/10.1007/978-3-031-63759-9_19

As consumers increasingly rely on online platforms to meet their shopping needs, the environmental impact of e-commerce activities is a growing concern. The energy consumption, transportation logistics, and server infrastructure required to maintain digital marketplaces contribute to greenhouse gas emissions, exacerbating the global climate crisis. While many of these aspects are beyond the control of e-commerce platform owners, they can still take steps to engage their business in environmentally responsible practices. It is critical to understand and mitigate the carbon footprint of e-shops to align digital commerce with broader sustainability goals. The carbon footprint is associated with every stage of the e-commerce lifecycle, from production and distribution to use and disposal. However, from the perspective of an individual e-commerce store, optimizing the site that customers use every day and choosing data centers that use renewable energy are two initial steps that can be taken.

The aim of this paper is to analyze the carbon footprint of a selected group of e-commerce sites and verify the algorithm proposed by the Sustainable Web Design (SWD) community group for calculating the carbon footprint of websites [9]. The methodology used in this research involves the collection and analysis of data from a representative sample of e-shop websites, spanning a variety of industries and business models. One hundred Polish online stores with the highest traffic were selected for the study, as this is the country with the highest carbon dioxide emissions per kWh of energy (835g/kWh in 2023) in Europe [7]. This means that changes made to Polish e-shops can have a much greater impact on the environment than similar changes made to e-shops in countries with much lower $CO_2$ emissions from electricity production. Such a conclusion is based on an estimate that the end-user devices are responsible for more than 50 percent of the energy consumed in the e-commerce [9]. Well-designed user interfaces can reduce the amount of data transferred and the time it takes to make an online purchase, reducing the environmental impact of using these devices.

The contribution of this paper is threefold. First, it presents the results of a study of the carbon footprint of a selected group of online stores, taking into account how they power the data centers they use. Second, it discusses the assumptions of the carbon footprint calculation algorithm adopted by the SDW community and suggests ways to improve it. Finally, the paper contributes to the discourse on sustainable e-commerce practices and provides insights that can guide the industry towards a greener and more environmentally responsible future.

## 2   Literature Review

In recent years, the rapid growth of e-commerce has significantly changed consumer behavior and business operations. As the digital landscape expands, concerns about its ecological implications have gained prominence. Understanding and mitigating the environmental impact of online shopping is critical to fostering a more sustainable future. One of the most popular directions is the concept of *digital decarbonization* [5]. Reducing the digital carbon footprint should be a critical part of any corporate sustainability strategy, based on the conscious reuse of knowledge and data and the promotion of digital best practices to minimize $CO_2$ emissions from data [13]. But minimizing your online carbon footprint is something everyone can do. Reducing the resolution of a streaming video or turning off the camera during an online conversation can reduce the resulting greenhouse gas emissions by up to 25 times [15].

Digitization can contribute to greenhouse gas emissions in several ways, some of which are directly related to the increased importance of e-commerce in the global economy. The energy used to collect, process, transmit and display information is a primary concern [12]. The quantity of data utilized by individuals is increasing annually. For instance, in 2020, the average European citizen used approximately 187.3 Gigabytes (GB) of data per year, which represents a yearly increase of over 30% and a nearly 300% increase over the course of five years [10]. However, it should be emphasized that the environmental impact of e-commerce is not limited to data transmission. Online stores are just one sales channel in a supply chain that includes manufacturers, suppliers, distributors and end customers. Comprehensive and consistent action is necessary to reduce the carbon footprint at each stage [3]. For example, $CO_2$ emissions related to online shopping can also be analyzed in terms of the packaging used and its environmental impact [1]. Packaging materials commonly used in standard shipments include wrapping paper, envelopes, cardboard boxes, plastic bags, woven bags, tape, and cushioning materials such as bubble wrap and styrofoam [4]. It is important to consider that some packaging is made from non-renewable materials when making a purchase or selecting a delivery method. Environmentally responsible practices may include the use of packaging made from renewable resources, such as naturally occurring cellulose materials or recycled materials such as paperboard [8]. On the other hand, logistics is undoubtedly crucial, as it accounts for approximately one-fifth of global CO2 emissions [16]. The choice of logistics model [19] can have a significant impact on the final outcome. Finally, it is important to consider artificial intelligence, which is increasingly used in e-commerce, but also has a significant carbon footprint. It is estimated that the ChatGPT tool alone produces 24 kg of *Carbon dioxide equivalent* ($CO_2e$) per day [11].

Reducing the carbon footprint of an e-commerce store can be a daunting task for business owners. However, a good starting point is to focus on reducing the amount of data sent during a customer visit and using data centers that rely on renewable energy sources.

## 3    Research on Carbon Footprint of E-Shop Websites

### 3.1    Methodology

The goal of the study was to verify the carbon footprint of TOP100 online stores in Poland with the highest traffic according to Ahref's stats [18].

For the purposes of the study, the assumptions suggested by the Sustainable Web Design community were adopted. The following assumptions were made for the $CO_2$ footprint calculations:

- Annual Internet Energy (AIE) = 1988 TWh and Annual End User Traffic (AEUT) = 2444 EB,
- Energy Intensity of Data Transfer (EIDT) = AIE/AEUT = 0.81 [kWh/GB],
- Carbon factor - global grid (CFgg) = 442 [g/kWh],
- Carbon factor - renewable energy source (CFres) = 50 [g/kWh],
- Entering the site requires transmission:
  - 100% of the data the first time,
  - Cache Efficiency (CE) estimated as 2% of the data each time a web page is downloaded, by using the device's cache,
- Returning Customers (RC) make up 25% of the total user base [6],
- Data centers (DCE) accounts for 15% of energy used.

The experimental research involved the following steps:

1. using the Google PageSpeed Insights API to get information about the total amount of traffic on the first visit to the site ($DTpV$ - Data Transfer per Visit),
2. determining whether the data centre used by the site uses renewable energy sources (DCres=1 if so, DCres=0 if not), based on the service available in the Green Web Foundation's API (https://www.thegreenwebfoundation.org/),
3. calculating the carbon footprint of the site.

The following formulas, expanding the SWD concept to include renewable energy sources in powering data centers, were used to calculate the carbon footprint:

$$EpV = DTpV * EIDT * RC + DTpV * EIDT * (1 - RC) * CE$$

where: $EpV$ - Energy per visit [kWh]

$$CO_2e = \begin{cases} EpV * CFgg, & \text{if } DCres = 0 \\ EpV * CFgg * (1 - DCE) + EpV * CFres * DCE, & \text{if } DCres = 1 \end{cases}$$

where: $CO_2e$ - Carbon dioxide equivalent [g]

For each online store studied, the DTpV determination was made for the home page only; sub-pages such as listings, product cards, fixed content were not checked. A detailed examination of these aspects of e-commerce sites will be the subject of future research.

## 3.2  Results

100 Polish on-line shops were selected for the survey, but 6 of them failed to retrieve information on the amount of data submitted during the first visit. Therefore, the carbon footprint analyses were performed on a set of 94 e-shops.

In addition to gathering information about the data being transferred, the datacenter being used was also verified. In this case, three options were possible:

1. standard datacenter,
2. datacenter using renewable energy sources,

**Table 1.** Summary of data centers used.

| Data center | Quantity | Percentage share | The most popular hosting |
|---|---|---|---|
| Standard | 19 | 20.2% | - |
| Using renewable energy sources | 42 | 44.7% | Akamai Technologies (18), e24cloud by Beyond (6) |
| Unknown (proxy) | 33 | 35.1% | Cloudflare (26), Amazon CloudFront (7) |

3. reverse proxy, not allowing to determine the actual hosting.

A summary of the data centers used by the analyzed e-shops is shown in Table 1. Due to the fact that there was a significant portion of hostings that could not be directly classified into either group one or group two, and therefore it was not possible to assign a unique value to the $DCres$ parameter, it was necessary to modify the original assumptions regarding the calculation of the $CO_2e$ index value.

The change was the addition of a third variant expression for calculating the value of $CO_2e$ in case it cannot be determined whether the hosting uses renewable energy sources (DCres=2):

$$EpV = EpV * CFgg * (1 - DCE) + EpV * CFunkn * DCE,$$

The $CFunkn$ value was calculated based on the $CFgg$ and $CFres$ values and the proportion of standard (31,1%) and "green" (68,9%) hosting among the e-stores included in the survey.

$$CFunkn = 0,311 * CFgg + 0,689 * CFres = 172,10[g/kWh]$$

Taking into account the above modification, it was possible to calculate $CO_2e$ values for 94 e-stores. The lowest index value was recorded for Bonito's online store, while the highest value was recorded for Agata Meble's store.

It can be observed that the $CO_2e$ value increases linearly for most e-commerce sites, but the results change sharply for about 10% of the worst sites.

In order to analyze this situation in detail, the surveyed stores were divided into quadrants labeled Q1, Q2, Q3, Q4 (Table 2).

**Table 2.** Values of $CO_2e$ by quadrants [kg/100k visits].

| Quadrant | Quantity | Average | Std. dev |
|---|---|---|---|
| Q1 | 24 | 3.740 | 0.673 |
| Q2 | 23 | 5.744 | 0.662 |
| Q3 | 24 | 7.688 | 0.661 |
| Q4 | 23 | 17.633 | 11.561 |

In quadrants Q1, Q2 and Q3, the standard deviation of the results is practically the same, with a practically linearly increasing mean value of the index. The case is different in the Q4 quadrant, where the standard deviation is close to 2/3 of the mean, and in the box plot three observations are even outside the whiskers. This means that the collection of e-commerce sites in Q4, which includes the worst sites in terms of carbon footprint, is the most diverse.

### 3.3    Discussion

The results of the analysis show that the largest Polish online stores differ significantly in terms of their carbon footprint, with the $CO_2e$ value of the worst being more than 16 times higher than that of the best. This indicates that there is great potential for reducing the negative environmental impact of e-commerce in Poland. It is worth noting that about 10% of the e-shops with the highest carbon value were responsible for such a wide spread of results. If these were omitted, the difference between the worst and best results would decrease by more than 3 times (from 1644% to 466%). This is still a large gap, however, and justifies the work to optimize websites through sustainability measures, such as reducing data transfer (e.g., by reducing the quality of image files), improving the user experience (e.g., by making it easier to search for products and content), using "green" data centers, and raising environmental awareness among customers.

However, the assumptions and simplifications made in the algorithm used must be kept in mind. While they are acceptable at a general level and provide a rough estimate of the carbon footprint of e-commerce websites, they should be improved for detailed analysis. The main changes to the $CO_2e$ calculation algorithm described above should include:

1. the location of end customers, due to the varying values of $CO_2$ emissions per kWh produced in each country,
2. end-user devices due to differences in power consumption,
3. the ratio of returning customers to new customers - this metric should be related to the length of time content is cached on end user devices,
4. detailed information about the data center in use, including energy intensity and location,
5. dynamic adjustment of the values of the parameters used in the calculations to energy production market changes.

With this approach, it would be possible to dynamically incorporate changing values into the calculation of the carbon footprint of e-commerce.

## 4    Future Research and Conclusions

This paper underscores the importance of conducting regular carbon footprint assessments for e-commerce websites and encourages the adoption of sustainable practices to minimize the ecological consequences associated with online shopping. E-commerce sites generate carbon emissions throughout their lifecycle, from the manufacturing of electronic devices and infrastructure to the operation of data centers and the delivery of goods. It is important to note that the carbon footprint of each e-commerce platform varies based on factors such as energy sources, transportation methods, and overall sustainability practices. Efforts can also be made to optimize the amount of data sent during online shopping, thereby reducing environmental costs, and one direction is to modify and customize the user interface. However, it is important to keep in mind that reducing the carbon footprint of an online store's website may be associated with a decrease in its perceived usability (e.g., due to a decrease in images quality). Such a situation can be a business problem, especially if the e-shop's customers are not engaged in sustainability. If problems arise, it may be worth considering implementing a solution that uses multi-variant user interfaces [20] to allow customers to choose between visual attractiveness and reduced environmental impact.

The results of the study conducted and described in this paper show that there is a wide variation between the carbon footprints of online stores. This means that there is potential for action to optimize environmental impact. This is crucial given the growing importance of e-commerce in the economy and the increasing traffic to online stores. It is necessary to remember, but also to make customers aware, that every time they visit a store's website, refresh it, browse products, etc., there is a carbon footprint. The number of visits, often in the millions per day, multiplies these values every day. The research also included a verification of the data centers used. Some are actively working to reduce their environmental impact. Future work will include expanding the algorithms for calculating the carbon footprint of e-commerce to include as many influencing factors as possible. A reliable and accurate mechanism for analyzing the environmental impact of e-commerce stores can provide a tool for designers to optimize the user interface, but also point in other directions to minimize $CO_2$ emissions. Such a solution could also provide an incentive for e-commerce customers to engage and contribute to eco-friendly activities. Its practical implementation would be a step toward a more environmentally responsible e-commerce sector that aligns business practices with the global sustainability imperative.

# References

1. Arora, T., Chirla, S., Singla, N., Gupta, L.: Product packaging by e-commerce platforms: impact of covid-19 and proposal for circular model to reduce the demand of virgin packaging. Circular Econ. Sustain. **3**, 1255–1273 (2022). https://doi.org/10.1007/s43615-022-00231-4
2. Barthel, M., et al.: Winning formulas for e-commerce growth. Tech. rep, Boston Consulting Group (BCG) (2023)
3. Brinken, J., Behrendt, F., Trojahn, S.: Comparing decarbonization potential of digital and green technologies. Sustain. Futures **6**, 100125 (2023)

4. Chueamuangphan, K.., Kashyap, P.., Visvanathan, C..: Packaging waste from E-Commerce: consumers' awareness and concern. In: Ghosh, Sadhan Kumar (ed.) Sustainable Waste Management: Policies and Case Studies, pp. 27–41. Springer, Singapore (2020). https://doi.org/10.1007/978-981-13-7071-7_3

5. Doleski, Oliver D.., Kaiser, Thomas, Metzger, Michael, Niessen, Stefan, Thiem, Sebastian: Digital Decarbonization. Springer, Wiesbaden (2022). https://doi.org/10.1007/978-3-658-33330-0

6. Dopson, E.: Ecommerce customer retention marketing: How to use emails, loyalty programs & communities to improve retention. Tech. rep., Shopify (2022). https://www.shopify.com/enterprise/ecommerce-customer-retention

7. Electricity Maps: https://app.electricitymaps.com/map Retrieved 20 Jan 2024

8. Escursell, S., Llorach-Massana, P., Roncero, M.B.: Sustainability in e-commerce packaging: a review. J. Cleaner Prod. **280**, 124314 (2021). https://doi.org/10.1016/j.jclepro.2020.124314

9. Estimating Digital Emissions: https://sustainablewebdesign.org/calculating-digital-emissions/

10. Farfan, J., Lohrmann, A.: Gone with the clouds: estimating the electricity and water footprint of digital data services in Europe. Energy Convers. Manage. **290**, 117225 (2023). https://doi.org/10.1016/j.enconman.2023.117225

11. Groes, K., Ludvigsen, A.: Chatgpt's electricity consumption. Technical report, medium.com (2023). https://towardsdatascience.com/chatgpts-electricity-consumption-7873483feac4

12. Jackson, T., Hodgkinson, I.R.: Is there a role for knowledge management in saving the planet from too much data? Knowl. Manage. Res. Pract. **21**(3), 427–435 (2023). https://doi.org/10.1080/14778238.2023.2192580

13. Jackson, T.W., Hodgkinson, I.R.: Keeping a lower profile: how firms can reduce their digital carbon footprints. J. Bus. Strateg. **44**(6), 363–370 (2022)

14. Masanet, E., Shehabi, A., Lei, N., Smith, S., Koomey, J.: Recalibrating global data center energy-use estimates. Science **367**(6481), 984–986 (2020)

15. Obringer, R., Rachunok, B., Maia-Silva, D., Arbabzadeh, M., Nateghi, R., Madani, K.: The overlooked environmental footprint of increasing internet use. Resour. Conserv. Recycl. **167**, 105389 (2021). https://doi.org/10.1016/j.resconrec.2020.105389

16. Ritchie, H.: Cars, planes, trains: where do co2 emissions from transport come from? Our World Data (2020). https://ourworldindata.org/co2-emissions-from-transport

17. Teuful, B., Sprus, C.: How digitalization acts as a driver of decarbonization. Tech. rep, EY (2020)

18. TOP100 online stores in Poland with the highest traffic: https://xann.pl/top100-e-commerce-w-polsce-czyli-sklepy-internetowe-z-najwiekszym-ruchem Retrieved 20 Jan 2024)

19. Wasilewski, Adam: Integration challenges for outsourcing of logistics processes in e-commerce. In: Huk, Maciej, Maleszka, Marcin, Szczerbicki, Edward (eds.) ACIIDS 2019. SCI, vol. 830, pp. 363–372. Springer, Cham (2020). https://doi.org/10.1007/978-3-030-14132-5_29

20. Wasilewski, A.: Functional framework for multivariant e-commerce user interfaces. J. Theor. Appl. Electron. Commer. Res. **19**(1), 412–430 (2024). https://doi.org/10.3390/jtaer19010022

# Evaluating R-CNN and YOLO V8 for Megalithic Monument Detection in Satellite Images

Daniel Marçal[4], Ariele Câmara[1]([✉]), João Oliveira[1,2], and Ana de Almeida[1,3]

[1] Instituto Universitário de Lisboa (ISCTE-IUL), ISTAR, Lisbon, Portugal
acaer@iscte-iul.pt
[2] Instituto de Telecomunicações, Lisbon, Portugal
[3] CISUC - Centre for Informatics and Systems of the University of Coimbra, Coimbra, Portugal
[4] Lisbon, Portugal

**Abstract.** Over recent years, archaeologists have started to use object detection methods in satellite images to search for potential archaeological sites. Within image object recognition, due to its ability to recognize objects with great accuracy, convolutional neural networks (CNN) are becoming increasingly popular. This study compares the performance of existing deep-learning algorithms for the detection of small megalithic monuments in satellite imagery, namely RCNN (Region-based Convolutional Neural Networks) and YOLO (You Only Look Once). Using a satellite image dataset and after adequate preprocessing, results showed that this is a feasible approach for archaeological image prospection, with RCNN achieving a remarkable precision of 93% in detecting these small monuments.

**Keywords:** object detection · satellite images · CNN · megalithic monuments · archaeology

## 1 Introduction

Object detection, a pivotal task in computer vision, has emerged as a crucial method for archaeologists to recognize specific monuments, thereby facilitating the prospection and the study of ancient societies. In the domain of satellite imagery, this task becomes especially challenging due to a myriad of factors [13], ranging from inherent spatial resolution constraints, where significant yet relatively small constructions like dolmens might be represented by a mere handful of pixels (around 15 pixels in this case), to issues of rotation invariance, with monuments appearing in any possible orientation. Additionally, accuracy is influenced by intraclass variations, where the visualization of the same object type, such as a megalithic monument, can vary based on environmental conditions, vegetation, shadow length, and soil type [6]. Beyond these technical challenges, the intensive task of data labelling is yet another challenging task. Nevertheless, deep learning presents a promising solution for object detection, and machine learning-based methods are becoming increasingly common, albeit in recognising easily detectable monuments [9].

L. Franco et al. (Eds.): ICCS 2024, LNCS 14834, pp. 162–170, 2024.
https://doi.org/10.1007/978-3-031-63759-9_20

Acknowledging the importance and intricacies of this problem, we delve into a performance analysis of different object detection pipelines to understand their capability for identifying small megalithic monuments in satellite images, aiming to provide a tool for helping archaeologists to recognise monuments that have been traditionally difficult to detect. For this purpose, with the help of an expert, we collected a customized dataset featuring high-resolution images containing known dolmen sites [3]. The new annotated dataset intends to benchmark the speed and accuracy of the pipelines developed, striving for optimal detection and localization of small heritage monuments in satellite images. The information regarding the locations of the analyzed monuments comprising our dataset is available on Zenodo [5].

To optimize the dolmens' detection and classification process using satellite imagery, we evaluate and compare two recent renowned algorithms, RCNN (Region-based Convolutional Neural Networks) and YOLO (You Only Look Once), benchmarking their performance metrics, including running time and accuracy.

This paper is organized into five distinct sections: after this introduction, a brief review of related works is presented. Next, we describe an exploration of object detection methodologies, followed by a discussion of our results. The work finishes with the presentation of conclusions and the discussion of probable implications.

## 2   Case Study Area Characterization

The archaeological data used in this study focus on the Mora and Arraiolos regions in southern Alentejo, Portugal (Fig. 1). Situated within the Ancient Massif, known for its granitic, schistose, quartzites, and other metamorphic rocks, these areas harbour significant clusters of known megalithic monuments, totalling 272 structures classified as dolmens according to the Portuguese *Portal do Arqueólogo* [14]. Dating back to the Neolithic/Chalcolithic period (4000/5000 BC), these dolmens primarily served as funerary sites, typically constructed from granite or schist, with diameters ranging from 2 to 5 m [3]. While some are visible above ground, others are buried or integrated into modern constructions, making recognition challenging. To our knowledge, there has

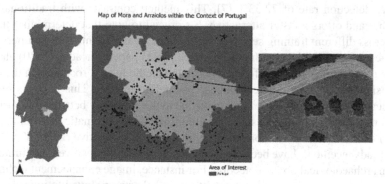

**Fig. 1.** Map of Portugal highlighting the regions of Mora and Arraiolos in detail, as well as the dolmens in these regions. On the right, a representation of dolmen Lapeira 1 is shown: an aerial view from Google Earth.

been no prior research on object recognition of these monuments in the analyzed region. This study stands out as it goes beyond simply detecting easily identifiable figures in satellite images.

## 3 Literature Review

Remote sensing has been utilized as an indispensable tool for archaeologists for decades, and recent advancements in Deep Learning (DL) present new opportunities to enhance archaeological research methodologies, notably in object recognition within satellite images [4]. Despite growing interest in machine learning for site identification, its adoption in archaeology remains limited due to its complex nature, demanding computational expertise. Among the prevailing neural network architectures in contemporary technology, Artificial Neural Networks (ANNs) stand out. Notably, CNNs have garnered significant attention and are the most rigorously explored among the different techniques. CNN architectures can generally be categorized into two groups based on their approach to object detection: one-stage detectors, such as You Look Only Once (YOLO), and two-stage detectors like Faster Region-based CNN. In the domain of archaeology, the primary use of Remote Sensing Images (RSI) revolves around detecting distinct ground structures, examples of which include burial mounds, tells, rectangular enclosures, charcoal burning platforms, and qanats [4].

In recent years, studies have increasingly utilized CNNs to recognize archaeological structures in RSIs. Among these architectures, the RCNN has been described as an ideal solution for high-precision object detection tasks. For instance, Caspari & Crespo (2019) employed a CNN to detect Early Iron Age tombs within the Eastern Central Asian steppes using optical satellite imagery. The authors' findings underlined the superior performance of CNNs in RSIs analysis, achieving an impressive accuracy of 0.99 (F1-score) in images without the presence of tombs, contrasting with a 0.91 score in their presence [8]. This study exemplifies the prowess of CNNs in precisely detecting archaeological landmarks even from satellite imagery. Another popular approach, YOLO, has also been utilized, demonstrating faster detection rates and reduced false positive detections. For example, Canedo et al. (2023) used YOLO to detect burial mounds, achieving a positive detection rate of 72.53% [7]. This method contrasts with traditional CNN approaches and offers a faster alternative. It's crucial to note that determining the best model across different training sets may vary, and other algorithms could outperform in distinct tests. In comparison with our approach in this paper, Caçador (2020) analyzed the same monuments using a different dataset and methodology, focusing on the hyperspectral signature of how dolmens appear in satellite images. This analysis revealed the challenge of discerning these signatures due to the similarity between the surrounding terrain and the monument itself. Despite utilizing panchromatic and multispectral images, identification was feasible, albeit with a high false positive rate of 87.2% [2].

While advancements have been made, challenges like false positives and limited data persist in archaeological object detection. For instance, image enhancement techniques, including rotation, flipping, and augmentation, have been employed to improve object detection in challenging environments and expand datasets for analysis [8]. Additionally, techniques such as Location-Based Ranking and Bagging have been utilized to mitigate false positives. Despite these improvements, automated results still require refinement to consistently match human expertise, highlighting ongoing challenges and the need for further advancements in the field. Current efforts are focused on detecting niches or those structures deemed 'easy' to recognize [9], but future research may explore analyzing smaller or less identifiable monuments as a potential area of interest.

## 4 Methodology

After collecting all the dolmen locations within the area to be covered, the next step was to obtain their satellite images. For this study, all the images were gathered from Google Earth Pro and corresponded to the same monuments analyzed in a previous work conducted by Câmara (2017), where the author performed a photo interpretation analysis [3]. These images were then extracted and saved in 8k resolution, providing higher-quality images. Our coverage features 62 dolmens visible from a software perspective. We collected five images for each dolmen, changing the monument's position within each image and, therefore, the surrounding background also changed, giving a total dataset of 310 different images. It is important to note that the low quantity of images of monuments for visualization derives from the fact that these are millennia-old structures, many of which are either destroyed or not visible in satellite images.

To test the algorithm's response to an image that does not contain any dolmen, two more images, very similar in background but devoid of dolmens were added to the test set. The data was split as follows: the test set comprises seventeen images, fifteen of which depict three different dolmens in different locations, and two images that, to the best of our knowledge, do not contain any dolmen. The remaining 285 images were used to train the algorithms. The random selection ensures a representative data distribution in training and validation subsets. In the subsequent step, the images underwent a preprocessing stage, including image labeling, enhancement, and augmentation, facilitated by Roboflow [12], augmenting the dataset to 855 images. Following a review of the state of the art [11], we determined that the optimal approach for augmentation in this setting was to introduce random values for various types of augmentation until identifying the most effective set of modifications for improved results. Non-colour-based and color-based augmentations such as cropping, rotation, hue, saturation, brightness, and exposure were employed. Contrast enhancement was applied using Roboflow's histogram and adaptive histogram equalization. These adjustments accentuate local details, making darker or lighter regions of the image more discernible. Such enhancement is especially valuable in images with a high vegetation index.

After pre-processing the images, we transitioned to the modeling phase to train the selected algorithms. Our choices included YOLO version 8, the most recent version at the time, and Fast R-CNN. YOLO, known for its efficiency in rapidly detecting objects within images, and Fast R-CNN, renowned for its high precision in object detection, were deemed suitable candidates for our archaeological structure detection project. In our experimentation, we systematically explored nine different architectures, leveraging auto-tuning in each experiment to fine-tune the parameters and hyperparameters for optimal performance on the training data. Using Fast R-CNN models, we opted to use ResNet-101 and ResNet-50 as backbone networks, without pre-training, exploring three different network structures: Feature Pyramid Network (FPN), Dilated Convolutional Network (DC), and Convolutional Network (C). We conducted experiments using two types of training schedules, namely 1x and 3x. These models were trained using the custom dataset created on Roboflow, which was then converted to COCO format, facilitating training with frameworks like Detectron2. Google Colab was used for training, and we chose to set 5000 iterations to minimize the total loss and approach the optimal learning rate for each algorithm. The training time for each Faster R-CNN algorithm was approximately 40 min, while YOLO required only half this time. This process consumed 10 GB of RAM and 8 GB of GPU memory in Colab. While the backbone networks were not pre-trained, using COCO-formatted data enabled us to leverage pre-trained weights for specific architectures to expedite convergence during training.

## 5  Results

For evaluation, the most common metrics were used. Precision (P) is calculated as the ratio of true positive (TP) predictions to the sum of true positives and false positives (FP): $P = \frac{TP}{TP+FP}$. F1-score, the harmonic mean of precision and recall(R), provides a balanced measure calculated by $F1 = \frac{2*P*R}{P+R}$, where R measures the model's ability to capture positive instances and is the ratio of true positive predictions to the sum of true positives and false negatives: $R = \frac{TP}{TP+FN}$. These metrics help assess the model's ability to detect objects while minimizing the false positives correctly, and the F1-score is specifically emphasized for its relevance in binary classification scenarios [1]. Table 1 presents the average precision and F1-score metrics that have been obtained for each of the trained models for the test set. Notably, the FasterRCNN model, utilizing a ResNet-50 backbone network with a DC network for structure and a 1x training schedule, achieved the best results for this test set, achieving a precision of 0.93 and an F1-score of 0.78. The choice of backbone architecture substantially impacts the model's ability to learn complex patterns, and these metrics provide insights into the strengths and weaknesses of each configuration.

**Table 1.** Performance results for all the architectures tested in terms of average precision and F1-score values.

| Model | Average Precision | F1 Score |
|---|---|---|
| R_50_FPN_3x | 0.67 | 0.51 |
| R_50_FPN_1x | 0.69 | 0.64 |
| R_50_DC_3x | 0.70 | 0.65 |
| R_50_DC_1x | **0.93** | **0.78** |
| R_50_C_3x | 0.61 | 0.57 |
| R_50_C_1x | 0.60 | 0.57 |

(*continued*)

**Table 1.** (*continued*)

| Model | Average Precision | F1 Score |
|---|---|---|
| R_101_FPN_3x | 0.71 | 0.64 |
| R_101_DC_3x | 0.74 | 0.71 |
| R_101_C_3x | 0.63 | 0.59 |
| YoloV8 | 0.79 | 0.71 |

The training of this particular type of algorithm involved monitoring various metrics to identify signs of overfitting or underfitting during the defined epochs. Figure 2 illustrates the losses and accuracy metrics throughout the training process. Particularly, the classification loss, which typically measures the divergence between predicted class probabilities and actual class labels, was scrutinized. The objective during training was to minimize this metric across epochs, aiming for optimal model performance. Figure 2A shows a notable decrease in classification loss that becomes more stable after 4000 epochs, indicating convergence towards an optimal solution. This observation suggests that the training process effectively learns the underlying patterns in the data, enabling the model to make accurate predictions. The loss stableness implies reduced sensitivity to minor fluctuations in the training data, indicative of a well-generalised model [10].

In the Region Proposal Network (RPN) of the faster RCNN algorithm (R_50_DC_1x), minimizing the classification loss allows the prediction of high objectness scores for anchors overlapping significantly with ground-truth object bounding boxes and low scores for anchors far distant from any object. This ensures that the model focuses on relevant regions likely to contain objects while ignoring background or irrelevant areas, which is particularly relevant given that each dolmen was tested against five different backgrounds [10]. Figure 2.B) illustrates that after approximately 2000 epochs, the model becomes more stable. This stableness suggests that the model has reached a point of diminishing returns regarding classification improvement. While this stabilisation is a positive sign, indicating the likely convergence of the RPN to a satisfactory level of classification accuracy, it's crucial to acknowledge that further training beyond this point may yield insignificant additional benefits and may even risk overfitting. Minimizing the localization loss is crucial because it ensures that the algorithm effectively learns to accurately predict the correct bounding box coordinates for the positive region proposals. Through improved training, the regression of the predicted bounding box coordinates for each positive anchor aligns more closely with the ground-truth bounding box. The successful training of the Faster R-CNN algorithm, exhibiting minimal losses and high accuracy, underscores the feasibility of implementing object detection models for small megalithic monuments in a rocky terrain through remote imagery, making this a feasible approach for automating and enhancing archaeological prospection work.

In Fig. 2.C), the analysis includes the localization loss in the RPN. However, the presence of numerous spikes suggests that there are instances where the model encounters

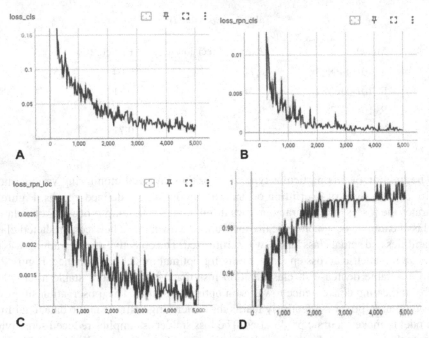

**Fig. 2.** Fig. 2 A-D depicts the plots of the train results in terms of loss and accuracy: (A) Classification Loss; (B) Classification loss in the Region Proposal Network; (C) Location loss in the Region Proposal Network; and (D) Classification Accuracy.

challenges in precisely pinpointing object locations. These spikes may be due to various factors, including complex object geometries, augmentations, or variations in image backgrounds. In Fig. 2.D), the classification accuracy for this algorithm was tracked by epoch, illustrating a consistent improvement trend. This means the model's growing adeptness in accurately classifying objects during training. While the Faster R-CNN algorithm, trained with minimal losses and high accuracy, demonstrates proficiency in object detection, it's noteworthy that YOLOV8, despite a lower confidence rate in true positive results, excels in minimizing false positive detections. This reduction in false positives is particularly advantageous for our overarching goal of providing a helpful tool for archaeologists in their prospection work. This trade-off bears practical implications, as Faster R-CNN models may be preferable for precise localization, whereas YOLO models could be advantageous in scenarios prioritizing false positive reduction. Moreover, the results underscore the influence of background contexts on confidence scores, emphasizing the importance of background diversity in training datasets to enhance the adaptability of object detection models in real-world scenarios. These findings stress the necessity of comprehensive evaluation considering environmental characteristics for robust detection performance.

## 6 Conclusions

The paper addresses the challenge of object detection in identifying dolmens in satellite imagery. Its primary contribution includes a set of 62 annotated high-quality images of dolmens in Portugal. Given data constraints, image augmentation and enhancement were crucial in increasing the dataset from 285 to 855 images, as well as highlighting the monuments, which can often be obscured by surrounding features. However, challenges persist due to the scarcity of expert-confirmed dolmen locations, resulting in a relatively small dataset. Evaluation of results highlighted the YOLOv8 model that, although showing lower confidence in true positives presented fewer false positives. Nevertheless, the Faster R-CNN model, despite the higher number of false positives, presents the lowest confidence rate in erroneous identifications.

Future research should prioritize the collection of a broader and diversified dataset for a more comprehensive evaluation assessment. Moreover, venturing into advanced or hybrid modeling techniques could improve accuracy in detecting dolmens in satellite images.

**Acknowledgments.** This work was partially supported by the Fundação para a Ciência e a Tecnologia, I.P. (FCT) through the ISTAR-Iscte project UIDB/04466/2020 and UIDP/04466/2020, through the scholarship UI/BD/151495/2021.

**Disclosure of Interests.** The authors have no competing interests to declare that are relevant to the content of this article.

## References

1. Müller, A.C., Guido, S.: Introduction to Machine Learning with Python A Guide for Data Scientists (2017)
2. Caçador, D.G.C.: Automatic recognition of megalithic objects in areas of interest in satellite imagery. ISCTE (2020)
3. Câmara, A.: A fotointerpretação como recurso de prospeção arqueológica. Chaves para a identificação e interpretação de monumentos megalíticos no Alentejo: aplicação nos concelhos de Mora e Arraiolos. Universidade de Évora (2017)
4. Câmara, A., et al.: Automated methods for image detection of cultural heritage: overviews and perspectives. Archaeol. Prospect. (2022). https://doi.org/10.1002/arp.1883
5. Câmara, A.: Data Description (ICCS). Zenodo (2024) https://doi.org/10.5281/zenodo.109 88490
6. Câmara, A., Batista, T.: Photo interpretation and geographic information systems for dolmen identification in Portugal: the case study of Mora and Arraiolos. Presented at the (2017). https://doi.org/10.23919/cisti.2017.7975890
7. Canedo, D., et al.: Uncovering archaeological sites in airborne LiDAR data with data-centric artificial intelligence. IEEE Access **11** (2023). https://doi.org/10.1109/ACCESS.2023.329 0305
8. Caspari, G., Crespo, P.: Convolutional neural networks for archaeological site detection – finding "princely" tombs. J. Archaeol. Sci. **110**, 104998 (2019). https://doi.org/10.1016/J. JAS.2019.104998

9. Davis, D.: Theoretical repositioning of automated remote sensing archaeology: shifting from features to ephemeral landscapes. J. Comput. Appl. Archaeol. **4**(1), 94–109 (2021). https://doi.org/10.5334/JCAA.72/METRICS/
10. Goodfellow, I., RegGoodfellow, I., Bengio, Y., Courville, A.: Regularization for Deep Learning. Deep Learning, pp. 216–261 (2016)
11. Guo, H., et al.: Dynamic low-light image enhancement for object detection via end-to-end training. In: Proceedings - International Conference on Pattern Recognition (2020). https://doi.org/10.1109/ICPR48806.2021.9412802
12. Roboflow: Introduction - Roboflow Docs. https://roboflow.com/. Accessed 18 Apr 2024
13. Tahir, A., et al.: Automatic target detection from satellite imagery using machine learning. Sensors **22**, 3 (2022). https://doi.org/10.3390/s22031147
14. Archaeologist's Portal. https://arqueologia.patrimoniocultural.pt/. Accessed 04 Mar 2024

# How to Facilitate Hybrid Model Development by Rebuilding the Demographic Simulation Model

Jacek Zabawa[✉] [iD]

Department of Operations Research and Business Intelligence, Faculty of Management,
Wroclaw University of Science and Technology, Wybrzeże Wyspiańskiego 27,
50-370 Wrocław, Poland
Jacek.Zabawa@pwr.edu.pl

**Abstract.** Demographic information can be used to analyze processes occurring in a wide range of human social activity: in the area of management, healthcare, social security systems. One of the leading methods of demographic modeling is continuous simulation based on the system dynamics approach. Reducing computation time for fast-running continuous model implementations enables efficient development of hybrid models. In the hybrid simulation model, the discrete-event simulation approach was used as one of the components. Technical and conceptual solutions to the observed problems were presented, experiments based on demographic data from the Wrocław region, Poland were performed and it was indicated that an effective hybrid simulation will allow to include additional cause-and-effect relationships in the models.

**Keywords:** Operations research · Decision support · Improving the simulation performance

## 1 Introduction

Knowledge of the state of demography, i.e. at least the number of people who can be classified into cohorts (groups distinguished on the basis of the age and sex) and trends in phenomena affecting these quantities, is useful for studying issues such as: prediction of premature mortality [2], support needs in certain age groups in the coming decades [8], demand for housing [9], economic reform scenario analysis [3], consumption levels study [7], long-term care expenditure analysis [10], methods of improving neonatal care [13], long-term care capacity analysis [1], changes in the pension system [15].

Among the approaches used to model demographic dependencies, simulation in accordance with the system dynamics (SD) approach (continuous simulation) is widely used. In addition to dividing the population into cohorts, this approach distinguishes main elements as input flows: births, immigrations and ageing/growing up and output flows: deaths, ageing/growing up, emigrations. Additional elements are: birth coefficients, death coefficients, transition time between cohorts. Studies using the SD approach are usually performed in accordance with the dynamic synthesis methodology presented e.g. in [13], i.e. creating a descriptive model (causal loop diagrams, CLD) and then building a computer system dynamics model (SDM), composed of flows (rates) and stocks (levels). CLD (a conceptual model) and SDM (a computational model) form the modeling and simulation cycle [4].

## 2    Conclusions from Previous Studies. Discovered Issues

This paper is one of a series of works on simulation modeling applications, i.e. the study of demographic changes and the use of the discrete-event system (DES) modeling approach to modeling the demand for hospital services [11], the prospects of replacing the SD approach with discrete rate modeling (DRM) in demographic research and how cohort grouping affects these prospects [16], studying the impact of demographic scenarios on the intensity of the demand for medical services [12].

The usefulness of our models will be determined by measuring the differences between historical and simulated cohorts. In the models built so far, it was possible to prevent the problems noticed in [6], by assigning an increase in cohorts in the model to the number of cohorts equal to the maximum age of people in the model, for each gender (2 * 105). However, we have observed problems in the use of models built on the Extendsim platform [5], such as: e.g. assigning initial states in stock blocks (Holding Tank) and slowing down the performance of models as the number of cohorts increases. The simulation run lasted too long (about 1 h) for our needs. We have decided to apply a DES to pass values between streams and resources. All of the above problems result in the fact that the discrete - continuous hybrid seemed difficult to implement.

Population data from 2006 were obtained from Statistics Poland [14] and concern the Wrocław region. The data used relate to the cohorts of human populations mostly 5 years wide (i.e. 17 cohorts from 0_5 to 80–84 years old) and one cohort supposedly infinite width (the model included a 20-year cohort for persons 85–104 years old).

We built separate models: SD and DES. They must therefore be run separately. DES has to wait for simulation results from SD model. It became necessary to construct an additional structure (interface between models). Purpose of SD is to save the state of the cohorts in a output file (e.g. a spreadsheet or a database file) at successive points in time that are important from the point of view of calculations.

The main task of the SD module in hybrid is to generate the status (values) of individual cohorts. The state of each cohort affects the parameters that control DES module, i.e. the flux of objects representing events, the number of which depends on the number (expressed in persons) of each cohort. An example would be the generation of objects (representing patients) with characteristics that may indicate particular types of diseases. It is assumed that these conditions may have a frequency depending on cohort.

Below is illustration of structure of hybrid simulation:

- Input: Description of demographic processes
- SD model
- Output of SD – cohort predictions
- Input: Description of the relationship between cohorts and studied phenomenon
- DES model
- Output of DES – objects (individuals with properties)
- Results analysis – data resource usage, finance, well-being etc.

## 3 Modeling Approach

The paper presents a description of the key places in the model built in the Extendsim simulation environment [5], where model is graphically created by placing blocks from various libraries into the working area, linking them using connectors, adjusting parameters and, to a lesser extent, creating programs in the ModL (similar to C) language. It seems that the proposed solutions to the outlined problems will also be useful when using tools with similar architecture and methods of operation (using the push-pull approach). In the author's view, Extendsim is distinguished by the intuitive structure of the models, however, obtaining results requires additional efforts of the modeller.

The following suggestions for model implementation were made:

- Supplement the simulation run with an additional year (step) preceding the initial year to assign baselines to all cohorts based on the values stored in the spreadsheet.
- Use 'Go' connectors (triggering calculations on demand) and careful select and arrange signals triggering calculations (inputs and outputs) - in Equation blocks.
- Use Equation blocks rather than elementary Math blocks (such as Add, Subtract, Multiply, Divide) when it does not reduce the readability of the models.
- Use of internal databases (increase the calculation speed) for intermediate results.
- Store initial data and state variables in the course of the simulation (dynamic birth rates) and results in spreadsheets (preferably a single.XLSX file), in separate sheets but with permanently assigned sheet columns for each cohort.
- During the test experiments, it seems useful to store the initial states of individual cohorts and the simulations results concerning the state of cohorts in one worksheet, in distant groups of rows (first the initial states of the cohorts, then the space for the simulation results).

## 4 Technical Solutions

For the sake of simplicity, we consider a simulation model for female cohorts. The male part of the model differs only in that male cohorts do not affect the birth stream and have their coefficients (except duration of residence in the cohort).

First, the initial values are loaded from the spreadsheet cells (see Fig. 1) selected by additional control module (Fig. 2).

The situation will be similar for saving the results, but the space will be shifted by several dozen (e.g. 20) rows down (the offset will depend on the size of the space reserved for input data when testing the model). The task of the lower part of the submodel is to determine which simulation step number we are dealing with; The block on the left generates auxiliary objects at fixed intervals (1 year). The first auxiliary object will be generated as soon as the simulation is run. The variable signal2 changes its value from 0 to 1 only when the second helper object arrives – this fixes the row representing the initial data "set back by 1" and correctly sets the initial data in the Holding Tank (HT) blocks (representing the term "level" used in the continuous approach). The value of the signal variable is equal to 1, of course, after the arrival of the second object in the control stream, but it is calculated and used as a double "message" running calculations in "cohort" blocks for the first year actually processed (the point is that the streams incoming and outgoing from the blocks implementing the "level" concept, i.e. the HT, are updated after assigning initial values to the HT blocks (via the Init connector) and update (resulted in) also saving the new value of the HT. Without the above described action, the result of the calculation would cause the state of the HT block to change only after the arrival of a new control object, despite the fact that the streams associated with the HT are already updated (by the first message).

Secondly, the number of live births in a given step of the simulation should be calculated (see Fig. 3). The birth rate stream is directly affected by the increase in the number of people (in this case, women) aged 0 (belonging to the F0_4 cohort).

| ◢ | A | C | D | E | F | G | H |
|---|---|---|---|---|---|---|---|
| 2 | | F_0_4 | F_5_9 | F_10_14 | F_15_19 | F_20_24 | F_25_29 |
| 3 | Year | 1 | 2 | 3 | 4 | 5 | 6 |
| 5 | 2006 | 24028 | 25019 | 29648 | 37431 | 51220 | 53727 |
| | Population | Ageing | BirthsCoef | Migrations | DeathsCoef | Births | |

**Fig. 1.** Actual data on the initial population – excerpt from worksheet

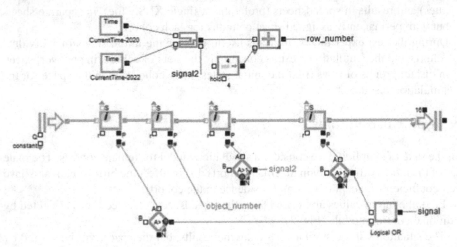

**Fig. 2.** Continuous model control submodule. It ensures that the initial values of the stock blocks (Holding Tank, outside the figure) and the inflow and outflow streams are correctly assigned

The non-standard settings of the Equation block are marked with red lines (see Fig. 3) on the input and output connectors. The markings denote that the calculations are performed only when a signal is sent to the block about a new, upcoming object in the object stream (a submodule built according to the DES approach, but this does not apply to the first object generated at the start of the simulation). The model contains a lot of other submodules, e.g. for the total inflow (birth + growing up + immigration).

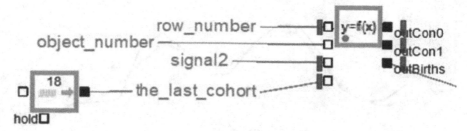

**Fig. 3.** Sub-module that calculates the birth rate in the first cohort (it uses data on the size of female cohorts and fertility rates (newborns, e.g. females))

## 5  Experiments

We want to test whether the model results are close to the real data. Therefore, deviations (differences) between the above values in individual cohorts will also be observed. From several tests, the long simulation time - 10 years was chosen to present (see Fig. 4). The actual data is the same as in the papers [11, 12, 16]. The data refers to the female population of the Wroclaw region (Poland), begins at 2006, was divided into cohorts 5 years wide (from cohort 0_4 to 80_84). The model is fed by size of individual cohorts, birth rates (for individual female cohorts), death rates, migration rates.

The dependence of the average percentage deviations on the number of years of simulation was determined (Fig. 4). A monotonic increase in deviations is visible. The simulation run (15 years) takes approx. 12 s (Intel Core i7–10700). Shapes of graphs of real cohort size distribution and model results (simulation) are similar (Fig. 5).

| Years of simulation | 1 | 2 | 3 | 4 | 5 | 6 | 7 | 8 | 9 | 10 | 11 | 12 | 13 | 14 |
|---|---|---|---|---|---|---|---|---|---|---|---|---|---|---|
| Average of % Error | 3,3 | 3,8 | 5,2 | 7,1 | 7,9 | 8,0 | 8,2 | 8,9 | 10,1 | 10,8 | 11,2 | 12,8 | 15,1 | 18,5 |

**Fig. 4.** The dependence of the average percentage deviation on the number of simulation length

## 6  Hybrid Simulation Model. Discussion of the Results

After providing fast calculations for continuous simulation, we will try to attach a DES module. Two cohorts were selected for the study: F_30_34 and F_35_36. The actual cohorts size and number of corresponding objects should be similar. However, the DES module is driven by the values of the current continuous simulation results, not by the actual size of the cohorts. It is noteworthy that the results of the simulation were more and more distant from the actual values (see Fig. 6).

The 15-year simulation lasted about 1 min and 10 s when the DES module was tested for the two cohorts mentioned above. Taking into account the previously given information about the length of the simulation time of the continuous module, we can conclude that obtaining results from the discrete part of the model requires much more computational time. We know that the time needed for calculations in a continuous

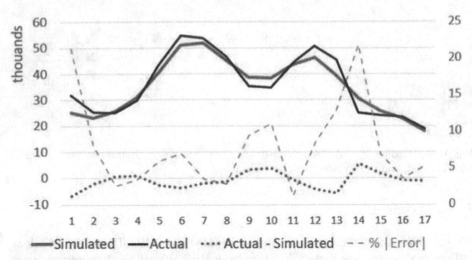

**Fig. 5.** Year 2016, the 10th year of simulation. Actual data, simulation results and deviations between them. The highest percentage deviations are in cohort no. 15 (F70_74), the deviations are irregular - an additional measure should be proposed

module does not depend on the size of the studied population, but only on the complexity of the model, i.e. the number of cohorts. In DES models, we will want to observe every object (person), even if the attributes assigned to it have stochastic values. We can assume that the larger the nominal numbers of people belonging to each cohort, the longer the time will be required to perform the calculations. As we can see in the example under consideration, even with a population of 40,000 people, the time of calculations in a continuous model is negligible in relation to the time needed for calculations in DES module, and this is the situation we wanted to achieve. In the models we used previously (210 cohorts), the calculation time was approx. 40 min, which made it particularly difficult to perform experiments for different variants – scenarios concerning, in particular, the values of birth and death rates and migration. We estimate that it will now be possible to include modules designed to model the above indicators so that they are a function of the current state of the cohorts or a function of their history. It should be noted that graphs (see Fig. 6) roughly reflect the dynamics of the actual states of the cohorts. However, as the simulation time horizon increases, the waveforms of the simulation results move away from the trajectory of the actual values and these are negative deviations (i.e., the simulation generates too few objects). These conclusions coincide with the observations on the results of the continuous model. We presume that they are due to the long, 5 years time span represented by the elementary cohort. Therefore, heterogeneities in the age distribution of individuals in a given elementary cohort are not taken into account.

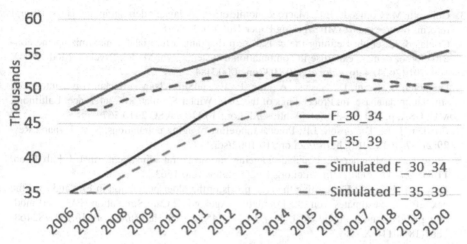

**Fig. 6.** A comparative chart of the actual size of cohorts and objects generated in subsequent years in DES module. Two female cohorts in the 4th decade of life. Negative deviations of the simulation results from the actual data are visible

## 7   Conclusions

Long calculation times and difficulties with the initial values assigning were observed in the past. These implementation shortcomings made it difficult to create effective hybrids of a continuous-discrete-event system. The architecture of the continuous model has been redesigned. Construction solutions were presented to overcome the current shortcomings of the implementation. A very high acceleration of calculations and a reduction in the simulation time were achieved. This enabled the implementation of an efficient and direct continuous-discrete connection (hybrid). The consistency of the state trajectories of the selected cohorts was also tested in terms of actual data and simulation results. The observed shortcomings prompt a modification of the model consisting in increasing the number of aggregated (composite) cohorts and shortening the time interval per elementary cohort. It corresponds to one of the recommendations [7].

**Acknowledgments.** ExtendSim blocks copyright © 2024 ANDRITZ Inc.

**Disclosure of Interests.** The authors have no competing interests to declare that are relevant to the content of this article.

## References

1. Ansah, J.P., et al.: Implications of long-term care capacity response policies for an aging population: a simulation analysis. Health Policy (New York) **116**(1), 105–113 (2014). https://doi.org///dx.doi.org/10.1016/j.healthpol.2014.01.006
2. Best, A., et al.: Premature mortality projections in the USA through 2030: a modelling study. Lancet Publ. Heal. **3**(8), E374-384 (2018). https://doi.org/10.1016/S2468-2667(18)30114-2

3. Colacelli, M., Corugedo, E.F.: Macroeconomic effects of Japan's demographics: can structural reforms reverse them? IMF Working Paper **18**(248), 1 (2018)
4. Crielaard, L., et al.: Refining the causal loop diagram: a tutorial for maximizing the contribution of domain expertise in computational system. Dyn. Model. Psychol. Meth. **29**(1), 169–201 (2024). https://doi.org/10.1037/met0000484
5. Diamond, B., Krahl, D., Nastasi, A., Tag, P.: ExtendSim advanced technology: -integrated simulation database. In: Proceedings of the 2010 Winter Simulation Conference, Baltimore, MD, USA, p. 100, 32–39 (2010). https://doi.org/10.1109/WSC.2010.5679178
6. Eberlein, R.L., Thompson, J.P.: Precise modeling of aging populations. Syst. Dynam. Rev. **29**(2), 87–101 (2013). https://doi.org/10.1002/sdr.1497
7. Giesecke, J., Meagher, G.: Population ageing and structural adjustment. Aust. J. Lab. Econ. **11**(3), 227–47 (2008). https://doi.org/10.26686/lew.v0i0.1663
8. Kingston, A., et al.: Forecasting the care needs of the older population in England over the next 20 years: estimates from the Population Ageing and Care Simulation (PACSim) modelling study. Lancet Public Heal. **3**(9), E447–E455 (2018). https://doi.org/10.1016/S2468-2667(18)30118-X
9. Lauf, S., et al.: Simulating demography and housing demand in an urban region under scenarios of growth and shrinkage. Environ. Plann. B. Plann. Des. **39**(2), 229–246 (2012). https://doi.org/10.1068/b36046t
10. McGrattan, E., et al.: On financing retirement, health, and long-term care in Japan. IMF Work. Reuters. **18**(249), 1–43 (2018). https://doi.org/10.5089/9781484384718.001
11. Mielczarek, B., Zabawa, J.: Modelling demographic changes using simulation: supportive analyses for socioeconomic studies. Socioecon. Plann. Sci. **74**, 100938 (2021). https://doi.org/10.1016/j.seps.2020.100938
12. Mielczarek, B., et al.: The impact of demographic trends on future hospital demand based on a hybrid simulation model. In: Proceedings of the 2018 Winter Simulation Conference, pp. 1476–1487. https://doi.org/10.1109/WSC.2018.8632317
13. Semwanga, A.R., Nakubulwa, S., Adam, T.: Applying a system dynamics modelling approach to explore policy options for improving neonatal health in Uganda. Health Res. Policy Syst. **14**, 35 (2016). https://doi.org/10.1186/s12961-016-0101-8
14. Statistics Poland Homepage. Główny Urząd Statystyczny. https://stat.gov.pl/en/
15. van Sonsbeek, J.: Micro simulations on the effects of ageing-related policy measures. Econ. Modell. **27**, 968–979 (2010). https://doi.org/10.1016/j.econmod.2010.05.004
16. Zabawa, J., Mielczarek, B.: An attempt to replace system dynamics with discrete rate modeling in demographic simulations. In: Pashinsky, M., et al. (eds.) ICCS 2021. LNCS, vol. 12745, pp. 269–283. Springer, Cham (2021). https://doi.org/10.1007/978-3-030-77970-2_21

# PCA Dimensionality Reduction
# for Categorical Data

Aleksander Denisiuk[(✉)] [iD]

University of Warmia and Mazury in Olsztyn, ul. Słoneczna 54,
10-710 Olsztyn, Poland
denisiuk@matman.uwm.edu.pl

**Abstract.** The purpose of the article is to develop a new dimensionality reduction algorithm for categorical data. We give a new geometric formulation of the PCA dimensionality reduction method for numerical data that can be effectively transferred to the case of categorical data with the Hamming metric.

**Keywords:** PCA · Hamming metric · weighted Hamming metric · categorical data · dimensionality reduction · classification

## 1 Introduction

One of the objectives of principal component analysis (PCA) is to reduce the dimension of the data space while retaining as much information as possible. The standard algorithm (see, for instance [12]) consists of calculating the eigenvectors of the covariant (correlation) matrix and using it as a new basis in the space of data. Coordinates of data vectors in this new basis are the principle components. The major principle component corresponds to the eigenvector with the largest eigenvalue, the minor component—to the eigenvector with the smallest eigenvalue. The dimensionality reduction involves discarding minor components.

However this algorithm is not applicable in case of categorical data, where the structure of linear space is not available.

In this article we propose a new interpretation of the PCA dimensionality reduction and transfer it to a set of categorical data. Namely, consider an affine transform of the data space that minimizes the total relative squared inner-class distance (Fig. 1). It turns out that the major principle component will be scaled with the minimal multiplier, while minor component with the maximal one.

The problem of dimensionality reduction can be brought to finding a scaling that minimizes the relative total squared distance. The direction with the most scaling multiplier corresponds to the minor component that can be dropped with a minimal information lost.

L. Franco et al. (Eds.): ICCS 2024, LNCS 14834, pp. 179–186, 2024.
https://doi.org/10.1007/978-3-031-63759-9_22

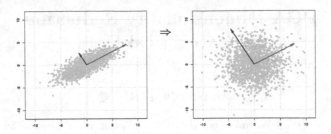

**Fig. 1.** Scaling that minimizes total relative inner-class squared distance

Such interpretation can be transferred to a space of categorical data with the weighted Hamming metric.

To prove the concept we perform numeric experiments on three datasets. Experiments show that discarding the features in order according to our method always results in the loss of a minimum amount of information. Discarding the features in reverse order results in maximal information loss.

The rest of the paper is organized as follows. In Sect. 2 we shortly recall basic related works paying main attention to recent works concerning non-numeric data. Section 3 contains our new interpretation of PCA dimensionality reduction for numeric data. This interpretation in transferred to categorical data in the Sect. 4. To verify our concept we performed numerical experiments that are described in the Sect. 5. Finally, we give some concluding remarks in the Sect. 6.

## 2    Related Works

Developed for numerical continuous data in the pioneering works of Pearson [18] and Hotelling [11], PCA has found many applications in many fields of data analysis. The method and recent developments for continuous numerical data are described in book [12] and review [13].

We focus on extensions of the PCA to discrete data. The most known is the correspondence analysis (see, for instance [12, section 5]) which deals with the principal components of the normalized contingency matrix.

In articles [7,10] the PCA was applied to a binary data. In case of ordinal data the authors of [14] suggested a variant of PCA based on Spearman's and Kendall's rank correlation coefficients. In our algorithm, the data features re not ordered and can have arbitrary cardinalities.

Let us also mention some recent papers that develop PCA for discrete data with additional complex structure as intervals or histograms [2,4,16]. In our parer we do not assume any additional structure defined on the data.

The PCA is often formulated as an optimization problem: maximizing dispersion of data projection, optimal approximation of the data with a linear manifold, finding projection that maximize the total inner-class squared distance. To the best of our knowledge the optimization problem suggested in this paper was not directly considered before. A characteristic feature that distinguishes our algorithm from other approaches: first we define a minor feature, and the other algorithms start by determining the most important feature.

The weighted Hamming metric itself was recently used for unsupervised [9] and supervised [8] metric learning for numeric-categorical data. Application of the weighted Hamming metric to the problem of dimensionality reduction of categorical data seems to be new.

## 3    Reformulation of PCA Dimensionality Reduction for Numerical Data

In this section we will show that minimizing of the total relative squared inner-class distance allows us to determine the minor principle component.

Assume that each data instance $x$ has $n$ numerical features, i.e. $x \in \mathbb{R}^n$ with the standard Euclidean distance. The distance is invariant with respect to translations and rotations. So, let us make two following assumptions:

1. The features are uncorrelated,
2. The data has a multivariate normal distribution centered at the origin.

That means that distribution function is as follows:

$$f(x) = \frac{\exp\left(-\frac{1}{2}x^T \Sigma^{-1} x\right)}{\sqrt{(2\pi)^n \det \Sigma}},$$

where $\Sigma$ is the correlation matrix, which in case of uncorrelated data is diagonal $\Sigma = \mathrm{diag}(\lambda_1, \ldots, \lambda_n)$, $\Sigma^{-1} = \mathrm{diag}(\lambda_1^{-1}, \ldots, \lambda_n^{-1})$, $\det \Sigma = \lambda_1 \cdots \lambda_n$, and $\lambda_i > 0$ for $i = 1, \ldots, n$. The total inner-class squared distance is

$$H = \int_{\mathbb{R}^n \times \mathbb{R}^n} \mathrm{dist}^2(x, y) f(x) f(y)\, dx\, dy,$$

where $dx = dx_1 \ldots dx_n$, $dy = dy_1 \ldots dy_n$.

We are to find a scaling $(x_1, \ldots, x_n) \mapsto (w_1 x_1, \ldots, w_n x_n)$, where $w_i \geq 0$ for $i = 1, \ldots, n$, that minimizes the total inner-class squared distance:

$$H(w) = \sum_{i=1}^n w_i^2 \int_{\mathbb{R}^n \times \mathbb{R}^n} (x_i - y_i)^2 f(x) f(y)\, dx\, dy. \tag{1}$$

Since the function $H(w)$ is homogenious we add a constraint on the weight $w$:

$$\sum_{i=1}^n w_i = 1. \tag{2}$$

The restriction of $H(w)$ to (2) is the *total relative inner-class squared distance*.

By the standard calculation the coefficient at $w_i^2$ in (1) equals $z_i = 2\lambda_i$. So, the minimization problem is as follows:

$$\begin{cases} \sum_{j=1}^n w_j^2 \lambda_j \to \min, \\ \sum_{j=1}^n w_j = 1, \quad w_j \geq 0 \text{ for } j = 1, \ldots, n. \end{cases}$$

One can solve it with the method of Lagrange multipliers:

$$w_i = \left(\sum_{j=1}^{n} \lambda_j^{-1}\right)^{-1} / \lambda_i, \quad i = 1, \ldots, n.$$

Specifically, the minor component (of minimal $\lambda_i$) has the maximal multiplier.

## 4  Dimensionality Reduction for Categorical Data

In this section we transfer considerations from the last section to the case of categorical data.

Assume that the dataset $\mathbf{X}$ of $M$ instances is given. Let each instance $x \in \mathbf{X}$ has $n$ categorical features of finite cardinalities $a_1, \ldots, a_n$ respectively, $x = (x_1, \ldots, x_n)$. We use the standard Hamming metric as a distance on $\mathbf{X}$:

$$\mathrm{dist}_h(x, y) = \sum_{i=1}^{n} \mathrm{diff}(x_i, y_i),$$

where $x, y \in \mathbf{X}$, and $\mathrm{diff}(\alpha, \beta) = \begin{cases} 1, & \text{if } \alpha \neq \beta, \\ 0, & \text{if } \alpha = \beta. \end{cases}$

Let us also make an assumption that the dataset is divided into $c$ classes, $\mathbf{X} = C_1 \cup \cdots \cup C_c$. The total inner-class distance is

$$G = \frac{1}{M^2} \sum_{k=1}^{c} \sum_{x,y \in C_k} \mathrm{dist}_h(x, y).$$

Do define a "scaling", let us introduce the weights vector $u = (u_1, \ldots, u_n) \in \mathbb{R}^n$, where $u_i \geq 0$ for $i = 1, \ldots, n$ and the weighted Hamming distance:

$$\mathrm{dist}_{h,u}(x, y) = \sum_{i=1}^{n} u_i \, \mathrm{diff}(x_i, y_i).$$

Consider minimization problem for the function

$$G(u) = \frac{1}{M^2} \sum_{k=1}^{c} \sum_{x,y \in C_k} \mathrm{dist}_{h,u}(x, y) = \sum_{i=1}^{n} u_i \left( \frac{1}{M^2} \sum_{k=1}^{c} \sum_{x,y \in C_k} \mathrm{diff}(x_i, y_i) \right) \quad (3)$$

with the following constraint on weights $u$:

$$\sum_{i=1}^{n} u_i = 1, \quad (4)$$

which is exactly the same as (2), and call the restriction of $G(u)$ to the hyperplane (4) the *total relative inner-class distance*.

Denoting coefficient at $u_i$ in (3) by $s_i$, we obtain the following minimization problem:

$$\begin{cases} \sum_{i=1}^{n} u_i s_i \to \min, \\ \sum_{i=1}^{n} u_i = 1, \quad u_j \geq 0 \text{ for } i = 1, \ldots, n, \end{cases} \quad (5)$$

**Algorithm 4.1.** Reduction of $m$ dimensions

---
**Require:** the dimension of dataset $\mathbf{X}$ is $n$, $n > m$
**Ensure:** the dimension of dataset $\mathbf{X}$ is $n - m$
   $s \leftarrow 0$
  **while** $s < m$ **do**
     solve the optimization problem (5) and discard the minor component
     $s \leftarrow s + 1$
  **end while**

---

which is known as the linear programming problem. The objective function reaches its optimal value at one of the vertices of the polytope defined by the constraint. Hence the optimal vector has the form $u_{\text{opt}} = (0, \ldots, 0, 1, 0, \ldots 0)$, i.e. all the coordinates but one are zeroes. The feature $k$ that corresponds to coordinate $u_{\text{opt},k} = 1$ is called *the minor component* and can be discarded first in dimensionality reduction.

Repeating this procedure after the feature $k$ reduction, one determines the next minor feature, and so on. We summarize the method in the algorithm 4.1.

## 5    Numerical Experiments

To illustrate the concept a few R scripts have been created. The code is available as a project on Gitlab at https://gitlab.com/adenisiuk/pca.

The purpose of our test is to show that the algorithm allows to reduce dimensionality while retaining as much information as possible. To do this we consider the classification problem. We classified data with complete set of features and then we discarded features one by one in different orders. First, according to the proposed algorithm: starting from the most minor feature, this order is referred as *pca* order (red line on the figures). Second order is reverse to the pca order: starting from most major features. We call this order *acp* order (blue line on the figures). We also performed two tests with random order of features discarding, the *sample* order (green lines on the figures).

Implementations of three classifiers: random forest, SVM and XGBoost in R were used in experiments: [5, 15, 17]. For the random forest classifier an average result for 100 tests is presented.

We use the $F_1$Score as a measure of classification accuracy. Some of datasets have more than two classes. In all the tests we define $F_1$Score as

$$F_1 \text{Score} = \frac{2 \cdot \text{total\_presicion} \cdot \text{total\_recall}}{\text{total\_presicion} + \text{total\_recall}},$$

where total\_presicion and total\_recall are the sums of respectively presicion and recall for all the classes.

We considered the following three datasets: the Car Evaluation [3], the Congressional Voting Records [6] and the Tic-Tac-Toe Endgame [1]. All the tested datasets were split into train (80%) and test (20%) parts.

One can see that in all the tests the *pca* order always gives the minimal information loss while the *acp* order gives the maximal lost.

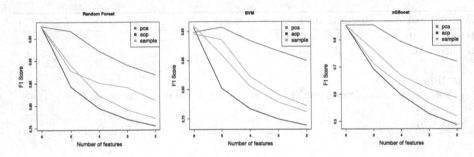

**Fig. 2.** Classification accuracy for the Car Evaluation dataset (Color figure online)

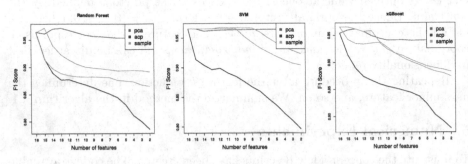

**Fig. 3.** Classification accuracy for the Congressional Voting Records dataset (Color figure online)

### 5.1 Car Evaluation Dataset

The dataset contains 1728 instances. Each instance has 6 features of cardinalities respectively 4, 4, 4, 3, 3, 3. The dataset is split into 4 classes [3].

The results of classification are presented at the Fig. 2. The *pca* order for this dataset is 3, 2, 1, 5, 4, 6. Orders sampled in tests were as follows: 4, 1, 3, 2, 6, 5 and 1, 5, 3, 6, 4, 2 for the random forest classifier, 1, 6, 4, 3, 5, 2 and 5, 4, 3, 1, 6, 2 for SVM, 2, 4, 1, 3, 5, 6 and 4, 2, 6, 3, 5, 1 for XGBoost.

One can observe the for all classifiers the *pca* order has minimal and the *acp* order has maximal accuracy loss when discarding features.

### 5.2 Congressional Voting Records Dataset

The dataset contains 435 instances. Each instance has 16 features of cardinalities 3. According to the data description we interpreted values "?" as the third option in voting. The dataset is split into 2 classes [6].

The results of classification are presented at the Fig. 3. The *pca* order for this dataset is 2, 10, 16, 11, 1, 15, 13, 6, 14, 9, 7, 12, 5, 8, 3, 4. Orders sampled in tests were as follows: 4, 16, 6, 10, 7, 1, 5, 14, 2, 3, 12, 13, 11, 9, 8, 15 and 10, 4, 8, 6, 16, 5, 2, 7, 15, 14, 9, 1, 12, 13, 11, 3 for the random forest classifier, 1, 11, 7, 9, 14, 2, 4, 13, 8, 10, 16, 6, 15, 5, 3, 12 and 9, 10, 5, 8, 12, 16, 2, 14, 4, 11, 15,

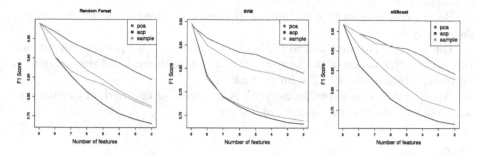

**Fig. 4.** Classification accuracy for the Tic-Tac-Toe Endgame (Color figure online)

13, 6, 7, 3, 1 for SVM, 12, 15, 7, 9, 4, 11, 1, 8, 14, 3, 16, 10, 6, 13, 2, 5 and 11, 16, 5, 7, 13, 4, 9, 14, 2, 6, 3, 8, 10, 15, 12, 1 for XGBoost.

Again, the *pca* order has minimal and the *acp* order has maximal accuracy loss when discarding features. Note also very stable classification for the *pca* order and drastic drop in accuracy after discarding the component 4, which is most significant according to our algorithm. This is especially noticeable at the *sample* orders for random forest classifier: the 4th feature was the first and the second one.

### 5.3    Tic-Tac-Toe Endgame Dataset

The dataset contains 958 instances. Each instance has 16 features, each one is of cardinality 3. The dataset is split into 2 classes [1].

The results of classification are presented at the Fig. 4. The *pca* order for this dataset is 2, 4, 8, 6, 9, 3, 7, 1, 5. Orders sampled in tests were as follows: 5, 6, 2, 3, 4, 8, 1, 9, 7 and 4, 9, 3, 6, 5, 1, 8, 7, 2 for the random forest classifier, 5, 4, 7, 3, 9, 2, 8, 1, 6 and 7, 8, 3, 6, 4, 9, 1, 2, 5 for SVM, 2, 8, 6, 3, 1, 7, 9, 5, 4 and 3, 4, 7, 5, 2, 1, 9, 6, 8 for XGBoost.

As in two previous experiments, the *pca* order has minimal and the *acp* order has maximal accuracy loss when discarding features. The accuracy graphs for this dataset is more monotonic than for the Congressional Voting Records. This is probably related to greater uncorrelatedness of the features. As in the previous example, let us notice drop in accuracy after discarding the feature 5: see two *sample* orders for the random forest and the SVM classifiers where this feature is the first one.

## 6    Conclusion and Future Work

In this article we propose a new geometric interpretation of the PCA dimensionality reduction algorithm and transfer it to categorical data. The algorithm involves determining and discarding minor features.

Numerical experiments confirm that the method allows to discard features with minimal lost of information, at least for the classification problem.

So, we can suggest this method of dimensionality reduction for categorical data analysis.

However, unlike the classical PCA, our method does not provide any quantitative characterization of the amount of information discarded. Developing such an interpretation is one of the future tasks.

Another direction of the future work is extension of the method to data that have both numerical and categorical features.

# References

1. Aha, D.: Tic-Tac-Toe Endgame. UCI Machine Learning Repository (1991)
2. Bock, H. H., Diday, E.: Analysis of Symbolic Data: Exploratory Methods for Extracting Statistical Information from Complex Data. Springer Science & Business Media (2012)
3. Bohanec, M.: Car Evaluation, UCI Machine Learning Repository (1997)
4. Brito, P.: Symbolic data analysis: another look at the interaction of data mining and statistics. Wiley Interdisc. Rev. Data Min. Knowl. Discovery 4(4), 281–295 (2014)
5. Chen, T., et al.: XGBoost: Extreme Gradient Boosting (2023)
6. Congressional Voting Records: UCI Machine Learning Repository (1987)
7. Cox, D.R.: The analysis of multivariate binary data. Appl. stat. 113–120 (1972)
8. Denisiuk, A.: Weighted hamming metric and KNN classification of nominal-continuous data. In: Mikyška, J., de Mulatier, C., Paszynski, M., Krzhizhanovskaya, V.V., Dongarra, J.J., Sloot, P.M. (eds.) Computational Science - ICCS 2023, pp. 306–313. Springer Nature Switzerland, Cham (2023)
9. Denisiuk, A., Grabowski, M.: Embedding of the hamming space into a sphere with weighted quadrance metric and c-means clustering of nominal-continuous data. Intell. Data Anal. 22(6), 1297001314 (2018)
10. Gower, J. C.: Some distance properties of latent root and vector methods used in multivariate analysis. Biometrika 53(3-4), 325–338 (1966)
11. Hotelling, H.: Analysis of a complex of statistical variables into principal components. J. Educ. Psychol. 24(6), 417 (1933)
12. Jolliffe, I. T.: Principal Component Analysis. Springer (2002)
13. Jolliffe, I.T., Cadima, J.: Principal component analysis: a review and recent developments. Philos. Trans. R. Soc. Math. Phys. Eng. Sci. 374(2065), 20150202 (2016)
14. Korhonen, P., Siljamäki, A.: Ordinal principal component analysis theory and an application. Comput. Stat. Data Anal. 26(4), 411–424 (1998)
15. Liaw, A., Wiener, M.: Classification and regression by randomForest. R News 2(3), 18–22 (2002)
16. Makosso-Kallyth, S.: Principal axes analysis of symbolic histogram variables. ASA Data Sci. J. 9(3), 188–200 (2016)
17. Meyer, D., Dimitriadou, E., Hornik, K., Weingessel, A., Leisch, F.: e1071: misc Functions of the Department of Statistics, Probability Theory Group (Formerly: E1071), TU Wien. R Package Version 1.7-12 (2022)
18. Pearson, K.: On lines and planes of closest fit to systems of points in space. London, Edinburgh Dublin Philos. Mag. J. Sci. 2(11), 559–572 (1901)

# EsmTemp - Transfer Learning Approach for Predicting Protein Thermostability

Adam Sułek[1(✉)], Jakub Jończyk[1,2], Patryk Orzechowski[1,3],
Ahmed Abdeen Hamed[1], and Marek Wodziński[4,5(✉)]

[1] Sano Centre for Computational Medicine, 30-072 Kraków, Poland
a.sulek@sanoscience.org
[2] Jagiellonian University Medical College, 30-688 Kraków, Poland
[3] University of Pennsylvania, Philadelphia, PA 19104, USA
[4] AGH University of Science and Technology, 32059 Kraków, Poland
wodzinski@agh.edu.pl
[5] University of Applied Sciences Western Switzerland, Sierre, Switzerland

**Abstract.** Protein thermostability is one of the most important features of bio-engineered proteins with significant scientific and industrial applications. Unfortunately, obtaining thermostable proteins is both expensive and complex. Recent advances in Protein Language Models (pLM) offer promising framework for sequence-to-sequence problems, especially in the realm of protein thermostability prediction. In this work, we present EsmTemp, a transfer learning model based on the ESM-2 pLM architecture. EsmTemp undergoes training on a meticulously curated dataset comprising 24,000 protein sequences with known melting temperatures. A rigorous evaluation, conducted through a 10-fold cross-validation, yields a coefficient of determination ($R^2$) of 0.70 and a mean absolute error of 4.3°C. These outcomes highlight how pLM has the potential to advance our understanding of protein thermostability and facilitate the rational design of enzymes for various applications.

**Keywords:** protein thermostability · protein language models · transfer learning · ESM-2 · artificial neural network

## 1 Introduction

Thermostable proteins are important for biotechnology, medicine, and pharmacy [8,18]. Key examples include the heat-resistant DNA polymerase employed in Polymerase Chain Reaction [16], proteases used in the synthesis of angiotensin-converting-enzyme inhibitors [2], or Ferritins harnessed for crafting nanocarriers. Enhanced enzyme heat resistance improves industrial processes, resulting in faster reactions, enhanced substrate solubility, and reduced contamination risks [12]. Thermostable proteins are often sourced from thermophilic organisms or engineered through mutagenesis and directed evolution. Despite their effectiveness, these methods have limitations, such as time investment and trial-and-error

© The Author(s), under exclusive license to Springer Nature Switzerland AG 2024
L. Franco et al. (Eds.): ICCS 2024, LNCS 14834, pp. 187–194, 2024.
https://doi.org/10.1007/978-3-031-63759-9_23

reliance. Determining the melting temperature (Tm) is a common method for assessing protein thermostability. It indicates the temperature at which half of the protein unfolds. However, accurate experimental Tm determination demands substantial quantities of high-purity proteins or cell cultures, making it labor-intensive and costly. Hence, there's a critical need to develop a more efficient and cost-effective method for predicting protein Tm.

The protein structure is determined by the sequence of amino acids, which directly impact their function and properties. Proteins with similar amino acid sequences not only share corresponding biochemical functions, but also exhibit similar characteristics, even if they are found in different organisms. On the other hand, single amino acid changes can profoundly alter protein structure, interactions, and properties, as exemplified by the critical impact of point mutations on the behavior of some proteins [17]. This observation is essential for predicting protein structures and properties using computational techniques. Over the past few years, remarkable progress has been made in the study of proteins, largely due to the utilization of machine learning (ML) approaches in the field of natural language processing (NLP). Through the translation of the amino acid sequence into tokenized vectors, models become capable of capturing pertinent protein features. Sequence embeddings serve as a versatile tool for predicting various properties of proteins, both at the global and local levels [15]. Protein Language Models (pLM) developed so far, such as Evoformer, ProteinBERT, ESM-1b and ProtGPT2 are successfully used in automated function or structure predictions [1,3,6,14]. Prediction of thermostability using ML and pLM offers insights for protein engineering.

Several classification algorithms and tools have been developed to distinguish proteins into thermostable and thermolabile based on their sequences. Support Vector Machines, Random Forests and Gradient Boosting have reported accuracy rates even up to 85% [9,20]. Although certain classification-based predictors have shown satisfactory performance, they are not a suitable solution for accurately determining the precise Tm value. ProTstab2, developed with the LightGBM framework and protein sequence featurization, stands out among the available regression models [19].

ProThermDB held about 10k entries on protein melting temperatures, serving as the main thermostability data repository [13]. The Meltome Atlas database enriched existing data by including Tm information for 48k proteins across 13 species and a human cell line dataset of 13k cases from 14 cell lines [5]. Still, the thermostability dataset is considerably smaller than >200M protein sequences stored in UniProt Knowledgebase, underscoring a substantial data deficit.

The introduction of pLM-based transfer learning for evaluating protein thermostability through sequence analysis was first presented in the DeepSTABp tool [7]. This model combines multilayer perceptron with a pre-trained transformer-based model, ProtTrans-XL. In addition to protein sequences, it considers measurement context (cell lysate or whole cell) and the source organism's Optimal Growth Temperature (OGT).

ESM-2 protein language models are constructed using the transformer architecture, specifically tailored for training on protein sequences [4,11]. Recognizing amino acid sequences as a unique language, transformer models, similar to BERT-like architectures, have demonstrated proficiency in capturing crucial features of proteins [1]. The input exclusively comprised protein sequences, encompassing the 20 standard amino acids. Our tool EsmTemp predicts Tm values for various proteins by leveraging pre-trained ESM2 (650M) embeddings and fine-tuning. By combining the data from Meltome Atlas and ProThermDB, we curated a comprehensive dataset characterized by a broad Tm value range and standardized experimental protocols. EsmTemp's effectiveness has been validated in down-stream task pertaining to protein thermostability, demonstrating its usability in amino acid sequence analyses. The code, the data and additional figures are provided here: https://github.com/SanoScience/esm_temp

## 2  Materials and Methods

The dataset combined data from ProThermDB and Meltome Atlas [5,13]. The aggregated data underwent preprocessing to ensure consistency and relevance for subsequent analysis. Experimental protocols were standardized by retaining only measurements conducted on cell lysates. The dataset was refined by excluding protein sequences with Tm values outside the range of 30°C to 98°C and lengths below 100 or exceeding 900 amino acids. After eliminating duplicate entries, the remaining sequences underwent clustering via the CD-HIT algorithm [10] to ensure that the training and test sets do not have identical or near identical examples. To minimize redundancy, we set a sequence similarity threshold of 0.7 and a minimum cluster size of 3, while keeping all other parameters at their default values. As a result of this process, we obtained a curated dataset comprising 24,472 distinct amino acid sequences. Each sequence is accompanied by its respective source organism and Tm value.

We utilized ESM-2, a model equipped with 650M parameters and 1280-dimensional embeddings for amino acids. This allowed us to represent each protein through pooling the embeddings from the amino acids, with predictions made using an output layer containing a single neuron [11]. The network weights were optimized using AdamW with a learning rate of 1e-4. MAE quantifies a model's accuracy in predicting continuous variables by assessing the average error magnitude. The model's performance was assessed with $(R^2)$, determining the variance in the dependent variable that can be predicted by the independent variable.

Due to GPU memory constraints (40GB, NVIDIA A100 units), a pseudo-batch approach was employed, where only 1 sequence was included in each batch. The gradient updates in a backward step were performed after accumulating the loss over a mini-batch containing 16 proteins. Our investigation involved gradually unfreezing three layers of the ESM-2 model to analyze the advantages of fine-tuning in transfer learning. This strategy aims to refine the pre-trained model's parameters for the Tm prediction task. We used 10-fold cross-validation to address training variations.

# 3 Results

In this study, three separate hypotheses were tested. First, we verified how data scaling affected the performance of the model. Secondly, we explored potential benefits from fine-tuning process on the range of unfrozen layers (Fig. 1). Thirdly, we evaluated the model's predictive performance both on a broad scale and within specific organisms.

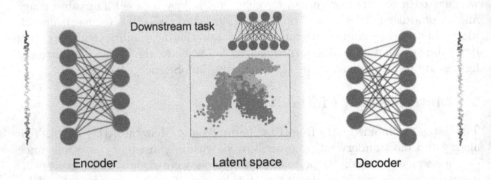

**Fig. 1.** Leveraging a transfer learning approach with the pLM autoencoder model involves utilizing a pre-trained latent space and its application to the downstream task.

Normalization techniques enhance the convergence and stability of optimization algorithms used in regression tasks. We examined how normalization and scaling methods, such as log scaling, square scaling, and min-max normalization, affect model prediction performance compared to non-scaling labels. Log and square scaling had no positive impact on the model's performance, as shown by Table 1's MAE and $R^2$ values. Normalizing the target variable consistently improved training outcomes, so we adopted min-max scaled labels for subsequent experiments.

The ESM-2 model has a complex architecture with multiple layers. By unfreezing specific layers, sequence embeddings can be fine-tuned to enhance model performance. Carefully controlling unfrozen layers is crucial to prevent gradient decreases and ensure effective learning. We used selective layer unfreezing in transfer learning framework and an extra Fully Connected layer to tailor the model for the study down-stream task. The effectiveness of ESM-2 in transfer learning for thermostability prediction was discerned by utilizing raw embeddings, which were subsequently passed as input to the regression task. We noticed a relatively strong correlation ($R^2$=0.64), though accompanied by a high mean absolute error. By unfreezing the final layers of ESM-2, it becomes possible to fine-tune not only the raw embeddings but also specific layers within the model, resulting in improved predictions. The results showed enhanced $R^2$ values and a corresponding decrease in MAE, as illustrated in Table 1.

**Table 1.** The results of sequence stability prediction with ESM-2 (650M parameters) model with different transfer learning and scaling procedures.

| Method | MAE | $R^2$ | Scaling Function |
|---|---|---|---|
| ESM-2 raw embeddings | $8.86 \pm 0.105$ | $0.64 \pm 0.012$ | Min-max |
| Unfrozen 0 | $4.90 \pm 0.061$ | $0.63 \pm 0.018$ | Min-max |
| Unfrozen 1 | $4.90 \pm 0.060$ | $0.63 \pm 0.015$ | Min-max |
| Unfrozen 2 | $4.42 \pm 0.096$ | $0.70 \pm 0.025$ | Min-max |
| Unfrozen 3 | $4.33 \pm 0.076$ | $0.70 \pm 0.018$ | Min-max |
| Unfrozen 3 | $4.57 \pm 0.051$ | $0.67 \pm 0.005$ | Non-scaling |
| Unfrozen 3 | $4.50 \pm 0.069$ | $0.68 \pm 0.019$ | Log |
| Unfrozen 3 | $4.54 \pm 0.076$ | $0.67 \pm 0.018$ | sqrt |

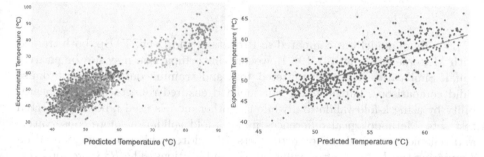

**Fig. 2.** The correlation analysis depicts the relationship between melting temperatures with the prediction for the best fine-tune ESM-2 model across specific organisms (A) and with a focus on fine-tuning exclusively on Human lysates (B).

In this study, we experimented with unfreezing up to three consecutive layers. We found that unfreezing three layers yielded the most favorable results, leading us to adopt this approach for subsequent experiments. The final model achieves MAE of $4.3°C$ and an $R^2$ score of $0.704$ on a 10-fold validation procedure (Table 1, Fig. 2A.). While the model prediction demonstrates a relatively strong correlation during global assessment of protein thermostability over the full temperature range, the visualization with color coding for each species reveals lower or even no correlation within the species (Table 2, Fig. 2A). To further explore the targeted applicability of EsmTemp, aiming specifically for improved correlation within organisms and potential applications for point mutations, we conducted fine-tuning within a single organism, utilizing the most effective strategies from previous experiments. It shows that the availability of more data may facilitate the training of a more accurate model within the species. Considering the extensive datasets on protein thermostability across diverse organisms, the variation in protein sequences, which can be attributed to their unique biological functions, presents difficulties in accessing information regarding their thermostability. The results summarized in Table 2 clearly showed that despite a similar temperature range and a very similar Tm distribution, the model is able to accurately capture the relationship between the amino acid sequence and thermostability only in the case of human proteins.

**Table 2.** The results of Tm prediction $R^2$ using the optimal fine-tune ESM-2 model for specific organisms.

| Lysate from organism | Correlation $R^2$ | Number of cases |
|---|---|---|
| Human | 0.49 | 7997 |
| Arabidopsis thaliana seedling | 0.26 | 2260 |
| Caenorhabditis elegans | 0.15 | 2917 |
| Mus musculus | 0.10 | 3870 |

EsmTemp model was compared to ProtStab2 and DeepSTABp, both trained with Meltome Atlas data [7,19]. However, this comparison may not be entirely equal. Similarities between unfiltered test and training data can compromise valid comparisons. Furthermore, our method ensured robustness and repeatability by using k-fold validation, while the other models used randomly selected test sets. Methodological differences make k-fold validation more conservative and reliable than single random test sets. To address this issue, we focused on comparing the $R^2$ correlation values in the publications. The $R^2$ score for ProtStab2 on a blind test subset was 0.58, while DeepSTABp performed better with a score of 0.80. While EsmTemp achieved an $R^2$ score of 0.70, surpassing ProtStab2, it did not demonstrate the same level of correlation as DeepSTABp.

## 4    Conclusions

This study examines the possibilities and constraints of employing transfer learning from ESM-2 embeddings and ESM-2 fine-tuning for the prediction of protein melting temperature (Tm). Protein language models are advanced neural networks that undergo training using vast databases of protein sequences such as UniProt. Typically, pLMs are trained by predicting missing amino acids or motifs in protein sequences, enhancing their understanding of protein properties. As a result, pLMs serve as valuable repositories of parameterized data, adept at addressing various tasks within their domain. Such knowledge transfer allows to obtain predictive models that are more effective than the original model training on a limited data resource. Our results demonstrate the model's accurate prediction of Tm for a vast landscape of proteins, with a mean absolute error of 4.3°C. Furthermore, our model does not necessitate external features, such as the source of the experiment or the OGT. The limitations of our approach lie in predicting the effects of point mutations on Tm, as our model relies on the global sequence information and does not account for the local structural features. Taking this into consideration, our future plans involve implementing more advanced feature extraction techniques, like graph convolutions, to enhance the precision and practicality of Tm prediction.

**Acknowledgements.** The publication was created within the project of the Minister of Science and Higher Education "Support for the activity of Centers of Excellence established in Poland under Horizon 2020" on the basis of the contract number MEiN/2023/DIR/3796. This project has received funding from the European Union's Horizon 2020 research and innovation programme under grant agreement No 857533. This publication is supported by Sano project carried out within the International Research Agendas programme of the Foundation for Polish Science, co-financed by the European Union under the European Regional Development Fund. This research was possible due to the support of PLGrid infrastructure grant plgsano4-gpu and PLG/2023/016261 on the Athena supercomputer cluster.

# References

1. Brandes, N., Ofer, D., Peleg, Y., Rappoport, N., Linial, M.: ProteinBERT: a universal deep-learning model of protein sequence and function. Bioinformatics **38**, 2102–2110 (2022). https://doi.org/10.1093/bioinformatics/btac020
2. Cheung, I.W.Y., Nakayama, S., Hsu, M.N.K., Samaranayaka, A.G.P., Li-Chan, E.C.Y.: Angiotensin-I converting enzyme inhibitory activity of hydrolysates from oat (*Avena sativa*) proteins by in silico and in vitro analyses. J. Agric. Food Chem. **57**, 9234–9242 (2009). https://doi.org/10.1021/jf9018245
3. Ferruz, N., Schmidt, S., Höcker, B.: ProtGPT2 is a deep unsupervised language model for protein design. Nat. Commun. **13**, 4348 (2022). https://doi.org/10.1038/s41467-022-32007-7
4. Hu, M., et al.: Exploring evolution-aware and-free protein language models as protein function predictors (2022). http://arxiv.org/abs/2206.06583
5. Jarzab, A., et al.: Meltome atlas-thermal proteome stability across the tree of life. Nat. Methods **17**, 495–503 (2020). https://doi.org/10.1038/s41592-020-0801-4
6. Jumper, J., et al.: Highly accurate protein structure prediction with AlphaFold. Nature **596**, 583–589 (2021). https://doi.org/10.1038/s41586-021-03819-2
7. Jung, F., Frey, K., Zimmer, D., Mühlhaus, T.: DeepSTABp: a deep learning approach for the prediction of thermal protein stability. Int. J. Mol. Sci. **24**, 7444 (2023). https://doi.org/10.3390/ijms24087444
8. Kaneko, H., Minagawa, H., Shimada, J.: Rational design of thermostable lactate oxidase by analyzing quaternary structure and prevention of deamidation. Biotech. Lett. **27**, 1777–1784 (2005). https://doi.org/10.1007/s10529-005-3555-2
9. Ku, T., et al.: Predicting melting temperature directly from protein sequences. Comput. Biol. Chem. **33**, 445–450 (2009). https://doi.org/10.1016/j.compbiolchem.2009.10.002
10. Li, W., Godzik, A.: Cd-hit: a fast program for clustering and comparing large sets of protein or nucleotide sequences. Bioinformatics **22**, 1658–1659 (2006). https://doi.org/10.1093/bioinformatics/btl158
11. Lin, Z., et al.: Evolutionary-scale prediction of atomic-level protein structure with a language model. Science **379**, 1123–1130 (2023). https://doi.org/10.1126/science.ade2574

12. Mesbah, N.: Editorial: enzymes from extreme environments, volume ii. Front. Bioeng. Biotechnol. **9**, 799426 (2021). https://doi.org/10.3389/fbioe.2021.799426

13. Nikam, R., Kulandaisamy, A., Harini, K., Sharma, D., Gromiha, M.: ProThermDB: thermodynamic database for proteins and mutants revisited after 15 years. Nucleic Acids Res. **49**, D420–D424 (2021). https://doi.org/10.1093/nar/gkaa1035

14. Rives, A., et al.: Biological structure and function emerge from scaling unsupervised learning to 250 million protein sequences. Proc. Natl. Acad. Sci. **118**, e2016239118 (2021). https://doi.org/10.1073/pnas.2016239118

15. Saar, K.L., et al.: Learning the molecular grammar of protein condensates from sequence determinants and embeddings. Proc. Natl. Acad. Sci. USA **118**(e2019053118) (2021). https://doi.org/10.1073/pnas.2019053118

16. Saiki, R.K., et al.: Primer-directed enzymatic amplification of DNA with a thermostable DNA polymerase. Science **239**(4839), 487–491 (1988). https://doi.org/10.1126/science.2448875

17. Schaefer, C., Rost, B.: Predict impact of single amino acid change upon protein structure. BMC Genomics **13**(S4), S4 (2012). https://doi.org/10.1186/1471-2164-13-S4-S4

18. Schilling, J., et al.: Thermostable designed ankyrin repeat proteins (DARPins) as building blocks for innovative drugs. J. Bio. Chem. **298**(1) (2022). https://doi.org/10.1016/j.jbc.2021.101403

19. Yang, Y., Zhao, J., Zeng, L., Vihinen, M.: ProTstab2 for prediction of protein thermal stabilities. Int. J. Mol. Sci. **23**, 10798 (2022). https://doi.org/10.3390/ijms231810798

20. Zhang, G., Fang, B.: Support vector machine for discrimination of thermophilic and mesophilic proteins based on amino acid composition. Protein Pept. Lett. **13**, 965–970 (2006). https://doi.org/10.2174/092986606778777560

# CrimeSeen: An Interactive Visualization Environment for Scenario Testing on Criminal Cocaine Networks

Frederike Oetker[1]($\boxtimes$), Liza A. S. Roelofsen[1], Rob G. Belleman[1], and Rick Quax[1,2]

[1] Computational Science Lab, University of Amsterdam, Amsterdam, The Netherlands
f.oetker@uva.nl
[2] Institute for Advanced Study, University of Amsterdam, Amsterdam, The Netherlands

**Abstract.** The resiliency of criminal networks against law enforcement interventions has driven researchers to investigate methods of creating accurate simulated criminal networks. Despite these efforts, insights reaching law enforcement agencies remain general and insufficient, warranting a new approach. Therefore, we created CrimeSeen - an interactive visualization and simulation environment for exploring criminal network dynamics using computational models. CrimeSeen empowers law enforcement agencies with the possibility to independently test specific scenarios and identify the most effective disruption strategy before deploying it. CrimeSeen comprises of three components: Citadel, a web-based network visualization and simulation tool serving as the interface; the model, defining rules for criminal network dynamics over time, with the Criminal Cocaine Replacement Model as the use-case in this project; and the simulator, connecting the model and interface and enhancing their functionality through transformations, triggers, and statistics. CrimeSeen was evaluated with sequential usability testing, revealing a positive trend in effectiveness and efficiency over time, with mean scores exceeding 80%. However, user satisfaction did not significantly change and remained below the average for web applications, prompting recommendations for future work.

**Keywords:** Criminal Networks · Human-Computer Interaction · Graph Visualization · Simulation · Interaction · Collaboration · Usability Testing · Computational Modelling

## 1 Introduction

Understanding criminal cocaine networks is challenging due to their secretive nature, posing difficulties for law enforcement's intervention efforts [1,2]. Existing approaches lack the tools necessary for effective interventions, particularly

**Supplementary Information** The online version contains supplementary material available at https://doi.org/10.1007/978-3-031-63759-9_24.

in identifying key figures like kingpins. Computational criminal network models offer promise in this regard, yet their complex nature presents accessibility challenges for law enforcement personnel without specialized training [3]. Collaboration among law enforcement agencies and stakeholders is essential for disrupting criminal networks, but integrating data from various sources is challenging due to privacy restrictions [3, 4]. CrimeSeen addresses these challenges by offering law enforcement accessible visualization and simulation tools for evaluating intervention scenarios collaboratively. By bridging the gap between researchers' models and law enforcement, CrimeSeen enables independent, efficient, and collaborative evaluation of intervention methods. Oetker et al. developed a comprehensive criminal network model of the Netherlands, but privacy laws limit data sharing, hindering the model's effectiveness [3, 4]. CrimeSeen aims to overcome these limitations by providing law enforcement with a platform to input real data and explore intervention scenarios independently. CrimeSeen, short for **CR**iminal **I**nvestigation and **M**odelling **E**nvironment for **S**cenario testing and **E**valuation using **E**xploratory interactive **N**etwork visualization, empowers law enforcement to evaluate intervention strategies effectively [3, 4]. It facilitates the visualization of different scenarios, helping domain experts identify the most effective interventions. Moreover, CrimeSeen fosters collaboration between researchers and law enforcement by providing a platform for independent and collaborative evaluation of intervention methods.

## 2   Background and Related Work

This project spans three critical research areas: criminal networks, Human-Computer Interaction (HCI), and usability testing. Criminal network research, particularly focused on drug networks in the Netherlands, is crucial for law enforcement to plan and execute interventions effectively, emphasizing the need for new disruption strategies [5, 6]. Criminal Network Theory (CNT) and Social Network Analysis (SNA) provide insights, but specific strategies must be tailored to the network structure to combat organized crime effectively [7]. In HCI, the design, evaluation, and implementation of interactive computing systems for human use are paramount, ensuring that regardless of system sophistication, usability requirements are met through iterative user-inclusive design processes [8, 9]. Usability testing, a common method for evaluating interactive systems, measures users' ability to efficiently accomplish goals, influencing their experience and future system usage [10]. To assess outcomes, metrics related to users' performance and usability experience are defined, drawing from established tasks in interactive visualization and network analysis within the criminal network context [11–14].

**Metrics.** Usability is measured across three dimensions [15]: effectiveness, efficiency, and satisfaction. Effectiveness pertains to users' ability to successfully complete tasks with the system [15, 16]. Efficiency quantifies the resources expended by users to complete tasks, whether absolutely in time units or relatively as a percentage of total task time spent [15]. Both dimensions necessitate well-defined task specifications for measurement. User satisfaction with a system is assessed via standardized questionnaires like the System Usability Scale (sus), providing a usability score between 0 and 100, requiring both qualitative and quantitative measures, encompassing issues from task completion to navigation [17–21].

**Collaborative and Interactive Visualization.** Collaborative visualization, a facet of Computer-Supported Cooperative Work (cscw), enhances collaboration among participants by employing shared visualizations, crucial for addressing complex problems requiring input from multiple individuals and levels of engagement [22, 23]. Interactive systems offer greater insight and visualization possibilities compared to static visualizations, addressing challenges like cluttering and scalability in network visualization [12]. However, current network visualization applications lack specialized interventions for law enforcement, necessitating an integrated system providing direct accessibility of criminal network models to law enforcement agencies [24].

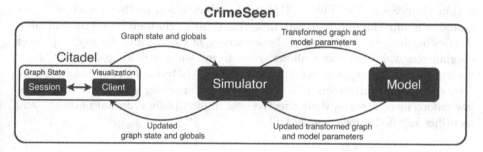

**Fig. 1.** The three main components of CrimeSeen, including their interactions in running a simulation. Citadel is further specified in the session containing the graph state and the client containing the visualization of the graph.

## 3  System Requirements

To attain direct accessibility to criminal network models, the created system has to meet several requirements. These requirements are further specified by domain experts through semi-structured interviews. CrimeSeen aims to provide: general accessibility (easy and intuitive usage, no specific skills or programming expertise required); collaboration (facilitating collaboration among multiple parties, viewing, and sharing information); network analysis (support for all analysis tasks, including SNA techniques for criminal networks, and basic graph analysis tasks); and high level of usability (ensuring adoption by law enforcement, enjoyable user experience, intuitive access to all functionalities).

# 4   CrimeSeen

CrimeSeen consists of three key components: the interface, the model, and the simulator (Fig. 1). The interface relies on Citadel, enabling easy model interaction without coding and enhancing accessibility for law enforcement [25]. Adaptations to Citadel, such as global variables, improve analytical capabilities and model integration [25]. The simulator facilitates compatibility with Citadel, using the ccrm as a demonstration, with a stateless step function approach ensuring system integrity and adaptability [26]. The model, exemplified by the ccrm, is crucial for scenario testing and must adhere to specific conditions for proper execution within CrimeSeen [26,27].

**Usage Scenario.** In CrimeSeen, law enforcement can execute intervention strategies tailored to the Criminal Cocaine Replacement Model (CCRM), aimed at disrupting criminal cocaine networks effectively [26]. Users initiate sessions and connect to the simulator, selecting either pre-existing networks or custom graphs compatible with the simulator and model requirements [26]. Once sessions are established, users explore intervention scenarios, guided by instructions within CrimeSeen [26]. The CCRM offers intervention methods such as automated or manual kingpin removal, targeting central criminals based on centrality metrics like in-degree, out-degree, betweenness, or closeness centrality [26]. After kingpin removal, CrimeSeen allows users to influence new kingpin selection by designating an agent as a "Kingpin Candidate," potentially proposing replacements to conclave members [26]. Through experimenting with these strategies, law enforcement assesses their effectiveness in disrupting criminal networks and identifies key nodes for removal [26].

# 5   Evaluation

We conducted three sequential usability testing sessions for CrimeSeen to gather feedback and enhance the tool. These sessions, supervised and conducted in a controlled environment, lasted about an hour each, with participants using personal laptops for familiarity. Participants from diverse backgrounds were recruited to evaluate accessibility, while tasks were designed to assess CrimeSeen's analysis capabilities and overall usability [28].

**Experimental Setup.** Participants engaged in an experimental setup involving a Qualtrics questionnaire with detailed instructions and sections covering network tasks, simulation running, demographics, and usability assessment. The questionnaire aimed to achieve specific goals, such as testing network understanding, gathering feedback on simulation dynamics, and evaluating usability through the SUS and open-ended feedback [15].

**Outcome Metrics.** To evaluate usability, the metrics of interest are: effectiveness, efficiency and satisfaction [15]. Effectiveness was measured as the percentage of the questions answered correctly. The formula used for the computation of the percentage score for the effectiveness of a participant was:

$$\text{Effectiveness}(j) = \frac{1}{N} \sum_{i=1}^{n} \frac{SC(i,j)}{l} \cdot 100 \tag{1}$$

where $i$ is the task index, $j$ the index of the sessions participant, $n$ the total number of sub-questions, $l$ the number of sub-questions the task has and N the number of main tasks participants were given. Since we did not define a desired speed for task completion time, we use relative efficiency, which is measured per participant using the following formula:

$$\text{Relative Efficiency}(j) = \frac{\sum_{i=1}^{N} n_{ij} t_{ij}}{\sum_{i=1}^{N} t_{ij}} \tag{2}$$

where $j$ is the index of the participant, $i$ is the task index, $N$ the number of tasks (11), $n_{ij}$ the score of task $i$ by participant $j$ and $t_{ij}$ the time spent on task $i$ by participant $j$.

Satisfaction was assessed using the System Usability Scale (sus [15,18]). To tailor the sus for our evaluation, minor adjustments were made to the questionnaire, detailed in A.9. The SUS score is determined by the sum of the answer scores from 0 to 4 where for the odd questions the score increases the more the user agreed with it and for even questions the score increases the more the user disagreed with it.

$$\text{SUS score} = \sum_{i=1}^{10} sc(i) \cdot 2.5 \tag{3}$$

where:

$$sc(i) = \begin{cases} q_{score} - 1 \text{ if } i \bmod 2 \neq 0 \\ 5 - q_{score} \text{ if } i \bmod 2 = 0 \end{cases} \tag{4}$$

Here, $q_{score}$ is the score for the question taken from the response's position on the Likert-scale from left (1) to right (5).

## 6   Results

The analysis of the results of the usability testing involving 32 participants focuses on the development of outcome metrics between test sessions, and the impact of collaboration and interaction test settings on these metrics.

**Usability.** Results from Table 1 show effectiveness, relative efficiency, and satisfaction evolution across test sessions. Effectiveness notably improves from below 50% to over 80%, with reduced score range. Relative efficiency improves between first and second sessions but less between second and third. Satisfaction scores vary across sessions without clear trend. Kruskal-Wallis H test confirms effectiveness and relative efficiency increases but not satisfaction. Kendall's test reveals strong relationships, especially between effectiveness and relative efficiency. Total time spent correlates strongly with task time and outcome metrics, increasing across sessions [28].

**Table 1.** The result of the Kruskal-Wallis H test for comparing the effectiveness, relative efficiency and satisfaction value per testing session. The H-statistics are shown including their p-value between brackets and significance indicated with $*\alpha < 0.05$, $**\alpha < 0.01$, $***\alpha < 0.001$.

|  | All Sessions | Session 1 and 2 | Session 1 and 3 | Session 2 and 3 |
|---|---|---|---|---|
| Effectiveness | 20.7 (0.0)*** | 9.61 (0.002)** | 15.66 (0.000)*** | 6.29 (0.012)* |
| Relative Efficiency | 16.91 (0.000)*** | 11.57 (0.001)*** | 13.15 (0.000)*** | 0.73 (0.391) |
| Satisfaction | 3.6 (0.166) | 3.31 (0.069) | 1.58 (0.209) | 0.58 (0.448) |

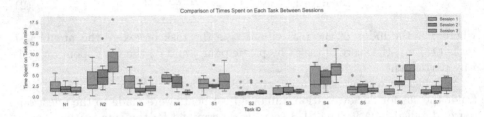

**Fig. 2.** The time spent in minutes on each task across sessions.

**Task Analysis.** Analyzing task performance across sessions, we focused on completion time and effectiveness (Fig. 3) [28]. Task completion times varied significantly, reflecting differences in complexity (for an overview of the tasks, refer to A.8 in Supplementary Materials). For instance, task N4 displayed decreased completion time across sessions, indicating improved efficiency likely due to enhanced familiarity and improved system usability. Concurrently, task effectiveness demonstrated a positive trend, with an increasing percentage of participants completing tasks correctly over sessions, as evident in Fig. 3. While some tasks, like N1 and S2, consistently showed high completion rates, others, like S7, remained below the industry standard even in the final session.

**Collaboration and Interaction.** Figure 2 shows lower outcome metric scores in the "with collaboration" setting compared to "without collaboration". Weak negative correlations between group participation and efficiency, satisfaction, and effectiveness were observed. Participants in groups tended to spend moderately more time on tasks and questionnaires, with no significant differences in medians between groups [28].

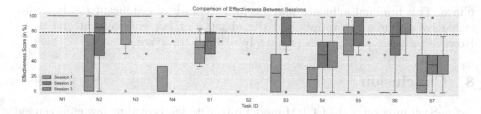

**Fig. 3.** The effectiveness percentages for each task across sessions. The dotted line represents a threshold for comparison with the industry standard which has a task completion rate of 78%.

**Table 2.** The result of the Kruskal-Wallis H test for comparing the effectiveness, relative efficiency and satisfaction value for the collaboration and interaction test setting. The H-statistics are shown including their p-value between brackets and if applicable, significance is indicated with $*\alpha < 0.05$, $**\alpha < 0.01$, $***\alpha < 0.001$.

|  | Collaboration | Interaction |
|---|---|---|
| Effectiveness | 0.13 (0.718) | 0.72 (0.396) |
| Relative Efficiency | 0.69 (0.403) | 0.01 (0.940) |
| Satisfaction | 1.62 (0.203) | 0.39 (0.533) |

## 7 Discussion

Sequential usability testing showed improvements in effectiveness and efficiency across sessions, while user satisfaction remained relatively stable (Fig. 2). Task completion rates were generally high, though some tasks proved challenging, indicating potential overestimation of difficulty. Usability issues persisted, highlighting CrimeSeen's ongoing development stage and the need for further enhancements. Participant demographics had minimal impact on outcomes, suggesting consistency in CrimeSeen's performance across users and devices (for an overview of the participants' demographics, see Figure S3 in the Supplementary Materials).

**Collaboration and Interaction.** The results from both collaborative and interaction settings showed no significant differences among specific test groups, as evident in Table 2. This unexpected finding may be due to small group sizes and the distribution of these settings across sessions, limiting the detection of significant effects. Additionally, the implementation of these settings may have influenced the outcomes, with participants primarily working individually despite the collaborative setting, leading to challenges in task coordination and differing time perspectives.

**Limitations.** The interaction setting solely involved engagement with the simulation, providing participants with similar test conditions until the simulation phase. This minimal influence may explain the lack of significant results, influenced by resource constraints and the absence of law enforcement end users. However, diverse participants effectively utilized the system.

## 8 Conclusion

CrimeSeen integrates the CCRM model into Citadel, providing law enforcement with a comprehensive network visualization and simulation tool. Through iterative evaluations, we identified usability issues and desired features, informing future developments focused on scalability, security, and statistical significance. As a prototype, CrimeSeen requires further development before deployment in law enforcement, with ongoing efforts aimed at refining and enhancing its capabilities. Ultimately, CrimeSeen enables independent simulation of disruption strategies and optimizes resource allocation for more effective interventions.

## References

1. Kleemans, E.R.: Organized crime and the visible hand: a theoretical critique on the economic analysis of organized crime. Criminol. Crim. Justice 13(5), 615–629 (2013). https://doi.org/10.1177/1748895812465296
2. Carrington, P.J.: Crime and social network analysis. In: The SAGE Handbook of Social Network Analysis, pp. 236–255. Sage, Thousand Oaks (2011). https://doi.org/10.4135/9781446294413
3. Oetker, F., Nespeca, V., Vis, T., Duijn, P., Quax, R.: Framework for Developing Quantitative Agent-Based Models Based on Qualitative Expert Knowledge: An Organised Crime Use Case (2023, unpublished manuscript)
4. Diviák, T.: Key aspects of covert networks data collection: problems, challenges, and opportunities. Soc. Netw. 69, 160–169 (2022). https://doi.org/10.1016/j.socnet.2019.10.002
5. Preece, J., Rogers, Y., Sharp, H., Benyon, D., Holland, S., Carey, T.: Human-Computer Interaction. Addison-Wesley Longman Ltd. (1994)
6. Sinha, G., Shahi, R., Shankar, M.: Human computer interaction. In: 2010 3rd International Conference on Emerging Trends in Engineering and Technology, pp. 1–4 (2010). https://doi.org/10.1109/ICETET.2010.85
7. Karray, F., Alemzadeh, M., Saleh, J.A., Arab, M.N.: Human-computer interaction: overview on state of the art. Int. J. Smart Sens. Intell. Syst. 1(1), 137–159 (2008). https://doi.org/10.21307/ijssis-2017-283
8. Issa, T., Isaias, P.: Sustainable Design: HCI, Usability and Environmental Concerns, 2nd edn. Springer, Cham (2022). https://doi.org/10.1007/978-1-4471-7513-1
9. Dumas, J.S., Fox, J.E.: Usability testing: current practice and future directions. In: Human-Computer Interaction, pp. 247–268. CRC Press (2009)
10. Niès, J., Pelayo, S.: From users involvement to users' needs understanding: a case study. Int. J. Med. Inform. 79(4), 76–82 (2010). https://doi.org/10.1016/j.ijmedinf.2009.06.007
11. Xu, J.J., Chen, H.: CrimeNet explorer: a framework for criminal network knowledge discovery. ACM Trans. Inf. Syst. 23(2), 201–226 (2005). https://doi.org/10.1145/1059981.1059984

12. Rasheed, A., Wiil, U.K., Niazi, M.: Evaluating PEVNET: a framework for visualization of criminal networks. In: Wang, L., Uesugi, S., Ting, I.-H., Okuhara, K., Wang, K. (eds.) MISNC 2015. CCIS, vol. 540, pp. 131–149. Springer, Heidelberg (2015). https://doi.org/10.1007/978-3-662-48319-0_11

13. Sparrow, M.K.: The application of network analysis to criminal intelligence: an assessment of the prospects. Soc. Netw. **13**(3), 251–274 (1991). https://doi.org/10.1016/0378-8733(91)90008-H

14. McAndrew, D.: The structural analysis of criminal networks. In: The Social Psychology of Crime, pp. 51–94. Routledge (2021)

15. Brooke, J.: SUS: a quick and dirty usability scale. In: Usability Evaluation in Industry, vol. 189, no. 194, pp. 4–7 (1996)

16. Sauro, J.: What Is A Good Task-Completion Rate? (2011). https://measuringu.com/task-completion/

17. Tullis, T., Stetson, J.: A comparison of questionnaires for assessing website usability. In: Proceedings of the Usability Professionals Association (UPA), Minneapolis, MN, USA, 7–11 June 2004. ResearchGate (2004)

18. Peres, S., Pham, T., Phillips, R.: Validation of the system usability scale (SUS). In: Proceedings of the Human Factors and Ergonomics Society Annual Meeting, vol. 57, pp. 192–196. SAGE Publications (2013). https://doi.org/10.1177/1541931213571043

19. Bangor, A., Kortum, P.T., Miller, J.T.: An empirical evaluation of the system usability scale. Int. J. Hum.-Comput. Interact. **24**, 574–594 (2008). https://doi.org/10.1080/10447310802205776

20. Ellis, G., Dix, A.: An explorative analysis of user evaluation studies in information visualisation. In: Bertini, E., Plaisant, C., Santucci, G. (eds.) Proceedings of the 2006 AVI Workshop on BEyond Time and Errors: Novel Evaluation Methods for Information Visualization, BELIV, pp. 1–7. AMC Press (2006). https://doi.org/10.1145/1168149.1168152

21. Tullis, T., Albert, B.: Measuring the User Experience: Collecting, Analyzing, and Presenting Usability Metrics, Interactive Technologies. Morgan Kaufmann, San Francisco (2008). https://doi.org/10.1016/B978-0-12-373558-4.X0001-5

22. Isenberg, P., Elmqvist, N., Scholtz, J., Cernea, D., Ma, K.-L., Hagen, H.: Collaborative visualization: definition, challenges, and research agenda. Inf. Vis. **10**(4), 310–326 (2011). https://doi.org/10.1177/1473871611412817. https://inria.hal.science/inria-00638540

23. Wulf, W.A.: The national collaboratory - a white paper. In: Lederberg, J., Uncaphar, K. (eds.) Towards a National Collaboratory: Report of an Invitational Workshop at the Rockefeller University, pp. 17–18. Rockefeller University, Washington, D.C. (1989)

24. Bertin, J.: Sémiologie Graphique: Les diagrammes, les réseaux, les cartes, Gauthier-Villars, Paris (1967). (Translation 1983. Semiology of Graphics by William J. Berg)

25. Roelofsen, L.A.S., Oetker, F., Van der Lely, M., Belleman, R.G., Quax, R.: CitadelPolice: an interactive visualization environment for scenario testing on criminal networks. In: EuroVis 2023 - Posters, pp. 17–19. The Eurographics Association (2023). https://doi.org/10.2312/evp.20231057

26. Oetker, F., Nespeca, V., Vis, T., Duijn, P., Sloot, P., Quax, R.: Framework for developing quantitative agent-based models based on qualitative expert knowledge: an organised crime use-case (2023). http://arxiv.org/abs/2308.00505. https://arxiv.org/abs/2308.00505

27. Kazil, J., Masad, D., Crooks, A.: Utilizing python for agent-based modeling: the mesa framework. In: Thomson, R., Bisgin, H., Dancy, C., Hyder, A., Hussain, M. (eds.) SBP-BRiMS 2020. LNCS, vol. 12268, pp. 308–317. Springer, Cham (2020). https://doi.org/10.1007/978-3-030-61255-9_30

28. Issa, T., Isaias, P.: Usability and human–computer interaction (HCI). In: Issa, T., Isaias, P. (eds.) Sustainable Design, pp. 23–40. Springer, London (2022). https://doi.org/10.1007/978-1-4471-7513-1_2

29. Van Der Lely, M.: A Novel Visualisation Paradigm Using AR Superimposition, Bachelor Thesis, University of Amsterdam (2022). https://scripties.uba.uva.nl/search?id=728606

30. Bangor, A., Kortum, P., Miller, J.: Determining what individual SUS scores mean: adding an adjective rating scale. J. Usability Stud. **4**(3), 114–123 (2009)

31. Shneiderman, B.: The eyes have it: a task by data type taxonomy for information visualizations. In: Proceedings 1996 IEEE Symposium on Visual Languages, pp. 336–343 (1996). https://doi.org/10.1109/VL.1996.545307

32. Kenton, W.: Value Networks: Definition, Benefits and Types (2022). https://www.investopedia.com/terms/v/value-network.asp

# Automatic Kernel Construction During the Neural Network Learning by Modified Fast Singular Value Decomposition

Norbert Jankowski[1]([✉]) and Grzegorz Dudek[2]

[1] Department of Informatics, Nicolaus Copernicus University, Toruń, Poland
norbert@umk.pl
[2] Faculty of Electrical Engineering, Technical University of Częstochowa,
Częstochowa, Poland
grzegorz.dudek@pcz.pl

**Abstract.** Thanks to the broad application fields, learning neural networks is still a more significant problem nowadays. Any attempt in the construction of faster learning algorithms is highly well come. This article presents a new way of learning neural networks with kernels with modified pseudo-inverse learning by modified SVD.

The new algorithm constructs the kernels during the learning and estimates the right number in the results. There is no longer a need to define their number of kernels before the learning. This means there is no need to test networks with a number of kernels that is too large, and the number of kernels is no longer a parameter in the selection process (in cross-validation).

The results show that the proposed algorithm constructs reasonable kernel bases, and final neural networks are accurate in classification.

**Keywords:** neural network learning · kernels · classification · singular value decomposition

## 1 Introduction

The classification or approximation problems are represented by a training dataset as a matrix $X$, which consists of learning vectors $\mathbf{x}_i$ ($\mathbf{x}_i \in R^n, i \in [1,\ldots,m]$), and each vector has corresponding $y_i \in \{0,\ldots,K\}$ (in binary case $y_i = \pm 1$).

We can look at a nonlinear model constructed as a linear combination of nonlinear functions:

$$F(\mathbf{x},\mathbf{w}) = \sum_{j=1}^{l} w_j g_j(\mathbf{x}) + w_0, \qquad (1)$$

which is a combination ($\mathbf{w}$) of kernels $g_j$. The above form is fully consistent with the Radial Basis Function Network (RBFN) [1] and the Extreme Learning

© The Author(s), under exclusive license to Springer Nature Switzerland AG 2024
L. Franco et al. (Eds.): ICCS 2024, LNCS 14834, pp. 205–212, 2024.
https://doi.org/10.1007/978-3-031-63759-9_25

Machine (ELM) [4,5] as well. The same form also applies to nonlinear Support Vector Machines [8].

The sigmoidal function was the original kernel used in the ELM, while in the RBFN the Gaussian kernel is usually chosen (although it is sometimes used in ELM as well [2,5]). [7] has introduced multilayer perceptron learned as a multi-layered ELM. It can be seen as a special case of deep learning—first layers are the layers of autoencoders and the final layer is a typical ELM.

The goal (the error function) for the neural networks can be defined by:

$$J_n(\mathbf{w}, G) = ||G\mathbf{w} - \mathbf{y}||^2 \tag{2}$$

where $G$ is defined by: $G = \begin{bmatrix} 1 & g_1(\mathbf{x}_1) & g_2(\mathbf{x}_1) & \cdots & g_l(\mathbf{x}_1) \\ 1 & g_1(\mathbf{x}_2) & g_2(\mathbf{x}_2) & \cdots & g_l(\mathbf{x}_2) \\ \cdots & & & & \\ 1 & g_1(\mathbf{x}_m) & g_2(\mathbf{x}_m) & \cdots & g_l(\mathbf{x}_m) \end{bmatrix}$.

To find the solution we have to compute the gradient:

$$\nabla J(\mathbf{w}, G) = 2G^T(G\mathbf{w} - \mathbf{y}). \tag{3}$$

After a few substitutions, we finally have:

$$\mathbf{w} = (G^T G)^{-1} G^T \mathbf{y} = G^\dagger \mathbf{y}. \tag{4}$$

where $G^\dagger$ is the Moore-Penrose pseudo-inverse matrix of $G$. The SVD can be used to compute the pseudo-inverse of $G$.

The costs of learning via SVD are relatively low—it is a sum of costs of construction of $G$ and costs of SVD. The complexity of construction of $G$ is $O(mln)$ (computation of single $G_{ij}$ is $O(n)$). The complexity of SVD of matrix $G$ is $O(ml^2)$.

In the above learning scheme, we see that for the learning to be based on kernels, the kernel has to be created earlier, while we do not know the correct number of kernels before the learning. Therefore, the main contribution of this article is to propose a way in which the kernels necessary to train the neural network are created and used during the training process, i.e. in the above scheme, it would be during the execution of the SVD procedure. This is presented in the next section. The last section presents analysis of empirical results.

## 2    Automatic Kernel Construction During the Pseudo-inverse Learning

The above algorithm can be sped up and simplified by changing two things. The first is related to the fact that the number of kernels $l$ as well as the $G$ matrix must be known before starting pseudo-inversion training. Note that the selection process of the number of kernels is typically done by repeated training and testing with a different number of kernels [2]. Here we propose that the $G$ matrix as well as the number of kernels will be determined in the modified SVD algorithm. The second change will consist in changing the fast version of the SVD determination to enable the dynamic creation of the $G$ matrix.

Thus, the starting point is the fast SVD algorithm proposed in [3]. The main idea of this algorithm lies in construction of matrix $Q$ of the following property:

$$G \approx QQ^T G \tag{5}$$

where $Q$ is matrix composed of orthonormal columns $(m \times k, \ k \leq l,)$. $Q$ is constructed by low-rank approximation of a matrix.

Now to compute the SVD of $G$ we have to: 1 – construct matrix $Q$, 2 – construct matrix $B$ such that $B = Q^T G$ ($B$ is $k \times l$), 3 – compute $SVD(B) = \tilde{U}\Sigma V^T$ and 4 – define the final matrix $U$ as $U = Q\tilde{U}$. In the next step final vector $\mathbf{w}$ can be computed as in Eq. 4.

## 2.1 Standard $Q$ Matrix Construction by Gaussian Randomization and Ortogonalization

The proposed construction of the matrix $Q$ in [3] was performed by iterative Gaussian randomization and orthogonalization. First let the matrix $H$ be defined by

$$\mathbf{h}^{(i)} = G\boldsymbol{\omega}^{(i)} \qquad i = 1, \ldots, k, \tag{6}$$

where $\boldsymbol{\omega}^{(i)}$ is a random Gaussian vector $1 \times n$, and $H \in \mathcal{M}_{m \times k}(\mathbb{R})$. The Gaussian random vector is drawn from a Gaussian distribution with a mean of 0 and variance equal to 1.

The next step is to use an orthonormalization algorithm to build an orthonormal matrix $Q$ from matrix $H$. The two steps, the randomization and the orthonormalization, can be performed iteratively.

The selection of $k$ as the number of columns in the matrix $Q$ is helped by the following lemma from [3]:

**Lemma 1.** Let $B \in \mathcal{M}_{m \times m}(\mathbb{R})$, $r$ be a positive integer and $\alpha > 1$. Draw an independent family $\{\boldsymbol{\omega}^{(i)} : i = 1, \ldots, r\}$ of standard Gaussian vectors. Then:

$$\|B\| \leq \alpha\sqrt{\frac{2}{\pi}} \max_{i=1,\ldots,r} \|B\boldsymbol{\omega}^{(i)}\|$$

with a probability of at least $1 - \alpha^{-r}$.

In the context of the above definition of the orthonormal matrix $Q$ and the above lemma, the direct conclusion is that the decomposition error is bounded as below:

$$\|(I - QQ^T)G\| \leq 10\sqrt{\frac{2}{\pi}} \max_{i=1,\ldots,r} \|(I - QQ^T)G\boldsymbol{\omega}^{(i)}\| \tag{7}$$

with a probability of at least $1 - 10^{-r}$. The detailed algorithm for construction of $Q$ is presented as Algorithm 4.2 in [3] on page 25.

The complexity of the above algorithm is $O(mlk)$ where $m \times l$ is the size of $G$ and $k$ is the number of columns in $Q$ ($k \leq l$). After the construction of $Q$, the SVD is calculated on the matrix $B = Q^T G$ and the matrix $B \in \mathcal{M}_{k \times l}(\mathbb{R})$. That means that the complexity of SVD on $B$ is $O(kl^2)$. This leads us to a final complexity of the fast version of SVD: $O(mlk)$.

## 2.2 New Way of Q Matrix Construction Directly from Learning Data

The concept of the algorithm that we want to publish is an attempt to avoid building kernel matrices $G$ that are too large. Then the complexity of building kernel matrices could change from $O(mnl)$ to $O(m^2n)$ when we have to insert kernels in too many data vectors. This also significantly impacts the final complexity of training via SVD, where the complexity will go from $O(ml^2)$ to $O(m^3)$. In the old procedure of creating the $Q$ matrix, unfortunately, we had to have a ready-to-use kernel matrix $G$. Training ends with poorer classification accuracy when the kernel matrix is too small. However, training on a large kernel matrix has high (sometimes unnecessarily) complexity. Thanks to this, the new $Q$ matrix creation procedure dynamically adjusts the number of kernels, i.e., the size of the kernel matrix increases dynamically in the new $Q$ matrix creation algorithm. As a result, in the process of creating the $Q$ matrix, the necessary number of kernels will be generated, and the further SVD learning process will take place on a possibly small matrix, reducing the complexity of both stages and removing the need to sample the appropriate number of kernels repeatedly. So, we do not want to build an array of 5000 kernels if 48 kernels are enough.

For this purpose, the new procedure for determining the $Q$ matrix will not work on the basis of the already prepared $G$ matrix, but on the basis of the $X$ data matrix, which is necessary to construct kernels, and the kernel columns of the $G$ matrix will be created as needed.

To make this process workable, random linear combinations of columns of matrix $G$ will gradually cover more and more of the target number of columns (kernels) of the final matrix $G$. Initially, the algorithm will use the starting amount of kernels $\nu = 16$. When the number of columns of the $Q$ matrix will increase from time to time algorithm will have to create new groups of kernels by increasing $\nu = 2\nu$. This scheme of increase prevents the increase in complexity due to the necessary matrix manipulations (increase = new memory allocations and copying). More precisely, when the number of columns of the $Q$ matrix reaches half the number of kernels, new kernels will be added to the $G$ matrix.

The algorithm of the $Q$ matrix calculation with the proposed changes can be found in Algorithm 1. It can be seen here that changes have been made mainly in two places related to generating a random combination of columns of the $G$ matrix, but also to the fact that the data matrix $X$ is the input for this algorithm. The original matrix $X$ is necessary for the construction of kernels. And this, in turn, we see in separate Algorithm 2. Here the new kernels are periodically added to the $G$ matrix and based on them, random combinations of columns of the current $G$ matrix are generated. In this procedure, when Gaussian

---

**Algorithm 1:** New way of Q matrix construction directly from $X$

---

1: **function** constructQ($X$)
2: **for** $i = 1$ to $r$ **do**
3: $\quad$ $\mathbf{h}^{(i)} = $ newRandComb($X$)
4: **end for**
5: $Q^{(0)} = []$, the $m \times 0$ matrix
6: $j = 0$
7: **while** $\left(\max_{k=1,\ldots,r} \|\mathbf{h}^{(j+k)}\|\right) >$
$\quad$ $\epsilon/(10\sqrt{2/\pi})$ **do**
8: $\quad$ $j = j + 1$
9: $\quad$ $\mathbf{h}^{(j)} = (\mathbf{I} - Q^{(j-1)}(Q^{(j-1)})^T)\mathbf{h}^{(j)}$
10: $\quad$ $q^{(j)} = \mathbf{h}^{(j)}/\|\mathbf{h}^{(j)}\|$

11: $\quad$ $Q^{(j)} = [Q^{(j-1)} q^{(j)}]$
12: $\quad$ $\mathbf{h} = $ newRandComb($X$)
13: $\quad$ $\mathbf{h}^{(j+r)} = (\mathbf{I} - Q^{(j)}(Q^{(j)})^T)\mathbf{h}$
14: $\quad$ **for**
$\quad\quad$ $i = (j+1), (j+2), \ldots, (j+r-1)$
$\quad\quad$ **do**
15: $\quad\quad$ $\mathbf{h}^{(i)} = \mathbf{h}^{(i)} - q^{(j)}\langle q^{(j)}, \mathbf{h}^{(i)}\rangle$
16: $\quad$ **end for**
17: **end while**
18: $Q = Q^{(j)}$
19: **return** $Q$

---

---

**Algorithm 2:** Random combination over current set of kernels

---

1: $\nu = 16$
2: $kCount = 0$
3: $z = 0$
4: $G = []$, the $m \times 0$ matrix
5: $Y = []$, the $m \times 0$ matrix
6:
7: **function** newRandComb($X$)
8: **if** $G$ is empty matrix **then**
9: $\quad$ $G = $ add a column of 1's
10: **end if**
11: **if** $kCount = 0 \lor z = kCount * 3/4$
$\quad$ **then**

12: $\quad$ add $\nu$ next kernel columns to $G$,
$\quad\quad$ but not more than $m$ in total
13: $\quad$ $\Omega = $ Gaussian random matrix
$\quad\quad$ (size: number of columns in $G \times \nu$)
14: $\quad$ $Y = [Y \ G\Omega]$
15: $\quad$ $kCount = kCount + \nu$
16: $\quad$ $\nu = 2 * \nu$
17: **end if**
18: $z = z + 1$
19: **return** $z$'s column of $Y$

---

kernels are created, they are placed in random instances of the $X$ data matrix (no instance is selected twice).

The algorithm working in this way may finish its work faster or slower depending on the data and performance of built kernels in the context of a given classification problem.

The final learning algorithm of a neural network with the above fast SVD algorithm is presented in Algorithm 3.

## 3 Empirical Analysis of Proposed Algorithm

In order to compare the above proposed algorithm, it was decided to select a number of different data sets from the UCI Machine Learning Repository [6]. The number of instances of selected datasets are summarized in Table 2. In all tests, we used 10-fold stratified cross-validation and all learning machines were trained on the same sets of data partitions.

---

**Algorithm 3:** Neural network learning with SVD

1: **function** networkLearnig $(X, \mathbf{y})$
2: $Q = \text{constructQ}(X)$
3: $[\tilde{U}, \Sigma, V^T] = \text{SVD}(Q^T G)$
4: $\mathbf{w} = V\Sigma^{-1}\tilde{U}^T Q^T \mathbf{y}$
5: **return** $\mathbf{w}$

---

To visualize the performance of all algorithms we present average accuracy for each benchmark dataset and for each learning machine. Additionally, we present the average reduction of dataset size in separate tables. *Ranks* are calculated for each machine for a given dataset. The ranks are calculated as follows: First, for a given benchmark dataset the averaged accuracies of all learning machines are sorted in descending order. The machine with the highest average accuracy is ranked 1. Then, the following machines in the accuracy order whose accuracies are not statistically different[1] from the result of the first machine are ranked 1, until a machine with a statistically different result is encountered. That machine starts the next rank group (2, 3, and so on), and an analogous process is repeated on the remaining (yet unranked) machines. Notice that each cell of the main part of Table 1 is in a form: $acc + std(rank)$, where $acc$ is average accuracy (for a given data set and given learning machine), $std$ is its standard deviation and $rank$ is the rank described just above. If a given cell of the table is in bold it means that this result is the best for given data set or not worse than the best one (rank 1 = winners).

Table 1 compares the average accuracies of the classification of the proposed algorithm that automatically determines kernels during modified SVD method with standard training of the neural network through SVD from 20, 200 and 2000 random Gaussian kernels, respectively.

As you can see, the proposed method is not always the best one, but always it achieves results close to the best without any manual selection of the number of kernels. It is worth comparing this data with the data from Table 2. This allows us to see how different the number of columns of the $Q$ matrix was needed to achieve convergence and this has a direct impact on the working time of the entire learning process.

Even for a large data set like shuttle-all, the averaged number of columns in the $Q$ matrix is only 48 out of 58,000 instances. This clearly shows that the size of the data set is not responsible for the number of necessary kernels, and the proposed method deals with it in an extremely interesting way. This is much simpler computationally than selection by many trials and testing, like via cross-validation. However, the proper selection of the number of kernels needs to start cross-validation at several points, which makes the process several times more time-consuming. Time consumption grows with the square of $l$ in the complexity of SVD and $O(mlk)$ complexity of $Q$ matrix construction. In conclusion, testing huge numbers of kernels in cross-validation provides huge costs. However, the

---

[1] We use the paired t-test to test the significance of statistical differences.

**Table 1.** Comparison of classification average accuracies for methods: auto-kernel SVD, SVD with 20, 200 and 2000 kernels.

| Dataset | Auto-kernel | SVD 20 | SVD 200 | SVD 2000 |
|---|---|---|---|---|
| cardiotocography-1 | **83.1 ± 2.3(1)** | 67.9±2.8(3) | 81.2±2.2(2) | **83.3 ± 2.3(1)** |
| cardiotocography-2 | 92.4±1.7(2) | 87.5±2.1(4) | 91.6±1.7(3) | **92.6 ± 1.8(1)** |
| chess-king-rook-vs-king-pawn | 99.3±0.54(2) | 84.5±4.5(4) | 97.2±1.1(3) | **99.4 ± 0.45(1)** |
| spambase | 91.9±1.2(3) | 87.4±1.8(4) | 92.4±1.3(2) | **92.7 ± 1.2(1)** |
| thyroid-disease | 94.5±0.89(3) | 93.8±0.28(4) | 94.7±0.5(2) | **95.4 ± 0.51(1)** |
| abalone | 26.2±1.9(2) | 25.7±2.1(3) | **26.3 ± 1.9(1)** | 26.4±1.8(1) |
| image | **94.4 ± 1.5(1)** | 89.1±2(3) | **94.5 ± 1.5(1)** | 94.2±1.6(2) |
| letter-recognition | **87.9 ± 0.76(1)** | 58.4±1.6(3) | 82.5±0.94(2) | **87.9 ± 0.75(1)** |
| magic04 | 86.1±0.68(2) | 82.4±0.93(4) | 86±0.69(3) | **86.3 ± 0.66(1)** |
| musk2 | **100 ± 0.04(1)** | 85.3±0.55(4) | 89.9±1(3) | 99.4±0.37(2) |
| nursery | **95.9 ± 0.43(1)** | 87.2±0.92(3) | 93.8±0.58(2) | **95.9 ± 0.49(1)** |
| sat-all | **90.1 ± 1(1)** | 83.8±1.1(4) | 88.5±1.3(3) | 90±1.1(2) |
| segmentation-all | **94.4 ± 1.5(1)** | 89.1±2(3) | **94.5 ± 1.5(1)** | 94.2±1.6(2) |
| SHUTTLE-all | 98.2±0.22(3) | 96.9±0.71(4) | 98.4±0.17(2) | **98.5 ± 0.17(1)** |
| Waveform | 84.3±1.5(2) | 83.7±1.7(3) | **86.4 ± 1.4(1)** | 84.3±1.6(2) |
| Mean Accuracy | 87.9±1.1 | 80.2±1.7 | 86.5±1.2 | 88±1.1 |
| Mean Rank | 1.73±0.21 | 3.53±0.14 | 2.07±0.21 | 1.33±0.13 |
| Wins[unique] | 7[2] | 0[0] | 4[1] | 10[6] |

**Table 2.** Reduction strength—the number of columns in matrix $Q$.

| Dataset | # Instances in dataset | Reduction Total | Fraction | Time Auto-kernel | CV-kernel |
|---|---|---|---|---|---|
| cardiotocography-1 | 2126 | 640 | 0.334 | 2.14 | 86.6 |
| cardiotocography-2 | 2126 | 637 | 0.333 | 2.46 | 83.8 |
| chess-king-rook-vs-king-pawn | 3196 | 2.3E+03 | 0.799 | 51.4 | 143 |
| spambase | 4601 | 2.98E+03 | 0.719 | 103 | 151 |
| thyroid-disease | 7200 | 162 | 0.0123 | 2 | 187 |
| abalone | 4177 | 46.4 | 0.0123 | 1.43 | 189 |
| image | 2310 | 137 | 0.0659 | 1.68 | 111 |
| letter-recognition | 20000 | 1.21E+03 | 0.067 | 45.7 | 996 |
| magic04 | 19020 | 206 | 0.012 | 7.95 | 1.03E+03 |
| musk2 | 6598 | 5.83E+03 | 0.981 | 888 | 609 |
| nursery | 12960 | 645 | 0.0553 | 19.5 | 849 |
| sat-all | 6435 | 2.19E+03 | 0.378 | 99.2 | 720 |
| segmentation-all | 2310 | 137 | 0.0659 | 15 | 415 |
| SHUTTLE-all | 58000 | 48 | 0.00092 | 14.8 | 4.58E+03 |
| Waveform | 5000 | 3.51E+03 | 0.779 | 259 | 1.35E+03 |

proposed algorithm immediately determines the correct number of kernels. This was proven by the next experiment (see Table 2 columns Reduction), which compares the CPU times of the proposed Auto-kernel algorithm and the SVD-based learning with cross-validation for the selection of kernel count. In the second algorithm, cross-validation is used to select kernel counts just between three values: 20, 200, and 2000. In many benchmarks, the Auto-kernel algorithm is several times faster than learning with CV-based selection of kernels, even in a simple three point selection. If we added 20,000 kernels to CV training as an additional optionpoint, learning would become terribly slower comparing to the Auto-kernel.

## 4 Summary

The process of selection of kernels for neural networks was never simple. The methods that were based on finding the right number of kernels required many learning processes and tests of the learned neural networks. Such a scheme of learning strongly uses the CPU. In the proposed algorithm, we modify a fast SVD method by automatically generating kernels and their appropriate number in the learning process, specifically in the sub-process of the fast SVD.

As can be seen, such learning is characterized by good classification quality for various data sets. What's more, the selection of the number of columns in the $Q$ matrix is quite impressive, which translates into shortening the learning process if the classifier does not have to use many kernels.

## References

1. Broomhead, D.S., Lowe, D.: Multivariable functional interpolation and adaptive networks. Complex Syst. **2**(3), 321–355 (1988)
2. Dudek, G.: A constructive approach to data-driven randomized learning for feedforward neural networks. Appl. Soft Comput. **112** (2021). https://doi.org/10.1016/j.asoc.2021.107797
3. Halko, N., Martinsson, P.G., Tropp, J.A.: Finding structure with randomness: probabilistic algorithms for constructing approximate matrix decompositions. SIAM Rev. **53**(2), 217–288 (2011)
4. Huang, G.B., Zhu, Q.Y., Siew, C.K.: Extreme learning machine: a new learning scheme of feedforward neural networks. In: International Joint Conference on Neural Networks, Budapest, Hungary, pp. 985–990. IEEE Press (2004)
5. Huang, G.B., Zhu, Q.Y., Siew, C.K.: Extreme learning machine: theory and applications. Neurocomputing **70**(1–3), 489–501 (2006)
6. Merz, C.J., Murphy, P.M.: UCI repository of machine learning databases (1998). http://www.ics.uci.edu/~mlearn/MLRepository.html
7. Tang, J., Deng, C., Member, S., Huang, G.B.: Extreme learning machine for multilayer perceptron. IEEE Trans. Neural Netw. Learn. Syst. **27**(4), 809–821 (2016)
8. Vapnik, V.: The Nature of Statistical Learning Theory. Springer, New York (1995)

# Multivariate Time Series Modelling with Neural SDE Driven by Jump Diffusion

Kirill Zakharov[✉][iD]

Research Center "Strong Artificial Intelligence in Industry", ITMO University,
Saint Petersburg 199034, Russia
kazakharov@itmo.ru

**Abstract.** Neural stochastic differential equations (neural SDEs) are effective for modelling complex dynamics in time series data, especially random behavior. We introduced JDFlow, a novel normalizing flow method to capture multivariate structures in time series data. The framework involves a latent process driven by a neural SDE based on the Merton jump diffusion model. By using maximum likelihood estimation to determine the intensity parameter of the Poisson process in neural SDE, we achieved better results in generating time series data compared to previous methods. We also proposed a new approach to assess synthetic time series quality using a Wasserstein-based similarity measure, which compares signature cross-section distributions of original and generated time series.

**Keywords:** neural stochastic differential equations · normalising flows · Merton jump diffusion · multivariate time series modelling · path signature

## 1 Introduction

In the ever-evolving landscape of data science and machine learning, the field of time series modeling has emerged as an essential and challenging area of research. Time series data, with its unique temporal dependencies and sequential patterns, finds applications in various fields such as finance, healthcare, and climate science, among others [1–3]. The accurate modeling of time series is crucial for creating robust models and understanding complex systems. One approach to modelling time series is through generative models [4], which have practical applications in anomaly detection [5] and data augmentation [6]. In this paper, we propose a novel method based on normalizing flows and neural SDEs for time series generation and modelling. Especially, we aim to create a model that can account the jumps in real markets, utilizing the Merton model [3] as the jump framework.

Normalising flows are a family of generative models with tractable density estimation. The main idea is to transform the initial complicated data distribution to a simple one, by composition of several functions $f_i$. There are some constraints to implement such architecture: i) each function $f_i$ should be a bijection (should exist an invertible transformation $g_i^{-1} = f_i$); ii) they should have an analytically closed form and must be differentiable; iii) the Jacobian determinant of $f_i$ should be easily calculable. The composition of the functions $f_i$ is called the *flow*. Functions $g_i = f_i^{-1}$ form the *forward* or *generative direction* and $f_i$ form the *backward* or *normalising direction* [4].

Stochastic differential equations (SDEs) are applied in various fields due to their ability to model systems, influenced by both deterministic and random factors. Here are some common applications: (i) in finance, SDEs determine the asset price dynamic, e.g., the famous Black-Scholes model for option pricing involves SDEs, by using geometric Brownian motion to determine stock prices; (ii) in physics, SDEs are used to model a system with random fluctuations. For example, the motion of particles in a fluid can be described by Langevin equations; (iii) SDEs are employed in modelling biological systems, e.g., population dynamics and the spread of diseases can be described using stochastic models.

In this article, we consider the Itô's type of *stochastic differential equations* [7] with additional jump component, which defined as (1).

## 2    Method

Figure 1 shows the general method's pipeline. The red section is dedicated to the latent process driven by neural SDE from the Sect. 2.1. The blue section is devoted to intensity estimation from the Sect. 2.2. Finally, the general generative model framework is discussed in the Sect. 2.3.

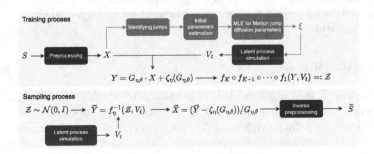

**Fig. 1.** General pipeline.

## 2.1    Latent Process and Neural SDE Formulation

Consider the time series $S$ of the length $T$ with dimension $M$. For a given $S$, we apply the preprocessing procedure, which scales the series in the range $[0, 1]$ and result it as $X \in \mathbb{R}^{M \times T}$. The main idea of our approach is to use the jump diffusion dynamic instead of the standard Itô's diffusion. To add the jump diffusion, we implement a *latent process* $V_t$. We suppose, that $V_t$ is square-integrable and it follows the jump diffusion dynamic (neural SDE [8]),

$$dV_t = \mu_\theta(V_t,t)dt + \sigma_\theta(V_t,t)dW_t + J_\theta(V_t,t)dP_t, \qquad (1)$$

where $W_t$ is the Wiener process, $P_t$ is the Poisson process, $\mu_\theta$, $\sigma_\theta$, $J_\theta$ are the drift, diffusion, and jump magnitude parameters, respectively. $W_t$ and $P_t$ are assumed to be independent. The functions $\mu,\sigma,J$ are parameterised by neural networks instead of using the predefined dynamic. The networks' parameters are stored in $\theta$ for short. The networks $\mu,\sigma,J$ take the vector $V_t \in \mathbb{R}^M$ and the vector $(t,\sin t,\cos t) \in \mathbb{R}^3$ as input, where time component is augmented by sine and cosine transformations, and produce the vector of the dimension $M$. For hidden layers we use fully connected layers, with hidden dimension $2^7$, and hyperbolic function, as activation function.

The initial value $V_0$ of the process $V_t$ is identified by another neural network $\varphi_\theta$, which depends on the random noise $Z_t \in \mathbb{R}^{1\times M}$ and some initial information $X_0$ about $X$ (we use $X$'s values at zero time point $X_0 \in \mathbb{R}^{M\times 1}$). The network $\varphi_\theta$ takes the vector $(Z_t, X_0^\top) \in \mathbb{R}^{2M}$ as input, and its outcome is the initial value $V_0 \in \mathbb{R}^M$.

It is complicated to find an analytic solution of the Eq. (1), because the functions $\mu,\sigma,J$ have complicated structure (they are the neural networks with huge amounts of parameters). But, it is possible to use the discrete scheme to approximate the solution. We choose the standard one so-called the *Euler-Maruyama scheme* [9]. Finally, we can represent the Eq. (1) as the following discrete scheme,

$$V_{t_{i+1}} = V_{t_i} + \mu_\theta(V_{t_i},t_i)\Delta t + \sigma_\theta(V_{t_i},t_i)\varepsilon\sqrt{\Delta t} + J_\theta(V_{t_i},t_i)P_{t_i}, \qquad (2)$$

where $\Delta t = \frac{T}{\tau}$, $\tau$ is the number of steps in the discrete scheme, $P_{t_i}$ is the Poisson random variable with intensity $\xi\Delta t$, and $\varepsilon \sim \mathcal{N}(0,1)$. We denote the solution block with discrete Euler-Maruyama scheme as $F_\theta(V_0,\Delta t,\mu_\theta,\sigma_\theta,J_\theta)$, which will produce the multivariate time series of the dimension $M$ and the length $\tau$. We refer $\tau$ as hyperparameter and recommend to choose it not greater than $T$.

## 2.2  Intensity Estimation with Merton Model

Consider the log return process $R_t = \log \frac{X_{t+dt}}{X_t}$. If there is a jump in the time interval $[t_i,t_{i+1})$, then we can observe it by $R_{t_i}$. We separate the dynamic of $R_t$ in two components $R^J$ and $R^D$, using the hyperparameter $\lambda$ (threshold of a jump presence in $R_t$). The $R_J$ is dedicated to observations with jumps and $R_D$ with the diffusion part. We use the Merton model to represent the jump diffusion process (MJD), because it is more natural to use the Gaussian distribution for jump amplitude in real markets.

To get the intensity parameter $\xi$, we use the maximum likelihood approach for MJD on the given data. For that, we define the initial parameters to achieve better convergence as in [3].

$$\hat{\mu}_D = \frac{2\mathbb{E}[R^D] + \mathrm{Var}[R^D]dt}{2dt}, \quad \hat{\sigma}_D^2 = \frac{\mathrm{Var}[R^D]}{dt}, \quad \hat{\mu}_J = \mathbb{E}[R^J] - \left(\hat{\mu}_D - \frac{\hat{\sigma}_D^2}{2}\right)dt,$$

$$\hat{\sigma}_J^2 = \mathrm{Var}[R^J] - \hat{\sigma}_D^2 dt, \quad \hat{\xi} = \frac{\#R^J}{T}, \tag{3}$$

where $\#R^J$ is the number of jumps (cardinality of a set $R^J$).

Finally, we maximise the likelihood function for the log return process,

$$\log L(R|\xi, \mu_D, \sigma_D^2, \mu_J, \sigma_J^2) =$$

$$= \sum_{i=1}^{T} \log\left(\sum_{k=0}^{\infty} \frac{(\xi dt)^k}{k!} e^{-\xi dt} \cdot \mathcal{N}\left(R_i \Big| (\mu_D - \frac{\sigma_D^2}{2})dt + \mu_J k, \ \sigma_D^2 dt + \sigma_J^2 k\right)\right), \tag{4}$$

where $\mathcal{N}(x|\mu, \sigma^2) = \frac{1}{\sqrt{2\pi\sigma^2}} e^{-(x-\mu)^2/2\sigma^2}$. The bounds for the optimisation task are $\xi \geq 0, \mu_D, \mu_J \in \mathbb{R}$, and $\sigma_D^2, \sigma_J^2 > 0$. The parameter $k$ is dedicated to a number of jumps in a single time period and it can be chosen empirically.

## 2.3   General Framework

For the given $V_0, \Delta t$, the solution for Eq. (1) is equal to $V_t \equiv V_t^\theta = F_\theta(V_0, \Delta t, \mu_\theta, \sigma_\theta, J_\theta) \in \mathbb{R}^{\tau \times M}$. After that, we compute the non-linearity term as $Y \equiv Y_{\eta,\theta} = G_{\eta,\theta} \cdot X + \zeta_\eta(G_{\eta,\theta})$, where $G_{\eta,\theta} = e^{-\psi_\eta(V_t^\top)} \in \mathbb{R}^{M \times T}$. The neural networks $\zeta_\eta$ and $\psi_\eta$ are dedicated to the same part of the model and its parameters are stored in $\eta$. The network $\psi_\eta$ for each $m \in \{1, \ldots, M\}$ takes the vector from $\mathbb{R}^\tau$ as input and produces the vector from $\mathbb{R}^T$. Further, we identify the normalising flow for $Y_{\eta,\theta}$ as a composition $f_\eta(Y, V_t) = f_K \circ f_{K-1} \circ \cdots \circ f_1(Y, V_t)$ and the generative flow as $\tilde{Y} \equiv f_\eta^{-1}(\mathcal{Z}, V_t) = f_1^{-1} \circ \cdots \circ f_{K-1}^{-1} \circ f_K^{-1}(\mathcal{Z}, V_t)$. The value $f_\eta(Y, V_t)$ should be approximately equals to $\mathcal{Z} \sim \mathcal{N}(0, I)$ by distribution.

We propose a new bijection function, which depends on the latent process $V_t$. The forward path of $f_i(Y, V_t)$ starts from separation of the process $Y \in \mathbb{R}^{M \times T}$ in two parts $Y_1 \in \mathbb{R}^{M \times d}$ and $Y_2 \in \mathbb{R}^{M \times (T-d)}$. Further, we calculate the term which depends on the latent process $\hat{p} = p_\eta(V_t^\top)$ to extract the information given on the latent dynamic. Then we implement the affine coupling layer [4] with following modifications,

$$\mathcal{Z}_1 = Y_1 + \hat{p},$$

$$\mathcal{Z}_2 = \Lambda_\eta(\mathcal{Z}_1) \odot g_\eta(\nu_\eta(\mathcal{Z}_1^\top)^\top) + Y_2 \odot e^{\Sigma_\eta(\mathcal{Z}_1)}, \tag{5}$$

where $\odot$ is an element-wise product. After that, we concatenate the outcomes as $\mathcal{Z} = (\mathcal{Z}_1, \mathcal{Z}_2) \in \mathbb{R}^{M \times T}$. The networks $\Lambda_\eta, \Sigma_\eta$ have the same structure with one recurrent layer and several linear layers with sigmoid activation between them. Due to we use the affine coupling layer, we also permute the component $\mathcal{Z}_1$ with $\mathcal{Z}_2$ after each flow.

Our novel idea is to use the transpose version of $\mathcal{Z}_1$ to extract the information from the multivariate dynamic, utilising the neural network $\nu_\eta$, and we transpose the result again to get the term structure dynamic. Then we apply the neural network $g_\eta$ to extract once again the time structure component, but in a new representation of $\mathcal{Z}_1$. The networks $\nu_\eta, g_\eta$ were constructed with two linear layers with sigmoid activation function between them. The flow constructed in this manner maintains the form of the Jacobian of a coupling flow.

Therefore, the log determinant of Jacobian is equal to the sum of elements $\Sigma_\eta(\mathcal{Z}_1)(j)$, where $j \in \{d+1, \ldots, T\}$. For the likelihood distribution we choose the standard Gaussian one.

We define the inverse flow as

$$\widetilde{Y}_2 = (\mathcal{Z}_2 - \Lambda_\eta(\mathcal{Z}_1) \odot g_\eta(\nu_\eta(\mathcal{Z}_1^\top)^\top)) \odot e^{-\Sigma_\eta(\mathcal{Z}_1)},$$
$$\widetilde{Y}_1 = \mathcal{Z}_1 - \hat{p}. \tag{6}$$

The final value $\widetilde{Y}$ achieved by concatenating components $(\widetilde{Y}_1, \widetilde{Y}_2)$. After the flow we implement the inverse procedure for non-linearity term as $\widetilde{X} = (\widetilde{Y} - \zeta_\eta(G_{\eta,\theta}))/e^{-\psi_\eta(V_\theta^\top)}$ by sampling the new latent process with discrete scheme (2). Finally, we get a synthetic multivariate time series $\widetilde{S}$, by applying the inverse preprocessing procedure to $\widetilde{X}$.

# 3 Experimental Study

## 3.1 Data Description

To evaluate the proposed model we use three multivariate time series. The first is dedicated to the jump diffusion so-called Merton model which driven by SDE in a risk-neutral measure as

$$dX_t = \left(r - \xi \cdot (e^{\mu_J + 0.5\sigma_J^2} - 1) - \frac{1}{2}\sigma_D^2\right)dt + \sigma_D dW_t + J dP_t. \tag{7}$$

As parameters we use $\xi = 15, \mu_J = 0, \sigma_J = 0.2, r = 0.04, \sigma_D = 0.6$. The length of the time series and the dimension are equal to 1000 and 10, respectively. By the Merton jump diffusion, we check the ability of the models to detect the complicated dynamics.

The second time series is the diffusion process, driven by DCL stochastic process [10], with $\theta = 1$ and $\delta = 2$. The time series length is equal to 500 and the dimension is 5. Using the DCL process, we wish to test the smoothness of the resulting series.

The third time series is derived from the kaggle stock dataset[1] to analyse the ability of the models to generate the realistic series for the market purposes. The multivariate attributes are formed by open, close, low, and high prices with 2645 total observations.

---

[1] https://www.kaggle.com/datasets/borismarjanovic/price-volume-data-for-all-us-stocks-etfs.

## 3.2   Quality Assessment

We employ three methods to compare results: Fourier Flow [11], fSDE [12] and PAR from SDV [13].

We choose the Jensen-Shannon divergence and $W_1$-distance to compare the distribution similarity. We also utilise the forecasting model to assess the prediction error, $MSE$. We use the local extrema QQ-plot, to evaluate the ability of the model to represent the local dynamic features [14]. To evaluate the multivariate similarity, we use t-SNE plots [15], which provides the dimension reduction with structure preserving.

We introduced a new method to assess global dynamic similarity by comparing signature distributions at different time points. By calculating path integrals and extracting cross sections over the interval [0, T], we can compare the distributions of real-world time series signatures with synthetic data. Using the $W_1$-distance metric, we visualize the differences in a plot for all models. The path integrals in the signatures play a key role in determining the time series dynamics uniquely [16]. This approach helps identify specific time periods where the synthetic data deviates from the real data.

## 3.3   Results

Figure 2 illustrates QQ-plots for local extrema. Our approach JDFlow (red color) is better to approximate the local structure, by matching the diagonal line, which refers to a real extrema quantile (black color). Fourier Flow is also close to the real time series, but PAR (orange color) and fSDE (green color) are too far from the diagonal. For example, PAR's synthetics differ in the first quantiles in the stock dataset, which means it has greater local minima values, than it has the initial time series.

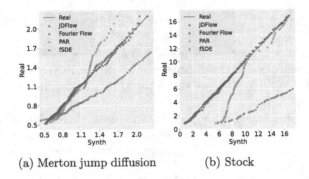

(a) Merton jump diffusion          (b) Stock

**Fig. 2.** QQ-plot for the local extrema comparison. (Color figure online)

Figure 3 shows the $W_1$-distance between cross sections of the signatures in different time periods. The best values should be close to a zero line (black color). PAR and fSDE models greatly differ from the real time series, especially in comparison with our approach JDFlow and Fourier Flow. For the Merton jump diffusion (Fig. 3a), the PAR and fSDE models have a $W_1$-distance that increases significantly over time. This means that their synthetics branch out too much over time. The same structure is observed in the stock data, only JDFlow and Fourier Flow accurately discover the initial data patterns.

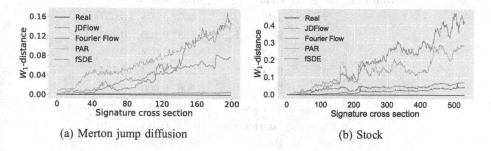

(a) Merton jump diffusion                    (b) Stock

**Fig. 3.** Signature cross section distribution comparison.

Figure 4 shows distinctions in multivariate structure. As a result, real-world time series, JDFlow, and Fourier Flow synthetics have similar multivariate structure. fSDE model repeats the local patterns of the source data, but globally behaves differently. The PAR synthesizer doesn't match the initial multivariate dynamic.

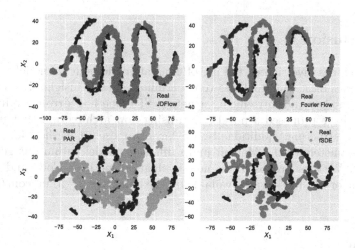

**Fig. 4.** t-SNE plot for stock.

The quantitative evaluation can be found in the Table 1. For all time series JDFlow has shown the best results in terms of distribution similarity (blue color for best results). When analysing $MSE$ metric for the forecasting task, JDFlow shows the best results and even decrease the initial time series prediction error.

**Table 1.** Quality assessment with quantitative metrics.

| Data | Models | $W_1$ | $D_{JS}$ | $MSE$ |
|---|---|---|---|---|
| Merton jump diffusion | Real | 0.000 | 0.0000 | 0.059 |
| | JDFlow | 0.003 | 0.005 | 0.057 |
| | Fourier Flow | 0.153 | 0.136 | 0.439 |
| | PAR | 0.093 | 0.078 | 0.288 |
| | fSDE | 2.275 | 0.303 | 2.181 |
| DCL process | Real | 0.000 | 0.000 | 0.019 |
| | JDFlow | 0.002 | 0.019 | 0.019 |
| | Fourier Flow | 0.094 | 0.345 | 0.101 |
| | PAR | 0.041 | 0.218 | 0.030 |
| | fSDE | 0.422 | 0.333 | 0.451 |
| Stock | Real | 0.000 | 0.000 | 12.655 |
| | JDFlow | 0.022 | 0.007 | 11.423 |
| | Fourier Flow | 0.144 | 0.010 | 12.569 |
| | PAR | 2.351 | 0.143 | 12.704 |
| | fSDE | 5.833 | 0.124 | 13.476 |

## 4 Conclusion

Our research has investigated the use of generative methods to model multivariate time series data. Our goal was to contribute valuable insights by evaluating generative models and introducing a neural SDE method driven by jump diffusion. We have proposed a novel normalising flow architecture that utilises a multivariate structure. Additionally, we have proposed a method for calibrating the parameters of the Merton jump diffusion model within our generative framework. We also proposed the new evaluation method. Code and training parameters for test models are available by the link https://github.com/kirillzx/JDFlow.

## References

1. Cohen, S.N., Reisinger, C., Wang, S.: Arbitrage-free neural-SDE market models. Appl. Math. Finance **30**(1), 1–46 (2023)
2. Esteban, C., Hyland, S.L., Rätsch, G.: Real-valued (medical) time series generation with recurrent conditional GANs. arXiv preprint arXiv:1706.02633 (2017)

3. Tang, F.: Merton jump-diffusion modeling of stock price data (2018)
4. Kobyzev, I., Prince, S.J., Brubaker, M.A.: Normalizing flows: an introduction and review of current methods. IEEE Trans. Pattern Anal. Mach. Intell. **43**(11), 3964–3979 (2020)
5. Geiger, A., Liu, D., Alnegheimish, S., Cuesta-Infante, A., Veeramachaneni, K.: Tadgan: time series anomaly detection using generative adversarial networks. In: 2020 IEEE International Conference on Big Data (Big Data), pp. 33–43. IEEE (2020)
6. Wen, Q., et al.: Time series data augmentation for deep learning: a survey. arXiv preprint arXiv:2002.12478 (2020)
7. Øksendal, B., Øksendal, B.: Stochastic Differential Equations. Springer, Heidelberg (2003). https://doi.org/10.1007/978-3-642-14394-6
8. Liu, X., Xiao, T., Si, S., Cao, Q., Kumar, S., Hsieh, C.J.: Neural SDE: stabilizing neural ode networks with stochastic noise. arXiv preprint arXiv:1906.02355 (2019)
9. Liu, W., Mao, X.: Strong convergence of the stopped euler-maruyama method for nonlinear stochastic differential equations. Appl. Math. Comput. **223**, 389–400 (2013)
10. Domingo, D., d'Onofrio, A., Flandoli, F.: Properties of bounded stochastic processes employed in biophysics. Stoch. Anal. Appl. **38**(2), 277–306 (2020)
11. Alaa, A., Chan, A.J., van der Schaar, M.: Generative time-series modeling with fourier flows. In: International Conference on Learning Representations (2020)
12. Hayashi, K., Nakagawa, K.: Fractional SDE-net: generation of time series data with long-term memory. In: 2022 IEEE 9th International Conference on Data Science and Advanced Analytics (DSAA), pp. 1–10. IEEE (2022)
13. Zhang, K., Patki, N., Veeramachaneni, K.: Sequential models in the synthetic data vault (2022)
14. Zakharov, K., Stavinova, E., Boukhanovsky, A.: Synthetic financial time series generation with regime clustering. J. Adv. Inf. Technol. **14**(6) (2023)
15. Yoon, J., Jarrett, D., Van der Schaar, M.: Time-series generative adversarial networks. In: Advances in Neural Information Processing Systems, vol. 32 (2019)
16. Morrill, J., Fermanian, A., Kidger, P., Lyons, T.: A generalised signature method for multivariate time series feature extraction. arXiv preprint arXiv:2006.00873 (2020)

# *S3LLM*: Large-*S*cale *S*cientific *S*oftware Understanding with *LLMs* Using Source, Metadata, and Document

Kareem Shaik[1], Dali Wang[2], Weijian Zheng[3], Qinglei Cao[4], Heng Fan[1], Peter Schwartz[2], and Yunhe Feng[1(✉)]

[1] University of North Texas, Denton, TX 76207, USA
kareembabashaik@my.unt.edu, {heng.fan,yunhe.feng}@unt.edu
[2] Oak Ridge National Laboratory, Oak Ridge, TN 37830, USA
{wangd,schwartzpd}@ornl.gov
[3] Argonne National Laboratory, Lemont, IL 60439, USA
wzheng@anl.gov
[4] Saint Louis University, St. Louis, MO 63103, USA
qinglei.cao@slu.edu

**Abstract.** The understanding of large-scale scientific software is a significant challenge due to its diverse codebase, extensive code length, and target computing architectures. The emergence of generative AI, specifically large language models (LLMs), provides novel pathways for understanding such complex scientific codes. This paper presents *S3LLM*, an LLM-based framework designed to enable the examination of source code, code metadata, and summarized information in conjunction with textual technical reports in an interactive, conversational manner through a user-friendly interface. *S3LLM* leverages open-source LLaMA-2 models to enhance code analysis through the automatic transformation of natural language queries into domain-specific language (DSL) queries. In addition, *S3LLM* is equipped to handle diverse metadata types, including DOT, SQL, and customized formats. Furthermore, *S3LLM* incorporates retrieval-augmented generation (RAG) and LangChain technologies to directly query extensive documents. *S3LLM* demonstrates the potential of using locally deployed open-source LLMs for the rapid understanding of large-scale scientific computing software, eliminating the need for extensive coding expertise and thereby making the process more efficient and effective. S3LLM is available at https://github.com/ResponsibleAILab/s3llm.

This manuscript has been authored by UT-Battelle, LLC under contract DE-AC05-00OR22725 with the US Department of Energy (DOE). The US government retains and the publisher, by accepting the article for publication, acknowledges that the US government retains a nonexclusive, paid-up, irrevocable, worldwide license to publish or reproduce the published form of this manuscript, or allow others to do so, for US government purposes. DOE will provide public access to these results of federally sponsored research in accordance with the DOE Public Access Plan (http://energy.gov/downloads/doe-public-access-plan).

L. Franco et al. (Eds.): ICCS 2024, LNCS 14834, pp. 222–230, 2024.
https://doi.org/10.1007/978-3-031-63759-9_27

**Keywords:** Large-Scale Scientific Software · Research Software Analysis · E3SM Land Model · Retrieval Augmented Generation (RAG) · LLM · LLaMA · ChatGPT

# 1 Introduction

Large-scale scientific computing software is crucial in various scientific fields, undergoing extensive development cycles that lead to the formation of intricate software libraries and ecosystems. This complexity stems from the lengthy development periods, ongoing extensions, and evolving development paradigms, making it imperative to provide users with insights into these computing tools. However, understanding such software is a challenging task for several reasons. First, large-scale scientific software often incorporates multiple programming languages, including older languages such as Fortran and Pascal, which poses a significant challenge for contemporary programmers trying to understand the code. Second, the large volume of scientific software, which may encompass millions of lines of code, makes comprehensively understanding each segment of the code an additional obstacle. Lastly, the documentation for these software systems is sometimes less than ideal, often lacking detailed explanations, which further complicates the task of gaining a thorough understanding of the software.

To enhance comprehension of large-scale scientific software, numerous tools have been devised to aid in code analysis and documentation. For instance, Doxygen [1] is capable of generating documentation from the source code and performing static code analysis for software source trees. Nonetheless, the available tools are primarily tailored for static code analysis and lack the capability to accommodate dynamic queries. Moreover, given the complexity inherent in large-scale scientific software, both developers and users may struggle to formulate queries in both instructed (e.g., textural documents) and structured formats (e.g., SQL). Therefore, it is imperative to devise methods for understanding and parsing large-scale scientific software that are both user-friendly and precise.

The emergence of generative AI, particularly large language models (LLMs), represents a new era in software comprehension and interaction. LLMs have shown remarkable capabilities across various tasks, including chatbot interactions [3,5,15], text summarization [4,13,14], and content creation [2,7,8], demonstrating their potential to vastly improve programming and documentation practices. Beyond these applications, LLMs offer promising solutions for understanding the complex landscapes of large-scale scientific software [12]. By leveraging LLMs, we can envision a future in which software comprehension is not only more accessible but also more intuitive, enabling users to query and interact with software in natural language. This paper introduces *S3LLM*, a novel framework that embodies this vision by providing a user-friendly interface for interacting with scientific computing software through natural language queries. *S3LLM* aims to bridge the gap between the world of complex scientific software and the diverse community of users and developers, fostering a deeper understanding and facilitating more effective use of these critical computational tools.

As opposed to most existing works on software understanding, the proposed *S3LLM* can handle various types of tasks for large-scale scientific software understanding—including source code query, metadata analysis, and text-based technical report understanding. *S3LLM* is capable of conducting queries over the information extracted from source code in diverse formats, such as DOT (graph description language)[1] and relational databases. By leveraging the few-shot learning capability of LLMs, *S3LLM* can also generate domain-specific language (DSL) queries, such as Feature Query Language (FQL) [17], to gather and extract software features through code analysis. Furthermore, *S3LLM* implements LangChain and retrieval-augmented generation (RAG) [9] schemes to enable text-based queries from technical reports and project summaries. More importantly, all the aforementioned interactions and inquiries facilitated by *S3LLM* are executed utilizing natural language.

The contributions outlined in this paper are summarized as follows:

- We conceptualized, designed, and implemented *S3LLM*, a novel framework that utilizes LLMs to improve the understanding of large-scale scientific software. This framework excels in analyzing source code, metadata, and textual technical reports, providing a holistic approach to software comprehension.
- *S3LLM* presents a user-friendly natural language–based interface that allows users, even those with limited programming knowledge, to easily query and gain insights into scientific software.
- Given the need to balance inference speed with the framework's computational demands, *S3LLM* provides three options featuring LLaMA-2 models with 7B, 13B, and 70B parameters, allowing users to choose the most appropriate model based on their specific requirements.
- Tests on the Energy Exascale Earth System Model (E3SM) [6] shows our model's effectiveness in analyzing source code, metadata, and documents.
- We contribute to the wider scientific computing community by releasing *S3LLM* as an open-source tool, ensuring broad accessibility and usefulness across a spectrum of scientific computing applications and research pursuits.

## 2 Method

### 2.1 Framework Overview

Figure 1 shows that *S3LLM* consists of three main components designed to process source code, code metadata, and textual technical documents. Central to all three components are LLaMA-2 models, which translate natural language into domain-specific language (DSL) queries or perform in-text information analysis and retrieval. Some components within *S3LLM* use the RAG methodology to improve LLM responses by referencing an external knowledge base, enhancing the model's output. *S3LLM* employs open-source LLaMA-2 models (7B, 13B, and 70B) for language understanding, providing a range of model sizes and ensuring reproducibility across different computational settings.

---

[1] https://en.wikipedia.org/wiki/DOT_(graph_description_language).

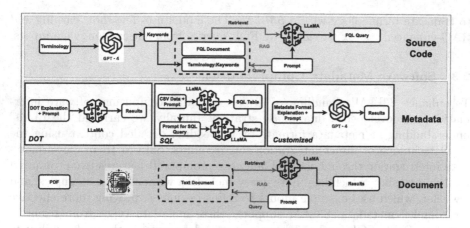

**Fig. 1.** Framework overview of *S3LLM*

## 2.2   Source Code Analysis

Large-scale scientific software like E3SM often involves over a million lines of source code, posing challenges for open-source LLMs due to the massive token count exceeding their context windows. Additionally, contemporary sophisticated source code analysis tools demand extensive programming knowledge or specialized domain expertise from users. To tackle these issues, *S3LLM* integrates the analytical strengths of existing tools with the natural language capabilities of LLMs, enabling source code queries in natural language without loading the entire codebase into the LLMs.

To demonstrate *S3LLM*'s strategy for LLM-based source code analysis, we use XScan as a key backend engine [16]. XScan, an integrated toolkit, extracts essential data from scientific code like lines of code, programming languages, library dependencies, and features of parallel software. While XScan simplifies tasks like line tallying, it employs FQL for deeper static code analysis. FQL, which handles queries like Library Utilization, Version Assessment, and Feature Enumeration, requires skills for precise query crafting. *S3LLM* bridges the gap by enabling accessible source code analysis with less need for extensive programming knowledge. Additionally, the core LLMs of *S3LLM* are versatile and can integrate with various source code analysis engines, not limited to XScan.

In transforming natural language into FQL queries, *S3LLM* focuses on three aspects: understanding FQL's purpose and syntax, grasping relevant terminologies like high-performance computing standards, and accurately translating natural language to FQL. To equip LLM models with a solid understanding of FQL, we integrated a foundational FQL document and query examples into the RAG framework as external resources. Addressing the challenge of converting specific terminologies into programming code keywords, initial efforts using open-source LLMs were inadequate. Consequently, GPT-4 was chosen for its ability to translate terminology within *S3LLM*. The model uses few-shot learning and RAG-enhanced contexts to generate FQL queries from natural language inputs.

## 2.3   Software Metadata Comprehension

To enhance S3LLM's handling of diverse scientific computing software information, we developed a dual-phase strategy. Initially, we enriched the LLMs' understanding of metadata formats by embedding detailed context using the RAG technique or direct integration into prompts. S3LLM then allows metadata interrogation through natural language prompts. It features two prompting modes: zero-shot, which lets users query metadata without prior examples, and few-shot, which uses example inputs to refine responses, proving more effective for complex metadata. S3LLM supports three metadata types from scientific software: DOT, SQL, and third-party specified formats, with the flexibility to expand to more in the future. For DOT, we enhance understanding by including a detailed format explanation in prompts. SQL queries are directly generated using LLaMA-2 models, bypassing traditional instruction-based prompts. For custom third-party formats, we showcased S3LLM's capabilities with SPEL, a toolkit for adapting E3SM models for GPU execution, demonstrating S3LLM's ability to handle specialized data formats through GPT-4. This broadens its utility across various metadata types.

## 2.4   Technical Document Interpretation

Large-scale scientific software is often accompanied by detailed documentation such as technical reports, user manuals, and research papers, which can be challenging to navigate. To efficiently extract information from these extensive texts, *S3LLM* utilizes a combination of RAG and LLMs to improve the accuracy of document-related queries. The RAG framework operates in three phases: document indexing, retrieval, and generation. It first segments external texts for the LLM's contextual analysis, then indexes and stores document embeddings. Upon receiving a query, it retrieves relevant embeddings to provide a context window for the LLM, which generates responses using both the query and the retrieved data. *S3LLM* incorporates the LangChain framework, an advanced tool for developing LLM applications, to implement RAG. LangChain's DocumentLoaders and Text Splitters organize and segment documents for query processing. VectorStore and Embeddings models, using *all-MiniLM-L6-v2* for embeddings and a FAISS-based index for efficient retrieval, maintain document embeddings. The Retriever component fetches relevant segments for inclusion in user prompts. Finally, the refined query, enhanced with retrieved data, is processed by LLaMA-2 to produce customized responses, demonstrating the effective integration of RAG to enhance document comprehension in scientific software.

# 3 Case Study

We deployed *S3LLM* on E3SM as a case study to demonstrate its effectiveness in analyzing source code, code metadata, and text-based technical reports. E3SM is a state-of-the-art Earth system modeling framework, developed and supported by the United States Department of Energy (DOE) [6].

## 3.1 Source Code Query Results

To illustrate the effectiveness of *S3LLM* in source code analysis, we assessed its ability to generate FQL queries across three distinct categories: Library Utilization Queries, Version Assessment Queries, and Feature Enumeration Queries. Through the use of crafted prompts, as demonstrated in the subsequent text box, *S3LLM* successfully generates the correct FQL query in certain instances. Specifically, for the Library Utilization Query, *S3LLM* produces the precise FQL query in a zero-shot manner. This accuracy is likely attributed to the pretraining of the LLaMA-2 model within *S3LLM* on RAG supporting documents that include FQL queries and HPC terminologies. Similarly, for Version Assessment and Feature Enumeration Query, "*Please Generate FQL query to Find the minimum version of MPI*" and "*Please Generate FQL query to List MPI process topology used in the Code*" can prompt *S3LLM* to generate accurate queries.

---

**Prompt for Library Utilization Query:**
*Please Generate FQL query to Check Whether OpenMP is used*

**Response:**
```
CHECK (omp task || end task ||  omp taskloop || omp taskloop simd
|| omp taskyield) WHERE (*)
```

---

## 3.2 Metadata Comprehension with *S3LLM*

In our study, we utilized three distinct forms of metadata to demonstrate the capabilities of *S3LLM* in enhancing metadata comprehension. These include DOT files produced by XScan [16], CSV files generated by SPEL [11][2], and bespoke metadata formats also created by SPEL, all derived from E3SM.

First, *S3LLM*'s efficacy was assessed on standardized DOT files, as produced by XScan and on a unique metadata format devised by SPEL. A DOT file, as illustrated in the box below, represents a directed graph that visualizes the interactions between caller and callee functions within E3SM, offering insights into the software's structural dynamics.

---

[2] A software toolkit for porting E3SM land models onto GPUs using OpenACC.

> **Prompt for Summarizing DOT File:**
> *DOT files can describe a directed graph using the keyword of "digraph". Inside a directed graph, each line starts with a module name, which is followed by the "::" symbol. For example, "allocationmod" is the module in "allocationmod::allocationinit -> clm_time_manager." Please list all the unique modules in the following DOT file describing directed graph G.*
>
> ```
> digraph G {
>     emi_datamod::emidcopy -> emi_datamod::emidallocatememory_real_3d
>     ...
>     firemod::firefluxes -> histfilemod::hist_addfld1d
> }
> ```
>
> **Response:**
> Sure! Here are the unique modules in the directed graph G described by the DOT file:
> 1. emi_datamod ... 17. verticalprofilemod   There are 17 unique modules in total.

SQL is widely utilized for managing data in relational databases, and language models can translate natural language into SQL queries efficiently. We demonstrated the SQL capabilities of *S3LLM* using two CSV files from SPEL [11]. The process involved two steps. Initially, *S3LLM* generated SQL statements to convert the subtables into SQL tables. Then, it created specific SQL queries. Our experimental results demonstrate *S3LLM*'s skill in handling and manipulating SQL tables for complex database tasks.

Additionally, a sequence of executions for `LakeTemperature` as delineated by SPEL [11] has been tested. The result offers an in-depth view of variable usage during the execution process, which is beneficial for code performance optimization (through asynchronous kernel launch) on GPUs. This structured approach to metadata analysis underscores *S3LLM*'s capability to navigate and elucidate complex metadata representations, enhancing comprehension and facilitating deeper insights into the software's operational mechanisms. Due to the space limitation, please refer to our GitHub repository for more details.

### 3.3 Document Parsing

This section demonstrates the effectiveness of combining the RAG framework and LangChain with LLMs, as described in Sect. 2.4, for extracting specific technical information from comprehensive scientific documents. We utilized a 28-page technical report on lake temperature (in PDF format), from the Community Land Model [10], as the testbed. The example experiment in the following colored box confirms *S3LLM*'s proficiency in extracting complex physical processes, such as phase changes, and its capability in both comprehension and explanation.

> **Prompt for Querying Technical Report of Lake Temperature:**
> *What is the significance of phase changes in the model, and how are they accounted for?*
>
> **Response:** Phase changes, such as freezing and melting, play a significant role in the atmosphere model, as they affect the energy balance at the land surface. The model accounts for phase changes by calculating heat diffusion and phase change in given layers based on temperature and moisture conditions. The available energy for melting is computed as function of temperature and moisture content.

# 4 Conclusion

This paper introduces *S3LLM*, a robust framework built on LLMs designed to understand the intricacies of large-scale scientific software. Through advanced techniques such as instruction-based prompting, integration with external queries, and the use of RAG and LangChain, *S3LLM* enhances the operational capabilities of pre-trained LLMs, and its effectiveness across various data formats was demonstrated herein. This framework not only underscores our current achievements but also lights the way for future advancements in generative AI and software engineering, particularly within scientific computing.

# References

1. Doxygen. https://www.doxygen.nl/. Accessed 01 Mar 2024
2. Acharya, A., Singh, B., Onoe, N.: LLM based generation of item-description for recommendation system. In: Proceedings of the 17th ACM RecSys, pp. 1204–1207 (2023)
3. Bill, D., Eriksson, T.: Fine-tuning a LLM using reinforcement learning from human feedback for a therapy chatbot application (2023)
4. Chang, Y., Lo, K., Goyal, T., Iyyer, M.: Booookscore: a systematic exploration of book-length summarization in the era of LLMs. arXiv preprint arXiv:2310.00785 (2023)
5. Freire, S.K., Wang, C., Niforatos, E.: Chatbots in knowledge-intensive contexts: comparing intent and LLM-based systems. arXiv preprint arXiv:2402.04955 (2024)
6. Golaz, J.-C., et al.: The DOE E3SM model version 2: overview of the physical model and initial model evaluation. J. Adv. Model. Earth Syst. **14**(12), e2022MS003156 (2022)
7. Huang, S., Zhao, J., Li, Y., Wang, L.: Learning preference model for LLMs via automatic preference data generation. In: The 2023 Conference on EMLP (2023)
8. Jury, B., Lorusso, A., Leinonen, J., Denny, P., Luxton-Reilly, A.: Evaluating LLM-generated worked examples in an introductory programming course. In: Proceedings of the 26th Australasian Computing Education Conference, pp. 77–86 (2024)
9. Lewis, P., et al.: Retrieval-augmented generation for knowledge-intensive NLP tasks. In: Advances in Neural Information Processing Systems, vol. 33, pp. 9459–9474 (2020)
10. Oleson, K.W., Lawrence, D.M., et al.: Technical description of version 4.0 of the community land model (CLM). NCAR Tech. Note NCAR/TN-478+ STR **257**, 1–257 (2010)
11. Schwartz, P., Wang, D., Yuan, F., Thornton, P.: SPEL: software tool for porting E3SM land model with openacc in a function unit test framework. In: 2022 Workshop on Accelerator Programming Using Directives (WACCPD), pp. 43–51. IEEE (2022)
12. Tsigkanos, C., Rani, P., Müller, S., Kehrer, T.: Variable discovery with large language models for metamorphic testing of scientific software. In: Mikyška, J., de Mulatier, C., Paszynski, M., Krzhizhanovskaya, V.V., Dongarra, J.J., Sloot, P.M. (eds.) ICCS 2023. LNCS, vol. 14073, pp. 321–335. Springer, Cham (2023). https://doi.org/10.1007/978-3-031-35995-8_23
13. Van Veen, D., Van Uden, C., et al.: Clinical text summarization: adapting large language models can outperform human experts. Res. Square (2023)

14. Zhang, T., Ladhak, F., et al.: Benchmarking large language models for news summarization. Trans. Assoc. Comput. Linguist. **12**, 39–57 (2024)
15. Zheng, L., Chiang, W.-L., et al.: Judging LLM-as-a-judge with MT-bench and chatbot arena. In: Advances in Neural Information Processing Systems, vol. 36 (2024)
16. Zheng, W., Wang, D., Song, F.: XScan: an integrated tool for understanding open source community-based scientific code. In: Rodrigues, J.M.F., et al. (eds.) ICCS 2019. LNCS, vol. 11536, pp. 226–237. Springer, Cham (2019). https://doi.org/10.1007/978-3-030-22734-0_17
17. Zheng, W., Wang, D., Song, F.: FQL: an extensible feature query language and toolkit on searching software characteristics for HPC applications. In: Tools and Techniques for High Performance Computing, pp. 129–142 (2020)

# A Novel Iterative Decoding for Iterated Codes Using Classical and Convolutional Neural Networks

Marek Blok[1] (ID) and Bartosz Czaplewski[2](✉) (ID)

[1] Artificial Intelligence Center of Excellence, Hapag-Lloyd Knowledge Center Sp. z o.o., Grunwaldzka 413, 80-309 Gdańsk, Poland
[2] Faculty of Electronics, Telecommunications and Informatics, Gdansk University of Technology, Gabriela Narutowicza 11/12, 80-233 Gdansk, Poland
bartosz.czaplewski@pg.edu.pl

**Abstract.** Forward error correction is crucial for communication, enabling error rate or required SNR reduction. Longer codes improve correction ratio. Iterated codes offer a solution for constructing long codes with a simple coder and decoder. However, a basic iterative code decoder cannot fully exploit the code's potential, as some error patterns within its correction capacity remain uncorrected. We propose two neural network-assisted decoders: one based on a classical neural network, and the second employing a convolutional neural network. Based on conducted research, we proposed an iterative neural network-based decoder. The resulting decoder demonstrated significantly improved overall performance, exceeding that of the classical decoder, proving the efficient application of neural networks in iterative code decoding.

**Keywords:** Forward Error Correction · Neural Networks · Iterated Codes

## 1 Introduction

Forward error correction (FEC) is an effective approach to error detection and correction. This protection is achieved by adding parity bits to the transmitted data. Longer codes offer better protection but typically require more complicated decoders. An interesting method of long code construction with simple decoder are iterated codes [1, 2]. They combine two or more simpler codes into one longer code. Using this approach, a message can be fit into 2D table and then encoded iteratively, i.e. the rows are encoded first, followed by the encoding of the columns. Decoding is a reverse process and is also iterative. The correction capacity of the iterated code significantly increases since the minimum Hamming distance of the code is a product of the distances of the component codes. Unfortunately, the classic, iterative, decoder is not able to correct all error patterns even if their weight is within the code's correction capacity. To address this, we explore the error correction capabilities of classical neural networks (NN) and convolutional neural networks (CNN) [3–7].

L. Franco et al. (Eds.): ICCS 2024, LNCS 14834, pp. 231–238, 2024.
https://doi.org/10.1007/978-3-031-63759-9_28

This paper aims to introduce an innovative decoder structure for iterated codes, utilizing NNs and CNNs to improve error correction effectiveness. To the best of the authors' knowledge, this paper represents the first published research on decoding iterated codes using machine learning techniques ([8] was the groundwork). The contribution comprises three aspects: 1. Custom datasets; 2. NN and CNN models trained for the considered iterated code; 3. NN-based and CNN-based iterated decoder, wherein the proposed NN and CNN are used iteratively. The CNN-based iterative decoder achieves higher error correction efficiency than the classic detector.

## 2 Related Work

There are publications that use machine learning for decoding different FEC codes: Hamming codes, Bose–Chaudhuri–Hocquenghem (BCH) codes, polar codes, turbo codes, Reed–Solomon (RS) codes, or Low-Density Parity-Check (LDPC) codes.

Article [9] introduced an RNN-based decoder for Hamming codes, offering the advantage of correcting noisy code words in non-binary form. In [10], the BCH codes were decoded with recurrent neural network (RNN). The interconnection of decoder parameters in iterations to form a RNN was implemented. The article [11] involves incorporating redundant symbols for synchronization in noisy channels. It enhances decoding precision by forwarding soft decisions from the synchronization module to the receiver's decoder. The next article [12] introduced successive cancellation and belief propagation decoding algorithms for polar codes. It discusses deep learning (DL) decoders, their principles and performance assessments. In [13] an architecture for turbo code decoding was proposed. This Deep Turbo decoder showed outstanding performance in both AWGN (Additive White Gaussian Noise) and non-AWGN channels. In [14], DNNs for improving belief propagation decoding were tested for BCH codes with cycle-reduced parity-check matrices. It provided guidelines for refining a model by incorporating domain knowledge into a DL model. In [15], the authors used the advantages of DL and conventional LDPC decoding with normalized min-sum by distributing the iterative decoding between check nodes and variable nodes in a forward propagation network. In [16], RNN-based decoder was tested on recursive systematic convolutional (RSC) codes. In [17], graph neural network (GNN) was utilized for LDPC and BCH codes. Unlike many DL-based decoding methods, that solution is scalable to arbitrary block lengths and minimalizes problem of dimensionality during training. Next study [18] explored the integration of NN into polar codes decoding. The authors showed a fresh approach to reducing complexity by leveraging constituent codes and introduced a scheduling scheme for decoding latency reduction. The recent work [19] focused on a CNN-based LDPC decoder tailored for optical fiber communication systems. Employing a correlated Gaussian noise model CNN can identify noise correlation, leading to enhanced decoding performance through iterative processing. The study [20] introduced shared graph neural network (SGNN) decoding algorithms, reducing parameters by half or three-quarters for BCH and LDPC codes, while maintaining a marginal degradation in performance.

The solutions proposed in this article stem from research initiated in [8], which first attempted to use NNs to decode iterated codes. The article presented two NNs trained on different datasets for error detection and correction, and an NN-assisted decoder for iterated codes leveraging both networks to decode iterated codes.

## 3  The Considered Iterated Code

Let's consider double-iterated code [1, 2] composed of two FEC codes capable of correcting a single error: a Hamming code (7,4) with the generator matrix $\mathbf{G}_1$ and a shortened Hamming code (6,3) with the generator matrix $\mathbf{G}_2$ (2). With $\mathbf{G}_i$ and information vector $\mathbf{x}$, a code word $\mathbf{c}_i$ for the $i$-th component code is obtained as in (1).

$$\mathbf{G}_1 = \begin{bmatrix} 1\,0\,0\,0\,1\,1\,0 \\ 0\,1\,0\,0\,1\,0\,1 \\ 0\,0\,1\,0\,0\,1\,1 \\ 0\,0\,0\,1\,1\,1\,1 \end{bmatrix}, \mathbf{G}_2 = \begin{bmatrix} 1\,0\,0\,1\,1\,0 \\ 0\,1\,0\,1\,0\,1 \\ 0\,0\,1\,0\,1\,1 \end{bmatrix} \tag{1}$$

$$\mathbf{c}_i = \mathbf{x} \cdot \mathbf{G}_i \tag{2}$$

The first code has 4 information bits and a length of 7, while the second has 3 information bits and a length of 6. We split the message into 12-bit information blocks formed as a $3 \times 4$ matrix. First, each row is encoded with first code $\mathbf{G}_1$ resulting in 3 rows of 7-bit words. Then, each column is encoded with the second code $\mathbf{G}_2$ resulting in 7 columns of 6-bit words. The resulting $6 \times 7$ matrix forms a 42-bit code word with 12 information bits. The code rate is 0.285, and the error-correcting capacity is 4.

The classical decoder also follows the iterated approach of the coder. The parity check matrices (3) of Hamming codes can be easily obtained from generator matrices.

$$\mathbf{H}_1 = \begin{bmatrix} 1\,1\,0\,1\,1\,0\,0 \\ 1\,0\,1\,1\,0\,1\,0 \\ 0\,1\,1\,1\,0\,0\,1 \end{bmatrix}, \mathbf{H}_2 = \begin{bmatrix} 1\,1\,0\,1\,0\,0 \\ 1\,0\,1\,0\,1\,0 \\ 0\,1\,1\,0\,0\,1 \end{bmatrix} \tag{3}$$

$$\mathbf{s}_i = \mathbf{H}_i \cdot \mathbf{y} \tag{4}$$

The received vector $\mathbf{y}$ is converted into $6 \times 7$ matrix and the decoder calculates the syndrome $\mathbf{s}_i$ for each row with $\mathbf{H}_1$ and each column with $\mathbf{H}_2$ as in (4). For both component codes, syndromes are binary vectors of length 3. A syndrome with all zeros means that there is no need for correction while the other possible values correspond to 7 error patterns that are used to correct errors in rows and columns accordingly.

## 4  Network Structures

The first concept was to use a NN. Since syndrome calculations and error correction equations used in the studied iterated code are not highly complex mathematical problems, it was initially assumed that such a simple solution would suffice. Various NN structures with different numbers of layers, neurons, and activation functions were examined. After a grid search, the structure depicted in Fig. 1 exhibited the best results. The proposed NN consists of an $1 \times 42$ input layer for a 42-bit received message, two fully connected layers (512 neurons each), and an $1 \times 42$ output layer for a 42-bit predicted error vector. ReLU activation functions are applied after the hidden layers, while a sigmoid activation function is used after the output layer. The total number of trainable parameters is 306218.

The second concept was to use a CNN that utilizes spatial properties of the code. The encoding of a 6 × 7 matrix is done by solving 2D parity-check equations which impact individual rows and columns. In the result, an error in a received message distorts only parity checks in the row and the column at which it is located. It was hypothesized that it is possible to use CNN to learn these spatial features using filters in the sizes of the rows and columns in the message matrix. Various CNN structures were examined in a grid search. The structure presented in Fig. 2 exhibited the best results. Here, the input layer size is 6 × 7 for a 42-bit received message. The data is passed to two parallel convolutional layers: one for scanning the rows of the message (512 horizontal filters, size of 1 × 7, ReLU activation) and the other for scanning the columns (512 vertical filters, size of 6 × 1, ReLU activation). No pooling layers were used due to the low data dimensionality and no dropout layers were employed because overfitting is not a concern given the nature of the data. Convolutional layers produce 512 6 × 1 and 512 1 × 7 feature maps, which are concatenated into a feature vector of length 6656 and passed to two fully connected layers (512 and 256 neurons, ReLU activation). The output layer utilizes a sigmoid activation function and outputs a 42-bit predicted error vector. The number of trainable parameters is 3558186.

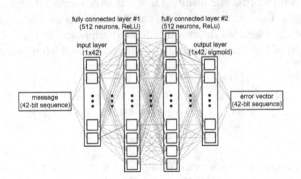

**Fig. 1.** The proposed NN model.

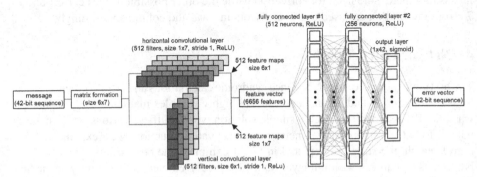

**Fig. 2.** The proposed CNN model.

# 5   Network Training

The training dataset consists of 872086 pairs of 42-bit messages and 42-bit error vectors. The number of errors in messages in the training dataset ranges from 1 to 4. This way, the networks can learn the actual behavior of the code because the code's error-correcting capability is equal to 4. The training dataset was utilized for supervised learning through a 10-fold cross-validation approach. Average validation loss and accuracy were calculated for the entire 10-fold cross-validation. The model with the lowest validation loss among all the folds was selected for the testing phase.

The test dataset contains 4912 pairs of 42-bit messages and 42-bit error vectors. The number of errors in messages in the test dataset ranges from 0 to 5. This way, we can verify the model's ability to predict zero error vectors or vectors with error weight exceeding the error-correcting capability of the code The test dataset was used for the final verification of the trained model. Total test loss, total test accuracy, and test accuracy for specific error weights were computed.

Learning parameters were assessed using the grid search method with the criterion of effective network training within a reasonable training time. Batch size was 128, solving algorithm was Adam (beta1 = 0.9, beta2 = 0.999, eps = 1e−08, decay = 0), initial learning rate was 0.001, learning rate schedule was ReduceLROnPlateau (mode = 'min', factor = 0.1), loss function was binary cross entropy (BCEWithLogitsLoss from the PyTorch), and max epochs was 200 with validation patience of 2. The implementation environment used Python 3.11.0, PyTorch 2.0.1 + cu118, Nvidia GeForce GTX 1080 Ti GPU, and CUDA 12.2.

# 6   The Proposed Neural Network-Based Iterative Decoder

To analyze the networks' performance in more detail, the classification accuracy was calculated additionally for datasets with messages containing from 0 up to 7 errors. For weights 0 and 1 all possible combinations of code words and error patterns have been tested (4096 and 172033 messages) while for larger weights 200000 randomly selected unique messages have been used in tests. We analyzed the error correction patterns for the classic decoder and the proposed networks as stand-alone detectors. Due to the limited number of pages, we are unable to include these results.

It was observed that the classic iterated decoder does not correct all the errors (misses 1, 2, or 4 errors) and at the same time adds additional errors (up to 3 new errors) when decoding messages with error vectors of weight 4. This pattern is different for NN and CNN networks, which are more likely to omit or add to the message just a single error, and almost never incorrectly detect all 4 errors. The neural networks working as stand-alone detectors almost always manage to detect some of the errors from the pattern, and in most cases when they fail and introduce additional errors, they add just one error. Thus, even if the proposed networks fail to correctly decode messages, the error pattern weight is decreased; thus, it is very likely that the network will be able to correct the message in the second attempt. These observations led us to the proposed neural network-based iterative decoder presented in Fig. 3.

In Fig. 4 the iterative neural network-based decoders are directly compared with the classic decoder. The number of correction attempts in the experiments has been limited to

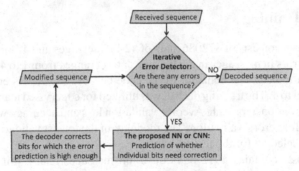

**Fig. 3.** The proposed neural network-based iterative decoder.

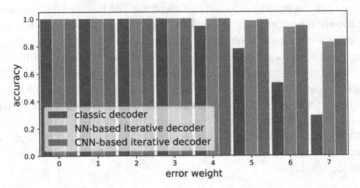

**Fig. 4.** Test accuracy results of the proposed neural network-based iterative decoders.

9. As we can see in Fig. 4, both NN and CNN architectures used iteratively significantly improve their performance with CNN achieving better results.

As we can see in Table 1, the NN-based iterative decoder almost perfectly corrected messages with error weights equal up to 3, only marginally being worse than the classic decoder in those cases, while beating it for larger error weights. These results have been even improved by the CNN-based iterative decoder which corrected all errors with the weight up to 3, provided almost perfect results for weights 4 and 5, and even showed very high performance for error vector weights of 6 and 7. Compared to the classic decoder, the results for the proposed CNN-based iterative decoder are a major improvement over a classical approach, especially in cases where the weight of the error vector exceeds the correction capacity of the code.

In most cases, the first attempt leads to the correct solution. Nevertheless, in the remaining cases, additional correction attempts are required. The distributions of the numbers of iterations for different error weights are presented in Table 2. The second correction attempt is usually sufficient with additional iterations needed for errors weights above 4. It can be observed that the CNN-based iterative decoder not only allows for improved accuracy but also reduces the number of required iterations.

**Table 1.** Test accuracy results of the iterative decoders with specific error vectors' weights.

| dataset error weight | classic decoder | NN-based decoder | CNN-based decoder |
|---|---|---|---|
| 0–2 | 1.00000 | 1.0 | 1.0 |
| 3 | 1.00000 | 0.999998 | 1.0 |
| 4 | 0.94565 | 0.998591 | 0.999861 |
| 5 | 0.78226 | 0.985696 | 0.990726 |
| 6 | 0.53211 | 0.936987 | 0.948255 |
| 7 | 0.29275 | 0.826528 | 0.845893 |

**Table 2.** Percentage of cases of decoding finished at a given iteration.

| Number of iterations | 1 | 2 | 3 | 4 | 5 | 6+ |
|---|---|---|---|---|---|---|
| NN-based iterative decoder | 69,36% | 24,67% | 4,43% | 1,02% | 0,25% | 0,27% |
| CNN-based iterative decoder | 71,70% | 24,70% | 3,17% | 0,36% | 0,04% | 0,03% |

## 7 Conclusions

It is possible to train a neural network to decode iterated codes. The analyzed problem relies on the network's ability to learn decoding rules from limited examples. The proposed networks, even if they fail to correct messages, typically decrease the number of errors. The proposed NN-based and CNN-based iterative decoders outperform the classic code decoder, with CNN-based iterative decoder proving particularly effective. For this study, we have selected an iterated code based on two different component codes, with a decoder not fully exploiting the code's capacity. These results prompt further research on different iterated codes or a wider range of FEC codes.

**Acknowledgments.** This study was funded by the statutory activities of Faculty of Electronics, Telecommunications, and Informatics of Gdańsk University of Technology, and by the research subsidy from the Polish Ministry of Science and Higher Education.

**Disclosure of Interests.** The authors have no competing interests to declare that are relevant to the content of this article.

## References

1. Elias, P.: Error free coding. IRE Trans. Inf. Theory **PGIT-4**, 29–37 (1954)
2. Lin, S., Costello, D.: Error Control Coding, 2nd edn. Prentice Hall, Upper Saddle River (2004)
3. Aggarwal, C.C.: Neural Networks and Deep Learning: A Textbook. Springer, Cham (2019)
4. Krohn, J., Beyleveld, G., Bassens, A.: Deep Learning Illustrated: A Visual, Interactive Guide to Artificial Intelligence. Addison-Wesley Data & Analytics Series. Addison-Wesley, Boston (2019)

5. Czaplewski, B., Dzwonkowski, M.: A novel approach exploiting properties of convolutional neural networks for vessel movement anomaly detection and classification. ISA Trans. **119**, 1–16 (2022)

6. Czaplewski, B., Dzwonkowski, M., Panas, D.: Convolutional neural networks for C. Elegans muscle age classification using only self-learned features. J. Telecommun. Inf. Technol. **4**, 85–96 (2022)

7. Czaplewski, B.: An improved convolutional neural network for steganalysis in the scenario of reuse of the stego-key. In: Tetko, I.V., Kůrková, V., Karpov, P., Theis, Fabian (eds.) ICANN 2019. LNCS, vol. 11729, pp. 81–92. Springer, Cham (2019). https://doi.org/10.1007/978-3-030-30508-6_7

8. Fiedot, O., Blok, M.: Decoding of iterated codes with the use of a neural network. In: Proceedings of the KRiT 2023 Conference on Radiocommunications and Teleinformatics, Cracow, Poland (2023)

9. Abdelbaki, H., Gelenbe, E., El-Khamy, S.E.: Random neural network decoder for error correcting codes. In: Proceedings of the International Joint Conference on Neural Networks, vol. 5, pp. 3241–3245 (1999)

10. Nachmani, E., Marciano, E., Lugosch L., Gross, W.J., Burshtein, D., Be'ery, Y.: Deep learning methods for improved decoding of linear codes. IEEE J. Sel. Top. Signal Process. **12**(1), 119–131 (2018)

11. Chadov, T.A., Erokhin, S.D., Tikhonyuk, A.I.: Machine learning approach on synchronization for FEC enabled channels. In: Proceeding of 2018 Systems of Signal Synchronization, Generating and Processing in Telecommunications, Minsk, Belarus (2018)

12. Gross, W.J., Doan, N., Mambou, E.N., Hashemi, S.A.: Deep learning techniques for decoding polar codes. In: Luo, F.-L. (ed.) Machine Learning for Future Wireless Communications. Wiley, Hoboken (2020)

13. Jiang, Y., Kim, H., Asnani, H., Kannan, S., Oh, S., Viswanath, P.: DEEPTURBO: deep turbo decoder. In: Proceedings of the IEEE 20th International Workshop on Signal Processing Advances in Wireless Communications SPAWC (2019)

14. Be'ery, I., Raviv, N., Raviv, T., Be'ery, Y.: Active deep decoding of linear codes. IEEE Trans. Commun. **68**(2), 728–736 (2020)

15. Wang, Q., Wang, S., Fang, H., Chen, L., Chen, L., Guo, Y.: A model-driven deep learning method for normalized min-sum LDPC decoding. In: Proceedings of the 2020 IEEE International Conference on Communications Workshops (ICC Workshops, Dublin, Ireland), pp. 1–6 (2020)

16. Devamane, S.B., Itagi, R.L.: Recurrent neural network based turbo decoding algorithms for different code rates. J. King Saud Univ. Comput. Inf. Sci. **34**, 2666–2679 (2022)

17. Cammerer, S., Hoydis, J., Aoudia, F.A., Keller, A.: Graph neural networks for channel decoding. In: Proceedings of 2022 IEEE Globecom Workshops (GC Wkshps), Rio de Janeiro, Brazil, pp. 486–491 (2022)

18. Zhou, L., Zhang, M., Chan, S., Kim, S.: Review and evaluation of belief propagation decoders for polar codes. Symmetry **14**(12), 2633 (2022)

19. Zhang, J., et al.: An iterative BP-CNN decoder for optical fiber communication systems. Opt. Lett. **48**(9), 2289–2292 (2023)

20. Wu, Q., Ng, B.K., Lam, C.-T., Cen, X., Liang, Y., Ma, Y.: Shared graph neural network for channel decoding. Appl. Sci. **13**(23), 12657 (2023)

# Towards a Generation of Digital Twins in Healthcare of Ischaemic and Haemorrhagic Stroke

Alfons G. Hoekstra[1]([✉]), Henk Marquering[2],
and on behalf of the GEMINI consortium

[1] Computational Science Lab, University of Amsterdam, Amsterdam, The Netherlands
a.g.hoesktra@uva.nl
[2] Amsterdam University Medical Center, Amsterdam, The Netherlands
h.a.marquering@amsterdamumc.nl
https://dth-gemini.eu/team/

**Abstract.** We introduce our approach towards development of Digital Twins in Healthcare for both ischemic and haemorrhagic stroke, in relation to aetiology and prevention, treatment, and disease progression. These models start their development as generic Digital Twins and in a series of steps are stratified to become fully patient-specific. The example of a Digital Twin in Healthcare for treatment of ischemic stroke is described, and an outlook on the need and applicability of such digital twin in relation to stroke is provided.

**Keywords:** Digital Twin in Healthcare · Stroke

## 1 Introduction

### 1.1 Context

Within the context of the recently started GEMINI project [1] our aim is to deliver validated multi-organ and multi-scale computational models for improved treatment and fundamental understanding of acute strokes, both haemorrhagic and haemorrhagic, and demonstrate the added benefit of these computational models for personalised disease management. This contribution describes our vision and approach towards fully personalized credible Digital Twins in Health for ischemic and haemorhaggic strokes.

### 1.2 Stroke

Neurological diseases are amongst the most challenging and expensive diseases around us, of which strokes are by far the most significant and expensive for society [2]. In the last decades, momentous progress has been realised in the understanding and treatment of stroke, resulting in some of the most effective novel treatment introductions in the whole field of medicine, namely the thrombectomy for intracranial large vessel occlusion stroke and endovascular treatment of intracranial aneurysms. However, we might

L. Franco et al. (Eds.): ICCS 2024, LNCS 14834, pp. 239–245, 2024.
https://doi.org/10.1007/978-3-031-63759-9_29

hit a block regarding treatment advancement, as the large parts of previous improvements are based on (1) treatment on a population level as the results of clinical trials and (2) decades of research with animal models. While valuable approaches, these efforts do not cover the patient-specific and multifaceted aspects of stroke (e.g., comorbidities, multi-organ involvement, the initiation of the stroke, and the subacute progression of the diseases) that are required to advance the field of stroke prevention, diagnosis, treatment, and monitoring, and improve patient care. Indeed, medical professionals currently lack patient-specific decision-making tools and well-established models that can accurately assist in the diagnosis and stratification of stroke patients for tailored treatments and device choice. Personalised computational modelling provides a viable, promising, and clinically required approach to improve patient management. On top of that, personalised computational modelling can address other (clinical) needs regarding stroke, such as fragmented knowledge from different medical specialties, evidence obtained in selected populations in protected environments, simplistic mono-organ disease models, and interdependencies between factors that are now considered independent [3].

### 1.3 Digital Twin in Healthcare

We follow the definitions as used in the coordination action EDITH, [4] where a Digital Twin in Healthcare (DTH) is defined as a computer simulation that predicts quantities of interest necessary to support decision-making within a specific context of use in healthcare. A DTH can be (1) generic: the predicted value is within the range of the values measured experimentally in a reference population, (2) population-specific: the predicted value is sufficiently close to some central property (e.g., mean, median) of the range of the values measured experimentally in the reference population, or (3) subject-specific: the predicted value is sufficiently close to the value measured experimentally in each individual in the reference population. Typically, model development starts generic, after which models are combined into workflows and turned into a generic DTH, which in rounds of development and stratification are moving to population-specific DTH and finally to fully personalised subject-specific DTH. This requires that the context of use becomes more and more specific, and that sufficient relevant clinical data is available to allow for validation of the DTHs.

## 2 Clinical Needs

### 2.1 Acute Ischaemic Stroke

An acute ischaemic stroke occurs when an artery that supplies blood to the brain is blocked by a blood clot. This results in a sudden loss of blood circulation to an area of the brain, resulting in loss of neurologic function, as the blockage-induced deprivation of oxygen and nutrients to the brain causes brain cell death. The effects of acute ischaemic stroke can be very severe: stroke is the second-leading cause of death (12% of total deaths) and the leading cause of serious long-term disability (5.7% of total disability-adjusted life years) globally [5]. On average, every 25 s someone has a stroke in Europe resulting in an estimated 1.3 million Europeans having a first stroke each year [6].

## 2.2 Intracranial Aneurysm

Intracranial Aneurysm (IA) is a disease of the vessel wall resulting in the deformation and enlargement of the vascular lumen. The aneurysms may rupture if the deformation process remains active, potentially resulting in haemorrhagic stroke and loss of blood circulation to the brain, resulting in a sudden headache in all cases and loss of consciousness or focal neurologic deficits. All vessels can be affected by an aneurysm, but structural and hemodynamic particularities of intracranial vessels justify the study of IAs as a distinct entity [7]. The major risk with IAs is their rupture resulting in a Sub-Arachnoid Haemorrhage (SAH) [9]. IAs are mostly quiescent and asymptomatic but when rupturing, IAs potentially induce severe brain damage and death. The prevalence of IAs is high (2 to 5%) and affects young people (median: 50 years) [8]. SAH occurs in 6.3 per 100,000 inhabitants per year in Europe [10].

## 2.3 Screening for Primary Stroke or IA Rupture Has Limited Value

It is important to note that for *ischaemic disease*, prevention of the primary event is nearly impossible. Although many behavioral characteristics have been associated with the prevalence of stroke (like obesity and smoking), the effects of addressing these are very small or adherence is complicated (e.g., to a healthy lifestyle). We focus on optimal treatment instead, since the right and swift treatment can have huge benefits for the patient's health. Here, time is crucial. Each hour without successful treatment, the brain loses as many neurons as it does in 3–4 years of ageing [6]. Moreover, effectiveness of stroke treatment is strongly reduced with delay [7, 9].

For IAs, theoretically, a complete screening of the population would allow detecting all IAs and possibly to prevent aneurysm rupture. This strategy, being very expensive, is currently only recommended for subjects with a positive family history for IA or SAH [11]. For the general population, screening of healthy subjects generates substantial stress, and preventive treatment does not offer a sufficient benefit in terms of quality of life, [11] which does not justify the screening of the full population. Nevertheless, the increase in the number of imaging facilities and the improving imaging quality results in an increase in incidental IA diagnosis. Too low a threshold for treatment exposes patients to unnecessary treatment-induced morbidity and mortality. Too frequent monitoring exposes patients to unnecessary stress and society to unnecessary costs. Careful selection of cases to be monitored and customisation of the monitoring protocol as well as adequate selection of an optimal treatment modality are major tools to prevent IA rupture and associated morbidity and mortality as well as care-induced stress, morbidity, mortality, and direct and indirect costs.

## 2.4 Need for Patient-Specific Models in Stroke

When patients are diagnosed with stroke, many initial questions arise. For *acute ischaemic stroke*, which treatment or treatment combinations are optimal for specific patients? What is the origin of the stroke (e.g., thrombus in the heart, thrombus coming from carotid plaques, or an origin in the brain)? Should thrombolytics be given? Which device or combination of devices should be used to achieve fast complete recanalization and reperfusion? Is there a risk for haemorrhagic transformation or early stroke recurrence? For *intracranial haemorrhage*, most of the improvement leverage is on prevention. What is the population at risk of having an intracranial aneurysm? Will a diagnosed aneurysm rupture, causing a haemorrhagic stroke? What are the different intracranial aneurysm treatment options to prevent IA rupture? What are the risks associated with treatments? How will the haemorrhage evolve into delayed cerebral infarction? Will there be a rebleed? What treatments are possible to limit the risks of delayed cerebral infarction? In the long run, should the patient's lifestyle be altered?

The answers to all these questions strongly depend on patient-specific factors and treatments available. In current clinical practice, treatment decisions are taken based on evidence generated in a relatively small number of clinical trials, so based on the effects on general or stratified populations and pragmatically assuming known relevant factors involved are independent. Validated and proven computational tools that enable a personalised patient-specific treatment and management of stroke patients integrating interdependencies on a knowledge base would be a significant step beyond current state of the art.

## 3   Towards DTH for Stroke

There is a clear clinical need for improved diagnosis of stroke patients, identification of patients at risk of a (secondary) event, and additional treatment and monitoring options. We have identified and prioritised these clear-cut clinical needs, resulting in a series of multi-scale multi-organ models to address them (see Fig. 1).

| | Ischaemic stroke | | Haemorrhagic stroke | |
|---|---|---|---|---|
| | Clinical need | GEMINI will model | Clinical need | GEMINI will model |
| Aetiology & Prevention | Prediction tool for stroke recurrence (long term) in patients +/- atrial fibrillation | Atrial fibrillation Plaque formation Thrombus characteristics | Identification tool patients at risk for IA rupture incl. gender differences | Aneurysm dome wall irregularity, Stability/growth at the aneurysm level and more focally |
| Treatment | Treatment allocation decision-making tool (thrombectomy +/- thrombolysis + timing) | Thrombolysis Endovascular thrombectomy | Treatment allocation decision-making tool (treatment modality + timing, gender differences) | Coil and stent placement and resulting effect on luminal flow and thrombosis |
| Disease progression | Prognosis and decision-making tool for management of patients +/- atrial fibrillation with subacute disease progression | Oedema formation Haemorrhagic transformation | Prognosis and decision-making tool management for SAH | Post diagnosis mortality & morbidity and probability of progression after UIA or SAH diagnosis and treatment |

**Fig. 1.** Clinical need addressed by GEMINI per acute stroke subtype and per disease stage. Areas that will be modelled per disease and disease stage are indicated. We address two distinctly different diseases with similar characteristics in the effect on patients: acute ischaemic stroke and subarachnoid haemorrhage. For both diseases, we model its aetiology & prevention, treatment, and subacute disease progression.

We will (further) develop and deliver a focused set of models at different stages of development to move into clinical practice addressing prevention, diagnosis, treatment, and monitoring of stroke. These models will be validated and combined into specific DTHs for ischaemic and haemorrhagic strokes. We start with further developing and integrating multi-scale (patho) physiological models that are then composed into population specific DTHs for ischaemic and for haemorrhagic strokes. Three selected DTHs, for treatment options for ischaemic and haemorrhagic strokes, and for risk of rupture of IAs, will then be fully personalized, and finally, the DTH for treatment of ischaemic stroke will be tested on clinical effect in a clinical trial. Figure 2, visualizing the work package structure of the GEMINI project, summarizes this approach.

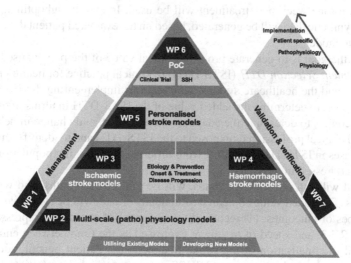

**Fig. 2.** The pyramid structure schematically depicts how higher work package number build upon the lower ones. WP3 and WP4 make use of the general (patho) physiology models of WP2, WP5 uses the population-based DTHs to build patient-specific DTHs, whereas WP6 takes one of the patient-specific DTHs to evaluate its value in clinical practice. All work packages interact with WP7 (validation and verification).

## 4  An Example, a DTH for Treatment of Acute Ischemic Stroke

Treatment options to remove the clot in an acute ischemic stroke patient are thrombolysis, thrombectomy, and aspiration. The INSIST project [12, 13] delivered fully validated models for thrombectomy [14] for M1 occlusions and for two different devices. Since simulating one case requires many hours of execution time on a high-end computer, which was not compatible with the intended use case (in silico stroke trials or personalized decision support in the acute phase), a computationally much more efficient surrogate model has been developed. This surrogate was trained with data from running thousands of instances of the fully mechanistic thrombectomy model [15]. The surrogate model takes as input details of the anatomy of the vasculature, the location of the clot, its size and some other parameters and returns a probability of a successful first-pass thrombectomy. More recently, a failure model calibrated with experimental tests on clot analogues was used to simulate thrombus fragmentation during a combined stent-retriever and aspiration thrombectomy procedure [15]. It is important to note that for the initial development and validation of the thrombectomy models in the INSIST project, dedicated benchtop experiments have been performed.

To fully personalize these models, we need to extract patient-specific data for model validation. Relevant input data for the treatment surrogates will be automatically extracted from a minimum of 250 patients (for each thrombectomy treatment). The surrogate ischaemic stroke treatment models will be evaluated on such personalised data. Metrics based on treatment success and patient outcome based on disease progression up to one-week post-treatment will be used. In case of suboptimal accuracy, additional synthetic data will be generated, based on the evaluated patient data, to update the surrogate model.

Finally, the goal is to generate proof of clinical value of the personalised *Ischemic Stroke Treatment Selection DTH* (ISTS-DTH) in clinical practice for health professionals, patients, and the healthcare system. This requires implementing the ISTS-DTH in clinical practice, to determine the added value of the ISTS-DTH in terms of established clinical outcomes, to determine the adaptation, perception, and change in behaviour of patients and clinical professionals exposed to the ISTS-DTH and to identify and mitigate potential biases in ISTS-DTH resulting from differences in training populations and the population addressed in the trial.

The trial will run in the Netherlands, will be run for 12 months, and will include 300 patients. The added value compared to standard-of-care will be evaluated according to three types of outcomes: (1) technical effectiveness of treatment, expressed as first pass eTICI 2C-3, and speed of treatment expressed as the time between entrance of a patient until the finalisation of the stroke treatment, (2) Patient benefit, addressed as functional outcome and measured as NIHSS approximately 1 day after stroke and the mRS 3 months after stroke, and (3) General quality of healthcare estimated as QALY (measure of how many years of life are lived in good health), DALYs (measure of lost healthy life years) and healthcare costs.

## 5  To Conclude

The GEMINI project aims to develop a series of DTHs relevant to strokes, for aetiology assessment and prevention, for treatments, and for disease progression. Our approach is to start with generic multiscale models of the basic physiology, e.g. brain perfusion and metabolism, and multiscale models of the pathologies, e.g. thrombosis or aneurysm development and rupture. These are then used to create a series of well-defined DTHs for stroke, some of which will then be fully personalized (for treatment of ischemic stroke, for treatment of IA, and for IA rupture risk estimation). Finally, one of these subject-specific DTHs, the DTH for treatment of acute ischemic stroke, will subsequently be tested in a randomized clinical trial as a decision support tool for the neuro-interventionist. The effect of using this DTH on the clinical endpoint, the status of the patient six months after treatment, will be assessed in this trial.

Towards a Generation of Digital Twins in Healthcare      245

**Acknowledgments.** We acknowledge funding from the European Union's Horizon Europe research and innovation programme under grant agreement No 101136438 (GEMINI project).

# References

1. https://dth-gemini.eu
2. Feigin, V.L., et al.: The global burden of neurological disorders: translating evidence into policy. Lancet. Neurol. **19**, 255–265 (2020)
3. Delucchi, M., et al.: Bayesian network analysis reveals the interplay of intracranial aneurysm rupture risk factors. Comput. Biol. Med. **147**, 105740 (2022)
4. https://www.edith-csa.eu
5. Feigin, V.L., et al.: Global, regional, and national burden of stroke and its risk factors, 1990–2019: a systematic analysis for the Global Burden of Disease Study 2019. Lancet Neurol. **20**, 795–820 (2021)
6. Truelsen, T., et al.: Stroke incidence and prevalence in Europe: a review of available data. Eur. J. Neurol. **13**, 581–598 (2006)
7. Portegies, M.L.P., Koudstaal, P.J., Ikram, M.A.: Cerebrovascular disease. Handb. Clin. Neurol. **138**, 239–261 (2016)
8. Vlak, M.H., et al.: Prevalence of unruptured intracranial aneurysms, with emphasis on sex, age, comorbidity, country, and time period: a systematic review and meta-analysis. Lancet. Neurol. **10**, 626–636 (2011)
9. van Donkelaar, C.E., et al.: Prediction of outcome after aneurysmal subarachnoid hemorrhage, development and validation of the SAFIRE grading scale. Stroke **50**, 837–844 (2019)
10. Etminan, N., et al.: Worldwide incidence of aneurysmal subarachnoid hemorrhage according to region, time period, blood pressure, and smoking prevalence in the population: a systematic review and meta-analysis. JAMA Neurol. **76**, 588–597 (2019)
11. Malhotra, A., et al.: MR angiography screening and surveillance for intracranial aneurysms in autosomal dominant polycystic kidney disease: a cost-effectiveness analysis. Radiology **291**(2) (2019)
12. Konduri, P.R., Marquering, H.A., van Bavel, E.E., Hoekstra, A., Majoie, C.B.L.M.: In-silico trials for treatment of acute ischemic stroke. Front. Neurol. **11**, 1–8 (2020). https://doi.org/10.3389/fneur.2020.558125
13. Luraghi, G., et al.: Applicability assessment of a stent-retriever thrombectomy finite-element model. Interface Focus **11**, 20190123 (2021)
14. Miller, C., et al.: In silico thrombectomy trials for acute ischemic stroke. Comput. Methods Programs Biomed. **228**, 107244 (2023)
15. Luraghi, G., et al.: A low dimensional surrogate model for a fast estimation of strain in the thrombus during a thrombectomy procedure. J. Mech. Behav. Biomed. Mater. **137**, 105577 (2023)

# A Working Week Simulation Approach to Forecast Personal Well-Being

Derek Groen[1]($\boxtimes$)(iD), Shivank Khullar[1](iD), Moqi Groen-Xu[2](iD),
and Rumyana Neykova[1](iD)

[1] Department of Computer Science, Brunel University London, Kingston Lane,
Uxbridge, UK
derek.groen@brunel.ac.uk
[2] School of Economics and Finance, Queen Mary University, London, UK

**Abstract.** Billions of people work every week. Forecasting which tasks gets done in a working week is important because it could help workers understand (i) whether their workload is manageable and (ii) which task scheduling approach helps to maximize the amount of work done and/or minimize the negative consequences of unfinished work. Here we present a working week simulation prototype, R2, and showcase how it can be used to forecast the working week for three archetypical workers. We show that R2 forecasts are sensitive to different task loads, task scheduling strategies and different levels of emerging work complications. We also highlight how R2 supports a new type of validation setting, namely that of user self-validation, and discuss the advantages and drawbacks of this new validation approach. We provide R2 as an online platform to allow users to create their own worker profile and task lists, and believe the tool could serve as a starting point for more in-depth research efforts on user-centric working week modelling.

**Keywords:** simulation · task management · working week · simulation development approach

## 1 Introduction

Many adults spend a non-negligible part of their weeks working. They have tasks that need to be done and face consequences, either in terms of professional or personal impact, when tasks are left unfinished [10]. The impact of the negative consequences depends on the perceived importance of the task, both from the perspective of the worker and of their possible superiors. The widespread prevalence of labour and its huge importance justifies the need to better understand how workers can operate effectively and how their well-being can be preserved. In this short paper we propose a simulation approach that focuses on modelling individual workers, and which is (perhaps unconventionally) intended to be eventually adopted by workers themselves. We sketch an

SK was supported by the Brunel Talent Marketplace (BTM).

early prototype of a workweek simulator, named R2, that may support workers, using their own assumptions, in forecasting their well-being and productivity at the end of a working week. The simulator requires workers to specify their workload in tasks, the expected impact on well-being for missing each task, and the expected availability and quality of working time throughout the week. Based on these provided assumptions, and a selected strategy for work task scheduling, the simulator then provides forecasts of which tasks are expected to be completed, and how much negative psychological events are expected from unfinished tasks and missed deadlines by the end of the working week.

## 1.1   Related Work

The work we present concerns individual workers and their well-being. In recent years, worker well-being has become a more prominent topic, particularly in the case of workers with family duties [7] and workers in the healthcare sector [9]. As an example of empirical research on the topic, Hafenbrack and Vohs quantified empirically that mindfulness meditation had a positive effect on task motivation but not necessarily task performance [4]. We also identified a range of investigations on the topic of structuring the working week from a managerial perspective. These efforts range from exploring interventions that aim to boost productivity [1,3] or well-being  [5,11] as well as guided design for job specifications [8]. However, a known complication for empirical studies on this topic is in terms of preserving the privacy of individual workers, and in terms of the reliability of externally reported worker well-being [2]. Within this paper we explore preset task scheduling strategies, which the user can manually select and apply. Although we focus mainly on the development approach of a user-centric working week simulation, it is also possible to focus on optimal task scheduling strategies. In this context the working week task allocation problem can be seen a variation of the job-shop problem [6], which has been extensively studied and for which many approaches exist [12,13]. And yet, the working week simulation as we present is different to a classic job shop problem in important ways. It is both simpler in that it features a single processing unit (the worker) and more complicated in that it has a probability for new complication tasks to be spawned at any point.

## 2   Approach

Our code R2 essentially relies on a variation of Discrete-Event Simulation, where a single worker agent attempts to "do its work" during the workweek. Within the conceptual model, we specifically simulate how a single worker: (i) has time availability during the workweek, (ii) tackles pending tasks from their task list using this available time and a predefined scheduling strategy, (iii) experiences complications [additional tasks] in some cases when performing the base tasks and (iv) experiences negative psychological events towards the end of the workweek due to tasks being completed after their deadline or not at all.

We provide a high-level overview of our model and our simulation development approach (SDA) in Fig. 1. Here, the main unit of work in the model is a `Task`, which has the following attributes: duration [hours], priority [1–5 scale], high_focus [true or false], cost_of_failure [1–5 intensity scale], deadline [day of the week or None], complication_probability [chance of triggering the emergence of a complication task], complication [the task created when a complication occurs]. The main simulation kernel propagates the simulation in time, taking into account the worker profile and the chosen task scheduling strategy to choose and complete tasks. Once the workweek simulation has completed, R2 will provide a range of diagnostic outputs. These include: number of tasks completed in the week, number of complication spawned in the week (these are extra tasks that may arise at a preset probability when certain tasks are done), a completion score (which is a sum of `task.priority`^2 for each completed task), a negative impact score (which is a sum of `task.cost_of_failure`^2 for each incomplete task) and a list of negative events which contains the number of negative psychological events triggered during the workweek. The list of negative events is categorized across the five possible `cost_of_faillure` scores, which we informally label as: effortless (1), inconvenient (2), annoying (3), painful (4), and desperate (5).

To enable users to reproduce our results and perform their own experimentation, we provide R2 as a free to use web service for anyone[1].

## 3   Exemplars

To illustrate the dynamics of R2 we provide simulation results for three exemplar workers: a manager, a software developer and a student. Although the exemplars are defined to be archetypal for these real-life roles, they are not intended to resemble real-world individuals. Below we provide a brief description for each exemplar, while the detailed input files can be found on Github[2]:

- Jordan is a manager overseeing various projects and teams. He has long working days with few periods dedicated to high-focus tasks and many meetings and administrative duties. Jordan's week is characterized by multitasking across numerous short-duration tasks, from team reviews to strategic planning. His schedule is predominantly filled with meetings and manageable tasks, although he has a few high focus planning tasks.
- Alexa is a software developer working at a tech startup. Her involves coding, fixing bugs, and meeting with the team. Alexa's week is filled with focused coding tasks, solving urgent problems, and participating in team meetings. She works typical office hours and does weekly sprint meetings early on the Monday morning.
- Mike, a university student, navigates a balanced academic week across 7 d, blending study with leisure and attending roughly 8 h of lectures. Tasks vary

---

[1] https://ratrace.streamlit.app.
[2] https://www.github.com/djgroen/ratrace-inputs.

**Model Architecture Overview**

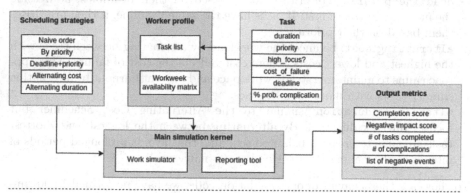

**Simulation Development Approach for Self-Validation**

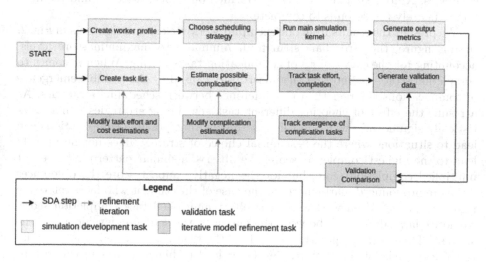

**Fig. 1.** Top: architecture overview of R2. Bottom: SDA overview of a self-validation with R2.

widely, from quick note-taking to intensive coursework, with a task cost mostly between 1–3, though crucial submissions may escalate to 4 or even 5, risking course failure.

In addition, we have implemented five basic task scheduling strategies to investigate to what extent these can affect the work outcomes for each of the three workers. These strategies are:

- **naive**: processes tasks in the order they are put in, without considering priorities, deadlines, or other attributes. This approach serves as a baseline.
- **priority**: prioritises tasks based on their priority value, with higher numerical values indicating higher priority. This strategy ensures that the most critical tasks are addressed first.

- `deadline+priority`: prioritises tasks based on their deadlines, addressing the most urgent tasks first. If tasks have the same deadline, it then prioritises them based on their priority.
- `alternating_cost`: balances the workload by alternating between tasks with the highest and lowest costs, where cost reflects the cost of failure. This approach aims to maintain a steady work pace and prevent burnout by alternating simpler tasks with more complex ones.
- `alternating_duration`: similar to the Alternating Cost Scheduler but focuses on task duration. By alternating between the longest and shortest tasks, it aims to create a balanced schedule that avoids prolonged periods of intense work, making the workload more manageable.

To showcase the behavior of our simulation code, we have performed 1,000 simulations for each combination of the three worker types with the five scheduling strategies. All our simulations were performed on a local desktop, and required approximately ten seconds to complete.

We present the results from our exemplar workweek simulation runs in Fig. 2. In this figure, the error bars span from minimum to maximum score levels, accounting for the uncertainty of complication tasks arising. When it comes to completion score, we find that the uncertainty caused by the possible emergence of complications is larger than the difference between scheduling strategies. As a result, the effect of choosing different task scheduling strategies is not "statistically significant", which in this case means that possible complications can lead to situations where the best initial choice of strategy does not necessarily lead to the highest completion score. We observe a similar pattern in the effort multiplied completion score, but for the negative impact score the differences are more pronounced, particularly in the case of the student who faces relatively tight deadlines. When we look at the effort multiplied task completion score, we notice large differences between the manager, the software developer and the student. This metric is greatly affected by the task list provided, and the relatively long tasks of the software developer lead to higher scores in this regard (the distribution of priority values is similar across the three exemplars).

In terms of negative impact, all three of our workers were given a very demanding task lists to demonstrate the behavior of the code. Given that missing out on a task with a cost of 5 (desperate) increases the negative impact score by 25, we can observe that the workers reach a negative impact exceeding that level in almost all cases. A notable exception is the student that adopts the alternating cost strategy, as in this case the negative impact remains remarkably limited. Overall, the results showcase that R2 is clearly sensitive to: (i) differences in task and effort properties of the three archetypes, (ii) different task scheduling strategies and (iii) probabilities of complication tasks during the week. As a result, we argue that our model can be adopted meaningfully by individual users for further experimentation and validation, following the SDA presented in Fig. 1. To further investigate the effect of complications on the results, we performed runs with differing complication probabilities, including runs with 0% and 100% probability. Here we find that complications increase negative scores

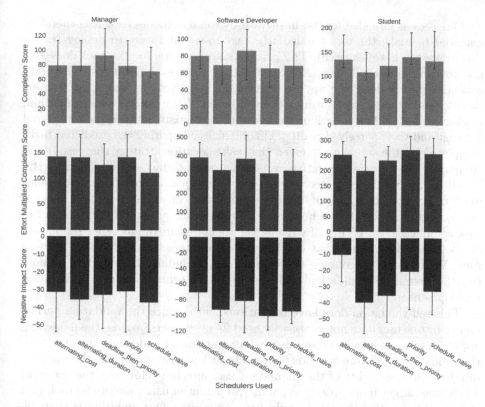

**Fig. 2.** Metrics Comparison Across Schedulers and Personas. Results of workweek simulations for the three exemplar workers: manager (left), software developer (middle) and student (right). Note: error bars indicate the full expected range of performance possible by the worker (maximum and minimum), accounting for the uncertainty of complication tasks arising.

by approximately 50 to 150% (manager), 20 to 150%(software developer) and -30% to 400%(student), depending on the scheduling strategy chosen. Here, the two alternating strategies in particular can lead to dramatic an counter-intuitive changes in the negative scores. For space reasons we provide our full result plots at: https://github.com/djgroen/r2_iccs2024_public.

# 4   Discussion

Within this paper we presented a new approach to simulation development, intended to be adopted by individual workers to better understand the dynamics of their working week. We showcased our prototype implementation by applying five different task scheduling strategies to three archetypal worker situations, and demonstrated that the simulation results are sensitive to factors such as the presence of complications, the nature of a worker's task list and availability, and the choice of task scheduling strategy.

Our code is intended for use in private by workers themselves. It is therefore essential to make the tool particularly easy-to-use and easy to deploy (hence the choice for a web service). The private setting of use for the tool does give rise to validation challenges. Although a large-scale evaluation test with data collection can be done, the way we defined the SDA (see Fig. 1 puts the workers themselves in charge of performing the validation. When doing such a validation, workers will use R2 for a working week and then compare between the initial tool forecast and their perceived reality. This includes comparing the predicted tasks completed vs. the actual ones (given the task allocation strategy they used) and comparing the perceived negative events incurred vs. the ones predicted by R2. Based on the validation outcome, the user can adjust assumptions in the simulation, for instance by modifying the duration or consequences of specific task types, modifying the effort schedule, or modifying the probability and impact of task complications. At time of writing, we are not aware of any other simulation development approaches that explicitly incorporate self-validation by the end user. We therefore encourage the community (including ourselves) to explore to which extent self-validation could be applied in other domains of simulation development.

This self-validation does have limitations: for instance the validation performance by one user does not necessarily hold for other users because the modelling approach relies so heavily on self-defined assumptions. In addition, the quality of the validation procedure depends on aspects such as intrinsic bias in the worker's mindset and the quality of the simulation user interface. However, the choice of workers as target users greatly expands the potential user base of the tool, and the possible positive impact it could have on society. Our ambition is that, by introducing this working week simulation and self-validation tool, we can help workers to learn about the task and effort dynamics in their working week, and their overall mental experience of it.

Lastly, there are many avenues that we have not yet explored. For future work, we'd like to (i) incorporate mechanisms where negative events incurred during the week could affect worked productivity in the remaining period, (ii) develop a user-friendly approach to define complications, (iii) explore how ad-hoc meetings or pre-emptive strategies of workload mitigation (e.g. requesting deadline extensions or scheduling overtime) affects completion scores and negative impact scores, and (iv) identify typical working weeks in different professions, in terms of the tasks given, their priority, cost and possible complications. Aspects (ii) and (iv) are directly relevant for the user, as both enhancements make it considerably easier for them to use the simulation platform.

# References

1. Attar, A., Gupta, A., Desai, D.: A study of various factors affecting labour productivity and methods to improve it. IOSR J. Mech. Civil Eng. (IOSR-JMCE) 1(3), 11–14 (2012)
2. Benjamin, D.J., Cooper, K., Heffetz, O., Kimball, M.S., Zhou, J.: Adjusting for Scale-use Heterogeneity in Self-reported Well-being. Tech. rep, National Bureau of Economic Research (2023)
3. Ghosh, J.: Distributed task scheduling and human resource distribution in industrial service solution production: a simulation application. Int. J. Prod. Res. 51(10), 2900–2914 (2013)
4. Hafenbrack, A.C., Vohs, K.D.: Mindfulness meditation impairs task motivation but not performance. Organ. Behav. Hum. Decis. Process. 147, 1–15 (2018)
5. Ivancevich, J.M., Lyon, H.L.: The shortened workweek: a field experiment. J. Appl. Psychol. 62(1), 34 (1977)
6. Manne, A.S.: On the job-shop scheduling problem. Oper. Res. 8(2), 219–223 (1960)
7. Obrenovic, B., Jianguo, D., Khudaykulov, A., Khan, M.A.S.: Work-family conflict impact on psychological safety and psychological well-being: a job performance model. Front. Psychol. 11, 475 (2020)
8. Parker, S., Wall, T.D.: Job and Work Design: Organizing Work To Promote Well-Being and Effectiveness, vol. 4. Sage (1998)
9. Søvold, L.E., Naslund, J.A., Kousoulis, A.A., Saxena, S., Qoronfleh, M.W., Grobler, C., Münter, L.: Prioritizing the mental health and well-being of healthcare workers: an urgent global public health priority. Front. Public Health 9, 679397 (2021)
10. Syrek, C.J., Weigelt, O., Peifer, C., Antoni, C.H.: Zeigarnik's sleepless nights: how unfinished tasks at the end of the week impair employee sleep on the weekend through rumination. J. Occup. Health Psychol. 22(2), 225 (2017)
11. Topp, J., Hille, J.H., Neumann, M., Mötefindt, D.: How a 4-day work week and remote work affect agile software development teams. In: International Conference on Lean and Agile Software Development., pp. 61–77. Springer, Cham (2022). https://doi.org/10.1007/978-3-030-94238-0_4
12. Türkyılmaz, A., Şenvar, Ö., Ünal, İ, Bulkan, S.: A research survey: heuristic approaches for solving multi objective flexible job shop problems. J. Intell. Manuf. 31, 1949–1983 (2020)
13. Xiong, H., Shi, S., Ren, D., Hu, J.: A survey of job shop scheduling problem: the types and models. Comput. Oper. Res. 142, 105731 (2022)

**Advances in High-Performance
Computational Earth Sciences:
Numerical Methods, Frameworks
and Applications**

# Low-Ordered Orthogonal Voxel Finite Element with INT8 Tensor Cores for GPU-Based Explicit Elastic Wave Propagation Analysis

Tsuyoshi Ichimura[1]([✉]), Kohei Fujita[1], Muneo Hori[2], and Maddegedara Lalith[1]

[1] Earthquake Research Institute and Department of Civil Engineering,
The University of Tokyo, Tokyo, Japan
`ichimura@eri.u-tokyo.ac.jp`
[2] Research Institute for Value-Added-Information Generation,
Japan Agency for Marine-Earth Science and Technology, Yokohama, Japan

**Abstract.** Faster explicit elastic wavefield simulations are required for large and complex three-dimensional media using a structured finite element method. Such wavefield simulations are suitable for GPUs, which have exhibited improved computational performance in recent years, and the use of GPUs is expected to speed up such simulations. However, available computational performance on GPUs is typically not fully exploited, and the conventional method involves some numerical dispersion. Thus, in this paper, we propose an explicit structured-mesh wavefield simulation method that uses INT8 Tensor Cores and reduces numerical dispersion to speed up computation on GPUs. The proposed method was implemented for GPUs, and its performance was evaluated in a simulation experiment of a real-world problem. The results demonstrate that the proposed method is 17.0 times faster than the conventional method.

**Keywords:** Explicit Wave Propagation Analysis · Finite Element · Tensor Core · GPU

## 1 Introduction

Explicit simulation of elastic wavefields, which can be computed without solving matrix equations, is often used to evaluate the dynamic response of large and complex three-dimensional media through sequential analysis and to estimate internal structures through optimization via many simulations. Thus, further speedup is desired for conducting such large and many-case analyses. In this paper, we focus on increasing the speed of explicit elastic wavefield simulations with structured finite elements, which are suitable for generating numerical models. In the standard finite element method, the generation of the finite element model can be a bottleneck in the analysis process. However, finite element models can be generated automatically for the structured finite element method by using cubic

© The Author(s), under exclusive license to Springer Nature Switzerland AG 2024
L. Franco et al. (Eds.): ICCS 2024, LNCS 14834, pp. 257–271, 2024.
https://doi.org/10.1007/978-3-031-63759-9_31

258 T. Ichimura et al.

elements that are finer than the elements used in standard finite element analysis and allowing some approximation of geometry, thereby making it suitable for analyzing large and complex media. Thus, it is used in seismic analysis [1, 2] and the estimation of the internal structure of structures [3]. There is a strong need for a method that can be executed many times in a short period because detailed simulations and model optimizations are desired in such analyses.

Explicit wavefield simulations using structured finite element methods, where continuous memory access is dominant, are well-suited for processing on GPUs and are expected to be executed quickly on recent GPUs with improved computational performance. However, the lumped mass matrix approximation is applied to the low-order structured finite elements that are conventionally used in such analyses, and this results in numerical dispersion, which increases computational cost due to the need to use smaller element sizes. In addition, some GPUs are equipped with acceleration mechanisms, e.g., Tensor Cores [4], and their use in deep learning and other applications is progressing. However, the use of Tensor Cores in physics-based simulations is limited because their high-speed operations are predicated on transforming operations into a somewhat special form, and they are only used in some cases (e.g., to obtain a coarse solution in implicit simulations [5, 6]. See [7] for a review of mixed-precision algorithms for linear numerical algebra in general). In particular, unlike the implicit solution method, the explicit solution method does not allow for processes that refine the coarse solution. Therefore, the development of such algorithms is challenging because the results must be equivalent to those of ordinary FP64 calculations while effectively using Tensor Cores. However, Tensor Cores have high computational performance; thus, further increases in processing speed can be expected if the hardware can be exploited effectively.

Therefore, this paper proposes an explicit structured finite element elastic wavefield simulation method that uses INT8 Tensor Cores, which is more accurate and faster than the conventional method. The remainder of this paper is organized as follows. In Sect. 2, we present an explicit structured finite element wavefield simulation method that solves the problems of conventional methods and is suitable for INT8 Tensor Core computation. Section 3 describes the implementation of the proposed method on a GPU and presents the details of the performance of the proposed method using a realistic analysis model. We show that the proposed method can realize an analysis that yields results that are equivalent to those of the conventional method at higher speed. Finally, the paper is concluded in Sect. 4.

## 2 Low-Ordered Orthogonal Voxel Finite Element with INT8 Tensor Cores

In this study, we target the governing equation of a linear dynamic elastic body:

$$\rho\ddot{\mathbf{u}} - (\nabla \cdot \mathbf{c} \cdot \nabla) \cdot \mathbf{u} = \mathbf{f}, \tag{1}$$

where $\rho$, $\mathbf{u}$, $\mathbf{c}$, and $\mathbf{f}$ denote the density, displacement, elasticity tensor, and body force, respectively. In addition, ($\dot{\phantom{x}}$) and $\nabla$ denote the temporal and spatial differential operators, respectively. In standard finite element analysis, the target domain is decomposed into small elements $e$, and the basis functions $\phi^\alpha$ are defined on these elements. By substituting $\mathbf{u} = \sum \mathbf{u}^\alpha \phi^\alpha$ into the functional in Eq. (1) and taking the stationary condition, the following linear equation is obtained for the unknown coefficients $\mathbf{u}^\alpha$:

$$\mathbf{Ku} + \mathbf{M\ddot{u}} = \mathbf{F}. \tag{2}$$

Here, $\mathbf{u}$ is a vector assembled using $\mathbf{u}^\alpha$. $\mathbf{K}$ and $\mathbf{M}$ are a stiffness matrix and mass matrix assembled using the element stiffness matrix $\mathbf{K}_e$ and the element mass matrix $\mathbf{M}_e$ obtained for each element, respectively.

In a typical structured finite element analysis, the domain of interest is divided into cubic subdomains of lengths $ds$ per side, and these are used as elements (hereafter referred to as voxel elements). As a result, a finite element model can be generated easily by simply covering a three-dimensional region with a structured grid of constant width $ds$ and evaluating the physical properties of each element. Thus, it is frequently used for the static and dynamic analysis of regions with complex geometry and property distributions or for geometry optimization. However, the object is modeled with cubes of uniform size; thus, it is necessary to reduce $ds$ to improve the accuracy of modeling geometry, which frequently results in a problem with large degrees of freedom. When using such voxel elements, low-order elements are frequently used, i.e., it is common to use basis functions

$$\phi^\beta = -\frac{1}{8}(r_1 + \bar{r}_1)(r_2 + \bar{r}_2)(r_3 + \bar{r}_3) \quad (\beta = 1, 2, ..., 8)$$

for node $\beta$, which is often used for hexahedral elements (Fig. 1 shows the definition of the local coordinates, local node numbering, and the definition of $\bar{r}_1$, $\bar{r}_2$, $\bar{r}_3$). However, in this case, the element mass matrix

$$\mathbf{M}_e = \rho(\phi^\beta \phi^{\beta'})_e,$$

becomes nondiagonal, which necessitates solving a large-scale matrix equation to obtain $\mathbf{u}$ in Eq. (2). Note that $(\ )_e$ indicates volume integration in each element. To solve this problem, an approximation that concentrates the off-diagonal terms to the diagonal terms (lumped mass matrix approximation) is frequently applied, which leads to a diagonal mass matrix $\mathbf{M}_e^{ap}$, where the off-diagonal terms are approximated as 0, and the diagonal terms are approximated as $\rho/8(1)_e$. Consequently, for example, if we approximate the acceleration using the central difference method with a time-step width of $dt$, Eq. (2) at time step $it$ becomes

$$\mathbf{Ku}^{it} + \mathbf{M}^{ap}(\mathbf{u}^{it+1} - 2\mathbf{u}^{it} + \mathbf{u}^{it-1})/dt^2 = \mathbf{F}^{it}, \tag{3}$$

which enables explicit computation of $\mathbf{u}^{it+1}$. This lumped matrix approximation is widely used when solving dynamic problems because eliminating the need to

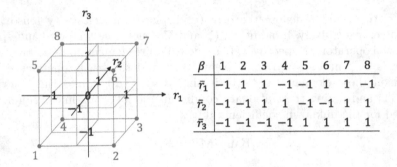

**Fig. 1.** Definition of a voxel element. For a cube of size $ds$ in each direction, a local coordinate $r_1r_2r_3$ is defined ($-1 \leq r_1 \leq 1$, $-1 \leq r_2 \leq 1$, $-1 \leq r_3 \leq 1$). The local node number ranging from 1–8 is defined on each node in the element. $\phi^\beta$ is the basis function for node $\beta$.

solve large matrix equations reduces the analysis cost significantly. This analysis method is referred to as VFEM in this paper.

In VFEM, the lumped mass matrix approximation causes numerical dispersion, which is particularly noticeable when the distance of the propagation of waves increases. To suppress such numerical dispersion, a VFEM using an orthogonal basis function has been proposed for the voxel elements [8]. Using the orthogonal basis functions, the mass matrix becomes diagonal without any approximation to the voxel elements. We formulate OVFEM using the Hellinger-Reissner functional:

$$J(\mathbf{u}, \sigma) = \int_V \int_T -\frac{1}{2}\rho \dot{\mathbf{u}} \cdot \dot{\mathbf{u}} - \frac{1}{2}\sigma : \mathbf{c}^{-1} : \sigma + \sigma : (\nabla \otimes \mathbf{u}) - \mathbf{u} \cdot \mathbf{f} \, dv dt. \quad (4)$$

Here, $\sigma$ denotes stress, which can be expressed as $\sigma = \sum \sigma^\beta \psi^\beta$ using the basis function $\psi^\beta$ and unknown coefficients $\sigma^\beta$. By taking the stationary condition of Eq. (4) for $\mathbf{u}^\alpha$ and $\sigma^\beta$, a discretized form of the governing equation can be obtained in the form of Eq. (2). Here, $\mathbf{K}$ and $\mathbf{M}$ become matrices assembled using the element stiffness matrix

$$\mathbf{K}_e^o = (\psi^\beta \nabla \phi^\alpha)_e^t \cdot (\psi^\beta \psi^{\beta'})_e^{-1} \mathbf{c} \cdot (\psi^\beta \nabla \phi^\alpha)_e, \quad (5)$$

and the element mass matrix

$$\mathbf{M}_e^o = \rho(\phi^\beta \phi^{\beta'})_e, \quad (6)$$

respectively. While the same local coordinates $-1 \leq r_1 \leq 1$, $-1 \leq r_2 \leq 1$, $-1 \leq r_3 \leq 1$ and node placement of VFEM (Fig. 1) are used, the following Heaviside function $H$ is used in OVFEM:

$$\phi^\beta = H(\bar{r}_1 r_1)H(\bar{r}_2 r_2)H(\bar{r}_3 r_3) \quad (\beta = 1, 2, ..., 8), \quad (7)$$

$$\psi^1 = 1, \psi^2 = r_1, \psi^3 = r_2, \psi^4 = r_3, \psi^5 = r_1 r_2, \psi^6 = r_2 r_3, \psi^7 = r_1 r_3. \quad (8)$$

As shown in the definition of Eq. (7), the element mass matrix of Eq. (6) becomes diagonal, and a stable solution is obtained because the discontinuous functions $H$ are used together with smooth polynomial functions. Thus, the use of $\mathbf{M}_e^{\varrho}$ allows us to solve Eq. (2) explicitly without approximation. OVFEM is expected to be more accurate than VFEM due to the suppression of numerical dispersion by the absence of the lumped mass approximation. This indicates that OVFEM may enable a coarser discretization size compared to VFEM, which can reduce the analysis costs of three-dimensional dynamic problems considerably.

Some GPUs are equipped with acceleration mechanisms that can perform certain forms of matrix products at high speed. For example, the A100 80 GB PCIe GPU [9] has Tensor Cores with peak performance of 624 TOPS for 8-bit integer matrix operations, which is significantly higher than the 9.7 TFLOPS peak performance of its FP64 cores. Therefore, effective use of Tensor Cores is expected to speed up analysis, and Tensor Cores are being used in deep learning and other applications. However, to apply Tensor Core operations efficiently, it is necessary to formulate an algorithm such that the amount of matrix operations of a particular form is dominant while guaranteeing the accuracy of the operations, which limits the applicability to only a few physics-based simulations. In particular, with the implicit method, Tensor Cores can be used as a substitute for part of the analysis by using Tensor Cores to obtain a coarse solution and then refining it to reduce the analysis cost; however, it is difficult to apply such a refining process to the explicit method. Thus, the part of the algorithm to be computed by Tensor Cores must demonstrate accuracy that is equivalent to that of FP64, and the difficulty in algorithm development lies in the fact that speedup must be achieved in consideration of the computing performance, data transfer, and other factors.

Therefore, we propose TCOVFEM, which is an OVFEM transformed to take advantage of the INT8 Tensor Core arithmetic performance with guaranteed arithmetic accuracy. Generally, solids can be treated as a linear isotropic dynamic elastic body in the most basic approximation; thus, the formulation is presented using a linear isotropic dynamic elastic body as an example. This method can be applied to the ultrasonic wave propagation analysis shown in the application examples presented in this paper and to the analysis of wavefields in other linear isotropic elastic bodies. The key is determining how to make $\mathbf{Ku}$, which has the highest analysis cost in Eq. (2), suitable for Tensor Core operations while guaranteeing FP64 accuracy. First, we consider the computation of $(\psi^\beta \nabla \phi^\alpha)_e$ among the operations given in Eq. (5). Note that there is arbitrariness in the selection of $\psi^\beta$ in Eq. (8); thus, we select a polynomial that follows the deformation mode of the cube, which leads to an expression within the range of low-precision numbers by proper normalization. In contrast, the computation of $(\psi^\beta \psi^{\beta'})_e^{-1}\mathbf{c}$ requires care. Generally, $\mathbf{c}$ does not fall within the range of low-precision numbers (even if it is linear isotropic); thus, it must be treated appropriately. In this section, we focus on the independence of the properties of linear isotropic elastic bodies and derive a formulation that falls within the range of 8-bit integers. For linear isotropic elastic bodies, $\mathbf{c}$ can generally be expressed by the product of the volume modulus $\kappa$, the shear modulus $G$, and a constant that fits into low-precision arithmetic. Here, by defining $\bar{\mathbf{D}}^\kappa = 3\frac{\partial(\psi^\beta\psi^{\beta'})_e^{-1}\mathbf{c}}{\partial\kappa}ds^3$ and $\bar{\mathbf{D}}^G = 3\frac{\partial(\psi^\beta\psi^{\beta'})_e^{-1}\mathbf{c}}{\partial G}ds^3$ as the contribution of $\kappa$ and $G$ to $(\psi^\beta\psi^{\beta'})_e^{-1}\mathbf{c}$, a constant $24 \times 48$ matrix

$$\mathbf{K}_e^{\text{INT8}} = \left(\mathbf{K}_e^\kappa, \bar{\mathbf{K}}_e^G\right),$$

where

$$\mathbf{K}_e^\kappa = 256/3\bar{\mathbf{B}}^t\bar{\mathbf{D}}^\kappa\bar{\mathbf{B}}ds^4,$$
$$\bar{\mathbf{K}}_e^G = 128\bar{\mathbf{B}}^t\bar{\mathbf{D}}^G\bar{\mathbf{B}}ds^4 - 128\mathbf{I},$$

can be constructed. Here, $\bar{\mathbf{B}}$ is an equivalent strain displacement matrix and $\mathbf{I}$ is a $24 \times 24$ identity matrix. Using this transformation, the core kernel, which has the highest analysis cost in finite element analyses, can be computed as follows:

$$\mathbf{K}_e^o\mathbf{u}_e = 1/256\kappa ds(\mathbf{K}_e^{\text{INT8}}\bar{\mathbf{u}}_e + 256/3G/\kappa\mathbf{u}_e). \tag{9}$$

Here, $\bar{\mathbf{u}}_e = \left(\mathbf{u}_e, 2/3G/\kappa\mathbf{u}_e\right)$.

In Eq. (9), we convert most of the analysis cost into operations that fall within the 8-bit integer range; however, to extract the full performance of Tensor Cores, it is necessary to implement Tensor Core computations in consideration of both the operation speed and the data transfer cost. Here, we present a concrete implementation of the fast Tensor Core-based computation of the dense matrix-vector product $\mathbf{K}_e^{\text{INT8}}\bar{\mathbf{u}}_e$. Here, $\mathbf{K}_e^{\text{INT8}}$ is a $24\times48$ constant matrix with values in the 8-bit integer range (-128 – 127), and $\bar{\mathbf{u}}_e$ is a vector with a length of 48 in FP64. Note that conversion between floating-point numbers and integers incurs a large cost; thus, we consider fast computation algorithms that minimize the number of variable conversions while reducing the data transfer cost of the computation.

First, to save the number of significant digits during integer arithmetic expansion, the element right-hand side vector is scaled as follows:

$$\bar{\mathbf{u}}_{es} = \frac{1}{s_e}\bar{\mathbf{u}}_e. \tag{10}$$

Here, $s_e = \max_{i=1,2,\dots,48}|\bar{u}_{ei}|$, where $\bar{u}_{ei}$ is the $i$-th component of $\bar{\mathbf{u}}_e$. This scales each component of $\bar{\mathbf{u}}_e$ to be within $\pm1$. Then, $\bar{\mathbf{u}}_{es}$ is expanded with the product of FP64 constants times integer value vectors:

$$\bar{\mathbf{u}}_{es} = \sum_{i=1}^{N}\frac{1}{a^i}\bar{\mathbf{u}}_{es\text{INT}(i)}, \tag{11}$$

where $N$ is the number of integer expansion stages, and $\bar{\mathbf{u}}_{es\text{INT}(i)}$ is a vector comprising 48 integers:

$$\bar{\mathbf{u}}_{es\text{INT}(1)} \underset{\text{INT}}{\Leftarrow} a\bar{\mathbf{u}}_{es}, \tag{12}$$

$$\bar{\mathbf{u}}_{es\text{INT}(2)} \underset{\text{INT}}{\Leftarrow} a\left\{a\bar{\mathbf{u}}_{es} - \bar{\mathbf{u}}_{es\text{INT}(1)}\right\}, \tag{13}$$

$$\bar{\mathbf{u}}_{es\text{INT}(3)} \underset{\text{INT}}{\Leftarrow} a\left\{a^2\bar{\mathbf{u}}_{es} - a\bar{\mathbf{u}}_{es\text{INT}(1)} - \bar{\mathbf{u}}_{es\text{INT}(2)}\right\}. \tag{14}$$

...

Here, $\underset{\text{INT}}{\Leftarrow}$ denotes truncation to integer values. Using these integer vectors, the element matrix-vector product is computed as follows:

$$\mathbf{K}_e^{\text{INT8}}\bar{\mathbf{u}}_e = s_e \sum_{i=1}^{N} \frac{1}{a^i} \mathbf{K}_e^{\text{INT8}}\bar{\mathbf{u}}_{es\text{INT}(i)}. \tag{15}$$

The above Eqs. (11)–(15) can be expanded and calculated with integers with an arbitrary number of bits. For example, using $a = 2^7$, all integer operations can be performed with 8-bit integers. However, in this case, Eqs. (12)–(14) must be calculated for the $N$ stages, which requires many conversions between floating-point variables and integers, and, therefore, reduces the computational performance. For example, $N = 8$ stages are required to realize 56-bit accuracy which is equivalent to FP64's 52 fraction bits, which requires many conversion operations. Thus, in this study, the number of conversions between the FP64 variables and integers is reduced using 64-bit integers to perform the conversion only once and by using a hierarchical expansion in which the 64-bit integer calculation is further expanded using 8-bit integer operations. In particular, Eqs. (11)–(12) are expanded using 64-bit integers with $a = 2^{56}$ and $N = 1$, and the main computation $\mathbf{K}_e^{\text{INT8}}\bar{\mathbf{u}}_{es\text{INT}(1)}$ (denoted $\mathbf{K}_e^{\text{INT8}}\bar{\mathbf{u}}_{es\text{INT64}(1)}$) is expanded using $M = 8$-stages of computations using 8-bit integers. Here, using $b = 2^7$, $\bar{\mathbf{u}}_{es\text{INT64}(1)}$ is expanded as

$$\bar{\mathbf{u}}_{es\text{INT64}(1)} = \sum_{j=1}^{M} b^{j-1}\bar{\mathbf{u}}_{es\text{INT8}(1,j)} \tag{16}$$

using 8-bit integers, and $\mathbf{K}_e^{\text{INT8}}\bar{\mathbf{u}}_{es\text{INT64}(1)}$ is computed as

$$\sum_{j=1}^{M} b^{j-1}\mathbf{K}_e^{\text{INT8}}\bar{\mathbf{u}}_{es\text{INT8}(1,j)}. \tag{17}$$

Figure 2 shows the computation flow of the matrix-vector products when directly converting data between FP64 and INT8 variables and when hierarchical FP64-INT64-INT8 conversion is used. While the data conversion and computation are looped $N$ times in the direct method, the proposed hierarchical method performs the data conversion outside of the $M$ loop. While the total computation accuracy is the same when $N = M$, the number of conversions between the integer and FP64 variables can be reduced by $N$-fold when using the hierarchical method.

Tensor Core operations are very fast when applied to matrix-matrix products of specific sizes; thus, we must transform the computation pattern of the matrix-vector products to matrix-matrix products suited for Tensor Cores. In addition, Tensor Core operations are very fast; thus, the cost of other operations must be suppressed to reduce time costs substantially. In particular, the data transfer from global memory and the data transfer between shared memory becomes a bottleneck. In the following, we describe how we map the matrix-vector computation to Tensor Cores and how we circumvent the data transfer bottlenecks.

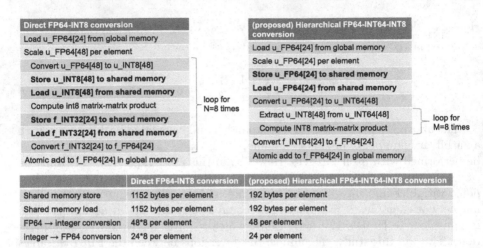

**Fig. 2.** Computation of matrix-vector products using INT8 Tensor Cores with direct FP64-INT8 conversion and the proposed hierarchical FP64-INT64-INT8 conversion. Here, 32 elements are computed using 32 threads per thread block.

- Transformation of matrix-vector products to matrix-matrix products: While $\mathbf{K}_e^{\text{INT8}}\bar{\mathbf{u}}_{es\text{INT8}(1,j)}$ is a matrix-vector product of size $(24\times48)\times48$, we can compute 32 elements in a single thread block as a matrix-matrix product of size $(24\times48)\times(48\times32)$. We compute this using INT8 Tensor Cores with $(n, m, k) = (8, 32, 16)$: C(8 × 32: INT32)=A(8 × 16: INT8)×B(16 × 32: INT8). As shown in Fig. 3, a $(24\times48)\times(48\times32)$ matrix-matrix multiplication is decomposed into $3 \times 3 = 9$ parts, and each part is computed using $(8 \times 16) \times (16 \times 32)$ matrix-matrix multiplication. Here, the matrix $\mathbf{K}_e^{\text{INT8}}$ is constant in the $j$-loop; thus, we can load the A fragment once and keep it in registers.
- Elimination of shared memory loads/stores by direct addition of results to global memory: As the register mapping of memory fragments for the matrices in Tensor Core operations is complex, an API is provided to exchange the input and output data between threads via shared memory. However, when this technique is used for the small-scale matrix-matrix multiplications targeted in this study, it frequently results in shared memory bottlenecks, which makes it difficult to fully utilize the available performance of the Tensor Cores. Thus, following the method presented in the literature [6], we skip the remapping process using shared memory and directly output the data to global memory. As shown in Fig. 4, rather than remapping the outputs from the Tensor Core operations and having each thread add element-wise results to the global memory, the results are added directly to the designated components in the global memory. Here, the scaling coefficient $s_e$ for each element is reflected in this procedure by sharing $s_e$ among the threads. Similarly, we can load the B fragments directly from the registers after conversion from INT64 value inputs by proper element mapping.

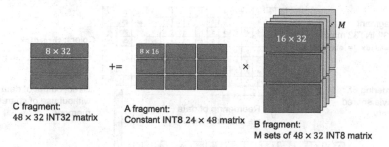

**Fig. 3.** Decomposition of $(24\times48)\times(48\times32)$ matrix-matrix product into nine $(8\times16)\times(16\times32)$ matrix-matrix products. Here, the A fragment can be reused throughout the $M$ sets of computations. Note that the results in the 32-bit integer C fragment are flushed every $M = 3$ stages to an INT64 buffer to avoid overflow.

– Improving cache reuse when accessing global memory variables: The matrix-vector product calculation requires reading the nodal data $\mathbf{u}_e$ from global memory and adding the nodal result $\mathbf{f}_e$ to the global vector $\mathbf{f}$. In this study, 32 elements belonging to a single thread block are arranged in the $x$-direction to facilitate sequential cache access in the $x$-direction, and then each thread block performs calculations continuously in the $z$-direction. This allows the reuse of nodal data ($\mathbf{u}$ and $\mathbf{f}$) in cache, which results in an approximately fourfold reduction in the volume of global memory accesses compared to the case when elements are ordered randomly.

By employing a hierarchical method that reduces the number of conversions between the floating-point and integer variables, as well as using the data access reduction methods described above, we can expect high increases in speed compared to the standard FP64 computations without reduction in computational accuracy.

## 3   Numerical Experiment

In the following, the efficacy of the proposed TCOVFEM is demonstrated through ultrasonic analysis using the model illustrated in Fig. 5, which is based on the literature in [10]. The model is a rebar with a radius of 15 mm embedded in concrete. An impulse force with a center time of $4.096 \times 10^{-4}$ s and a center frequency of 112.5 kHz with a bandpass of 100–125 kHz is applied in the $z$-direction from the input point, and its response is calculated and stored for $8.192 \times 10^{-4}$ s at the observation points shown in Table 1. The size of the model is $324 \times 128 \times 384$ mm and assumes a cylindrical rebar of radius 15 mm centered at $x = 160$, $z = 100$ mm that is penetrating the model in the $y$-direction. The four corner points at the bottom ($z = 0$) of the model are fixed in three directions ($x$, $y$, and $z$). Rayleigh damping (100–125 kHz) is used to compute attenuation in the simulation. The following time history simulations were performed on a

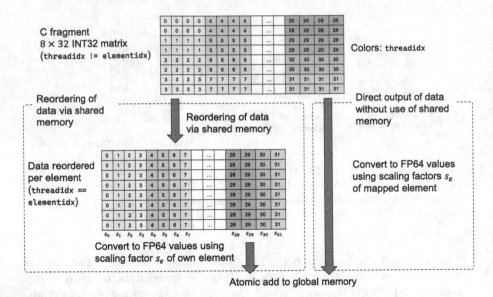

**Fig. 4.** Elimination of shared memory loads/stores by direct addition of results to global memory. Note that one out of the three C fragments is shown. 32 elements are computed using 32 threads per thread block.

single A100 80 GB PCIe GPU [9], and except for the INT8 Tensor Core operation in the proposed TCOVFEM, all other operations were computed and stored using FP64 variables.

First, a reference solution was generated using an implicit finite element analysis with second-order tetrahedral elements without the lumped mass matrix approximation (hereafter referred to as CFEM), and we compared the waveforms at the observation points to confirm the accuracy of the proposed TCOVFEM. As shown in Fig. 5, the finite element mesh used for the reference solution based on CFEM is sufficiently fine. For example, the area near the rebar is discretized with an element size of approximately 1 mm (i.e., the nodal spacing becomes approximately 0.5 mm as second-order tetrahedral elements are used). We used a model with 2-mm voxel elements for the proposed TCOVFEM, and to compare the accuracy with the conventional VFEM method, we conducted VFEM analysis for cases with seven element sizes (2.0, 1.5, 1.33, 1.2, 1.09, 1.0, and 0.5 mm). Table 2 shows the error from the reference solution at all observation points, which is defined as follows for each analysis case.

$$Err = \frac{1}{n_c} \sum_{i=1}^{n_c} \frac{\sum_{j=1}^{n_t}(u_{i,j}^{\mathrm{obs}} - u_{i,j}^{\mathrm{ref}})^2}{\sum_{j=1}^{n_t}(u_{i,j}^{\mathrm{ref}})^2} \tag{18}$$

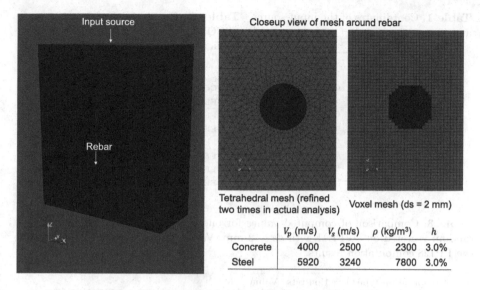

| | $V_p$ (m/s) | $V_s$ (m/s) | $\rho$ (kg/m³) | $h$ |
|---|---|---|---|---|
| Concrete | 4000 | 2500 | 2300 | 3.0% |
| Steel | 5920 | 3240 | 7800 | 3.0% |

**Fig. 5.** Model used for the numerical experiment.

Here, $n_c$ denotes the number of observation channels (eight observation points $\times (x, y, z)$-components $= 24$ channels), $n_t$ denotes the number of time steps, and $u_{i,j}^{obs}$ and $u_{i,j}^{ref}$ denote the computed and reference waves at time step $j$ for channel $i$, respectively. The results demonstrate that the proposed TCOVFEM can compute results with high accuracy compared to the conventional method. The error of CFEM using a coarser mesh (the length of an element edge is twice that of the mesh used in the reference solution) was 0.00022, which confirms that the reference solution computed using CFEM with the fine mesh is converged sufficiently. A comparison of the proposed TCOVFEM with the conventional VFEM indicates that the proposed method is more accurate than the VFEM in the case with the same discretization width. Here, the proposed TCOVFEM with $ds = 2$ mm achieved accuracy that is equivalent to the VFEM with $ds = 1.2$ mm. Figure 6, which visualizes the response in the $y$- and $z$-directions at observation point 1, also confirms that the proposed TCOVFEM with $ds = 2$ mm is considerably more accurate than VFEM at $ds = 2$ mm (even when viewed at individual observation points).

Next, we review the effect of INT8 Tensor Core operations on calculation accuracy. Here, we compare the accuracy when computing Eq. (9) without Tensor Cores using FP128, FP64, and FP32 operations, and computing Eq. (9) with INT8 Tensor Cores. Specifically, the calculations for multiplying the global stiffness matrix of the $ds = 2$ mm cubic element model by random vectors with double precision values were computed with FP128, FP64, and FP32 operations, and these calculations were compared with the results computed using INT8 Tensor Core operations. Note that FP128 has 112 fraction bits, FP64 has 52 fraction bits, and FP32 has 23 fraction bits. In contrast, as the resolution of

**Table 1.** Coordinates of the input source and observation points.

| | $x$ (mm) | $y$ (mm) | $z$ (mm) |
|---|---|---|---|
| Source point | 156 | 72 | 384 |
| Obs. point 1 | 26 | 60 | 384 |
| Obs. point 2 | 60 | 60 | 384 |
| Obs. point 3 | 108 | 60 | 384 |
| Obs. point 4 | 144 | 60 | 384 |
| Obs. point 5 | 180 | 60 | 384 |
| Obs. point 6 | 216 | 60 | 384 |
| Obs. point 7 | 264 | 60 | 384 |
| cre Obs. point 8 | 300 | 60 | 384 |

**Table 2.** Comparison of relative error (TCOVFEM: proposed method; VFEM: explicit standard voxel FEM).

| | $ds$ (mm) | $dt$ (s) | $Err$ |
|---|---|---|---|
| TCOVFEM | 2.000 | $5.0 \times 10^{-8}$ | **0.03947** |
| VFEM | 2.000 | $5.0 \times 10^{-8}$ | 0.28057 |
| VFEM | 1.500 | $2.5 \times 10^{-8}$ | 0.09579 |
| VFEM | 1.333 | $2.5 \times 10^{-8}$ | 0.05743 |
| VFEM | 1.200 | $2.5 \times 10^{-8}$ | **0.03602** |
| VFEM | 1.091 | $2.5 \times 10^{-8}$ | 0.02343 |
| VFEM | 1.000 | $2.5 \times 10^{-8}$ | 0.01570 |
| VFEM | 0.500 | $1.25 \times 10^{-8}$ | 0.00104 |

**Table 3.** Comparison of computed values for one component of **Ku** computed for a random vector **u** on OVFEM with $ds = 2$ mm. Values in bold letters are identical to the FP128 computation results.

| Computation type | Fraction bits | Value |
|---|---|---|
| FP128 | 112 | **507813.69059255961682786791**0192902549 |
| FP64 | 52 | **507813.6905925**632 |
| FP32 | 23 | **507813.**750 |
| INT8 ($M = 4$) | 28 | **507813.**7802133318 |
| INT8 ($M = 8$) | 56 | **507813.6905925**595 |

an 8-bit integer is 7 bits, the INT8 Tensor Core calculation is equivalent to 28 fraction bits for $N = 4$ stages and 56 fraction bits for $M = 8$ stages. Thus, it is expected that accuracy equivalent to that of FP32 can be obtained with $M = 4$ stages and higher than that of FP64 with $M = 8$ stages in the INT8 calculation. The results shown in Table 3 demonstrate that the accuracy of the INT8 operation increases with the number of stages, and that the INT8 operations with $M = 8$ stages are equivalent to or higher than that of FP64. In addition, the INT8 operations with $M = 4$ stages are equivalent to that of FP32. Thus, we use $M = 8$ stages in the proposed TCOVFEM throughout the rest of this paper.

Finally, we compare the elapsed time for the entire time series analysis and the matrix-vector product kernel (Table 4). In the following, we refer to $\mathbf{Ku}^{it}$ in

$$\mathbf{u}^{it+1} \Leftarrow 2\mathbf{u}^{it} - \mathbf{u}^{it-1} + dt^2 \mathbf{M}^{-1}(\mathbf{F}^{it} + \mathbf{Ku}^{it}), \tag{19}$$

which is obtained by transforming Eq. (3), as the matrix-vector product kernel, and the rest as other computations. We first compare the proposed TCOVFEM with $M = 8$ stages (TCOVFEM INT8 ($M = 8$)) with OVFEM without Tensor Cores (OVFEM FP64), which computes Eq. (9) as is with FP64 variables on the $ds = 2$ mm model. In the model with $ds = 2$ mm, the other computations took 4.62 s. By excluding this, we can evaluate the improved calculation time of the matrix-vector product operation when using the Tensor Cores. The matrix-vector product calculation was $43.3/9.62 = 4.5$-fold faster using the Tensor Cores,

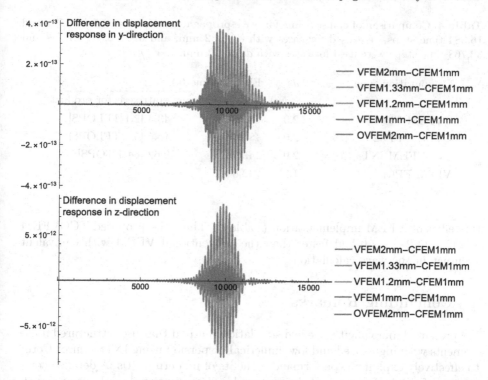

**Fig. 6.** Differences in time history displacement response at observation point 1 when compared with the reference solution ($y$- and $z$-directions with $dt = 5 \times 10^{-8}$ s).

which corresponds to 64.4 TOPS in computational performance. In contrast, the OVFEM matrix-vector product calculation using FP32 (without Tensor Cores), which had lower accuracy (Table 3), took 15.5 s. Thus, the proposed TCOVFEM, which demonstrates computational accuracy that is equivalent to that of FP64 computations, is 1.6 times faster than the FP32 implementation. This improved calculation time for the matrix-vector product resulted in a $48.3/14.2 = 3.39$-fold speedup of the overall time history simulation compared to the FP64-based OVFEM without Tensor Cores. We also compared the proposed method with the conventional VFEM. Note that VFEM uses the element-by-element (EBE) method in the explicit time integration code, which calculates the matrix-vector product in an element-by-element manner without keeping the entire stiffness matrix in memory (all calculations are performed in FP64). Here, the VFEM EBE (8-point integration) implementation in the paper [11], which achieves high computational performance as a regular EBE-based VFEM, was used for comparison. As shown in Table 2, elements with $ds = 1.2$ mm are required for the VFEM to obtain accuracy that is equivalent to that of the proposed TCOVFEM with $ds = 2$ mm sized elements. Consequently, since the number of elements is $(2.0/1.2)^3 = 4.63$ times greater, and the number of time steps is twice as large, the amount of operations increases such that the analysis takes 242.8 s even using

**Table 4.** Comparison of elapsed time for wave propagation. Here, $dt = 5 \times 10^{-8}$ s and 16,384 time steps were used for cases with $ds = 2$ mm, and $dt = 2.5 \times 10^{-8}$ s and 32,768 time steps were used for cases with $ds = 1.2$ mm.

| Computation type | $ds$ | Elapsed time (s) | |
|---|---|---|---|
| | (mm) | Time-step loop | Matrix-vector product |
| OVFEM FP64 | 2.0 | 48.3 | 43.3 [2.1 TFLOPS] |
| OVFEM FP32 | 2.0 | 18.9 | 15.5 [5.9 TFLOPS] |
| TCOVFEM INT8 ($M = 8$) | 2.0 | 14.2 | 9.62 [64.4 TOPS] |
| VFEM FP64 | 1.2 | 242.8 | - |

this efficient VFEM implementation (Table 4). Thus, the proposed TCOVFEM is $242.8/14.2 = 17.0$-fold faster than the conventional VFEM with equivalent accuracy in the wave calculation.

## 4    Concluding Remarks

We presented an explicit wavefield simulation method that uses structured finite elements with high speed and low numerical dispersion using INT8 Tensor Cores to effectively exploit the performance benefits of modern GPUs. A detailed comparison of the performance of the proposed and conventional methods on a real-world problem demonstrates that the proposed method utilizes GPU performance efficiently and that the proposed TCOVFEM (INT8 ($M = 8$)) is 17.0-fold faster than the conventional VFEM. Tensor Cores, which are an acceleration mechanism in modern GPUs, are used in deep learning and other applications due to their high computing performance. However, the application of Tensor Cores in physics-based simulations remains limited because it must be brought into a specific format and data transfer must be performed without losing computing performance. In addition, the use of Tensor Cores to obtain a coarse solution and then refine it, as in the implicit method, is difficult in the explicit method, thereby making it even more challenging. In this context, the formulation and implementation presented in this paper, which exploits the performance advantages of Tensor Cores in explicit physics-based simulations, are general and expected to provide insights into accelerating other applications. The conversion of floating-point matrix-vector computations to small integer-based matrix-matrix computations is expected to be accelerated in other GPU architectures with acceleration mechanisms for integer-based matrix-matrix multiplication. Note that the purpose of this study was to demonstrate the basic performance of the proposed method; thus, the analysis was performed on only a single GPU. However, we expect that the proposed method can easily be extended to solve large-scale problems using multiple GPUs.

# References

1. Koketsu, K., Fujiwara, H., Ikegami, Y.: Finite-element simulation of seismic ground motion with a Voxel Mesh. Pure Appl. Geophys. **161**(11), 2183–2198 (2004)
2. Bielak, J., Ghattas, O., Kim, E.J.: Parallel octree-based finite element method for large-scale earthquake ground motion simulation. Comput. Model. Eng. Sci. **10**(2), 99–112 (2005)
3. Nakahata, K., Amano, Y., Ogi, K., Mizukami, K., Saitoh, T.: Three-dimensional ultrasonic wave simulation in laminated CFRP using elastic parameters determined from wavefield data, Composites Part B: Engineering, 176 (2019)
4. NVIDIA Tensor Cores. https://www.nvidia.com/en-us/data-center/tensor-cores/. Accessed 28 Feb 2024
5. Ichimura, T., et al.: 416-PFLOPS Fast Scalable Implicit Solver on Low-Ordered Unstructured Finite Elements Accelerated by 1.10-ExaFLOPS Kernel with Reformulated AI-Like Algorithm: For Equation-Based Earthquake Modeling, SC19 (the International Conference for High Performance Computing, Networking, Storage, and Analysis), Research Poster (2019)
6. Yamaguchi, T., et al.: Low-order finite element solver with small matrix-matrix multiplication accelerated by AI-specific hardware for crustal deformation computation, In: Proceedings of the Platform for Advanced Scientific Computing Conference (PASC'20). Association for Computing Machinery, 16:1–11 (2020)
7. Higham, N.J., Mary, T.: Mixed precision algorithms in numerical linear algebra. Acta Num. **31**, 347–414 (2022). https://doi.org/10.1017/S0962492922000022
8. Ichimura, T., Hori, M., Wijerathne, M.L.L.: Linear finite elements with orthogonal discontinuous basis functions for explicit earthquake ground motion modeling. Int. J. Numer. Meth. Eng. **86**, 286–300 (2011)
9. NVIDIA Ampere architecture whitepaper. https://www.nvidia.com/content/dam/en-zz/Solutions/Data-Center/nvidia-ampere-architecture-whitepaper.pdf. Accessed 27 Feb 2024
10. Dinh, K., Tran, K., Gucunski, N., Ferraro, C. C., Nguyen, T.: Imaging Concrete Structures with Ultrasonic Shear Waves–Technology Development and Demonstration of Capabilities. Infrastructures **8**(53) (2023)
11. Ichimura, T., Fujita, K., Hori, M., Maddegedara, L., Ueda, N., Kikuchi, Y.: A fast scalable iterative implicit solver with green's function-based neural networks. In: IEEE/ACM 11th Workshop on Latest Advances in Scalable Algorithms for Large-Scale Systems (ScalA) (2020)

# Development of an Estimation Method for the Seismic Motion Reproducibility of a Three-Dimensional Ground Structure Model by Combining Surface-Observed Seismic Motion and Three-Dimensional Seismic Motion Analysis

Tsuyoshi Ichimura[1]([✉]), Kohei Fujita[1], Ryota Kusakabe[1], Hiroyuki Fujiwara[2], Muneo Hori[3], and Maddegedara Lalith[1]

[1] Earthquake Research Institute and Department of Civil Engineering, The University of Tokyo, Tokyo, Japan
ichimura@eri.u-tokyo.ac.jp
[2] Multi-hazard Risk Assessment Research Division, National Research Institute for Earth Science and Disaster Resilience, Tsukuba, Japan
[3] Research Institute for Value-Added-Information Generation, Japan Agency for Marine-Earth Science and Technology, Yokohama, Japan

**Abstract.** The ground structure can substantially influence seismic ground motion underscoring the need to develop a ground structure model with sufficient reliability in terms of ground motion estimation for earthquake damage mitigation. While many methods for generating ground structure models have been proposed and used in practice, there remains room for enhancing their reliability. In this study, amid many candidate 3D ground structure models generated from geotechnical engineering knowledge, we propose a method for selecting a credible 3D ground structure model capable of reproducing observed earthquake ground motion, utilizing seismic ground motion data solely observed at the ground surface and employing 3D seismic ground motion analysis. Through a numerical experiment, we illustrate the efficacy of this approach. By conducting $10^2$–$10^3$ cases of fast 3D seismic wave propagation analyses using graphic processing units (GPUs), we demonstrate that a credible 3D ground structure model is selected according to the quantity of seismic motion information. We show the effectiveness of the proposed method by showing that the accuracy of seismic motions using ground structure models that were selected from the pool of candidate models is higher than that using ground structure models that were not selected from the pool of candidate models.

**Keywords:** Three-dimensional Ground Structure Model · Three-dimensional Seismic Motion Analysis · Finite Element Method

L. Franco et al. (Eds.): ICCS 2024, LNCS 14834, pp. 272–287, 2024.
https://doi.org/10.1007/978-3-031-63759-9_32

# 1    Introduction

Structures on or within the ground can experience substantial impacts from the ground structure during seismic events [1] [2]. For instance, soft ground atop the hard ground, coupled with the characteristics of seismic motion and the ground structure, can lead to localized amplification of seismic motion, resulting in damage. Enhancing the reliability of ground structure assessments is required to enable accurate evaluation of structural behavior during seismic events and mitigate earthquake-induced damage effectively.

Many methods for estimating ground structures have been proposed and employed in practical applications. For instance, methods directly assessing *in situ* geotechnical properties, such as borehole testing, yield reliable outcomes. Nonetheless, due to the limited number of measurement points, interpolation becomes necessary to estimate the spatial extent of the ground structure. In the construction of 3D ground structure models, interpolation methods based on geotechnical engineering knowledge, such as inverse distance weighting methods [3], curvature minimization principles [4], and Kriging methods [5], are employed. These interpolations rely on various assumptions, resulting in multiple 3D ground structure models, necessitating additional evaluation to determine the reliability of each model.

Such evaluation can utilize the observed ground motion of small amplitudes. For instance, the reliability of a candidate ground structure model can be assessed by comparing the phase velocity derived from long-term microtremor observations [6] or by comparing the Green's functions obtained via seismic interferometry [7]. Alternatively, the quality of 3D ground structure models can be evaluated by directly utilizing observed seismic ground motions of small amplitudes, which typically exhibit stronger signal strengths. For example, [8] demonstrates the feasibility of evaluating a 3D ground structure model using observed seismic ground motions of small amplitudes. However, the method in [8] assumes that seismic motions input to the 3D ground structure model are observed at underground observation points, making it challenging to apply to sites where underground observation points cannot be installed.

Building upon the above insights, we propose a method that selects a credible ground structure model from many generated 3D ground structure models using small amplitude seismic motions solely observed at the ground surface. We illustrate the effectiveness of the method through numerical experiments by conducting $10^2$–$10^3$ cases of fast 3D seismic wave propagation analysis on GPUs, and show that we can select a credible ground structure model based on the amount of seismic motion information observed at the ground surface. Additionally, we show that the selected 3D ground structure model can be utilized to evaluate ground motion with sufficient accuracy.

## 2  Method

We introduce a method for extracting a credible 3D ground structure model from a pool of candidate 3D ground structure models using small amplitude seismic motions observed at the ground surface. The general setting, depicted on the left of Fig. 1, involves inputting seismic waves to a 3D ground structure comprising soft and hard soil, with resulting seismic ground motions observed at the designated points on the surface. Notably, the 3D ground structure and the input seismic waves are typically unknown.

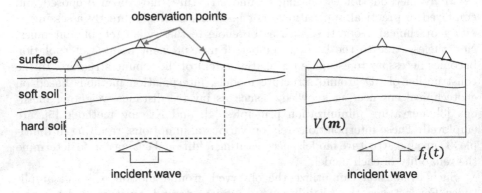

**Fig. 1.** Target system (left) and its numerical analysis model (right).

Next, we elucidate the numerical analysis model for the target system, as depicted on the right side of Fig. 1. This study uses the cartesian coordinate system $x_1, x_2, x_3$ for simplicity. The ground enclosed by the dashed line in the left panel of Fig. 1 is designated as $V(\mathbf{m})$, defined by a set of model parameters $\mathbf{m}$. An external force, represented by the input velocity wave $f_i(t)$, is applied to the bottom of $V(\mathbf{m})$, and its dynamic response is analyzed to yield the time-history response $obs_i^k(t)$ at each observation point $k$ on the ground surface. To compute the small amplitude seismic response of the ground to an earthquake, we model $V(\mathbf{m})$ as linearly elastic and solve the governing equations for a linear dynamic elastic body

$$d_i(c_{ijkl}, d_k u_l) = \rho \ddot{u}_j. \tag{1}$$

Here, $c_{ijkl}$, $u_j$, and $\rho$ represent the elasticity tensor, displacement in the $j$-th direction, and density, respectively. The notation ( ¨ ) denotes second-order derivatives in time, and $d_i$ indicates the differentiation in the $i$-th direction. Note that stress-free, non-reflective, and semi-infinite absorbing boundary conditions are enforced on the model's top, bottom, and side surfaces. We aim to develop a method that can evaluate $V(\mathbf{m})$ and $f_i(t)$ that is consistent with $obs_i^k(t)$.

We describe the parametrization employed for this evaluation. First, we suppose that the input velocity wave can be described as follows:

$$f_i(t) = \sum_j c_{ij} p(t - (j-1)\Delta t).$$

We utilize unit impulse waves $p(t)$, for instance, trigonometric functions, to represent the input velocity wave. Here, $c_{ij}$ represents unknown scalar values assessed through error minimization, and $\Delta t$ denotes the time-stepping stride for discretization. While arbitrary $p(t)$ and $\Delta t$ are permissible, we configure them to facilitate the reconstruction of the input wave within the target frequency range. Moreover, we denote the Green's function of the $i$-th directional velocity response at observation point $k$ when $p(t)$ is input in the $j$-th direction at the bottom of $V(\mathbf{m})$, as $G_{ij}^k(\mathbf{m}, t)$. The $i$-th directional velocity at observation point $k$ for $f_j(t)$ can thus be expressed as

$$U_i^k(\mathbf{m}, t) = \sum_{j=1}^{3} \sum_{l} G_{ij}^k(\mathbf{m}, t - (l-1)\Delta t) c_{jl}.$$

Utilizing the observed ground velocity in the $i$-th direction at the $k$-th observation point $(obs_i^k(t))$, the error can be assessed as follows:

$$ERR(\mathbf{m}) = \frac{1}{3n_{obs}} \sum_{k=1}^{n_{obs}} \sum_{i=1}^{3} \frac{\sqrt{\int (U_i^k(\mathbf{m}, t) - obs_i^k(t))^2 dt}}{\sqrt{\int obs_i^k(t)^2 dt}}. \tag{2}$$

With the parameter settings outlined above, we compute $c_{ij}$ that minimizes the error from the observed seismic motion $obs_i^k(t)$ for each candidate $l$-th 3D ground structure model characterized by model parameters $\mathbf{m}_l$. First, we compute $G_{ij}^k(\mathbf{m}_l, t)$ for the model parameters $\mathbf{m}_l$. Subsequently, by considering the stationary condition of $ERR$, which is a second-order function of $c_{ij}$, the coefficients $c_{ij}$ can be determined by solving:

$$\mathbf{Ac} = \mathbf{b}.$$

Here, $\mathbf{A}$, $\mathbf{b}$ represent a constant matrix and a constant vector, respectively, while $\mathbf{c}$ denotes an unknown vector with components $c_{ij}$.

By applying singular value decomposition to $\mathbf{A}$ and discarding negligible singular values, we construct a pseudoinverse matrix and compute $\mathbf{c}$ and $ERR$. Note that the dimension of $\mathbf{A}$ is small because this computation can be performed independently for each trial, and thus the cost of this computation can be kept small enough. That is, if the seismic ground motions contain sufficient information (i.e., the number of observation points and the number of events is adequate) and are suitably constrained by the model-specific time-history Green's functions, it is possible to assess the credibility of various 3D ground structure models by comparing the error $ERR$ obtained through attempts to reproduce observed seismic ground motions using $\mathbf{m}_l$.

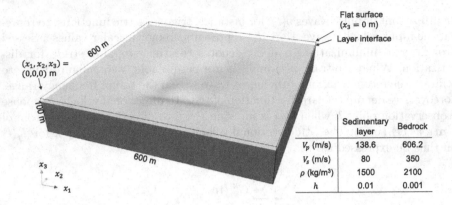

**Fig. 2.** Reference ground model. The model comprises two layers, with a flat surface and a sedimentary layer with varying thickness. The thickness of the sedimentary layer is illustrated in Fig. 7a).

## 3  Numerical Experiment

We conducted a numerical experiment to demonstrate the efficacy of the proposed method in selecting a credible 3D ground structure model from among geotechnically estimated ground models, using seismic ground motions observed at the ground surface. Here, a credible ground model can reproduce the seismic ground motions observed at the surface. Specifically, the method described in Sect. 2 estimates the seismic ground motion that best fits the observed seismic ground motion using a given 3D ground structure model, albeit with introduced errors. This analysis in estimating seismic ground motion that best fits the observed ground motion for a given 3D ground structure model is conducted each time an earthquake is observed. Discarding ground models that inadequately reproduce seismic ground motions can identify a credible ground model capable of reproducing observed seismic ground motions at a specific site.

First, we set a reference model to serve as the correct solution for the numerical experiment. The reference model comprises a two-layer structure comprising a sedimentary layer and bedrock, which imitates a real ground structure. It spans 600, 600, and 100 m in the $x_1$, $x_2$, and $x_3$ directions, respectively (see Fig. 2). The ground surface is flat at an elevation of 0 m, with the thickness of the first layer specified as depicted in Fig. 7a). The physical properties of each layer are given in Fig. 2. This ground structure emulates valley floor lowlands formed by sediment accumulation from river meandering. During earthquakes, local ground amplification occurs, emphasizing the importance of accurately assessing the ground structure for evaluating seismic ground motion and making effective earthquake mitigation strategies. Assuming that the reference model is situated at KiK-net [9] station IBRH19, we simulate a scenario where earthquakes listed in Table 1 occur sequentially, and seismic motions are observed. To simulate this scenario, actual ground motions observed at underground stations of IBRH19 were inputted as forces from the bottom of the reference ground model, and

**Table 1.** Properties of the earthquakes used in the numerical experiment

| Event # | Data time (JST) | Epicenter | Depth (km) | Magnitude |
|---|---|---|---|---|
| 1 | 2023/02/25 22:27 | 42.755N 145.075E | 63 | 6.0 |
| 2 | 2023/03/27 00:04 | 38.307N 141.615E | 60 | 5.3 |
| 3 | 2023/05/11 04:16 | 35.170N 140.185E | 40 | 5.2 |
| 4 | 2023/07/29 19:34 | 36.347N 139.958E | 77 | 4.6 |
| 5 | 2023/12/22 01:11 | 35.238N 141.137E | 10 | 4.8 |
| 6 | 2024/01/28 08:59 | 35.6N 140.0E | 80 | 4.8 |

pseudo-observed seismic ground motions were recorded at the ground surface. The analysis was performed up to 2.5 Hz, which is the frequency range considered to have a significant impact on structural damage. Semi-infinite absorbing boundary conditions are applied to the sides and bottom of the numerical model.

Next, a set of candidate ground structure models is generated using geotechnical engineering methods. While many methods generating ground structure models exist, we assume that the physical properties of the first and second layers, along with the boundary location between them, are determined from a simple ground survey conducted at the 120 survey points illustrated in Fig. 7a) (i.e., **m** represents model parameters describing the boundary shape between the first and second layers). The geometry of the boundary between the first and second layers within the target area of $600 \times 600$ m is estimated based on information regarding the boundary layer location at these 120 survey points, creating a set of candidate ground structure models. Although many methods for estimating the layer boundary geometry under these conditions exist, we employ an inverse distance weighting method. This method assigns weights to observed values obtained near the evaluation point based on the inverse of the distance, yielding an interpolated result. Utilizing the inverse distance weighting method, the layer thickness $z$ at any point in the domain is evaluated as follows:

$$z(x_1, x_2) = \sum_{i=1}^{M} \frac{\bar{z}_i/d_i^q}{\sum_{i=1}^{M} 1/d_i^q}, \text{ where } d_i = \sqrt{(x_1 - \bar{x}_1^i)^2 + (x_2 - \bar{x}_2^i)^2},$$

where $\bar{z}_i$ denotes the layer thickness at the $i$-th measurement point located at $(\bar{x}_1^i, \bar{x}_2^i)$, $M$ represents the number of measurement points nearest to the target position $(x_1, x_2)$ utilized for interpolation, and $q$ is a positive constant parameter. The values of $M$ and $q$ are arbitrary. In this context, we designate $M = i$ ($i = 1, 2, ..., 50$) and $q = 0.1i$ ($i = 1, 2, ...., 40$) to generate a total of $50 \times 40 = 2000$ 3D ground structure models. Subsequently, from the pool of candidate 3D ground structure models, we select 236 models that exhibit substantial variations in the layer boundary geometry, based on a 3D ground structure model generated using commonly used parameters (i.e., $M = 20$ and $q = 2.0$). Note that Laplace smoothing was applied five times to mitigate steep slopes in certain areas.

Utilizing the method described in Sect. 2 and the pseudo-observed seismic ground motions described above, we attempt to identify credible 3D ground structure models by reproducing the observed seismic ground motions for each of the 236 obtained models. First, we consider a scenario where only one observation point is available on the ground surface at $(x_1, x_2) = (300, 300)$ m. The analysis conducted for each candidate model is the same as the analysis using the reference model (i.e., the target frequency is set up to 2.5 Hz), with semi-infinite absorbing boundary conditions applied to the sides and bottom of the model (the same applies to the other analysis cases in this Section). Figure 3a) illustrates the errors obtained for each ground structure model across each earthquake event. Although the error varies depending on the model, it remains very small regardless of the ground structure model in the case of a single observation point (from the definition of $ERR$ in Eq. (2), $ERR = 0.05$ means that the waveform is matched within an error of about 5% of its amplitude). This indicates that the waveform constraints are insufficient when the number of observation points is limited. In other words, despite the complexity of the time history Green's function in a 3D medium, observed waveforms can be accurately reconstructed regardless of the model used by imposing errors on the estimated input waveform. Indeed, the estimated input waveform obtained using model000083 (i.e., model number 83 of the 236 candidate models), which exhibits a small error, substantially differs from the true waveform in amplitude and phase characteristics (Fig. 4). Conversely, increasing the number of observation points becomes imperative to leverage model-specific time-history constraints in Green's functions for accurately reconstructing both observed and input waveforms. This constraint can be leveraged to diminish the ability to reproduce the observed waveforms depending on the ground structure model, thereby facilitating the selection of the appropriate ground structure model.

Next we employ nine observation points on the ground surface $(x_1, x_2) = (100 + 200i, 100 + 200j)$ m, where $i, j = 0, 1, 2$. Figure 5a) illustrates the error obtained for each model across each earthquake event. Compared with the single observation point case, the information derived from the pseudo-observed earthquake ground motion adequately constrains the ground model. Consequently, models exhibiting small error levels are consistently identified irrespective of the earthquake event. The comparison with the historical maximum error depicted in Fig. 5b) reveals that the credible ground model remains consistently selected even as the number of experienced earthquakes and the quantity of information increases. Furthermore, we augment the information by increasing the number of surface observation points to 25, located at $(x_1, x_2) = (100 + 100i, 100 + 100j)$ m, where $i, j = 0, 1, ...., 4$. The error estimation is shown in Fig. 6a). As in the 9-point scenario, models with small errors are systematically chosen. The comparison with the historical maximum error exhibited in Fig. 6b) underscores the continued stable selection of the credible ground model even with the heightened number of experienced earthquakes and augmented information. Comparing the 9-point case with the 25-point case, it is evident that the increase in information

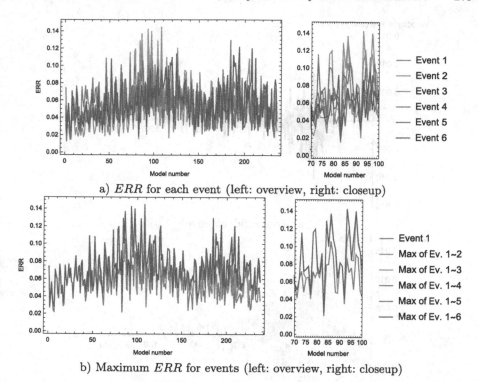

a) $ERR$ for each event (left: overview, right: closeup)

b) Maximum $ERR$ for events (left: overview, right: closeup)

**Fig. 3.** Estimated $ERR$ using one observation point

facilitates the stable and systematic extraction of ground models characterized by smaller errors.

Finally, we checked the extracted results. Figures 7a,b, and c) display the layer thickness distribution of the reference ground model, model000018 (with a minimum error of 0.388 for the case with 25 observation points), and model000228 (with error of 0.781 for the case with 25 observation points). Model000018, with the smallest error, closely resembles the reference ground model, indicating the extraction of a plausible 3D ground structure model through this method. While no model among the candidate ground models exactly matches the reference model, model000018 was selected due to its minimal difference compared to the target wavelength used for analysis. Conversely, model000228, with a large error, substantially deviates from the reference model. Next, Fig. 8 illustrates the $x_1$ component of the input seismic motion estimated by model000018 and model000228 using pseudo-observed seismic ground motion at 25 observation points during earthquake event #1. The input waveform estimated using model000018, evaluated to have a small error, closely resembles the

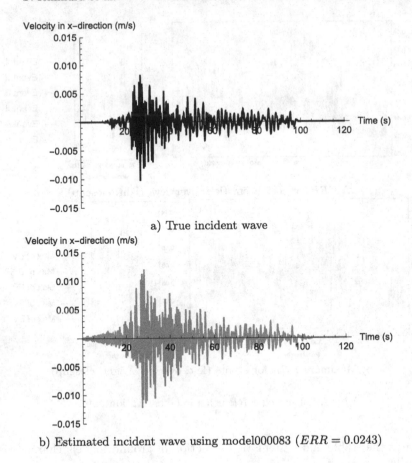

a) True incident wave

b) Estimated incident wave using model000083 ($ERR = 0.0243$)

**Fig. 4.** Estimated incident wave using one observation point for event #1. Estimation accuracy is low even if $ERR$ is low.

true input wave, while the waveform estimated using model000228 differs substantially from the true input wave. This underscores that sufficient observed seismic ground motions enable the Green's function constraint of the time history in the 3D medium obtained from a 3D wavefield analysis, ensuring estimation performance for both observed and input seismic motions when using a plausible model. Finally, Fig. 9 depicts the ground motion distribution using the reference model, model000018, and model000228. The responses of the reference model and model000018 closely align regardless of the location from the observation points, whereas the response of model000228 differs markedly. As shown in Fig. 10, model000018, selected as the most credible 3D ground structure model in this study, can estimate not only the time series response at the observation point ($(x_1, x_2) = (300, 300)$ m) but also the time series response at a point which is located away from the observation points ($(x_1, x_2) = (350, 350)$ m). This indicates that a consistent ground model that matches the observed

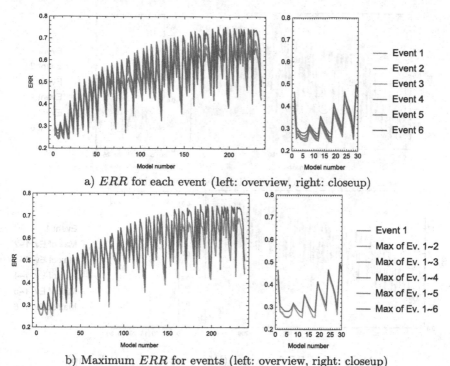

a) $ERR$ for each event (left: overview, right: closeup)

b) Maximum $ERR$ for events (left: overview, right: closeup)

**Fig. 5.** Estimated $ERR$ using nine observation points

seismic ground motion has been extracted. These results underscore the significance of selecting a plausible 3D ground structure model, as demonstrated in this study, for mitigating earthquake damage, since even 3D ground structure models estimated by geotechnical engineering methods exhibit substantial differences in their performance in reproducing seismic ground motions.

As described above, it is evident that a 3D ground structure model demonstrating high performance in reproducing observed seismic motions can be extracted by using seismic motion observations at the ground surface. Achieving this necessitates multiple 3D seismic response analyses. In this study, finite element models are generated for the 236 ground models with varying layer thicknesses, and for each model, the Green's function response of Eq. (1) to the unit impulse wave $p(t)$ in each of the $x_1$, $x_2$, and $x_3$ directions are computed, resulting in a total of $236 \times 3 = 708$ instances of 3D wave field analysis. An automatic method for generating high-quality finite element models is required to facilitate such a large number of analyses. Moreover, a fast implicit time integration-based 3D seismic analysis method capable of performing stable calculations, even on finite element models with small local elements that are essential to faithfully model the complex ground geometry, is required. Consequently, in this study, finite element models comprising second-order tetrahedral elements are auto-

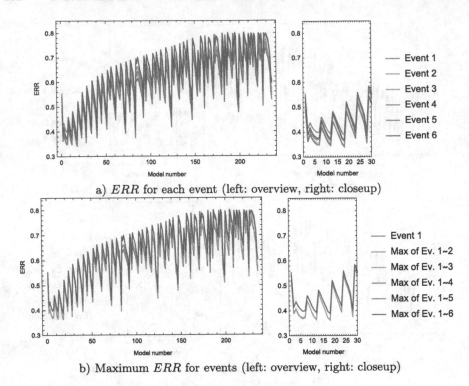

a) $ERR$ for each event (left: overview, right: closeup)

b) Maximum $ERR$ for events (left: overview, right: closeup)

**Fig. 6.** Estimated $ERR$ using 25 observation points

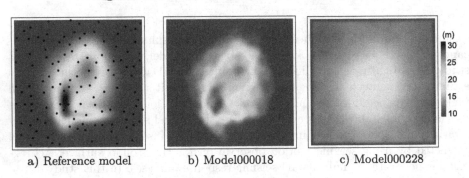

a) Reference model       b) Model000018       c) Model000228

**Fig. 7.** Thickness of sedimentary layers for each model. The position of 120 boring log points is also indicated in a).

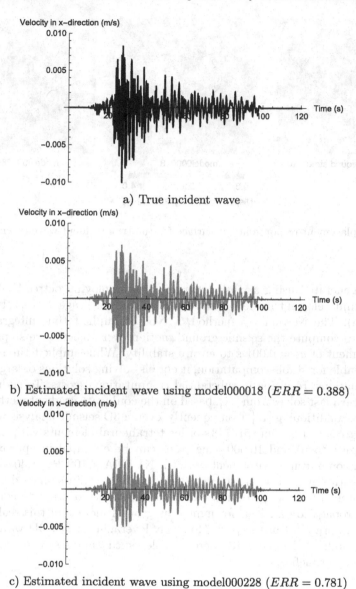

a) True incident wave

b) Estimated incident wave using model000018 ($ERR = 0.388$)

c) Estimated incident wave using model000228 ($ERR = 0.781$)

**Fig. 8.** Estimated incident wave using 25 observation points for event #1. The incident wave can be estimated accurately by selecting a model with a low $ERR$.

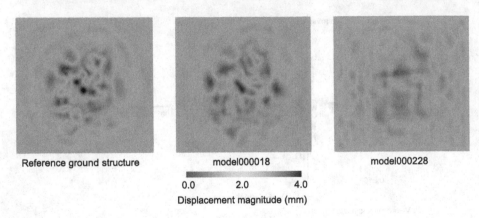

Reference ground structure          model000018          model000228

0.0          2.0          4.0

Displacement magnitude (mm)

**Fig. 9.** Displacement response at the surface for the true incident wave (event #1) at $t = 50$ s

matically generated using a mesh generator [10] employing octree background cells (the finite element models utilized in this study average 2,320,401 degrees of freedom). The Newmark-$\beta$ method, a type of implicit time integration, is employed to compute the seismic ground motion over 16,000 time steps with a time increment of $dt = 0.001$ s to ensure stability. While implicit time integration is suitable for stable computation, it entails solving solutions of large sparse matrix equations, resulting in substantial computational costs. To address this issue, we use a fast solver from [11], based on the conjugate gradient method with variable preconditioning [12]. Consequently, even a 3D seismic analysis with 2.32 million degrees of freedom (516748 s-order tetrahedral elements with minimum element size of 5 m) and 16,000 time steps can be executed in approximately 270 s on a computing environment with an NVIDIA A100 PCIe 40 GB GPU [13]. The entire evaluation can be completed stably within a short duration of approximately 1.5 h using 44 A100 GPUs. Due to time constraints, conducting 3D wave propagation analysis for numerous ground models of this scale would have been impractical in the past. However, leveraging a fast 3D wave propagation method, as demonstrated here, enables such analyses to be conducted within feasible timeframes.

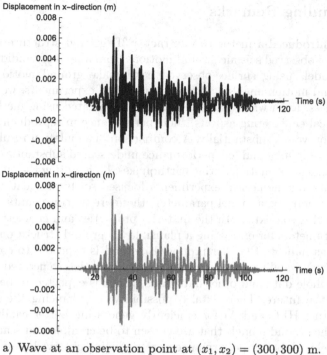

a) Wave at an observation point at $(x_1, x_2) = (300, 300)$ m.

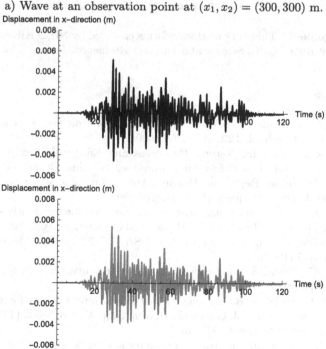

b) Wave at a point on the surface at $(x_1, x_2) = (350, 350)$ m, which is far from the observation points.

**Fig. 10.** Displacement time-history response for the true incident wave (event #1) using the reference model (black) and model000018 (blue). (Color figure online)

## 4   Concluding Remarks

This study introduced a method to extract a 3D ground structure model that can reproduce observed seismic ground motions from a pool of candidate ground structure models using surface-observed earthquake ground motions and 3D seismic ground motion analysis. Through numerical experiments, we illustrated its effectiveness. Even when ground models are generated using methods based on geotechnical engineering aspects, their performance in reproducing observed ground motion varies substantially. A comparison of simulation results between models exhibiting high and low performance underscored substantial differences in seismic damage estimation during earthquakes, highlighting the efficacy of our method. While our numerical experiments focused solely on treating the layer boundary geometry as a model parameter, there are no constraints on parameterization in this context. Both the material properties and geometry can serve as model parameters for extracting a plausible 3D ground structure model. The above findings indicate that our proposed method is expected to contribute to enhancing the reliability of 3D ground structure models generated by supplementing borehole data and other databases, which is expected to become more common in the future. There is also possibility in enhancing the method by utilizing further HPC and AI for efficiently generating better candidate models around the ground models that are chosen to be credible and searching for a ground model with better performance in reproducing observed ground motions.

**Acknowledgement.** This work used waveforms provided by NIED KiK-net, National Research Institute for Earth Science and Disaster Resilience, Japan.

## References

1. Liang, J., Sun, S.: Site effects on seismic behavior of pipelines: a review. ASME J. Pressure Vessel Technol. **122**, 469–475 (2000)
2. Advisory Committee for Natural Resources and Energy, Urban Area Thermal Energy Committee, Gas Safety Subcommittee, Working Group for Earthquake Disaster Prevention: Report on Disaster Mitigation for Gas Supply in View of Great East Japan Earthquake (in Japanese) (2012)
3. Shepard, D.: A two-dimensional interpolation function for irregularly-spaced data. In: Proceedings of the 1968 23rd ACM National Conference (ACM '68). Association for Computing Machinery, New York, NY, USA, 517–524. https://doi.org/10.1145/800186.810616 (1968)
4. Briggs, I. C.: Machine contouring using minimum curvature. Geophysics **39**(1), 39–48 (1974). https://doi.org/10.1190/1.1440410
5. Oliver, M.A., Webster, R.: Kriging: a method of interpolation for geographical information systems. Int. J. Geograph. Inform. Syst. **4**(3), 313–332 (1990). https://doi.org/10.1080/02693799008941549
6. Aki, K.: Space and time spectra of stationary stochastic waves with special reference to microtremors. Bull. Erthq. Res. Inst **35**, 415–456 (1957)
7. Wapenaar, K., Fokkema, J.: Green's function representation for seismic interferometry Geophysics **71**(4) SI33–SI46 (2006)

8. Yamaguchi, T., Ichimura, T., Fujita, K., Hori, M., Wijerathne, L., Ueda, N.: Data-driven approach to inversion analysis of three-dimensional inner soil structure via wave propagation analysis. In: Krzhizhanovskaya, V.V., Závodszky, G., Lees, M.H., Dongarra, J.J., Sloot, P.M.A., Brissos, S., Teixeira, J. (eds.) ICCS 2020. LNCS, vol. 12139, pp. 3–17. Springer, Cham (2020). https://doi.org/10.1007/978-3-030-50420-5_1

9. National Research Institute for Earth Science and Disaster Resilience: NIED K-NET, KiK-net, National Research Institute for Earth Science and Disaster Resilience. https://doi.org/10.17598/NIED.0004(2019)

10. Ichimura, T., Hori, M., Bielak, J.: A hybrid multiresolution meshing technique for finite element three-dimensional earthquake ground motion modelling in basins including topography. Geophys. J. Int. **177**(3), 1221–1232 (2009)

11. Kusakabe, R., Fujita, K., Ichimura, T., Yamaguchi, T., Hori, M., Wijerathne, L.: Development of regional simulation of seismic ground-motion and induced lique-faction enhanced by GPU computing. Earthquake Eng. Struct. Dyn. **50**, 197–213 (2021)

12. Ichimura, T., et al.: Physics-Based Urban Earthquake Simulation Enhanced by 10.7 BlnDOF × 30 K Time-Step Unstructured FE Non-Linear Seismic Wave Simulation, SC '14: Proceedings of the International Conference for High Performance Computing, Networking, Storage and Analysis, New Orleans, LA, USA, 2014, pp. 15–26 (2014)

13. NVIDIA Ampere architecture whitepaper. https://www.nvidia.com/content/dam/en-zz/Solutions/Data-Center/nvidia-ampere-architecture-whitepaper.pdf. Accessed 27 Feb 2024

# Numerical Studies on Coupled Stokes-Transport Systems for Mantle Convection

Ponsuganth Ilangovan$^{(\boxtimes)}$ [ID], Eugenio D'Ascoli, Nils Kohl [ID], and Marcus Mohr [ID]

Department of Earth and Environmental Sciences, LMU Munich, Munich, Germany
p.ilango@lmu.de

**Abstract.** Accurate retrodictions of the past evolution of convection in the Earth's mantle are crucial to obtain a qualitative understanding of this central mechanism behind impactful geological events on our planet. They require highly resolved simulations and therefore extremely scalable numerical methods. This paper applies the massively parallel matrix-free finite element framework HyTeG to approximate solutions to stationary Stokes systems and time-dependent, coupled convection problems. It summarizes the underlying mathematical model and verifies the implementation through semi-analytical setups and community benchmarks. The numerical results agree with the expected outcomes from the literature.

**Keywords:** Mantle Convection · Finite Elements · Matrix-Free Methods

## 1 Introduction

Mantle convection is the dominant mechanism of heat transfer from Earth's hot outer core to the cold lithosphere. This movement is the major driving force for plate tectonics and, thus, finally the trigger of earthquakes and other geological events, such as mountain building and back-arc volcanism, [9]. Modelling this mechanism opens up pathways towards answering fundamental geophysical questions [5]. A model that accurately represents mantle convection can be used inversely, with the adjoint method, to determine the past state of our planet from present-day observations, especially seismic data. Such retrodictions can help us to narrow the range of physical parameters of the mantle and identifying the location of oil deposits, among other things, [15].

On geological time scales the mantle behaves like a highly viscous fluid. The governing equations, thus, involve the Stokes system, which is a simplification of the Navier-Stokes equations for nearly vanishing Reynolds numbers [16]. Buoyancy, resulting from thermo-chemical density differences in the mantle, is the primary driving force of convection. In addition to the motion of the mantle,

L. Franco et al. (Eds.): ICCS 2024, LNCS 14834, pp. 288–302, 2024.
https://doi.org/10.1007/978-3-031-63759-9_33

one also needs to consider the evolution of heat, [7], which can be described by an advection-diffusion equation. Standard formulations can be found in literature for both compressible and incompressible models, [23, 25].

Accurate predictions require highly resolved simulations that can only be executed on massively parallel computers using extremely scalable numerical methods of optimal complexity. Resolving Earth's mantle globally with a resolution of 1 km requires meshes with trillions $(10^{12})$ of cells and yields linear systems of corresponding size. This is crucial to resolve small-scale features such as rising plumes and subducting slabs, but also to capture sharp viscosity changes between the lithosphere and underlying asthenosphere (4–5 orders of magnitude), and the resulting short wavelength asthenosphere dynamics [4].

We therefore employ the massively parallel matrix-free finite element software framework HyTeG [18, 21] to discretize the governing equations and solve the corresponding discrete problems. Its extreme-scalability has been demonstrated in, e.g., [19, 20].

The central focus of this paper is the evaluation and verification of the accuracy of the software through simulation of various benchmarks from mantle convection. We consider semi-analytic solutions to the stationary Stokes problem involving the relevant boundary conditions [22], two community benchmarks for time-dependent settings [3, 26], and eventually present results from a forward simulation of a compressible model. All development is driven by the TERRA-NEO project [1], which is an effort to create a scalable and accurate Earth model for the Geodynamics community.

## 2     Model and Formulation

On geologic time-scales the Earth's mantle behaves like a fluid. Its motion can, thus, be described by the Navier-Stokes equations. However, convection in the mantle is characterised by a very small Reynolds number $\mathcal{O}(10^{-15})$, i.e. the ratio of inertial to viscous forces, and an Ekmann number of $\mathcal{O}(10^{9})$, i.e. the ratio of viscous to Coriolis forces. Together this allows to neglect inertial and Coriolis forces and assume an instantaneous balance of viscous and buoyancy forces, resulting in a quasi-static flow field, [27]. Thus, one arrives at the momentum part of the Stokes equations

$$- \nabla \cdot \tau + \nabla p = \rho \mathbf{g} \tag{1}$$

with the deviatoric stress tensor $\tau$, pressure $p$, density $\rho$ and gravitational acceleration $\mathbf{g}$. Using the strain-rate tensor $\dot{\epsilon} = \frac{1}{2}\left(\nabla \mathbf{u} + (\nabla \mathbf{u})^{\top}\right)$ one can re-write the stress as

$$\tau = 2\eta\left(\dot{\epsilon} - \frac{1}{3}(\nabla \cdot \mathbf{u})\delta_{ij}\right) = \eta\left(\nabla \mathbf{u} + (\nabla \mathbf{u})^{\top}\right) - \frac{2}{3}\eta(\nabla \cdot \mathbf{u})I$$

where $\eta$ is the dynamic viscosity, $I$ the identity tensor and the second term on the right-hand side will vanish in an incompressible model. Buoyancy forces result from changes in local density, which are primarily driven by deviations in temperature $T$. The latter needs to satisfy a time-dependent advection-diffusion equation of the form

$$\frac{\partial T}{\partial t} = \nabla \cdot (k\nabla T) - (\mathbf{u} \cdot \nabla)T + f_T(x, T, p, \mathbf{u}) \ . \tag{2}$$

The term $f_T$ encompasses various possible heat sources, such as shear, adiabatic or radiogenic heating. The system then needs to be closed by selecting an appropriate equation of state to couple density to pressure and temperature. It is known that the hydrostatic density of the mantle increases by a factor of about two from the surface to the core-mantle boundary (CMB). Hence, a purely incompressible flow model will not be exact. A common approach in Geodynamics is to select a reference state $(\bar{p}, \bar{T})$, from which one derives $\bar{\rho}$, and express quantities of interest as deviations from this reference state, e.g. $T = \bar{T} + T'$. Commonly these reference states are only depth-dependent and time invariant. Buoyancy forces arise from deviations $\rho'$ to the radial background model $\bar{\rho}$, but are assumed to be much smaller than the latter. They can, thus, be neglected in the continuity equation, which avoids pressure waves in the model, but must, of course, be considered in the momentum equation, through $T'$. In total one commonly employs the truncated anelastic liquid approximation, see e.g. [11, 16] and uses the following system of non-dimensional equations

$$-\nabla \cdot \tau + \nabla p' = -\mathrm{Ra}\bar{\rho}\alpha g T' \tag{3}$$

$$\nabla \cdot (\bar{\rho}\mathbf{u}) = 0 \tag{4}$$

$$\bar{\rho}c_p \frac{DT'}{Dt} + \mathrm{Di}\bar{\rho}\alpha w(T' + T_s) - \nabla \cdot (k\nabla T) = \frac{\mathrm{Di}}{\mathrm{Ra}}\Phi \tag{5}$$

where $w = -g \cdot \mathbf{u}$, Ra and Di are the Rayleigh and dissipation number respectively, $\Phi = \tau : \dot{\epsilon}$ describes shear heating and $\frac{D}{Dt}(\cdot)$ is the material derivative while $c_p$, $k$, $\alpha$, and $T_s$ are the non-dimensionalised coefficients of heat capacity, diffusivity, thermal expansivity and the surface temperature.

In a typical, so-called *mantle circulation model*, Earth's mantle is modeled by a thick spherical shell. Temperature is fixed on the top and bottom by imposing Dirichlet boundary conditions. The tangential components of the velocity are taken from paleo-reconstructions of the movement of tectonic plates, while one requires the radial component to vanish. So one has a *no outflow* condition. At the CMB one also requires no outflow and combines this with the constraint of vanishing shear-stress, as the rocky mantle slides freely on the molten iron of the outer core. Formally one requires

$$\begin{aligned}\mathbf{u} \cdot \mathbf{n} &= 0 && \text{on } \Gamma_{\text{surf}} \cup \Gamma_{\text{CMB}} \\ \mathbf{u} \cdot \mathbf{t} &= \mathbf{u}_{\text{plate}} \cdot \mathbf{t} && \text{on } \Gamma_{\text{surf}} \\ \mathbf{t} \cdot \tau \cdot \mathbf{n} &= 0 && \text{on } \Gamma_{\text{CMB}}\end{aligned}$$

with $\mathbf{n}$ being the normal vector in a point and $\mathbf{t}$ any vector in the tangential plane. The combination of vanishing shear stress with no outflow constitutes a freeslip boundary condition.

# 3   Software Framework – HyTeG

In this contribution we consider the solution of mantle convection problems using the finite element framework HyTeG (Hybrid Tetrahedral Grids) [18,21]. HyTeG enables the solution of partial differential equations at the extreme-scale through massively parallel matrix-free geometric multigrid methods. The mesh hierarchy is constructed based on an unstructured triangular or tetrahedral base mesh by successive steps of uniform refinement. Figure 1 illustrates the regular refinement of a single coarse grid element. This results in an $L$-level hierarchy $\mathcal{T}_\ell$, $\ell = 0, \ldots, L-1$, with $\mathcal{T}_0$ being the unstructured coarse mesh. The compute kernels, then, take advantage of the resulting block-structured domain partitioning, while the mesh hierarchy supports construction of geometric multigrid solvers. Details on the refinement and finite element data structures are presented in [18].

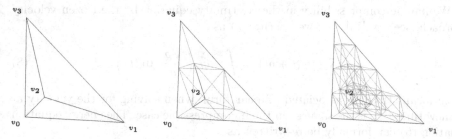

(a) *coarse grid tetrahedron*  (b) *one refinement iteration* (c) *two refinement iterations*

**Fig. 1.** Uniform refinement of a single coarse grid element.

HyTeG's extreme scalability and performance was demonstrated in, e.g., [20], where Stokes systems with more than a trillion ($> 10^{12}$) unknowns were solved on up to $147,456$ parallel processes. Such problem sizes are only feasible due to extremely memory-efficient solvers with optimal time complexity such as the employed matrix-free multigrid methods. For strongly advection-dominated flows, that are relevant in mantle convection models, HyTeG implements a massively parallel Eulerian-Lagrangian method that handles the advective terms through a particle-based implementation of the modified method of characteristics (MMOC) [19].

# 4   Finite Element Discretization

We discretise the Stokes equations, equations (3) and (4), of the convection model by means of $P_2 - P_1$ Taylor-Hood elements to approximate velocity **u** and pressure $p$. This gives us the discrete weak formulation of the problem as: find

$(\mathbf{u}_h, p_h) \in \mathcal{U}_h \times \mathcal{P}_h$ such that

$$\int_\Omega \boldsymbol{\tau} : \nabla \mathbf{v}_h - \int_\Omega p_h \nabla \cdot \mathbf{v}_h = \int_\Omega -\mathrm{Ra}\bar{\rho}\alpha T' \mathbf{g} \cdot \mathbf{v}_h \qquad (6)$$

$$-\int_\Omega q_h \nabla \cdot (\bar{\rho}\mathbf{u}_h) = 0 \qquad (7)$$

holds for all $(\mathbf{v}_h, q_h) \in \mathcal{U}_h \times \mathcal{P}_h$, with

$$\mathcal{U}_h = \left(P_2 \cap H_0^1\right)^3 , \quad \mathcal{P}_h = P_1 \cap H^0 .$$

Here $H_0^1$ contains all functions from the standard Sobolev space $H^1$, whose trace equals zero on the Dirichlet part $\Gamma_D$ of the domain boundary. Uniqueness of pressure within $\mathcal{P}_h$ is enforced by requiring that the average over all degrees of freedom (DoFs) vanishes.

We handle compressibility in the continuity equation by the frozen velocity approach, see e.g. [14], i.e., we rewrite (7) as

$$\int_\Omega q_h \nabla \cdot \mathbf{u}_h(t + \delta t) = \int_\Omega -q_h \left(\frac{\nabla \bar{\rho}}{\bar{\rho}} \cdot \mathbf{u}_h(t)\right) \qquad (8)$$

which means that the associated bilinear form, when solving for the velocity at the new timestep, is the same as in the incompressible case. The energy equation equation (5) can formally be re-written as

$$\frac{\partial T'}{\partial t} + \mathbf{u} \cdot \nabla T = \mathcal{F}(t, T', \mathbf{u}) , \qquad (9)$$

where $\mathcal{F}$ includes diffusion. We approximate $T$ with $P_2$ elements and apply a splitting approach, where the advective component is resolved with the MMOC [19]. Here virtual particles corresponding to the DoFs of $T_h(t + \delta t)$ are advected back in time along characteristics to obtain their departure points $\boldsymbol{x}_{\mathrm{dept}}$ at time $t$. A Runge-Kutta scheme of order 4 is used to solve the resulting ordinary differential equations (ODEs). With this, the energy equation can then be semi-discretised in time as

$$\frac{T'(\boldsymbol{x}, t + \delta t) - \hat{T}'(\boldsymbol{x}, t)}{\delta t} \approx \Theta \mathcal{F}(\boldsymbol{x}, t + \delta t, T', \mathbf{u}_h) + (1 - \Theta)\mathcal{F}(\boldsymbol{x}, t, \hat{T}', \mathbf{u}_h) \quad (10)$$

where $\hat{T}'(\boldsymbol{x}, t) = T'(\boldsymbol{x}_{\mathrm{dept}}, t)$. For a full derivation and other variants available in HyTeG see [19].

With this we proceed in solving the coupled system (3) to (5) according to Algorithm 1. For each step at time $t$, the $\delta t$ for the energy equation is computed with the Courant–Friedrichs–Lewy (CFL) condition based on the velocity field at time $t$. After obtaining the temperature field at time $(t + \delta t)$ by solving (9) and (10), the corresponding velocity field at $(t + \delta t)$ is obtained by solving (6) and (7). Under the assumption that a single Picard type iteration is enough to couple the Stokes and energy equation, we perform a single solve of each in every

---

**Algorithm 1:** Coupled timestepping algorithm

---

1  initialise $T'$;
2  solve Stokes system (6) and (7) for $\mathbf{u}_h$;
3  $t \leftarrow t_0$;
4  **while** $t < t_{end}$ **do**
5  $\quad$ calculate timestep size $\delta t$ using $\mathbf{u}_h(t)$ for the CFL condition;
6  $\quad$ execute advection step by calculating $\mathbf{x}_{\text{dept}}$ and evaluating $\hat{T}'$;
7  $\quad$ solve (10) for $T'$ (with $\Theta = 1.0$);
8  $\quad$ solve (6) and (7) for $\mathbf{u}_h$ (using new $T'$);
9  $\quad$ $t \leftarrow t + \delta t$;
10 **end**

---

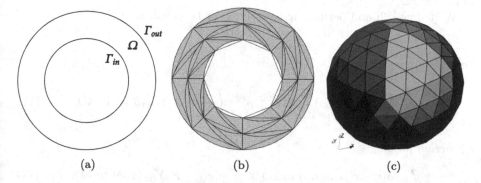

(a)  (b)  (c)

**Fig. 2.** Visualizations of domain and mesh. Figure 2a: schematic sketch of the annulus domain with inner and outer boundaries $\Gamma_{\text{in}}$ and $\Gamma_{\text{out}}$. Figure 2b: computational coarse mesh $\mathcal{T}_0$ without projection (green) overlayed by the projected mesh after refinement to $\mathcal{T}_3$ (macro-element boundaries in red, micro-elements in grey). Figure 2c: computational coarse mesh $\mathcal{T}_0$ of the spherical shell (without projection). (Color figure online)

timestep. Note that the MMOC is not formally bound to the CFL. In [19] it was demonstrated that, with a more sophisticated splitting approach, $\delta t$'s exceeding this condition are permissible.

The geometry of the Earth's mantle is approximated in our models in 2D by an annulus and in 3D by a thick spherical shell. The corresponding meshes are illustrated in Figs. 2b and 2c. We apply projections from the computational polyhedral meshes to the non-polyhedral physical domains to exactly capture the curvature of the boundaries. These so called blending maps are incorporated into the finite element integrals, see [2] for details.

## 5  Results and Discussion

In this section, we present results for a number of standard benchmarks used in the Geodynamics community, before demonstrating applicability of our code to model convection in an Earth-like setting.

## 5.1    Stationary Benchmarks – Stokes System

Here we assess the convergence of the computed finite element approximation to analytical solutions of the Stokes system (1) and (4). To approximate the mantle, we define $\Omega$ (annulus in 2D, thick spherical shell in 3D) with inner radius $r_{\min} = 1.22$ and outer radius $r_{\max} = 2.22$, thereby maintaining a ratio of $\frac{1.22}{2.22} \simeq 0.55$, which is close to the actual ratio for Earth, while also maintaining the non-dimensional thickness of the mantle to be $\Delta r = r_{\max} - r_{\min} = 1.0$. The forcing function, driving the flow, models a density anomaly, which from (1) becomes $-g\rho'\hat{\mathbf{r}}$, with $\bar{\rho} = 0$ and $\hat{\mathbf{r}}$ is a unit vector pointing outward (opposite to gravity). For the annulus, the forcing is based on a cosine term, while for the spherical shell it is based on the spherical harmonics function $Y_{\ell m}$ of degree $\ell$ and order $m$.

We follow [22] and perform experiments for two choices of $\rho'$:

*Smooth forcing*

$$\rho' = \left(\frac{r}{r_{\max}}\right)^k \cos(n\phi) \text{ in 2D} , \quad \rho' = \left(\frac{r}{r_{\max}}\right)^k Y_{\ell m}(\theta, \phi) \text{ in 3D}, \qquad (11)$$

*$\delta$-function forcing*

$$\rho' = \delta(r - r') \cos(n\phi) \text{ in 2D}, \quad \rho' = \delta(r - r') Y_{\ell m}(\theta, \phi) \text{ in 3D}, \qquad (12)$$

where $k, n$ are constants while $\phi$ refers to the angle with the $x$-axis in 2D and $\theta, \phi$ refer to the co-latitude and longitude respectively in 3D. In the $\delta$-function forcing case, the 3D volume integral in the finite element discrete form over the right-hand side of (1) reduces to a surface integral

$$\int_\Omega -g\rho'\hat{\mathbf{r}} \cdot \mathbf{v}_h d\Omega = \int_\Omega -g\delta(r - r')Y_{\ell m}\hat{\mathbf{r}} \cdot \mathbf{v}_h d\Omega = \int_{\Gamma'} -gY_{\ell m}\hat{\mathbf{r}} \cdot \mathbf{v}_h d\Gamma',$$

where $\Gamma'$ is the spherical surface of radius $r'$. We design our mesh in such a way that $\Gamma'$ is discretised by tetrahedral faces. The analogous reduction and meshing applies to the 2D case.

Two sets of boundary conditions are considered, a freeslip boundary condition on the inner boundary and a noslip condition on the outer boundary, denoted by noslip-freeslip ($\Gamma_{in} = \Gamma_{FS}, \Gamma_{out} = \Gamma_D$, see Fig. 2a), and another type where only freeslip is considered on both the boundaries, denoted by freeslip-freeslip ($\Gamma_{in} = \Gamma_{FS}, \Gamma_{out} = \Gamma_{FS}$, see Fig. 2a). The analytical solutions for the freeslip-freeslip case are available through the Python package *assess* from [22], where the general form of the solution has been derived for the annulus and spherical shell. Hence, we use the same form and derive the respective coefficients for noslip-freeslip boundary condition setup by means of computer algebra.

In the 2D annulus mesh, as shown in Fig. 2b, at $\mathcal{T}_0$, we consider 8 layers in tangential direction and 2 layers in radial direction at the coarsest level. For the

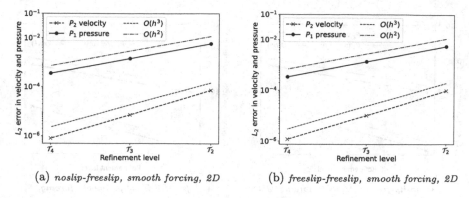

**Fig. 3.** $L_2$-errors of velocity and pressure, smooth forcing on the annulus with $n = 2, k = 2$ in (11).

3D shell mesh, we use the icosahedral meshing approach [8], which results in ten diamond shaped sections. Figure 2c shows a top view of the North pole, where five sections meet. The resolution of the base mesh is defined by two parameters. One is the number of divisions along the curved edges of each section, the other the number of layers in radial direction. We use as starting mesh for our convergence study $\mathcal{T}_2$, which contains $\simeq 10^2$ triangles with $\simeq 3 \times 10^4$ DoFs for the annulus and $\simeq 10^5$ tetrahedrons with $\simeq 3 \times 10^7$ DoFs for the spherical shell.

For this study, we use the Minres solver for the Stokes system with a lumped inverse mass matrix preconditioner [10] for the Schur complement. At the freeslip boundary, the velocity field is projected such that the normal component is zero at every step of the iterative solver.

Let $\mathbf{u}_h$ be the discrete solution on level $\ell$. Then we approximate the velocity error $\mathbf{e} = \mathbf{u} - \mathbf{u}_h$ and its $L_2$-norm by interpolating both $\mathbf{u}$ and $\mathbf{u}_h$ in the FE space for $\mathcal{T}_{\ell+1}$ and evaluating it there, similar for pressure. With smooth forcing, for both noslip-freeslip and freeslip-freeslip, we obtain the theoretically expected $L_2$ convergence rates under refinement [12], i.e. cubic, $\mathcal{O}(h^3)$, for velocity and quadratic, $\mathcal{O}(h^2)$, for pressure, for both the annulus and spherical shell as can be seen in Figs. 3 and 4. For the $\delta$-function forcing cases, the convergence speed deteriorates. We observe only $\mathcal{O}(h^{1.5})$ for velocity and $\mathcal{O}(h^{0.5})$ for pressure. See Figs. 5 and 6. This behaviour is consistent with [22]. The analytical pressure solution for the $\delta$-function forcing is a discontinuous function. As the $P_1$ finite element can only represent continuous functions, we should expect to see an impact on its convergence and in turn for the velocity as well. The convergence properties can be regained by choosing a discontinuous pressure element. A more detailed discussion and an example with an enriched $P_2$-element for velocity and a discontiunous $P_1$ element for pressure can be found in [22].

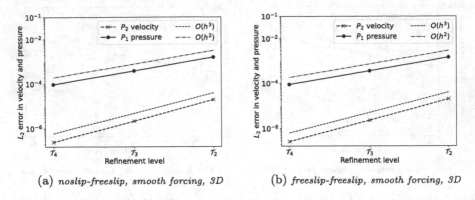

**Fig. 4.** $L_2$-errors of velocity and pressure, smooth forcing on the spherical shell with $\ell = 2, m = 2, k = 2$ in (11).

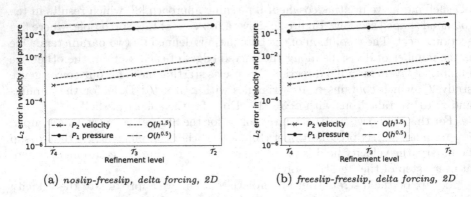

**Fig. 5.** $L_2$-errors of velocity and pressure, delta forcing on annulus with $n = 2$ in (12).

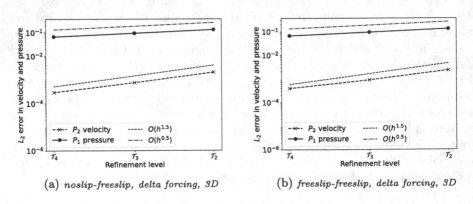

**Fig. 6.** $L_2$-errors of velocity and pressure, delta forcing on spherical shell with $\ell = 2, m = 2$ in (12).

## 5.2    Time Dependent Benchmarks – Unit Square

In this section, we present a benchmark from [3], specifically case 1a, where a suite of software codes for modelling convection processes were compared. The problem considers an isoviscous, bottom heated model on a unit square. The idea is to prescribe a sinusoidal perturbation of temperature given by

$$T(x, y, t = 0) = (1 - y) + A \cos(\pi x) \sin(\pi y), \text{ with } A = 0.05 , \qquad (13)$$

while imposing freeslip boundary conditions for the velocity on all four walls of the box. For the temperature field, a Dirichlet boundary condition of $T = 0$ and $T = 1$ is imposed at the top and bottom respectively, while zero flux is imposed at the vertical sides. The perturbation combined with the freeslip boundary condition induces a single convection cell in the square and heat advects and diffuses to reach a steady state. Once this state is reached, the Nusselt number is calculated at the top boundary. The latter gives an estimate of the amount of heat transported by advection and diffusion to that of pure diffusion.

Equations (3) to (5) are solved according to Algorithm 1, but we consider the total temperature $T$ instead of $T'$ with no internal heating, $\text{Di} = 0$, $c_p = 1$ and an isoviscous, incompressible model with constant density $\bar{\rho} = 1$, $\eta = 1$ and $\mathbf{g} = [0, -1]^{\top}$. The experiments are performed with $\text{Ra} = 10^4$ and the Nusselt number (Nu) for the top boundary of the unit square is calculated via

$$\text{Nu} = \frac{\int_0^1 \frac{\partial T(x, 1)}{\partial y} dx}{\int_0^1 T(x, 0) dx} , \qquad (14)$$

as in [3]. To verify our implementation, we choose our mesh based on the range of resolutions in [3] and compare the computed Nusselt numbers to the values obtained by other codes. With refinement, our computed results converge to the values reported in the benchmark and stay in the range of values predicted by other codes, see Fig. 7.

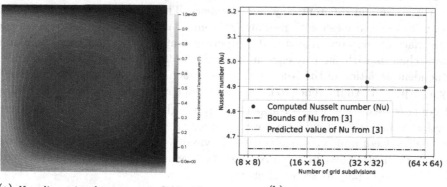

(a) *Non-dimensional temperature field with isolines*

(b) *Nu vs Number of DoFs*

**Fig. 7.** Convection on a square after reaching steady state.

## 5.3    Time Dependent Benchmarks – Thick Spherical Shell

In addition to the previous 2D benchmarks, we now perform computations on a thick spherical shell and verify our implementation through the community benchmark described in [6, 26]. It involves simulation of convection behaviour on a thick spherical shell by imposing freeslip boundary conditions for the velocity field on both $\Gamma_{in}$ and $\Gamma_{out}$, while a Dirichlet boundary condition is specified for the temperature field with $T = 0$ on $\Gamma_{out}$ and $T = 1$ on $\Gamma_{in}$. The initial condition for the temperature is given by

$$T(\boldsymbol{x}, 0) = T_c(\boldsymbol{x}) + A Y_{\ell m}(\theta, \phi) \sin(\pi(r - r_{min})),$$

with $A = 0.04$, $(\ell, m) = (3, 2)$ and the background profile,

$$T_c(\boldsymbol{x}) = \left( \frac{r_{min} r_{max}}{r} \right) - r_{min}.$$

The temperature deviations from the background profile due to this specific spherical harmonic induces four plumes to rise up and stabilize with time in a tetrahedral symmetry around the center. When the Rayleigh number increases, the convective vigour increases and the plumes get thinner. This can be seen from larger velocity magnitudes and higher Nusselt numbers, which vary proportional to the third root of the Rayleigh number [9], $Nu \simeq Ra^{\frac{1}{3}}$. For a spherical surface $\Gamma$ at radius $r$, the Nusselt number is calculated by

$$Nu = \frac{\displaystyle\int_\Gamma \frac{\partial T}{\partial r} d\Gamma}{\displaystyle\int_\Gamma \frac{\partial T_c}{\partial r} d\Gamma}. \tag{15}$$

Ideally the Nusselt number must be the same for $r \in [r_{min}, r_{max}]$. We choose $r = r_{min}$ for the evaluation.

Equations (3) to (5) are solved according to Algorithm 1, but we consider the total temperature $T$ instead of $T'$ with $Di = 0$, $c_p = 1$, constant density $\bar{\rho} = 1$ with $\mathbf{g} = -\hat{\mathbf{r}}$ and different Rayleigh numbers. The experiments are performed on an icosahedral mesh with roughly $10^6$ temperature DoFs. Figure 8 shows the iso-surface of the non-dimensional temperature at $T = 0.5$ and how the plume shape changes with the Rayleigh number. Figure 9a shows the Rayleigh number dependent variation of the Nusselt number, which obeys the cubic proportionality as expected. An inference to note is that [26] concludes with a proportionality of $Nu \simeq (Ra/Ra_{crit})^{1/4}$, where $Ra_{crit}$ is the Rayleigh number at the onset of convection, which is constant for a given temperature dependent viscosity law. But for the Rayleigh number range and the isoviscous case that we have considered, a cubic proportionality between Nu and Ra is seen which also corroborates with the values from [26], as seen in Fig. 9a.

## 5.4    Mantle Convection

Here we show results from a model based on the Truncated Anelastic Liquid Approximation (TALA) formulation which is currently under development. This is being done with the software framework HyTeG under the TERRANEO project, the goal of which is to create a scalable and accurate Earth model which can be used for various geophysical applications.

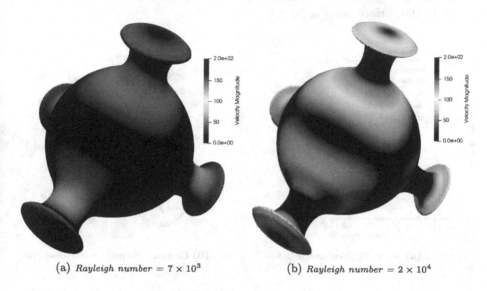

(a) *Rayleigh number* = $7 \times 10^3$          (b) *Rayleigh number* = $2 \times 10^4$

**Fig. 8.** Iso-surface of non-dimensional temperature at $T = 0.5$ coloured by magnitude of velocity for comparison between Rayleigh numbers after reaching steady state

Equations (3) to (5) are solved according to Algorithm 1. An implicit Euler scheme with $\Theta = 1.0$ in (10) is used for the time discretization of the diffusive and other forcing terms in (5). The Stokes system is solved with a monolithic geometric multigrid method that uses an inexact Uzawa method for relaxation, see [20] for details. The base mesh $\mathcal{T}_0$ is a thick spherical shell represented by an icosahedral mesh composed of $\simeq 10^3$ tetrahedrons. This is taken as the coarsest level for the V-cycle of the multigrid algorithm and $\mathcal{T}_4$ for the finest level, which then contains $\simeq 3 \times 10^7$ DoFs for the Stokes system and $\simeq 1 \times 10^7$ DoFs for temperature. The model is isoviscous with $\eta = 1$, compressible with the radially varying density profile given by

$$\bar{\rho} = \bar{\rho}_s \exp \left( \frac{\mathrm{Di}(r_{\max} - r)}{\mathrm{Gr}(r_{\max} - r_{\min})} \right) , \qquad (16)$$

where $\bar{\rho}_s = 1$ here and Gr is the Grüneisen parameter [17]. The values considered are Ra $= 10^5$ and $T_s = 0$ while all other nondimensional parameters are set to 1 and no-slip boundary conditions are considered both on the surface ($r_{max}$) and CMB ($r_{min}$). Around the CMB, we prescribe the initial temperature deviation

$$T'(\boldsymbol{x}, 0) = T'_0 \exp\left(-A\frac{r - r_{min}}{r_{max} - r_{min}}\right) Y_{\ell m}(\theta, \phi) \qquad (17)$$

with scaling factors $T'_0$ and $A$, and a spherical harmonics function of order $\ell = 6$ and degree $m = 4$, which then induces plumes according to the harmonic that rise up to the surface, see Fig. 9b.

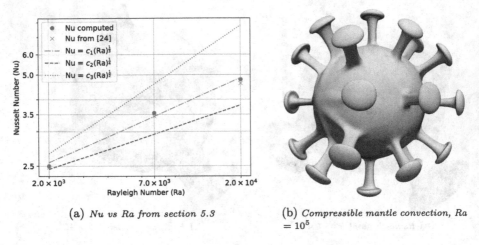

(a) *Nu vs Ra from section 5.3*

(b) *Compressible mantle convection, Ra* $= 10^5$

**Fig. 9.** Variation of Nusselt number with respect to Rayleigh number for the tetrahedral symmetry case from Sect. 5.3 (left), Compressible Mantle convection model with adiabatic density profile showing non-dimensional temperature iso-surfaces (right)

The results are obtained from running the model on our in-house cluster TETHYS-3g [24] which is composed of 24 AMD EPYC 7662 processors for a total of 1536 cores connected via 100 GBit/s Infiniband HDR. For the export of 3D data, the ADIOS2 package [13], which is developed for IO operations especially on supercomputers, is employed in our framework.

## 6   Conclusion

In this paper, we have verified the implementation of a model for mantle convection simulation in the HyTeG finite element framework against different semi-analytical setups and community benchmarks. The agreement of our numerical results with the literature demonstrates its applicability for Geophysical applications. First, we showed the numerical convergence of our approximation to the solution of a Stokes system for cases with different boundary conditions and different forcing terms. Then we showcased the capability of our framework to solve transport equations and verified it through time-dependent benchmark problems. Finally, we showed results for a model based on the TALA formulation. Future work will include benchmarking against temperature dependent viscosity cases and nonlinear rheologies.

**Acknowledgements.** Funding was provided by the German Research Foundation (DFG) research training group UPLIFT 440512084 (GRK 2698) and the German Ministry of Research and Education (BMBF) via its SCALEXA initiative (project CoMPS). Computing resources were provided by the Institute of Geophysics of LMU Munich [24] funded by DFG 495931446.

# References

1. Bauer, S., et al.: TerraNeo—Mantle convection beyond a trillion degrees of freedom. In: Bungartz, H.-J., Reiz, S., Uekermann, B., Neumann, P., Nagel, W.E. (eds.) Software for Exascale Computing - SPPEXA 2016-2019. LNCSE, vol. 136, pp. 569–610. Springer, Cham (2020). https://doi.org/10.1007/978-3-030-47956-5_19
2. Bauer, S., Mohr, M., Rüde, U., Weismüller, J., Wittmann, M., Wohlmuth, B.: A two-scale approach for efficient on-the-fly operator assembly in massively parallel high performance multigrid codes. Appl. Numer. Math. **122**, 14–38 (2017). https://doi.org/10.1016/j.apnum.2017.07.006
3. Blankenbach, B., et al.: A benchmark comparison for mantle convection codes. Geophys. J. Int. **98**(1), 23–38 (1989). https://doi.org/10.1111/j.1365-246X.1989.tb05511.x
4. Brown, H., Colli, L., Bunge, H.P.: Asthenospheric flow through the Izanagi-Pacific slab window and its influence on dynamic topography and intraplate volcanism in East Asia. Front. Earth Sci. **10**, 889907 (2022). https://doi.org/10.3389/feart.2022.889907
5. Bunge, H.P., Richards, M., Baumgardner, J.: A sensitivity study of three-dimensional spherical mantle convection at $10^8$ Rayleigh number: Effects of depth-dependent viscosity, heating mode, and an endothermic phase change. JGR: Solid Earth **102**(B6), 11991–12007 (1997). https://doi.org/10.1029/96JB03806
6. Burstedde, C., et al.: Large-scale adaptive mantle convection simulation. Geophys. J. Int. **192**(3), 889–906 (2013). https://doi.org/10.1093/gji/ggs070
7. Christensen, U., Yuen, D.: Layered convection induced by phase transitions. JGR: Solid Earth **90**(B12), 10291–10300 (1985). https://doi.org/10.1029/JB090iB12p10291
8. Davies, D.R., Davies, J.H., Bollada, P.C., Hassan, O., Morgan, K., Nithiarasu, P.: A hierarchical mesh refinement technique for global 3-D spherical mantle convection modelling. Geosci. Model Dev. **6**(4), 1095–1107 (2013). https://doi.org/10.5194/gmd-6-1095-2013
9. Davies, G.: Dynamic Earth: Plates. Cambridge University Press, Plumes and Mantle Convection (1999)
10. Elman, H., Silvester, D., Wathen, A.: Finite Elements and Fast Iterative Solvers: with Applications in Incompressible Fluid Dynamics. Oxford University Press (2014)
11. Gassmöller, R., Dannberg, J., Bangerth, W., Heister, T., Myhill, R.: On formulations of compressible mantle convection. Geophys. J. Int. **221**(2), 1264–1280 (2020). https://doi.org/10.1093/gji/ggaa078

12. Girault, V., Raviart, P.A.: Incompressible mixed finite element methods for solving the stokes problem. In: Girault, V., Raviart, P.A. (eds.) Finite Element Methods for Navier-Stokes Equations: Theory and Algorithms, pp. 193–277. Springer (1986). https://doi.org/10.1007/978-3-642-61623-5_3

13. Godoy, W.F., Podhorszki, N., et al.: ADIOS 2: the adaptable input output system. A framework for high-performance data management. SoftwareX **12**, 100561 (2020). https://doi.org/10.1016/j.softx.2020.100561

14. Heister, T., Dannberg, J., Gassmöller, R., Bangerth, W.: High accuracy mantle convection simulation through modern numerical methods - II: realistic models and problems. Geophys. J. Int. **210**(2), 833–851 (2017). https://doi.org/10.1093/gji/ggx195

15. Horbach, A., Bunge, H.P., Oeser, J.: The adjoint method in geodynamics: derivation from a general operator formulation and application to the initial condition problem in a high resolution mantle circulation model. Int. J. Geomath. **5**(2), 163–194 (2014). https://doi.org/10.1007/s13137-014-0061-5

16. Jarvis, G., Mckenzie, D.: Convection in a compressible fluid with infinite Prandtl number. J. Fluid Mech. **96**(3), 515–583 (1980). https://doi.org/10.1017/S002211208000225X

17. King, S.D., et al.: A community benchmark for 2-D Cartesian compressible convection in the Earth's mantle. Geophys. J. Int. **180**(1), 73–87 (2010). https://doi.org/10.1111/j.1365-246X.2009.04413.x

18. Kohl, N., Bauer, D., Böhm, F., Rüde, U.: Fundamental data structures for matrix-free finite elements on hybrid tetrahedral grids. Int. J. Parallel Emergent Distrib. Syst. **39**(1), 51–74 (2024). https://doi.org/10.1080/17445760.2023.2266875

19. Kohl, N., Mohr, M., Eibl, S., Rüde, U.: A massively parallel eulerian-lagrangian method for advection-dominated transport in viscous fluids. SIAM J. Sci. Comput. **44**(3), C260–C285 (2022). https://doi.org/10.1137/21M1402510

20. Kohl, N., Rüde, U.: Textbook efficiency: massively parallel matrix-free multigrid for the stokes system. SIAM J. Sci. Comput. **44**(2), C124–C155 (2022). https://doi.org/10.1137/20M1376005

21. Kohl, N., Thönnes, D., Drzisga, D., Bartuschat, D., Rüde, U.: The HyTeG finite-element software framework for scalable multigrid solvers. Int. J. Parallel Emergent Distrib. Syst. **34**(5), 477–496 (2019). https://doi.org/10.1080/17445760.2018.1506453

22. Kramer, S.C., Davies, D.R., Wilson, C.R.: Analytical solutions for mantle flow in cylindrical and spherical shells. Geosci. Model Dev. **14**(4), 1899–1919 (2021). https://doi.org/10.5194/gmd-14-1899-2021

23. Oberbeck, A.: Über die Wärmeleitung der Flüssigkeiten bei Berücksichtigung der Strömungen infolge von Temperaturdifferenzen. Ann. Phys. **243**(6), 271–292 (1879). https://doi.org/10.1002/andp.18792430606

24. Oeser, J., Bunge, H.-P., Mohr, M.: Cluster design in the earth sciences Tethys. In: Gerndt, M., Kranzlmüller, D. (eds.) HPCC 2006. LNCS, vol. 4208, pp. 31–40. Springer, Heidelberg (2006). https://doi.org/10.1007/11847366_4

25. Oxburgh, E.R., Turcotte, D.L.: Mechanisms of continental drift. Rep. Prog. Phys. **41**(8), 1249 (1978). https://doi.org/10.1088/0034-4885/41/8/003

26. Ratcliff, J.T., Schubert, G., Zebib, A.: Steady tetrahedral and cubic patterns of spherical shell convection with temperature-dependent viscosity. JGR: Solid Earth **101**(B11), 25473–25484 (1996). https://doi.org/10.1029/96JB02097

27. Ricard, Y.: Physics of Mantle Convection. In: Bercovici, D. (ed.) Mantle Dynamics, Treatise on Geophysics, vol. 7, pp. 23–71. Elsevier, 2nd edn. (2015). https://doi.org/10.1016/B978-0-444-53802-4.00127-5

# The Research Repository for Data and Diagnostics (R2D2): An Online Database Software System for High Performance Computing and Cloud-Based Satellite Data Assimilation Workflows

Eric J. Lingerfelt[1]([✉]), Tariq J. Hamzey[2], Maryam Abdi-Oskouei[1], Jérôme Barré[1], Fábio Diniz[1], Clémentine Gas[1], Ashley Griffin[1], Dominikus Heinzeller[1], Stephen Herbener[1], Evan Parker[1], Benjamin Ruston[1], Christian Sampson[1], Travis Sluka[1], Kristin Smith[1], and Yannick Trémolet[1]

[1] University Corporation for Atmospheric Research, Boulder, CO 80307, USA
eric2@ucar.edu
[2] Science Systems and Applications, Inc., Lanham, MD 20706, USA
tariq.hamzey@nasa.gov

**Abstract.** The Joint Center for Satellite Data Assimilation (JCSDA) is a multi-agency research center established to improve the quantitative use of satellite data in atmosphere, ocean, climate and environmental analysis and prediction systems. At the JCSDA, scientists and software engineers within the Joint Effort for Data Assimilation Integration (JEDI) are developing a unified data assimilation framework for research and operational use. To harness the full potential of ever-increasing volumes of data from new and evolving Earth observation systems, a new online database software system has been developed and deployed by the JCSDA. As a core component of JEDI, the Research Repository for Data and Diagnostics (R2D2) performs data registration, management, and configuration services for data assimilation computational workflows. We present an overview of R2D2's distributed system of data stores, SQL data model, intuitive python API, and user support efforts. In addition, we will detail R2D2's utilization by environmental prediction applications developed by the JCSDA and its partners.

**Keywords:** earth system data assimilation · fair principles · earth observations · sql data model · skylab · joint effort for data assimilation integration · jedi · high performance computing · data store · r2d2 · research repository for data and diagnostics · ioda · interface for observation data access · saas

## 1 Introduction

Computational data assimilation (DA) in numerical weather prediction is the process of combining observational data with model simulations to estimate the current state of the atmosphere, ocean, and climate. In pursuit of higher accuracy and finer resolutions, DA models rely on ever-increasing volumes of observational data, particularly

L. Franco et al. (Eds.): ICCS 2024, LNCS 14834, pp. 303–310, 2024.
https://doi.org/10.1007/978-3-031-63759-9_34

from new satellites [1]. A robust database system is thus crucial for efficient storage and retrieval of observations, both historical and in real-time. Managing model input and output is equally crucial for accessing past experiments and reanalysis of earlier time periods. Environmental modeling centers spend decades developing in-house database solutions for operational use and remote data retrieval. The European Centre for Medium-Range Weather Forecasts (ECMWF) [2] conceived the Meteorological Archival Retrieval System (MARS) in 1985, which currently manages petabytes of data [3]. In 2001, the National Centers for Environmental Prediction (NCEP) [4] developed the National Operational Model Archive and Distribution System (NOMADS) [5] based on the Open-Source Project for a Network Data Access Protocol (OPeNDAP) [6]. Together with its partner organizations, the Joint Center for Satellite Data Assimilation (JCSDA) [7] has developed a completely original software system called the Research Repository for Data and Diagnostics (R2D2) that performs data management, registration, and configuration services for DA workflows. As a central component of the Joint Effort for Data assimilation Integration (JEDI) project and its flagship SkyLab application [8, 9], R2D2 satisfies these requirements by offering a centralized, model agnostic platform for handling diverse datasets. Moving toward a unified DA framework, R2D2 provides a generic data solution for DA scientists and is not restricted to any one organization or data storage platform. In the past two years, R2D2 has been developed into a highly sophisticated data management system with internal knowledge of numerous high performance computing (HPC) systems throughout the United States, all of which are interconnected by a centralized cloud database. This paper details R2D2's architecture and data holdings, its utilization across partner organizations, and its near-term development objectives that are bringing JEDI's premise of a unified DA framework to fruition.

## 2 Distributed Data Access

Accessible from any place at any time, R2D2 relieves data assimilation scientists from the technical hurdles of data ingestion and restoration by linking the physical locations of data via a network of interconnected, mirrored data hubs established at several high performance computing centers, including the NASA Center for Climate Simulation [10], NCAR-Wyoming Supercomputing Center [11], Space Science and Engineering Center [12], and Mississippi State University's High Performance Computing Collaboratory [13]. Cloud-based data storage services provided by Amazon Web Services' Simple Storage Service (AWS S3) [14], Microsoft Azure's Blob Storage [15], and the Google Cloud [16] are also utilized as R2D2 data hubs. Each data hub is a collection of R2D2 data stores where each is physically instantiated as a posix directory or a cloud object storage container such as an AWS bucket (see Fig. 1). The data stores are populated by a subdirectory tree partitioned by the type of data file stored and the valid date for the file. In effect, these data type and date subdirectories are used as hash buckets for consumed data files indexed by the R2D2 service. Permissions for read and write operations to these data stores are managed in three ways: (1) Data stores are identified as requiring administrative write permission by R2D2, (2) R2D2 users are identified as either regular or administrative users where administrative users are granted modification access to all data stores, (3) Administrative permissions are set via role accounts at high performance computing centers and authorization policies executed on cloud platforms. R2D2 data stores are characterized by the level of data protection and the total size of the data. "Experiments" data stores provide data file storage for all R2D2 users and are made available on all data hubs. "Archive" data stores require administrative write permission and provide all the input files needed for a variety of JSCDA workflows. Mirrored across all data hubs, "archives" supply a variety of input data types including observation and bias correction files as well as deterministic (i.e., non-ensemble) forecast and analysis files. "Ensemble" data stores are like "archives", but they are not mirrored due to their large disk footprint which can range from 10 to 100 TB. These data stores contain model-specific, ensemble forecast input files and currently make up over 90% of R2D2's total data holdings. Computing resources must be registered in R2D2 to gain access to the collection of R2D2's data hubs. All compute resources are assigned one local data hub and one or more cloud-based data hubs.

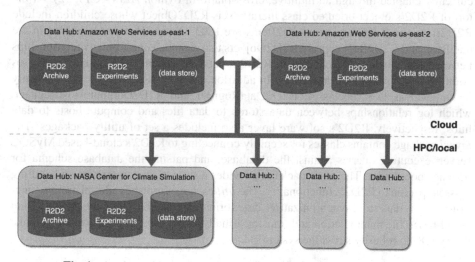

**Fig. 1.** A schematic representation of R2D2's data hubs and data stores.

## 3  SQL Design and Implementation

R2D2 persists knowledge through an SQL relational database designed to serve as an information indexer that stores data file attributes along with metadata concerning experiments, data vocabularies, data hubs, and supported compute host configurations. R2D2's SQL schema represents related but unique and non-overlapping categorizations of metadata and data files. To optimize internal efficiency and logical consistency, R2D2's MySQL database utilizes three design components: normalization, orthogonality, and extensibility. R2D2 tables are normalized to the third normal form (3NF) [17]. Orthogonality is ensured with the distinction of R2D2 *data*, *index*, and *register* tables, where *data* and *index* types are called *items*. A *data item* encapsulates all unique identifiers for the physical data files stored in the system. An *index item* is not associated with a data file; it stores metadata records only. *Register* tables link *data* and *index items* to physical infrastructure by logging one-to-many relationships between data stores and data files and between compute hosts and data hubs. All tables are granted a unique index by the MySQL server, and all *index* items have human-readable, unique *name* values. Lastly, R2D2 promotes extensibility by dynamically extracting the SQL schema at runtime, enabling the Python-based abstraction layer to be completely *item*-agnostic. Hence, extensions to the database design are automatically reflected at runtime and do not require source code updates.

## 4   Software Architecture

Secure access to this centralized database server and R2D2's end user functionality is currently enabled through an intuitive, cross-platform Python API (see Fig. 2). At the top of R2D2's object-oriented class hierarchy is R2D2Object whose children include R2D2Item and R2D2Register. R2D2Items are R2D2Objects that can be stored in the database, and R2D2Registers are R2D2Objects that establish one-to-many relationships between R2D2Items. As reflected in R2D2's database schema, R2D2Item subclasses come in two flavors: *data* and *index*. In addition, the R2D2Register class is currently instantiated as two subclasses, R2D2DataRegister and R2D2ComputeHostRegister, which log relationships between data stores to data files and compute hosts to data hubs, respectively. R2D2's software layer also includes a set of utility packages. The *mysql* package contains classes for securely connecting to R2D2's cloud-based MySQL server, executing queries against the database, and parsing the database schema for instance population. The *error* package provides a suite of custom exception handling classes specific to R2D2's functionality. The *util* package furnishes administrative functions such as data store synchronization, date formatting methods, logging, and local and cloud-based file manipulation and transfer. Finally, the *test* package utilizes the cmake utility [18] to robustly stress the system.

## 5   Adherence to FAIR

State-of-the-art numerical weather prediction models are impacted by Big Data issues brought on by an increasing amount of incoming observational data, an explosion of the number of points as the model grid shrinks, and an increase in the frequency of forecast cycling. FAIR principles [19] "emphasize machine-actionability because humans increasingly rely on computational support to deal with data as a result of the increase in volume, complexity, and creation speed of data" [20]. R2D2 follows these guidelines in several ways. The system enhances "findability" of data by assigning metadata (and data) a globally unique and persistent identifier from a searchable resource. Identifiers utilized by R2D2 are minted by the MySQL server. It enables "accessibility" by making metadata (and data) retrievable by their identifier using a standardized, authenticated communications protocol that is open, free, and universally implementable. Another JCSDA-supported software product, the Interface for Observation Data Access (IODA)

[21], enables "interoperability" and "reusability". By implementing multiple data storage backends such as Network Common Data Form (netCDF) [22], Hierarchical Data Format (HDF5) [23], and Observation DataBase (ODB) [24], IODA allows observational data to seamlessly integrate with other data, applications, and workflows while being well-described using standard vocabularies for data replication and combination.

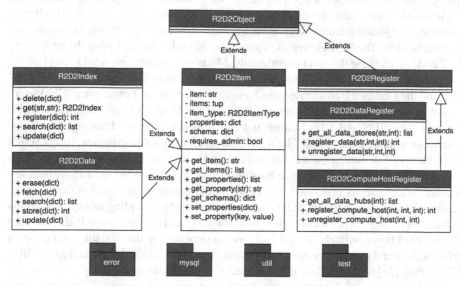

**Fig. 2.** A UML diagram of R2D2's core python API.

## 6   End User Support

Documentation provides a key interface between users and developers of a system [25]. Good documentation and user support encourages system adoption, increases end-user satisfaction with the product, and avoids time-intensive troubleshooting that can snowball into project delays and emergency support to partner agencies or other users [26]. For R2D2, three documents have proven key for ease of use (and lack of troubleshooting requests to the R2D2 team): tutorial, key reference tables, and the FAQ. The tutorial documentation covers system prerequisites and configurations, installation instructions, and step-by-step examples of how to use R2D2. Code snippets at each step ensure that the instructions are clear and easy to follow. Key lookup was formerly one of the main causes of confusion and support requests in R2D2. This problem was resolved by a new document containing lookup tables of required keys, key types, possible values for keys with an enumerated set of possible values, and which tables can only be changed by an R2D2 administrator. Use-instructions accompanying this document include code snippets and step-by-step examples to ensure clear communication. The FAQ document addresses the most frequent questions from end users, including inquiries targeted at data search and ingestion tasks. Collating assorted requests into one publicly accessible document has cut down on the time spent troubleshooting by the R2D2 team and made it easy for users to find answers to most of their questions, increasing ease of use.

## 7 Utilization

The JCSDA has developed a comprehensive workflow called SkyLab to facilitate the execution of diverse data assimilation scenarios. These scenarios can involve different instruments, algorithm choices such as background error covariance or cost function, and different models. The SkyLab workflow serves as the orchestrator, driving the underlying JEDI code to run these different experiments. Models currently supported in SkyLab are the Unified Forecast System (UFS) [27], the Model for Prediction Across Scales (MPAS) [28], and the Goddard Earth Observing System - Composition Forecasting (GEOS-CF) system [29], and work is in progress to integrate the Modular Ocean Model (MOM6) [30] and the GEOS and UFS Marine and Aerosols components. Existing workflows, called *suites*, range from simple HofX [31] calculations to more complex, fully cycling ensemble and deterministic data assimilation and forecasts. The individual tasks composing these *suites* involve data file handling using R2D2, execution of JEDI binaries, or other miscellaneous tasks such as extracting performance information from logs or creating visualization artifacts. R2D2 supports SkyLab by providing storage and access to data, and through experiment registration, tracking, and configuration. R2D2 supplies JCSDA workflows with the default runtime arguments for the scalable execution of JEDI binaries and other processing tasks using common job schedulers such as SLURM [32] or PBS [33]. All experiments registered with R2D2 are granted a globally unique identifier consisting of four alphanumeric characters and the name of the supervising research or operational center. This identification scheme allows centers to track, categorize, and interrelate results associated with each experiment. In addition to the JCSDA, NASA's Global Modeling and Assimilation Office (GMAO) [34], a JEDI partner, is also leveraging R2D2 in new data assimilation toolkits to handle experimental data storage. Its current data holdings include approximately 12K files comprising 250 GB of bias correction, forecast, and observational data. As new toolkits are developed and tested using R2D2's database solution, the GMAO is continually evaluating how best to expand R2D2's scope for both experimental and production workflows. Since its initial release in early 2023, R2D2 has registered over 9K experiments and stored 17M data files distributed across 9 data hubs and 40 data stores for 50 end user accounts. The results of a single computational cycle of a production JEDI-UFS experiment managed in R2D2 requires storage and tracking of approximately 50K files and 30 TB of data.

## 8 Conclusion

The Research Repository for Data and Diagnostics database software system attempts to address the requirements of Big Data management, access, and curation for high performance computational data assimilation experiments. Its multi-tier system comprises a series of distributed data hubs interconnected and indexed via an extendable SQL database schema and a dynamic, intuitive python API. R2D2 adheres to FAIR principles by deploying a system managing "findable" and "accessible" data files and artifacts.

Multiple user support products, such as tutorials, reference tables, API examples, and a FAQ, have been created to provide in-depth instruction and easy adoption by new users. Organizations such as the JCSDA and NASA's GMAO are currently applying the R2D2 system for world-class modeling and data assimilation. Current development paths include a redesign of R2D2 into a scalable, flexible client/server system by utilizing the OpenAPI 3.0 Specification [35] for RESTful web services [36], Swagger tooling [37], and the Flask micro web framework [38].

**Acknowledgments.** The Joint Center for Satellite Data Assimilation is a partnership between NOAA, NASA, the US Navy, and the US Air Force, committed to improving and accelerating the quantitative use of research and operational satellite data in weather, ocean, climate and environmental analysis and prediction systems. University Corporation for Atmospheric Research (UCAR) is hosting the core team of our multi-agency research center.

**Disclosure of Interests.** The authors have no competing interests to declare that are relevant to the content of this article.

# References

1. McCarty, W., et al.: MERRA-2 input observations: summary and assessment. In: Koster, R. (ed.) NASA Technical Report Series on Global Modeling and Data Assimilation, vol. 46 (2016)
2. ECMWF Homepage. https://www.ecmwf.int/
3. Raoult, B.: Architecture of the new MARS server. In: Sixth Workshop on Meteorological Operational Systems, Reading, United Kingdom (1997)
4. NCEP Homepage. https://www.weather.gov/ncep/
5. Rutledge, G.K., Alpert, J., Ebisuzaki, W.: NOMADS: a climate and weather model archive at the National Oceanic and Atmospheric Administration. Bull. Am. Meteorol. Soc. **87**(3), 327–341 (2006)
6. OPeNDAP Homepage. https://www.opendap.org/
7. JCSDA Homepage. https://www.jcsda.org. Accessed 16 Feb 2024
8. Heinzeller, D., Abdi-Oskouei, M., Herbener, S., Lingerfelt, E., Trémolet, Y., Auligné, T.: The Joint Effort for Data assimilation Integration (JEDI): a unified data assimilation framework for Earth system prediction supported by NOAA, NASA, U.S. Navy, U.S. Air Force, and UK Met Office. In: EGU General Assembly 2023, EGU23-10697, Vienna, Austria (2023)
9. Herbener, S., et al.: Accelerating research to operations with the joint effort for data assimilation integration (JEDI) and Skylab systems. In: 104th AMS Annual Meeting, Baltimore, MD (2024)
10. NCCS Homepage. https://www.nccs.nasa.gov. Accessed 16 Feb 2024
11. NWSC Homepage. https://www2.cisl.ucar.edu/ncar-wyoming-supercomputing-center. Accessed 16 Feb 2024
12. SSEC Homepage. https://www.ssec.wisc.edu. Accessed 16 Feb 2024
13. MSU-HPC2 Homepage. https://www.hpc.msstate.edu. Accessed 16 Feb 2024
14. AWS S3 Homepage. https://aws.amazon.com/s3. Accessed 16 Feb 2024
15. Microsoft Azure Blob Homepage. https://azure.microsoft.com/en-us/products/storage/blobs. Accessed 16 Feb 2024
16. Google Cloud Storage Homepage. https://cloud.google.com/storage. Accessed 16 Feb 2024

17. Codd, E.F.: Further normalization of the data base relational model. In: Research Report/RJ/IBM, **RJ909** (1971)

18. Wilke, J., Arndt, D., Lebrun-Grandie, D., Madsen, J., Ellingwood, N., Trott, C.: Modern problems require modern solutions: how modern CMake supports modern C++ in Kokkos. In: International Workshop on Performance, Portability & Productivity in HPC (2020)

19. Stall, S., et al.: Make scientific data FAIR. Nature **570**, 27–29 (2019)

20. FAIR Principles. https://www.go-fair.org/fair-principles. Accessed 16 Feb 2024

21. Kim, J., et al.: NOAA-NCEP next generation global ocean data assimilation system (NG-GODAS). Research activities in Earth system modelling. Working Group on Numerical Experimentation. Report 51, 8–05 (2021)

22. Rew, R., Davis, G.: NetCDF: an interface for scientific data access. IEEE Comput. Graphics Appl. **10**(4), 76–82 (1990)

23. Folk, M., Heber, G., Koziol, Q., Pourmal, E., Robinson, D.: An overview of the HDF5 technology suite and its applications. In: AD 2011, Proceedings of the EDBT/ICDT 2011 Workshop on Array Databases, pp. 36–47 (2011)

24. Fouilloux, A.: ODB (Observational DataBase) and its usage at ECMWF. In: Twelfth Workshop on Meteorological Operational Systems (2009)

25. Silva, L., Unterkalmsteiner, M., Wnuk, K.: Towards identifying and minimizing customer-facing documentation debt. In: 2023 ACM/IEEE International Conference on Technical Debt (TechDebt), Melbourne, Australia, pp. 72–81 (2023)

26. Burnett, J., et al.: Ten simple rules for creating a scientific web application. PLoS Comput. Biol. **17**(12), e1009574 (2021)

27. UFS Homepage. https://epic.noaa.gov/get-code/ufs-weather-model. Accessed 16 Feb 2024

28. MPAS Homepage. https://www.mmm.ucar.edu/models/mpas. Accessed 16 Feb 2024

29. GMAO Weather Analysis & Prediction. https://gmao.gsfc.nasa.gov/weather_prediction. Accessed 16 Feb 2024

30. MOM6 Homepage. https://www.gfdl.noaa.gov/mom-ocean-model. Accessed 16 Feb 2024

31. HofX applications in OOPS. https://jointcenterforsatellitedataassimilation-jedi-docs.readthedocs-hosted.com/en/latest/inside/jedi-components/oops/applications/hofx.html. Accessed 16 Feb 2024

32. Jette, M., Wickberg, T.: Architecture of the Slurm workload manager. In: Klusáček, D., Corbalán, J., Rodrigo, G.P. (eds.) Job Scheduling Strategies for Parallel Processing. LNCS, vol. 14283, pp. 3–23. Springer, Cham (2023). https://doi.org/10.1007/978-3-031-43943-8_1

33. Jones, J.: PBS: portable batch system. In: Sterling, T. (ed.) Beowulf Cluster Computing with Linux, MIT Press, Cambridge (2001)

34. GMAO Homepage. https://gmao.gsfc.nasa.gov. Accessed 16 Feb 2024

35. OpenAPI Specification Homepage. https://swagger.io/resources/open-api/

36. What is a RESTful API? Homepage. https://aws.amazon.com/what-is/restful-api/

37. Swagger Tools Homepage. https://swagger.io/tools/

38. Flask Homepage. https://flask.palletsprojects.com/en/3.0.x/

# Robustness and Accuracy in Pipelined Bi-conjugate Gradient Stabilized Methods

Mykhailo Havdiak[1]([✉]), José I. Aliaga[2], and Roman Iakymchuk[3,4]

[1] Ivan Franko National University of Lviv, Lviv, Ukraine
`mykhailo.havdiak@lnu.edu.ua`
[2] Universitat Jaime I, Castelló, Spain
`aliaga@icc.uji.es`
[3] Umeå University, Umeå, Sweden
`riakymch@cs.umu.se`
[4] Uppsala University, Uppsala, Sweden

**Abstract.** In this article, we propose an accuracy-assuring technique for finding a solution for unsymmetric linear systems. Such problems are related to different areas such as image processing, computer vision, and computational fluid dynamics. Parallel implementation of Krylov subspace methods speeds up finding approximate solutions for linear systems. In this context, the refined approach in pipelined BiCGStab enhances scalability on distributed memory machines, yielding to substantial speed improvements compared to the standard BiCGStab method. However, it's worth noting that the pipelined BiCGStab algorithm sacrifices some accuracy, which is stabilized with the residual replacement technique. This paper aims to address this issue by employing the ExBLAS-based reproducible approach. We validate the idea on a set of matrices from the SuiteSparse Matrix Collection.

**Keywords:** Krylov subspace methods · BiCGStab · Residual replacement · Numerical reliability · ExBLAS · HPC

## 1 Introduction

Krylov subspace methods form a powerful class of iterative techniques for solving large linear systems arising in diverse scientific and engineering applications. These methods are particularly well-suited for problems where the coefficient matrix is sparse and both symmetric or non-symmetric. Such methods are applicable and often used in image denoising, data compression, inverse problems, and other areas. The Conjugate Gradient (CG) method, introduced in [6], is one of the earliest members of this well-known class of iterative solvers. However, CG is limited to solving symmetric and positive definite (SPD) systems. In contrast, the Bi-Conjugate Gradient [3] (BiCG) method extends its applicability to more general classes of non-symmetric and indefinite linear systems. Additionally, the Conjugate Gradient Squared [8] (CGS) method provides an alternative approach.

© The Author(s), under exclusive license to Springer Nature Switzerland AG 2024
L. Franco et al. (Eds.): ICCS 2024, LNCS 14834, pp. 311–319, 2024.
https://doi.org/10.1007/978-3-031-63759-9_35

The BiConjugate Gradient Stabilized (BiCGStab) method [9] was introduced as a smoother converging version of both BiCG and CGS methods. Preconditioning is usually incorporated in real implementations of these methods in order to accelerate the convergence of the methods and improve their numerical features.

These classical Krylov subspace methods have been actively discussed and optimized. For instance, optimizations have been explored for a specific class of hepta-diagonal sparse matrices on GPUs, as well as the implementation of the pipelined Bi-Conjugate Gradient Stabilized method (p-BiCGStab) [1] to overlap (hide) communication and computation. The pipelined methods, in particular, introduced more operations compared to the original ones and with that impacted the convergence as the computer residual deviated from the true one. As a remedy, the residual replacement technique was proposed [1] to numerically stabilize convergence with a strong emphasis on mathematical aspects.

The purpose of this initial study is to explore the possibility of avoiding residual replacement in pipelined Krylov-type methods [1] with the help of accurate and reproducible computations via the ExBLAS approach [4,5]. As a test case, we use the pipelined BiCGStab method.

## 2    Reproducibility of BiCGStab and Pipelined BiCGStab

BiCGStab was developed to solve non-symmetric linear systems while avoiding the often irregular convergence patterns of the CGS method. In BiCGStab, minimizing a residual vector promotes smoother convergence. However, when the Generalized Minimal Residual method (GMRES) [7] stagnates, preventing the expansion of the Krylov subspace, BiCGStab may fail to proceed effectively.

In the light of the conventional BiCGStab algorithm, see Algorithm 1, introduced by Van der Vorst, Cools and Vanroose proposed an optimization known as pipelined BiCGStab (p-BiCGStab) [1], see Algorithm 2. This optimization entails two primary phases within the pipelining framework. Firstly, in what is termed the 'communication-avoiding' phase, the standard Krylov algorithm undergoes a transformation into a mathematically equivalent form with the reduced global synchronization points. This reduction is accomplished by merging the global reduction phases of various dot products scattered throughout the algorithm into a single global communication phase. Subsequently, in the 'communication-hiding' phase, the algorithm is further refined to overlap the remaining global reduction phases with the sparse matrix-vector product and application of the preconditioner. This strategic restructuring effectively mitigates the typical communication bottleneck by concealing communication time behind productive computational tasks. Although, the methods are mathematically equivalent they may lead to different numerical results and convergence patterns due to the non-associativity of floating-point operations.

The BiCGStab (Algorithm 1) and p-BiCGStab (Algorithm 2) methods will serve as the primary methods utilized throughout this article, although we mainly focus on the p-BiCGStab. Due to the non-associativity of finite precision floating-point operations, the mathematical equivalent of these two methods can show large divergence while implemented in parallel environments especially for the tolerance $10^{-6}$ and below. To stabilize this deviation in p-BiCGStab, the residual replacement technique was proposed [1]. This technique resets the residuals $r_i$, along with the auxiliary variables $w_i$, $s_i$, and $z_i$, to their original values every $k$ iterations. In the preconditioned version, this process also updates $\bar{r}_i = M^{-1} r_i$ and $\bar{s}_i = M^{-1} s_i$, where $M$ is the preconditioning operator. We refer to [1] for more details.

| **Algorithm 1.** BiCGStab [9] | **Algorithm 2.** Pipelined BiCGStab [1] |
|---|---|
| **function** BiCGSTAB(A, b, $x_0$) | **function** P-BiCGSTAB(A, b, $x_0$) |
| $\quad r_0 := b - Ax_0$ | $\quad r_0 := b - Ax_0$ |
| $\quad p_0 := r_0$ | $\quad w_0 := A - r_0$ |
| $\quad$ **for** i = 0, 1, 2, ... **do** | $\quad t_0 := Aw_0$ |
| $\quad\quad s_i := Ap_i$ | $\quad a_0 := (r_0, r_0)/(r_0, w_0)$ |
| $\quad\quad a_i := (r_0, r_i)/(r_0, s_i)$ | $\quad \beta_{-1} := 0$ |
| $\quad\quad q_i := r_i - a_i s_i$ | $\quad$ **for** i = 0, 1, 2, ... **do** |
| $\quad\quad y_i := Aq_i$ | $\quad\quad p_i := r_i + \beta_{i-1}(p_{i-1} - w_{i-1}s_{i-1})$ |
| $\quad\quad w_i := (q_i, y_i)/(y_i, y_i)$ | $\quad\quad s_i := w_i + \beta_{i-1}(s_{i-1} - w_{i-1}z_{i-1})$ |
| $\quad\quad x_{i+1} := x_i + a_i p_i + w_i q_i$ | $\quad\quad z_i := t_i + \beta_{i-1}(z_{i-1} - w_{i-1}v_{i-1})$ |
| $\quad\quad r_{i+1} := q_i - w_i y_i$ | $\quad\quad q_i := r_i - a_i s_i;\quad y_i := w_i - a_i z_i$ |
| $\quad\quad \beta_i :=$ | $\quad\quad v_i := Az_i;\quad w_i := (q_i, y_i)/(y_i, y_i)$ |
| $(a_i/w_i)(r_0, r_{i+1})/(r_0, r_i)$ | $\quad\quad x_{i+1} := x_i + a_i p_i + w_i q_i$ |
| $\quad\quad p_{i+1} := r_{i+1} + \beta_i(p_i - w_i s_i)$ | $\quad\quad r_{i+1} := q_i - w_i y_i$ |
| $\quad$ **end for** | $\quad\quad w_{i+1} := y_i - w_i(t_i - a_i v_i);\quad t_{i+1} := Aw_{i+1}$ |
| **end function** | $\quad\quad \beta_i := (a_i/w_i)(r_0, r_{i+1})/(r_0, r_i)$ |
| | $\quad\quad a_{i+1} := (r_0, r_{i+1})/((r_0, w_{i+1}) +$ |
| | $\beta_i(r_0, s_i) - \beta_i w_i(r_0, z_i))$ |
| | $\quad$ **end for** |
| | **end function** |

In [5], we proposed to ensure the reproducibility and accuracy of the pure MPI implementation of the preconditioned BiCGStab method via the ExBLAS approach. ExBLAS combines together long accumulator and floating-point expansions into algorithmic solutions as well as efficiently tunes and implements them on various architectures. ExBLAS aims to provide new algorithms and implementations for fundamental linear algebra operations (like those included in the BLAS library), that deliver reproducible and accurate results with small or without losses to their performance on modern parallel architectures such as desktop and server processors, Intel Xeon Phi co-processors, and GPU accelerators. We construct our approach in such a way that it is independent of data partitioning, order of computations, thread scheduling, or reduction tree schemes. Instead of using the residual replacement technique, we propose to exhibit the benefits of the ExBLAS approach in the pipelined BiCGStab method.

## 3   Experimental Results

This section presents a series of numerical experiments to evaluate the convergence, performance, and accuracy of the BiCGStab methods, including the reproducible with ExBLAS. The results include comparisons between BiCGStab, pipelined BiCGStab (p-BiCGStab), p-BiCGStab with ExBLAS, and p-BiCGStab with the residual replacement technique across various matrices from the Suite Sparse Matrix Collection [2]. In the experiments, IEEE754 double-precision arithmetic was utilized, and we run on nodes at HPC2N with the dual 14-core Intel Xeon Gold 6132 (Skylake) @2.60GHz interconnected via EDR Infiniband.

The SuiteSparse Matrix Collection is a comprehensive repository of sparse matrices widely used for benchmarking and testing numerical algorithms in the field of computational mathematics. It allows for robust and repeatable experiments, as performance results with artificially generated matrices can be misleading. Hence, repeatable experiments are crucial for ensuring the reliability of algorithm evaluations. The collection encompasses a diverse range of matrices representing real-world problems from various disciplines.

Table 1 presents the comparative performance of four iterative BiCGStab methods, namely BiCGStab, pipelined BiCGStab (p-BiCGStab), p-BiCGStab with ExBLAS, and pipelined BiCGStab with residual replacement (p-BiCGStab-RR), across a selection of sparse matrices from the SuiteSparse Matrix Collection. Each cell in the table represents the number of iterations required to

**Table 1.** Number of iterations for the BiCGStab-like methods on a set of the SuiteSparse matrices without precondition. The initial estimate is a zero vector $x_0$. The best-performing method is highlighted in bold.

| Problem | BiCGStab | | p-BiCGStab | | p-BiCGStabExBLAS | | p-BiCGStabRR | |
|---|---|---|---|---|---|---|---|---|
| | $10^{-6}$ | $10^{-9}$ | $10^{-6}$ | $10^{-9}$ | $10^{-6}$ | $10^{-9}$ | $10^{-6}$ | $10^{-9}$ |
| 1138_bus | 30 | 151 | **27** | 130 | 35 | **108** | **27** | 130 |
| add32 | **38** | 74 | **38** | 68 | 38 | **69** | 38 | **68** |
| bcsstk13 | 545 | − | 520 | 2258 | 350 | 3273 | **195** | **403** |
| bcsstk14 | 149 | 461 | 44 | 459 | **43** | **433** | 44 | − |
| bcsstk18 | 405 | 2806 | **261** | 1284 | 366 | **1274** | 309 | − |
| bcsstk27 | 283 | **958** | 335 | 2107 | **279** | 1477 | 335 | 2107 |
| bfwa782 | 99 | 647 | 74 | **448** | **54** | 463 | 115 | 576 |
| cdde6 | 36 | 122 | **34** | 121 | **34** | **115** | **34** | 388 |
| msc01050 | **29** | 61 | 30 | **47** | **28** | 60 | 30 | **47** |
| msc04515 | 123 | **257** | **96** | 275 | **98** | 308 | 263 | 334 |
| orsreg_1 | **21** | 161 | 22 | 168 | **20** | **106** | 22 | 371 |
| pde2961 | **100** | **166** | 111 | 278 | 134 | 287 | 170 | 683 |
| plat1919 | **79** | **132** | 99 | 185 | 84 | 179 | 87 | 250 |
| rdb3200l | 31 | 193 | 31 | 223 | 31 | **171** | 31 | 567 |
| saylr4 | 28 | 73 | 28 | 74 | 28 | **69** | 28 | 74 |
| sherman3 | 34 | 501 | 33 | 314 | **26** | 400 | 33 | **271** |
| utm5940 | 99 | 592 | 97 | **420** | **18** | 603 | 20 | **419** |

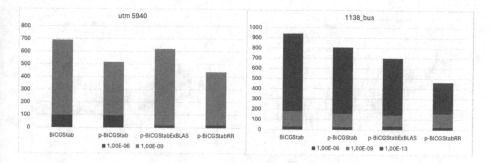

**Fig. 1.** Number of iterations required by various BiCGStab-like methods to achieve a specified tolerance $(10^{-6}, 10^{-9}, 10^{-13})$. p-BiCGStabRR stands for the pipelined version of the BiCGStab method with residual replacement; P-BiCGStabExBLAS refers to the method with ExBLAS.

achieve convergence for a specific method and a given matrix, with convergence thresholds set at $10^{-6}$ and $10^{-9}$. The results demonstrate varying convergence behavior among the methods across different matrices, providing insights into their respective efficiency in solving sparse linear systems.

When summarizing the findings, several notable observations come to light. Firstly, for $\varepsilon = 10^{-6}$, all methods demonstrate comparable performance in numerous cases. However, when considering a higher tolerance, $\varepsilon = 10^{-9}$, the p-BiCGStabExBLAS method consistently outperforms p-BiCGStab across the majority of matrices, owing to its enhanced accuracy. Moreover, the p-BiCGStab with residual replacement strategy (p-BiCGStabRR) method generally exhibits superior convergence rates compared to the p-BiCGStab method. Yet, increasing the tolerance also reveals instances where the classic BiCGStab method proves more efficient in terms of iterations, although it may encounter convergence issues for certain matrices. Notably, for the specific bcsstk13 problem, p-BiCGStabRR demonstrates the most favorable convergence characteristics for both $tol = 10^{-6}$, $tol = 10^{-9}$. Despite these advantages, there are scenarios where p-BiCGStabRR fails to converge, namely bcsstk14 and bcsstk18. The accuracy of the p-BiCGStabRR is highly contingent to the specific problem context and parameter choices, leading to variability in its effectiveness. This method may exhibit convergence speed under certain conditions while performing poorly under others, highlighting the sensitivity of its outcomes to these factors. The method requires multiple runs to determine the optimal step and the best place for applying residual replacement. A less optimal parameter choice can result in more iterations compared to the pipelined-BiCGStab method.

In Fig. 1, the utm5940 case highlights an interesting trend: the ExBLAS version performed the best for epsilon $10^{-6}$, yet with an increase to $10^{-9}$, it required slightly more iterations compared to other methods. Overall, p-BiCGStabExBLAS demonstrates good constant performance in terms of iterations. When examining p-BiCGStabRR, it's evident that for certain examples, it exhibits the lowest iteration count. However, there are instances for higher

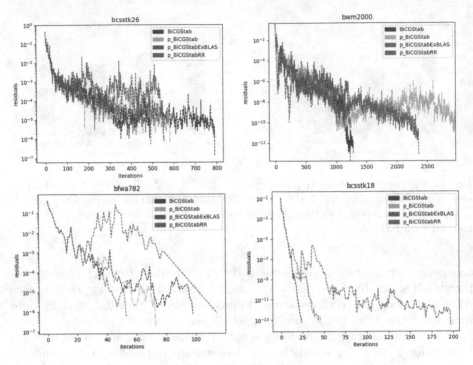

**Fig. 2.** Residual history of the four BiCGStab-like methods; $tol = 10^{-13}$.

tolerance the method results in significantly higher iteration counts compared to other methods.

Figure 2 provides the convergence history of the four BiCGStab-like methods. We can observe the performance characteristics of the methods on a particular problem instance, where the tolerance is set to $10^{-13}$. The pipelined BiCGStab method with ExBLAS consistently outperforms the regular pipelined variant in terms of iterations across a wide range of scenarios. p-BiCGStabExBLAS exhibits a reduced occurrence of spikes compared to p-BiCGStabRR, suggesting a smoother and more stable performance profile. This difference highlights the potential of ExBLAS to offer improved reliability and predictability in computational outcomes, namely results and iterations.

Following this, we evaluate the four considered methods using an increased number of processes. Subsequently, we present Tables 2 and 3 illustrating the outcomes achieved by the aforementioned methods across different process counts. A notable observation from both Table 2 and Table 3 is the consistency in the number of iterations required for the ExBLAS implementation across different numbers of processes. Thus, increasing the number of processes does not lead to a faster solution for pipelined BiCGStab method. The pipelined BiCGStabRR as indicated in Table 1 demonstrates its best outcome of 195 iterations for the bcsstk13 matrix.

**Table 2.** Number of iterations required for BiCGStab, p-BiCGStab, and p-BiCGStabExBLAS by varying numbers of processes ($nXX$) for $\varepsilon = 10^{-6}$.

| $\varepsilon = 10^{-6}$ | BiCGStab | | | p-BiCGStab | | | p-BiCGStabExBLAS | | |
|---|---|---|---|---|---|---|---|---|---|
| | $n01$ | $n08$ | $n16$ | $n01$ | $n08$ | $n16$ | $n01$ | $n08$ | $n16$ |
| bcsstk26 | 791 | 235 | 599 | 583 | 528 | 683 | 493 | 493 | 493 |
| bwm2000 | 37 | 37 | 37 | 37 | 37 | 37 | 37 | 37 | 37 |
| bfwa782 | 99 | 79 | 85 | 74 | 104 | 59 | 54 | 54 | 54 |
| bcsstk18 | 405 | 414 | 363 | 261 | 305 | 356 | 366 | 366 | 366 |

Table 4 illustrates the iteration counts for the BiCGStab, p-BiCGStab, and p-BiCGStabExBLAS methods across different numbers of processes. The method with residual replacement technique finds an approximation to the solution only on a single process. Additionally, the table presents the execution times for each scenario. p-BiCGStabExBLAS requires more time, attributed to its higher accuracy. Nonetheless, the overhead associated with p-BiCGStabExBLAS diminishes as the number of processes increases, dropping from 2.6x on a single process to 1.87x on 16 processes.

Figure 3 illustrates the benefits of using pipelined methods within a parallel environment, emphasizing their efficiency and scalability. Certainly, the scale is small but the gain starts to be visible on 16 processes, four per each of four nodes; we used only few processes per node to highlight the benefit. In this test, we focus on two large matrices: s3dkq4m2 with $4,427,725$ non-zero elements and Queen_4147 with $316,548,962$ non-zero elements. Larger-dimensional problems tend to demonstrate better strong scalability in parallel environments, using the existing potential of available resources especially on four and eight processes. Conversely, employing 16 processes for the s3dkq4m2 matrix with fewer non-zero elements did not yield significant improvements for BiCGStab and p-BiCGStab. Additionally, p-BiCGStabExBLAS demonstrates strong scalability for both problems due to more flops imposed by the ExBLAS approach. Overhead for s3dkq4m2 varied from 3.41x to 4.8x, and matrix Queen_4147 from 1.96x to 5.2x. With the increase in the number of processes from 4 to 16, the execution time for p-BiCGStabExBLAS was reduced by more than 2x.

**Table 3.** Number of iterations required for BiCGStab, p-BiCGStab, and p-BiCGStabExBLAS by varying the numbers of processes ($nXX$) for $tol = 10^{-9}$.

| Matrix | BiCGStab | | | p-BiCGStab | | | p-BiCGStabExBLAS | | |
|---|---|---|---|---|---|---|---|---|---|
| | $n01$ | $n08$ | $n16$ | $n01$ | $n08$ | $n16$ | $n01$ | $n08$ | $n16$ |
| bwm2000 | 1268 | 1267 | 1120 | 663 | 1131 | 1024 | 1232 | 1232 | 1232 |
| bfwa782 | 647 | 528 | 531 | 448 | 476 | 498 | 464 | 463 | 463 |
| bcsstk18 | 2806 | 2819 | 2286 | 1284 | 1811 | 2318 | 1274 | 1274 | 1274 |

**Table 4.** Number of iterations and time required for the BiCGStab-like methods for the `bcsstk13` matrix. Tolerance is set to $tol = 10^{-6}$.

| Method | $n01$ | | $n08$ | | $n16$ | |
|---|---|---|---|---|---|---|
| | iter | time | iter | time | iter | time |
| BiCGStab | 545 | $2.034 \times 10^{-1}$ | 520 | $4.8236 \times 10^{-2}$ | 544 | $3.94 \times 10^{-2}$ |
| p-BiCGStab | 520 | $2.0495 \times 10^{-1}$ | 482 | $4.457 \times 10^{-2}$ | 394 | $3.106 \times 10^{-2}$ |
| p-BiCGStabE | 350 | $5.321 \times 10^{-1}$ | 350 | $8.976 \times 10^{-2}$ | 350 | $5.819 \times 10^{-2}$ |
| p-BiCGStabRR | 195 | $6.198 \times 10^{-2}$ | – | – | – | – |

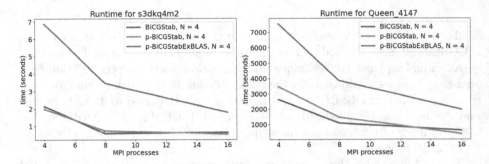

**Fig. 3.** Runtime comparison of BiCGStab-like methods on two matrices: s3dkq4m2 with $4,427,725$ nnz $tol = 10^{-9}$; Queen_4147 with $316,548,962$ nnz and $tol = 10^{-6}$. We used four nodes with $1, 2$, and $4$ MPI processes each.

## 4    Conclusion

In this study, we investigated the robustness and accuracy of the pipelined Biconjugate Gradient Stabilized using the ExBLAS approach as not only an accurate and reproducible solution but also as an alternative to the residual replacement technique. Our analysis focused on evaluating the convergence behavior of the method across a set of matrices from the SuiteSparse Matrix Collection. Through the numerical experiments, we demonstrated that the pipelined BiCGStab method with ExbLAS, consistently outperforms the conventional pipelined BiCGStab approach in terms of convergence rates and numerical reliability. Overall, this study emphasizes the importance of considering algorithmic refinements and numerical stability enhancements to achieve reliable and efficient solutions for challenging computational problems. The results underscore notable performance disparities among the assessed methods. Specifically, BiCGStab demonstrates better stability compared to p-BiCGStab, showcasing its reliability in solving linear systems. The residual replacement strategy is expected to address the stability of the pipelined method, bringing it closer to the robustness exhibited by BiCGStab. Although its performance is inferior to ExBLAS implementation. This suggests that the ExBLAS implementation capitalizes on the advantages of the pipelined method version while maintaining stability as in BiCGStab.

As a future work, we shall conduct theoretical study of the ExBLAS approach as a possible replacement for the residual replacement, which requires some empirical trials to get the right step. Furthermore, we plan to carry out an exhaustive study with more matrices from the SuiteSparse Matrix Collection as well as the real applications like the ones from the EU-funded EuroHPC JU Center of Excellence in Exascale CFD (CEEC)[1], where the last author contributes.

# References

1. Cools, S., Vanroose, W.: The communication-hiding pipelined BiCGstab method for the parallel solution of large unsymmetric linear systems. ParCo **65**, 1–20 (2017). https://doi.org/10.1016/j.parco.2017.04.005
2. Davis, T.A., Hu, Y.: The university of Florida sparse matrix collection. ACM TOMS **38**(1), 1–25 (2011). https://doi.org/10.1145/2049662.2049663
3. Fletcher, R.: Conjugate gradient methods for indefinite systems. In: Watson, G.A. (ed.) Numerical Analysis. LNM, vol. 506, pp. 73–89. Springer, Heidelberg (1976). https://doi.org/10.1007/BFb0080116
4. Iakymchuk, R., Collange, S., Defour, D., Graillat, S.: ExBLAS: reproducible and accurate BLAS library. In: NRE at SC15 (2015). https://hal.science/hal-01202396v3
5. Iakymchuk, R., Graillat, S., Aliaga, J.I.: General framework for re-assuring numerical reliability in parallel Krylov solvers: a case of bicgstab methods. IJHPCA **38**, 17–33 (2024). https://doi.org/10.1177/10943420231207642
6. Hestenes, M.R., Stiefel, E.: Methods of conjugate gradients for solving linear systems. NBS **49**, 409–435 (1952). https://doi.org/10.6028/jres.049.044
7. Saad, Y., Schultz, M.H.: GMRES: a generalized minimal residual algorithm for solving nonsymmetric linear systems. SIAM J Sci. Stat. Comp. **7**, 856–869 (1986). https://doi.org/10.1137/0907058
8. Sonneveld, P.: CGS, a fast Lanczos-type solver for nonsymmetric linear systems. SIAM J. Sci. Stat. Comp, **10**(1), 36–52 (1989). https://doi.org/10.1137/0910004
9. van der Vorst, H.A.: Bi-CGSTAB: a fast and smoothly converging variant of Bi-CG for the solution of nonsymmetric linear systems. SIAM J. Sci. Stat. Comp. **13**(2), 631–644 (1992). https://doi.org/10.1137/0913035

[1] This work was partially supported by CEEC under grant agreement No 10109339.

**Artificial Intelligence
and High-Performance Computing
for Advanced Simulations**

# Investigating Guiding Information for Adaptive Collocation Point Sampling in PINNs

Jose Florido[1]([✉])[ID], He Wang[2][ID], Amirul Khan[1][ID], and Peter K. Jimack[1][ID]

[1] University of Leeds, Leeds LS2 9JT, UK
{mn17jilf,A.Khan,P.K.Jimack}@leeds.ac.uk
[2] University College London, London WC1E 6BT, UK
he_wang@ucl.ac.uk

**Abstract.** Physics-informed neural networks (PINNs) provide a means of obtaining approximate solutions of partial differential equations and systems through the minimisation of an objective function which includes the evaluation of a residual function at a set of collocation points within the domain. The quality of a PINNs solution depends upon numerous parameters, including the number and distribution of these collocation points. In this paper we consider a number of strategies for selecting these points and investigate their impact on the overall accuracy of the method. In particular, we suggest that no single approach is likely to be "optimal" but we show how a number of important metrics can have an impact in improving the quality of the results obtained when using a fixed number of residual evaluations. We illustrate these approaches through the use of two benchmark test problems: Burgers' equation and the Allen-Cahn equation.

**Keywords:** Partial differential equations · Deep learning · Physics-informed neural networks · Adaptivity

## 1 Introduction

### 1.1 Context

This paper is concerned with the mechanism by which Physics-informed neural networks (PINNs) [1,2], allow us to introduce our prior knowledge of physics into a machine learning (ML) algorithm. PINNs are an ML approach to the solution of systems of partial differential equations (PDEs), where there is limited or noisy ground truth data. In this context, the PDEs provide a mathematical model of the physics, which is captured in the method through the inclusion of loss terms that are evaluated by calculating PDE residuals at collocation points throughout the domain. This point-based, meshless approach offers greater flexibility compared to meshed methods in traditional numerical models - but also poses an interesting problem regarding the number and distribution of said collocation points.

© The Author(s), under exclusive license to Springer Nature Switzerland AG 2024
L. Franco et al. (Eds.): ICCS 2024, LNCS 14834, pp. 323–337, 2024.
https://doi.org/10.1007/978-3-031-63759-9_36

The original implementation of PINNs uniformly distributes a fixed number of collocation points randomly throughout the domain. However, in certain problems some areas are intrinsically harder to learn than others - and this purely random approach can lead to slow or inefficient training. Different collocation points contain different amounts of information for learning, which indicates there is an underlying distribution of locations for collocation points that can maximise the learning - which is mostly likely non-uniform and problem dependent. Such a proposition suggests that adjusting the distribution of points, by biasing towards features of interest, may be beneficial in terms of efficiency. It may also be necessary for practical reasons to minimise the risk of the network getting trapped in local minima during training [3]. However, manually choosing this distribution of collocation points can be an arduous task and requires *a priori* knowledge of the solution, or at least of which areas of the domain will be "most important". The alternative is to increase the number of points used globally which, whilst potentially effective, adds significantly to the computational cost.

To find the optimal or a good distribution of the collocation points, there are two common approaches to explicitly or implicitly approximate this distribution. The first approach is to have fixed collocation points but weight them differently, i.e. *adaptive weighting*. Examples of this are [4,5]'s Self-Adaptive PINNs, a concept extended by [6]'s DASA point weighing, and more recently [7]'s Loss-Attentional PINNs. The residual-based attention scheme in [8] works similarly to the above, weighting the influence of specific collocation points in the domain to ensure key collocation points aren't overlooked.

The other approach is to refine the locations of collocation points, i.e. *adaptive resampling*. [9] first presents this type of adaptive refinement based on residual information. Other sampling approaches include [10–12]. [13]'s formulation for resampling is a more general version of some of the previous approaches, and systematically compares adaptive collocation resampling methods to fixed approaches. Also noteworthy is [14]'s implementation of a cosine-annealing strategy for restarting training from a uniform distribution when optimisation stalls. Whilst the above focus on the collocation points where the PDEs are evaluated, [15] also considers optimising the selection of experimental points (where the data is available for supervised learning problems) at the same time.

From the sampling point of view, existing methods essentially seek the ideal distribution of collocation points, via either explicit parameterisation [9], or implicit approximation [10]. The information exploited up to date is mostly focused on the loss function [10,12,13], e.g. information such as local PDE residual at collocation points, or gradients of the loss term. Deviating from existing literature, we investigate into a different category of information that proves to be useful in parametrising the ideal collocation points. This information is the geometric information of the estimated solution, e.g. their spatial and temporal derivatives. The intuition behind is the solution geometry reveals intrinsic information about the PDE. For instance, stiffness regions are harder to learn [7]. By introducing the solution geometry, our aim is to shed a deeper light into the complexity behind seeking an "ideal" distribution, and into the interplay between the effectiveness of different point sampling methods, the complexity of the problem and the computational cost.

## 1.2   PINNs

The process of training a PINN is similar to that for a regular NN: once a topology and a set of weights is defined we seek to find values of these weights that minimise a prescribed loss function. As discussed below, we will consider a fully unsupervised learning task for which the loss function is obtained from the sum of all of the residuals of the PDEs at all of the selected collocation points. The significance of the number and location of the collocation points is substantial therefore, since they directly impact on the loss function which is to be minimised. The crux of adaptive resampling is that a fixed number of points are moved based upon some criteria (that we investigate), with a view to this modified loss function being a better objective for the overall minimisation . Previous investigations of this approach have redistributed the collocation points based upon a probability density function (PDF) that was formed using the values of the residuals at the current points or the derivatives of the loss function with respect to the location of the collocation points. In this work, we consider alternative criteria based upon mixed derivatives of the estimated solution and of the residual.

The loss function $\mathcal{L}(\theta)$ of a PINN can be typically characterised as the sum of the individual losses due to data fitting ($\mathcal{L}_{data}$), enforcing the governing PDE ($\mathcal{L}_{PDE}$), and weakly enforcing boundary conditions ($\mathcal{L}_{BCs}$) as follows:

$$\mathcal{L}(\theta) = w_{data}\mathcal{L}(\theta)_{data} + w_{PDE}\mathcal{L}(\theta)_{PDE} + w_{BC}\mathcal{L}(\theta)_{BCs}$$

Here, the relative impact of each loss term can be controlled through the weight, $w$, whilst $\theta$ represents the trainable network parameters. For the one-dimensionsal time-dependent problems that we look at, we will consider a spatial dimension $x$, a temporal dimension $t$ and solve for the dependent variable $u(x,t)$. The PDE loss term is obtained by evaluating the PDE residual ($f$, say) for every collocation point in the set of selected collocation points $\mathrm{x} \in T$:

$$\mathcal{L}_{PDE} = \frac{1}{|T|} \sum_{\mathrm{x} \in T} |f(\mathrm{x}; u; u_x, u_t, u_{xx}; \theta)|^2$$

It is important to distribute these points throughout the domain however evidence suggests that it may also be beneficial to identify and place a greater proportion of points in particular regions of importance, so that the loss term especially includes these areas. This can ensure that optimising the cost function improves how well the PDE is satisfied around these locations, and therefore hopefully improves the solution in those areas.

For forward problems, PINNs can be trained without the need for data, and boundary conditions can be enforced in a 'hard' manner. This is the approach that we take here in order to focus exclusively on the PDE residual contribution to the loss. We strongly impose the boundary conditions by applying an output transformation to the results of the neural network and we treat the PINN as an unsupervised problem with no input data. As a result, $\mathcal{L}_{PDE}$ becomes the only term to optimise in our loss function.

Throughout this paper, for consistency with other studies in the literature [13], a simple feedforward NN is used consisting of 3 intermediate layers of 64 nodes each, with a tanh activation function after each layer. The training consists of 15,000 initial steps using the ADAM optimiser (with learning rate of 0.001), followed by 1000 steps using L-BFGS [13] before beginning the resampling process for adaptive methods. For these methods, the points are then redistributed at this stage and then the training continues with 1000 steps of ADAM and 1000 steps of L-BFGS, repeating until the number of resamples specified are completed.

## 2  Adaptive Resampling

### 2.1  Probability Density Functions

As mentioned in the introduction, for many problems PINNs will give more accurate results when the collocation points are distributed following an appropriate, problem-dependent distribution. The process of rearranging the collocation points can be automated by using information gathered during the training process. This approach has been shown to work well in the literature using the local residual as a guiding metric [9,13]; but this paper also proposes alternatives based on the spatial and temporal derivatives of both the residuals and estimates of the solution. These proposed alternatives are assessed against both uniform, fixed distributions and existing adaptive redistribution methods, and shown to be beneficial in many cases.

The method that we use to resample the points is consistent with [13] for comparison, which chooses the next set of collocation points $\mathcal{T}$ from a fine grid of random points $\mathcal{X}$. All points in $\mathcal{X}$ are assigned a probability $(P(\mathcal{X}))$, and the prescribed number of collocation points are chosen according to the normalised PDF $\hat{P}(\mathcal{X})$. These are obtained from Eqs. 1 and 2:

$$P(\mathcal{X}) = \frac{Y(\mathcal{X})^k}{\overline{Y(\mathcal{X})^k}} + c \tag{1}$$

$$\hat{P}(\mathcal{X}) = \frac{P(\mathcal{X})}{\|P(\mathcal{X})\|_1} \tag{2}$$

Here $Y(\mathcal{X})$ is the vector containing the magnitude of the information source chosen (e.g. the current residual or its derivatives, or some derivatives of the current estimate of the solution), $\overline{Y(\mathcal{X})^k}$ is the mean value across all points in $\mathcal{X}$ and $k, c$ are constant hyper-parameters that affect the resampling behaviour. A visualisation of the distribution of points obtained using the PDE residual as $Y(\mathcal{X})$ for solving the Burgers' Equation is shown in Fig. 1

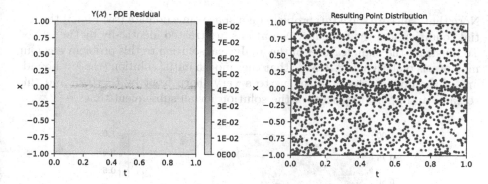

**Fig. 1.** $Y(\mathcal{X})$ = PDE Residual shown at every point in $\mathcal{X}$ (left); and the corresponding points selected for $T$ (right).

As a baseline, we also compare the performance of the adaptive methods that we consider against two fixed distributions: a uniform random distribution and an alternative uniform, pseudo-random, distribution made with a Hammersley sampling algorithm [16]. For these fixed distributions, instead of resampling after a given number of steps, the training is continued until an equivalent amount of training steps are carried out.

We consider two cases for the adaptive method for which $Y$ is based upon the current local residuals: one where the initial distribution of points before any resamples is randomly generated $(PDE, R)$, and the other where the initial distribution follows Hammersley sampling $(PDE, H)$. We also consider the use of the current solution estimates as an information source, defining $Y$ based upon the mixed (spatial and temporal) second derivative of the current solution. We will refer to this as the local geometric curvature of $u$, and denote this case as "$U_{xt}$" in the subsequent text. We also consider a version of $Y$ that is based upon the mixed second derivative of the local residual values, which will subsequently be denoted as "$PDE_{xt}$".

## 2.2   Problem Definition

The main benchmark that we initially use in order to assess the different sampling strategies that we consider is the 1D Burgers' Equation, given by:

$$uu_x + u_t = \nu\, u_{xx}, \quad x \in [-1, 1], \quad t \in [0, 1], \tag{3}$$

where the magnitude of $\nu$ determines the relative effect of diffusion. Dirichlet boundary conditions are assigned:

$$u(-1, t) = u(1, t) = 0,$$

as is the following initial condition:

$$u(x, 0) = -\sin(\pi x).$$

Note that, since we solve this problem on a space-time domain (see Fig. 2) the initial condition and the boundary conditions are treated identically by the PINN. To illustrate the main features of the analytical solution to this problem a colour map of $u$ is plotted in Fig. 2. Note that the smooth initial solution (the left boundary of the domain shown) steepens to a very sharp front by $t = 0.25$ and this remains an important feature of the solution for all subsequent times.

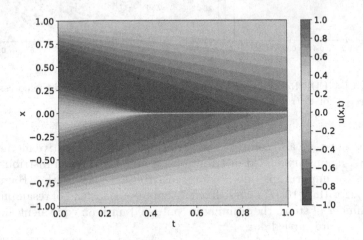

**Fig. 2.** Burgers' Equation contour plot of example solution with $\nu = \frac{\pi}{100}$

As discussed in the previous section the boundary and initial conditions may be imposed either weakly or strongly when using PINNs. Here we opt for the latter by applying an output transformation to the $u$ that is output by the network:

$$output = -\sin(\pi * x) + (1 - x^2)ut$$

For this case; the PDE residual when used as an information source is therefore given by:

$$PDE = uu_x + u_t - \nu\, u_{xx} \tag{4}$$

Similarly,

$$U_{xt} = \frac{\partial^2}{\partial x \partial t} u(x, t) \tag{5}$$

and

$$PDE_{xt} = \frac{\partial^2}{\partial x \partial t}(uu_x + u_t - \nu\, u_{xx}) \tag{6}$$

To measure and compare the accuracy of the different methods considered an $L^2$ relative error metric is used at the end of training. This compares the prediction $u$ to ground truth $u_{gt}$ as follows:

$$L^2 = \frac{\sqrt{\sum(u(i) - u_{gt}(i))^2}}{\sqrt{\sum u_{gt}(i)^2}} \,.$$

Note that the sums in this last expression are over every point $i$ in a very fine background grid which contains many more points than the number of collocation points used to evaluate $u$. Furthermore, due to the stochastic nature of ML training, a minimum of 20 repeats with different seeds were carried out for each case investigated throughout this paper; with the quoted $L^2$ errors always representing the average of these.

## 2.3   Results: Default Case

For this first test problem we have applied six different solution strategies using $N = 2000$ collocation points. In each case we follow the training regime of [13], described in Sect. 1.2 (15,000 initial steps of ADAM followed by 1000 steps of L-BFGS and then resampling every 2000 steps (1000 ADAM/1000 L-BFGS). Figure 3 plots the error against the number of resamples taken for each of the approaches considered, including the two constant point sets based upon a uniform random distribution and a distribution based upon the Hammersley sampling algorithm.

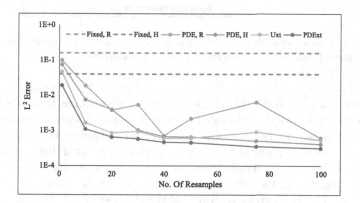

**Fig. 3.** Error against number of resamples solving Burgers' Equation with described default parameters $v = \frac{\pi}{100}; N = 2000$

The initial distribution is seen to have an effect in both the fixed and residual-based methods, with the use of the Hammersley sampling algorithm resulting in consistently lower error. The local residual method with random initialisation ($PDE, R$) exhibited the least robustness and greatest variability, which can be seen from the fact increasing the number of resamples did not consistently lower the average error between runs. Generally, more resamples does result in improving accuracy, with error descending below 0.1% for above 30 resamples for the most of the adaptive resampling methods.

For the same case we also look at the error achieved using different numbers of collocation points. This was obtained for 100 rounds of resampling and is plotted in Fig. 4. This figure suggests that the threshold for the number of points required to obtain results of a given accuracy varies from method to method. Even for the most accurate method ($PDE_{xt}$), the error seems to plateau however between 0.01% and 0.1%, which the other adaptive methods eventually reach at the default 2000 points. Nevertheless, this clearly suggests that the $PDE_{xt}$ approach could allow similar accuracy to other techniques but at a significantly lower cost.

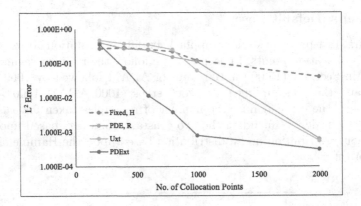

**Fig. 4.** Error against number of collocation points for different resampling methods solving Burgers' Equation with described default parameters $v = \frac{\pi}{100}$; 100 resamples

In conclusion, based on this single test case, we observe that the achieved error of a PINN is heavily influenced by the chosen collocation point distribution, echoing what has been seen previously (see Introduction). Furthermore, the adaptive resampling methods easily outperform the fixed distributions, with both of the derivative-based redistribution methods performing quite well (especially when considering the mixed second derivative of the residual). In the next section we explore the extent to which this behaviour generalises by changing multiple aspects of the problem under consideration, including the initial conditions and the magnitude of the diffusion term, and by considering a second benchmark PDE: the Allen-Cahn Equation.

## 3    Results: Other Cases

### 3.1    Alternate Initial Conditions

In this example we contrast the same methods as in the previous section to once again solve Burgers' Equation, but this time we vary the initial conditions, and therefore the entire evolution of the solution of the PDE. For each of the four cases considered we compute a high resolution numerical solution to represent the ground truth and use this to assess the error against an increasing number of adaptive resamples of the collocation points. The corresponding results are plotted in Fig. 5.

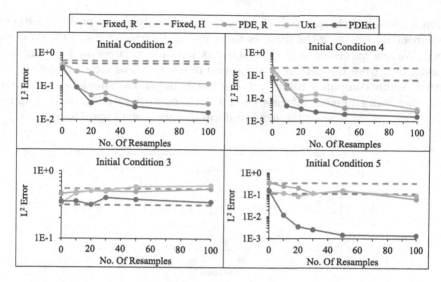

**Fig. 5.** Error versus number of resamples. Initial conditions 2 and 3 are combinations of randomly weighted sin curves of different frequency; with 2 being made up of three low frequency curves and 3 being a higher frequency combination of four curves. Initial conditions 4 and 5 are $sin(2\pi)$ and $1.5sin(\pi)$ respectively.

Inspection of Fig. 5 suggests that for three of the four initial conditions (2, 4 and 5), $PDE_{xt}$ clearly outperforms the other methods (as it did for initial condition 1 in the previous section). Not only does it deliver a smaller error after the last resample (the maximum amount of training investigated) but the error curve lies beneath the others at each stage in the adaptive process. This gap is most noticeable in the 5th case, which corresponds to an initial condition with a larger amplitude, thus leading to a faster shock formation and a sharper shock. In this case it far outperforms the next best method. Furthermore, in each case other than initial condition 3, the adaptive collocation point algorithms all performed significantly better than using fixed collocation points. Unsurprisingly, of the two fixed point approaches Hammersley always outperforms the uniform random distribution.

The obvious outlier in Fig. 5 is initial condition 3. Here all of the methods to quite poorly in terms of reducing the error and least worst result is obtained using fixed collocation points based upon a Hammersley sampling. Further investigation is required to better understand this result, which shows that the PINN performance (or, at least, the ease with which the PINN can be trained) is highly dependent upon the problem being solved.

## 3.2   Adjusting PDE Parameters

For the following example, we investigate adjusting the value of the $\nu$ parameter in the PDE; which affects the magnitude of the diffusion term $\nu\,u_{xx}$. Lower values increase the shock sharpness, increasing the complexity of the problem without fundamentally changing the shape of the solution. Increasing $\nu$ sees the opposite effect. We use values of $\nu$ of 0.01 and 0.001, approximately multiplying and dividing by 3 from the default $\nu = \frac{\pi}{100}$ used in the previous section. The ground truth is again computed using high-resolution numerical simulations and the error when using different information sources to control the adaptive resampling is again compared against the fixed uniform methods: see Figs. 6 and 7.

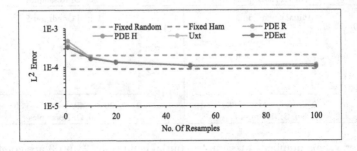

**Fig. 6.** Error against number of resamples, for Burgers' Equation with $\nu = 0.01$

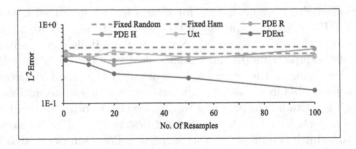

**Fig. 7.** Error against number of resamples, for Burgers' Equation with $\nu = 0.001$

For the simpler case, $\nu = 0.01$ (Fig. 6), all adaptive methods behave similarly, delivering a small error after relatively few resamples. Comparing the fixed methods, the Hammerlsey distribution is again consistently better than a random fixed distribution. Furthermore, in this specific example, it again outperforms the adaptive sampling methods. Unlike case 3 of Fig. 5 however, the accuracy of all of these PINNs solutions is high. One possible explanation for the Hammersley approach being the best in this case is that, unlike the resampling methods, it uses the L-BFGS optimisation algorithm for the majority of training steps (as opposed to an equal split between ADAM and L-BFGS for the adaptive methods). Overall, It is clear that by making the solution of the problem smoother (through increasing $\nu$) all of the methods do well and therefore use of the simplest approach, based upon a fixed distribution of collocation points, can be appropriate.

For the more challenging case of $\nu = 0.001$ (Fig. 7), all methods struggle to converge to an accurate solution using $N = 2000$ collocation points. The least worst approach, with an $L^2$ error of 14%, is again the adaptive method based upon $PDE_{xt}$. Based upon this set of results we can make the opposite proposition: that for complex problems, the choice of adaptive sampling method may not be the limiting factor to accuracy, and an increase to either the number of collocation points or the network complexity may be needed to increase accuracy.

We test this proposition by undertaking a study to look at the impact that changing the number of collocation points has on the solutions obtained in the $\nu = 0.01$ and $\nu = 0.001$ cases. In Table 1 we consider the simpler $\nu = 0.01$ case, observing the change in performance when the number of collocation points is halved to 1000. Conversely, in Table 2, the more complex case with the sharper shock ($\nu = 0.001$), we double the number of collocation points to $N = 4000$ and observe whether there are any significant increases in accuracy.

**Table 1.** Effect of changing $N$; $L^2$ Error at 100 resamples for $\nu = 0.01$

| $N$ | Fixed, R | Fixed, H | $PDE, R$ | $PDE, H$ | $U_{xt}$ | $PDE_{xt}$ |
|---|---|---|---|---|---|---|
| 2000 | 2.019E-04 | 8.95E-05 | 1.024E-04 | 1.152E-04 | 1.089E-04 | 1.055E-04 |
| 1000 | 3.089E-03 | 3.925E-04 | 1.503E-04 | 1.867E-04 | 1.458E-04 | 1.258E-04 |
| Error Change | 1429.9% | 338.4% | 46.8% | 62.1% | 33.9% | 19.3% |

In Table 1, decreasing the number of collocation points from 2000 to 1000 increases the error as expected. However, there is a big discrepancy in how much the different methods are affected. A positive % signifying increase in error, the fixed methods are substantially less effective with fewer points. The opposite is true for adaptive methods, especially for $PDE_{xt}$ which only sees an increase in error of 19.3% compared to the 3-fold increase of the fixed Hammersley distribution of points.

**Table 2.** Effect of changing $N$; $L^2$ Error at 100 resamples for $\nu = 0.001$

| $N$ | Fixed, R | Fixed, H | $PDE, R$ | $PDE, H$ | $U_{xt}$ | $PDE_{xt}$ |
|---|---|---|---|---|---|---|
| 2000 | 4.209E-01 | 5.093E-01 | 3.925E-01 | 4.780E-01 | 3.797E-01 | 1.425E-01 |
| 4000 | 4.049E-01 | 3.836E-01 | 1.858E-01 | 1.996E-01 | 2.591E-01 | 3.660E-02 |
| Error Change | −3.9% | −32.8% | −111.2% | −139.5% | −46.6% | −289.3% |

In Table 2, we are this time looking at the decrease in error as a result of doubling the number of collocation points in the case of $\nu = 0.001$. As anticipated, the error does decrease in each column (signified by a negative increase) however this improvement is relatively small in most cases, meaning that the overall errors are still not competitive, even with this increase in resources. The biggest decreases are seen for the $PDE, H$ and especially $PDE_{xt}$, which is the only method to achieve an error under 10%.

### 3.3    An Alternative PDE: Allen-Cahn

Having presented several variations of the Burgers' Equation problem, our next step is to consider a different PDE entirely and observe the results. The Allen-Cahn equation is chosen for this due to its complex transient behaviours which are known to be challenging to capture computationally. The problem is defined as follows:

$$\frac{\partial u}{\partial t} = D\frac{\partial^2 u}{\partial x^2} + 5(u - u^3), \quad x \in [-1, 1], \quad t \in [0, 1] \tag{7}$$

We take D = 0.001 and use the initial condition

$$u(x, 0) = x^2 \cos(\pi x) \, ,$$

and boundary conditions

$$u(-1, t) = u(1, t) = -1 \, .$$

For this problem we again compute a high-resolution numerical solution for the ground truth and consider the $L^2$ error as the number of resamples is increased. Based on the observations of the previous subsections, we undertake these tests for three distinct numbers of collocation points: $N = 500, 1000, 2000$. The same resampling methods as in Sect. 2 are used, using a minimum of 20 repeats per case. The results for $N = 500$ are shown below in Fig. 8. For this case, the adaptive methods are again consistently superior to using fixed collocation points. Similar to the Burger's equation solutions with a large $\nu = 0.01$, there are no clear advantages between $PDE_{xt}$, and the approaches based upon the local residual (though using $U_{xt}$ is not so good).

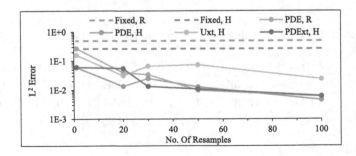

**Fig. 8.** Error against number of resamples solving the Allen-Cahn Equation with $N = 500$

For $N \geq 1000$, a similar pattern is observed to that shown in Fig. 6, where there is little difference between resampling and fixed methods. In this case, the increased number of points again makes the problem simple enough to solve with fixed collocation strategies, so that the benefits previously observed from using adaptive resampling methods disappear. In fact a fixed pseudo-random Hammersley distribution performs better than the other curves: perhaps again benefiting from more L-BFGS optimiser steps.

For the Allen-Cahn Equation case, the behaviour and importance of sampling in these methods varies significantly depending on the number of points used. The less sharp shocks in this solution could also be influencing the effectiveness of using derivatives of the solution as redistribution criteria. Nevertheless, the adaptive methods clearly outperform the fixed distribution of collocation points for the case $N = 500$, again suggesting that results of a prescribed accuracy can be obtained with fewer collocation points when adaptive sampling is permitted.

# 4    Discussion and Conclusion

In this paper, to better explore the problem of collocation point sampling in PINNs we have examined the two benchmark problems of Burgers' Equation and the Allen-Cahn Equation. We have used novel information sources based on derivatives of the local residual and the solution estimate to guide the resampling process, which has been kept consistent with other approaches in literature, and compared the results to other resampling methods and fixed random and pseudo-random distributions.

Overall, we have observed that using adaptive point distribution typically allows more accurate PINNs solutions to be obtained than using fixed point distributions for small numbers of collocation points. Using fewer collocation points is potentially important since the training cost of a PINN grows in proportion to this number (all other factors being equal). Furthermore, we have observed that defining the probability distribution for the collocation point locations based upon the mixed second derivative of the residual with respect to the independent variables typically outperforms, often significantly, using the residual alone, as in [13].

For Burgers' equation with a smooth initial condition that leads to the formation of a single "shock", and a diffusion parameter such that this features is sufficiently sharp, we obtained the smallest errors using this approach (Fig. 3). Equally importantly, we observed that we could reach such errors with significantly fewer collocation points than for any of the other techniques that we compared against (Fig. 4). When considering other initial conditions and other diffusion parameters the relative performance of the adaptive approaches was seen to depend upon the complexity of the specific solution in each case. Typically, there appears to be a value of $N$ above which the use of a uniform (Hammersley) distribution of points is preferred - however this value can be relatively large for the more complex problems (and is relatively small when the solution is smoothly varying everywhere). When considering numbers of points below this critical value of $N$ we invariably find that using the mixed derivative of the residual is the best strategy for defining the PDF for drawing the collocation point locations.

Consideration of a different PDE, the Allen-Cahn equation in this case, demonstrates similar behaviour. Specifically, for a given problem there is a value of $N$ below which the adaptive sampling approach is preferable (and above which a uniform Hammersley distribution suffices). In the particular examples considered here, defining the PDF based upon the mixed derivative of the residual is still reliable and robust, however its improvement relative to using the residual alone is less clear. Future work will extend the exploration of these resampling methods to other PDEs beyond Allen-Cahn, helping assess the generalizability of our findings. As our approach does not explicitly constrain the type of domain we can study, we should also be able to consider domains with higher spatial dimensions and more complex shapes.

In terms of computational cost, the PINNs network we have considered is not competitive with standard discretization methods for solving individual forward problems. This is primarily due to the high computational cost of training with full gradient descent methods like L_BFGS. As for the cost of the resampling strategies we have discussed, these only contribute to a small overhead (estimated at 3% in [13]) to the overall computational cost of PINNs training.

Considering other common problems in ML, the biggest challenge is determining what constitutes an appropriate training regime. We did not observe overfitting to occur, even with fixed sampling methods. However, the choice of optimisation algorithm and training regime was still impactful. The use of faster stochastic gradient descent methods like ADAM helped initialise training but had a limited ceiling to accuracy, hence the need for the more expensive L_BFGS. Whilst we were not able to fully explore the impact of different training regimes within the scope of this paper, efficiency could feasibly be improved by further optimising this via methods such as cosine annealing [14]; or by optimising the architecture employed for solving problems of varying complexity.

In summary, we have shown that a number of different strategies are viable for selecting the location of PINNs collocation points with resampling methods. In particular, we have shown that resampling can be guided not only by the local residuals, but also by the spatial and temporal derivatives of the residuals and even of the solution estimates themselves. Furthermore, we have observed that a number of different factors impact the accuracy of the PINN method, making it difficult to analyse any single aspect (such as collocation point selection) in isolation. Whilst we believe that our contribution begins to shed some light on the interplay between the location of the collocation points and the ability of a PINN to learn the PDE solution it is clear that further research is desirable in order to be able to propose robust implementations that support reliable training or provide accuracy guarantees.

**Acknowledgments.** The first author gratefully acknowledges financial support via the EPSRC Centre for Doctoral Training in Fluid Dynamics at Leeds (EP/S022732/1).

**Disclosure of Interests.** The authors have no competing interests to declare that are relevant to the content of this article.

# References

1. Lagaris, I., Likas, A., Fotiadis, D.: Artificial neural networks for solving ordinary and partial differential equations. IEEE Trans. Neural Netw. **9**(5), 987–1000 (1998)
2. Raissi, M., Perdikaris, P., Karniadakis, G.: Physics-informed neural networks: a deep learning framework for solving forward and inverse problems involving non-linear partial differential equations. J. Comput. Phys. **378**, 686–707 (2019)
3. Wang, Y., Lai, C., Gomez-Serrano, J., Buckmaster, T.: Asymptotic self-similar blow-up profile for three-dimensional axisymmetric Euler equations using neural networks. Phys. Rev. Lett. **130**, 244002 (2023)
4. Li, W., Zhang C., Wang, C., et al.: Revisiting PINNs: a Generative Adversarial Physics-Informed Neural Networks and Point-Weighting Method (2022). arXiv Preprint, arXiv:2205.08754
5. McClenny, L., Braga-Neto, U.: Self-adaptive physics-informed neural networks using a soft attention mechanism. J. Comput. Phys. **474**, 111722 (2022)
6. Zhang, G., Yang, H., Zhu, F., et al.: Dasa-Pinns: Differentiable adversarial self-adaptive pointwise weighting scheme for physicsinformed neural networks, SSRN (2023)
7. Song, Y., Wang, H., Yang, H.: Loss-attentional physics-informed neural networks. J. Comput. Phys. **501**, 112781 (2024)
8. Anagnostopoulos, S., Toscano, J., Karniadakis, G., et al.: Residual-based attention and connection to information bottleneck theory in PINNs (2023). arXiv Preprint, arXiv:2307.00379
9. Lu, L., Meng, X., Karniadakis, G., et al.: DeepXDE: a deep learning library for solving differential equations. SIAM Rev. **63**, 208–228 (2021)
10. Nabian, M., Gladstone, R., Meidani, H.: Efficient training of physics-informed neural networks via importance sampling. Comput.-Aided Civil Infrastruct. Eng. **36**(8), 962–977 (2021)
11. Hanna, J., Aguado, J., Borzacchiello, D., et al.: Residual-based adaptivity for two-phase flow simulation in porous media using physics-informed neural networks. Comput. Methods Appl. Mech. Eng. **396**, 115100 (2022)
12. Gao, Z., Yan, L., Zhou, T.: Failure-informed adaptive sampling for PINNs. SIAM J. Sci. Comput. **45**, A1971–A1994 (2023)
13. Wu, C., Zhu, M., Lu, L., et al.: A comprehensive study of non-adaptive and residual-based adaptive sampling for physics-informed neural networks. Comput. Methods Appl. Mech. Eng. **403**, 115671 (2023)
14. Subramanian, S., Kirby, R., Gholami, A., et al.: Adaptive Self-Supervision Algorithms for Physics-Informed Neural Networks. ECAI 2023, vol. 372 (2023)
15. Lau, G., Hemachandra, A., Low, B., et al.: PINNACLE: PINN Adaptive ColLocation and Experimental points selection. In: The Twelfth International Conference on Learning Representations (2024)
16. Hammersley, J.M., Handscomb, D.C.: Monte Carlo methods, methuen & co. Ltd., London 40 (1964): 32

# Estimating Soil Hydraulic Parameters for Unsaturated Flow Using Physics-Informed Neural Networks

Sai Karthikeya Vemuri$^{(\boxtimes)}$ [iD], Tim Büchner[iD], and Joachim Denzler[iD]

Computer Vision Group, Friedrich Schiller University Jena, 07743 Jena, Germany
`sai.karthikeya.vemuri@uni-jena.de`

**Abstract.** Water movement in soil is essential for weather monitoring, prediction of natural disasters, and agricultural water management. Richardson-Richards' equation (RRE) is the characteristic partial differential equation for studying soil water movement. RRE is a non-linear PDE involving water potential, hydraulic conductivity, and volumetric water content. This equation has underlying non-linear parametric relationships called water retention curves (WRCs) and hydraulic conductivity functions (HCFs). This two-level non-linearity makes the problem of unsaturated water flow of soils challenging to solve. Physics-Informed Neural Networks (PINNs) offer a powerful paradigm to combine physics in data-driven techniques. From noisy or sparse observations of one variable (water potential), we use PINNs to learn the complete system, estimate the parameters of the underlying model, and further facilitate the prediction of infiltration and discharge. We employ training on RRE, WRC, HCF, and measured values to resolve two-level non-linearity directly instead of explicitly deriving water potential or volumetric water content-based formulations. The parameters to be estimated are made trainable with initialized values. We take water potential data from simulations and use this data to solve the inverse problem with PINN and compare estimated parameters, volumetric water content, and hydraulic conductivity with actual values. We chose different types of parametric relationships and wetting conditions to show the approach's effectiveness.

**Keywords:** Physics-Informed Neural Networks · Richardson-Richards equation · Computational Hydrology

## 1 Introduction

Understanding the movement of water in near-surface levels of soil is of utmost importance for agricultural water management, prediction of floods, and microbial activities inside the soil. This understanding is heavily dependent on our capacity to model these hydrological processes. A partial differential equation called *Richardson-Richards equation* [28] (RRE) is at the heart of modeling water movement in soils. This non-linear partial differential equation (PDE) represents

© The Author(s), under exclusive license to Springer Nature Switzerland AG 2024
L. Franco et al. (Eds.): ICCS 2024, LNCS 14834, pp. 338–351, 2024.
https://doi.org/10.1007/978-3-031-63759-9_37

water movement in porous media. It involves the relationship between three quantities, namely pressure head or water potential ($\psi$), volumetric water content ($\theta$), and hydraulic conductivity ($K$) [7]. To make the RRE solvable, two parameterized relations called water retention curve (WRC) and hydraulic conductivity function (HCF) are used. The parameters used in these relationships characterize the hydraulic properties of soil. This double non-linearity makes solving RRE difficult and is of utmost interest to mathematicians and physicists. A detailed review of the history and significance of RRE in computational sciences is provided in the review by [7]. They referred to the equation as arguably one of the most difficult equations to reliably and accurately solve in all hydro-sciences.

Machine learning and deep learning are used in many applications to model and simulate physical processes. A new and exciting paradigm for using deep learning for physics involves solving problems related to differential equations, providing a robust framework to integrate physical knowledge into data-driven deep learning systems. Physics-Informed Neural Networks (PINNs) were introduced in the seminal works of Raissi et al. [24,25], showing the use of deep learning to solve forward and inverse problems involving differential equations.

Applying PINNs to solve problems involving RRE effectively has greatly interested the scientific community. Tartakovsky et al. [32] used PINNs and estimated saturated and unsaturated hydraulic conductivity and pressure from saturated hydraulic conductivity measurements. Bandai et al. [3] introduced a PINN-based framework for the inverse solution of the Richards equation to estimate both the water retention curve (WRC) and the hydraulic conductivity function (HCF). This involved training three neural networks, with one approximating the solution, the Richards PDE, regarding the water content and the remaining two describing, respectively, the WRC and HCF. Depina et al. [13] addressed the inverse problem by deriving individual formulations for potential-based and volumetric water content-based extensions of the classic mixed form of RRE, while Chen et al. [5] investigated the impacts of the loss weights and random state on the performance of the RRE-solving PINNs and possible solutions to mitigate such impacts.

We aim to complement and build upon this body of work in this study. We use PINNs to solve the inverse problem of one-dimensional RRE. Specifically, we use data on water potential ($\psi$) to estimate the remaining two variables, $\theta$ and K, thereby learning the complete system. This is done by inferring the parameters in WRC and HCF, which characterize the hydraulic properties of soil. We do this for different types of parametric relationships: Gardner [9] and van-Genuchten [10]. Depina et al. [13] use different formulations to solve the inverse problem using data from other variables. In contrast, our approach uses a single general mixed formulation that can handle water potential and volumetric water content measurements without rewriting the RRE. Additionally, Bandai et al. [3] use three separate neural networks to capture the RRE, WRCs, and HCFs, while our single neural network finds a representation by simultaneously resolving all of them by training on a multi-objective loss. This is illustrated in the Fig. 1.

## 2   Theoretical Background

This section offers an overview of the Richardson-Richards equation (RRE) with Physics-Informed Neural Networks (PINNs) [24,25,28], combining computational hydrology and machine learning. The RRE is fundamental for modeling water movement in unsaturated soils and is essential for understanding soil water dynamics. PINNs integrate deep learning to solve complex differential equations by enforcing physical laws. We provide a concise overview highlighting their combination and synergy to alleviate current challenges.

### 2.1   The Richardson-Richards Equation

Richardson-Richards equation [28] is used extensively for modeling water movement in unsaturated porous media. The equation applies when assuming isothermal water transport (heat exchange is negligible) and when soil hydraulic properties are homogeneous and isotropic, indicating uniform behavior across the soil. The general mixed formulation of the equation is given as

$$\frac{\partial \theta(\psi)}{\partial t} = \frac{\partial}{\partial z}\left[K(\theta)\left(\frac{\partial \psi}{\partial z} + 1\right)\right], \tag{1}$$

where $\theta[L^3 L^{-3}]$ is the volumetric water content, $t[T]$ is the time, $z[L]$ is the depth, $K[LT^{-1}]$ is the hydraulic conductivity and $\psi[L]$ is the water potential (hydraulic pressure). The measurement units of all the quantities are given in square brackets.

Since the number of unknowns is greater than the number of equations in Eq. 1, additional equations are required to solve the equation completely. These two additional equations are $K(\theta)$ and $\theta(\psi)$. The function $\theta(\psi)$ is called the water retention curve (WRC), and $K(\theta)$ is called the hydraulic conductivity function (HCF). These non-linear parametric relationships are used to solve the Eq. 1. They are of different types. Some of the most used ones in the community are relations given by Gardner [9], given by Equation 2 and van-Genuchten [10], given by Equation 3

$$\theta = \theta_r + (\theta_s - \theta_r)\,e^{\alpha\psi},$$
$$K = K_s e^{\alpha\psi}, \tag{2}$$

$$\theta = \theta_r + \frac{\theta_s - \theta_r}{(1 + (-\alpha\psi)^n)^m},$$
$$K = K_s S_e^l \left(1 - \left(1 - S_e^{1/m}\right)^m\right)^2. \tag{3}$$

The parameters given in Equation 2 and Equation 3 characterize the hydraulic properties of soil. The parameter $\theta_r[L^3 L^{-3}]$ is the residual water content, $\theta_s[L^3 L^{-3}]$ is the saturated water content, $\alpha[L^{-1}]$ is the pore-size distribution parameter and $K_s[LT^{-1}]$ is the saturated hydraulic conductivity. There are additional quantities in van-Genuchten relationships. $l$ is the tortuosity parameter, and $n, m$ are shape parameters. The intermediate quantity $S_e$ in Equation 3 is the effective saturation given by

$$S_e = \frac{\theta - \theta_r}{\theta_s - \theta_r}. \tag{4}$$

A more detailed review of the numerical solutions and the significance of RRE is provided in [7]. This study focuses on the three parameters common in both relationships: $\alpha$, $K_s$, $\theta_s$.

## 2.2   Physics-Informed Neural Networks

Physics-Informed Neural Networks (PINNs) provide a framework to include rules of physics as differential equations in neural networks. PINNs leverage the fact that deep neural networks are universal functional approximations. Backpropagation enables the calculation of derivatives of any order between the input and output of deep neural networks. Raissi et al. [25] showed that loss function can be constructed using combinations of higher order derivatives, enabling the neural network to learn from physics. To illustrate the concept, consider a general differential equation of the form

$$u_t + \mathcal{D}[u] = 0, t \in [0, T], \boldsymbol{x} \in \Omega, \tag{5}$$

where $u(\boldsymbol{x}, t)$ is the solution of the differential equation in the domain $[\Omega XT]$. A neural network, represented by $u_w$, with trainable parameters (weights and biases) $\boldsymbol{w}$ approximates the latent, hidden solution $u$. The loss function to penalize the behaviour of $u_w$ according to Eq. 5 is given as

$$\mathcal{L}(w) = \frac{1}{N} \sum_{i=1}^{N} \left| \frac{\partial u_w}{\partial t}(t^i, \boldsymbol{x}^i) + \mathcal{D}[u_w](t^i, \boldsymbol{x}^i) \right|^2, \tag{6}$$

where $N$ is the total number of sampled points in the domain $[\Omega XT]$. When trained with this loss function, the neural network is penalized on all these points such that the parameters $\theta$ are optimized to learn the behavior according to Eq. 5. This framework of informing physics through loss function can be used for various applications where there is a need to enforce physics-based constraints. It can solve PDEs by adding additional loss functions for boundary and initial conditions. It can solve inverse problems using data samples and the physics constraint where parameters are trainable. This flexibility has made PINNs useful in various scientific and engineering disciplines [6,22,26,30]. We recommend referring to Raissi et al. [25] and Karniadakis et al. [15] for a detailed overview of PINNs.

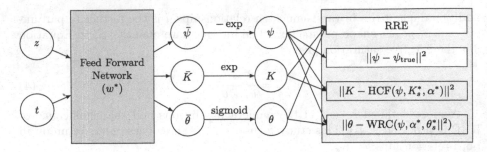

**Fig. 1.** Our proposed Physics-Informed Neural Network is designed to solve the Richardson-Richards equation, utilizing depth ($z$) and time ($t$) as input and giving $\psi, K, \theta$ as output. Specifically, the model can be trained using the observations of $\psi$, with the self-referential terms (highlighted in green). All learnable parameters are denoted by $^*$. Please note that we additionally integrate knowledge about each output's value range to enforce reasonable results. (Color figure online)

## 2.3 Inverse Problem of RRE

An inverse problem focuses on determining the unknown parameters and variables that govern a system based on already-known observations. This process entails using available data to work backward to identify the underlying factors or conditions that could have resulted in the observed outcomes. The system to solve Richard-Richardson-Equation consists of the three variables $\psi$, $\theta$, and $K$. As described before, we solely observe $\psi$ and have to uncover the remaining variables. However, knowing that $\theta$ and $K$ can be described by hydraulic quantities of soil (either based on Gardner [9] or Genuchten [10]), we can introduce learnable parameters $\alpha^*$, $K_s^*$, and $\theta_s^*$ to learn them on-the-fly. Given this setup, we could estimate the hydraulic behavior and type of soil given only the water potential $\psi$ observations.

We construct our Physics-Informed Neural Network as a standard feed-forward network. The water potential values $\psi$ are recorded at a specific time ($t$) and depth ($z$); hence, we utilize them as input for training the PINN. Our model predicts the estimated water potential $\psi$, hydraulic conductivity $K$, and the volumetric water content $\theta$. This model is trained on RRE, as given in Eq. 1, along with samples of $\psi$. This model is also trained on the WRCs and HCFs to enforce the representation of the complete system. These loss terms are self-referential, i.e., one output of the neural network is used to compare to another. For example, As shown in Fig. 1 (highlighted in green), in the third loss term, the output of the neural network $K$ is compared to HCF, a function of $\psi$, also an output of the neural network. The same applies to the fourth loss term where $\theta$ and WRC are compared. Due to this, we found the model is prone to exploding gradients.

We leverage existent physical knowledge of the $\psi$, $K$, and $\theta$ to alleviate this effect during the model's training. We follow current best practices [3,19] and restrict the possible value ranges for the model's output neurons through specific last-layer activation functions. We enforce $\psi$ as always negative using a negative exponential function attached to the output neuron [3]. For $K$, we use an exponential function to ensure non-negative values. Lastly, knowing that $\theta$ can only be in $[0, 1]$, we utilize a *sigmoid* function to constrain the value range. Therefore, the initial model outputs, denoted as $\bar{\psi}$, $\bar{K}$, and $\bar{\theta}$, represent intermediate representations of the actual desired parameters. As shown by Bandai et al. [3], these stabilize the training significantly and enable the model to learn the complete system. As both *exponential* and *sigmoid* functions are computationally stable using backward passes, the gradients used for the optimization terms do not explode. The full self-regularization and knowledge integration approach is visualized in Fig. 1.

Overall, the model minimizes its parameters $w^*$ and the hydraulic soil parameters $\alpha^*$, $K_s^*$, and $\theta_s^*$ by solving the following optimization problem:

$$\min_{w^*, \alpha^*, \theta_s^*, K_s^*} \|\mathrm{RRE}(\psi, \theta, K))\|^2 + \|\psi - \psi_{true}\|^2 +$$
$$\|K - \mathrm{HCF}(\psi, K_s^*, \alpha^*))\|^2 + \|\theta - \mathrm{WRC}(\psi, \alpha^*, \theta_s^*))\|^2 . \quad (7)$$

The model has to find a viable solution from the solution space defined by the following summarized objectives to predict the required parameters effectively:

1. The main target is solving the mixed formulation of the RRE based on Eq. 1, such that valid solutions arise.
2. Further, minimizing the squared error between the predicted $\psi$ and the measured values $\psi_{\text{true}}$ for a specific time $(t)$ and depth $(z)$ value adds to the total loss term.
3. The parametric relationships, WRC and HCF, based on either Gardner [9] or van-Genuchten [10], are used as loss terms. During the learning, the values of $\alpha^*, \theta_s^*$, and $K_s^*$ have to be updated to fit a suitable solution space. The initial values for these parameters are discussed in the experimental section.
4. Lastly, we avoid loss term weighting as in [20,33,35] to not interfere with output transforms used for value range restrictions.

## 3   Experiments

We first introduce our general experimental setup to validate our proposed model architecture and self-regularization approach to ensure comparable results. In our main experiments[1], we apply our approach to different types of parametric relationships for soil hydraulic properties.

### 3.1   Experimental Setup

To test the effectiveness of our proposed approach, we build our experiments on simulations of the Richard-Richardson-Equation [28]. We fix hydraulic quantities of soil parameters $(\alpha, \theta_s, K_s)$ and simulate RRE for certain depths until a

---

[1] The code is available at https://github.com/cvjena/InverseRRE.

certain time. We follow Ireson [12] to ensure correct simulations by setting an appropriate boundary and initial conditions. Accordingly, this is done for Gardner [9] and van-Genuchten [10]. The training data consists of 5000 samples of $\psi$ from the simulations,

We evaluate the capabilities of our model, thus our approach, based on the following criteria:

1. The complete map of the water potential $\psi$ is correctly estimated.
2. The learnable parameters describing the soil hydraulic properties $(\alpha^*, \theta_s^*, K_s^*)$ converge towards the selected true parameters $(\alpha, \theta_s, K_s)$.
3. The PINN output values of $\theta$ and $K$ in the domain are close to that of the numerical simulations.
4. We investigate the qualitative results by comparing the predicted values and simulation over depth $(z)$ and time $(t)$.

We use the same network architecture and learning hyper-parameters in all experiments shown. As we do not weigh the individual loss terms, we can refrain from additional hyper-parameter searches and focus solely on the general capabilities of our method. All models optimize the same optimization term in Eq. 7.

Our feed-forward PINN is constructed and trained with the DeepXDE library [18]. We utilize eight hidden layers, each having 50 neurons followed by $SiLU$ activations, as prescribed in [2], Therefore, we require fewer parameters and only a single network compared to Bandai et al. [3] to solve the Richards-Richardson-Equation. All models are trained with Adam as optimizer [16] over 40000 epochs with a constant learning rate of $1e^{-4}$. All experiments are repeated for ten independent runs, and single training takes around one hour on an Nvidia GeForce GTX 1080 Ti.

## 3.2   Gardner Simulation

First, we solve the inverse problem of RRE with parametric relationships given by Gardner as given in the Equation 2. For this purpose, we use the simulation code by Bandai et al. [3], which is based on analytical solutions given by Srivastava et al. [31]. We simulate for depth $z = 10\,\text{cm}$ and time $t = 10\,\text{h}$, and the parameters

**Table 1.** The table compares original parameters and estimated parameters by PINN for RRE using Gardner's parametric relationships [9]. Additionally, we display the mean prediction errors between predicted variables by PINN with original simulated values. The given values are over ten independent trials.

| Parameter | Original | Prediction | RRE Parameter | Prediction Error |
|---:|---:|---|---:|---|
| $\alpha$ | 1.00 | $1.01 \pm 0.07$ | $\psi$ | $0.0361 \pm 0.0075$ |
| $\theta_s$ | 0.40 | $0.42 \pm 0.15$ | $K$ | $0.0367 \pm 0.0138$ |
| $K_s$ | 1.00 | $1.02 \pm 0.18$ | $\theta$ | $0.0287 \pm 0.0258$ |

|                                  |                            |
|:--------------------------------:|:--------------------------:|
| (a) Hydraulic Soil Parameters    | (b) Simulation Data        |

are chosen to be $\alpha = 1.0$, $\theta_s = 0.4$, $K_s = 1.0$. The initial condition $\psi(0, t)$ is taken as parabolic. The simulation results of all variables are shown in Fig. 2 (middle column). The details about architecture and training are described in Subsect. 2.3.

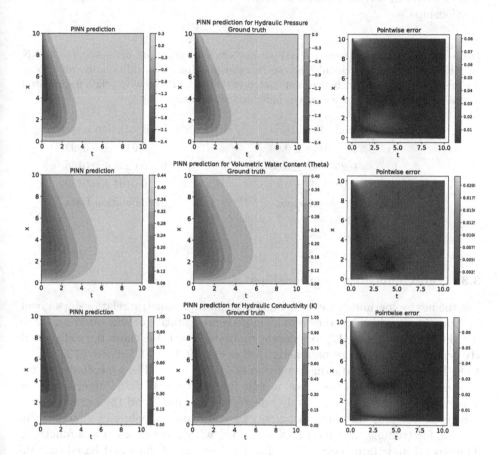

**Fig. 2.** Prediction of $\psi$, $\theta$, and K using PINN for the inverse problem of RRE using Gardner relationships. The first column shows the PINN predictions, and the second column shows ground truth data from simulations [3]. The predictions are close to the ground truth, and the corresponding pointwise errors are shown in the third column.

We train the PINN using the Adam optimizer for 40000 epochs with a learning rate $1e^{-4}$. In the first row of Fig. 2, we show the predicted values of $\psi$ as the first step to ensure PINN has learned from the data correctly. The $\theta$ and

K predictions are shown in the second and third rows of the Fig. 2. These figures show that the predicted values of PINN are accurate with reference solutions from simulations. Further, the estimated parameters are also close to the original values from the simulation, as shown in the Table 1. This shows the effectiveness of PINN in solving the inverse problem of RRE with Gardner's parametric relationships.

**Table 2.** The table compares original parameters and estimated parameters by PINN for RRE using van-Genuchten's parametric relationships [31]. Additionally, we display the mean prediction errors between predicted variables by PINN with original simulated values. The given values are over ten independent trials.

| Parameter | Original | Prediction |
|---|---|---|
| $\alpha$ | 0.79 | $0.78 \pm 0.06$ |
| $\theta_s$ | 0.25 | $0.28 \pm 0.08$ |
| $K_s$ | 1.08 | $1.07 \pm 0.13$ |

(a) Soil hydraulic parameters

| RRE variable | Prediction error |
|---|---|
| $\psi$ | $0.0198 \pm 0.0093$ |
| $K$ | $0.0238 \pm 0.0146$ |
| $\theta$ | $0.0104 \pm 0.0139$ |

(b) Simulation Data

### 3.3   Van-Genuchten Relationships

For the next experiment, we used van-Genuchten parametric relationships given by Equation 3, which are known to be more representative of real-world soils. We use the work by Ireson [12] to generate simulated data. We chose the soil type of Hygiene sandstone whose parameters are $\alpha = 0.79$, $\theta_s = 0.25$, $\theta_r = 0153$, $K_s = 1.08$. The simulation's starting condition is $\psi(0, z) = -z$. The generated data for $\psi$ is shown in the Fig. 3 (first row, middle figure). One thousand observations of $\psi$ across the domain are randomly selected for training, and the basic training procedure remains the same as before; we use sampled points of $\psi$ across the domain for training, and the PINN is trained with three types of loss functions. The second and third terms of loss function are now formulated based on van-Genuchten parametric relationships. We set the value of 0.1 as the starting value of trainable parameters. The PINN is trained and predicted volumetric water content, $\theta$ and hydraulic conductivity, K values are shown in the second and third rows of the Fig. 3 and the estimated parameters in the Table 2. We see that the predictions are accurate with the ground truth, showing the effectiveness of PINN in solving the inverse problem of RRE with van-Genuchten parametric relationships.

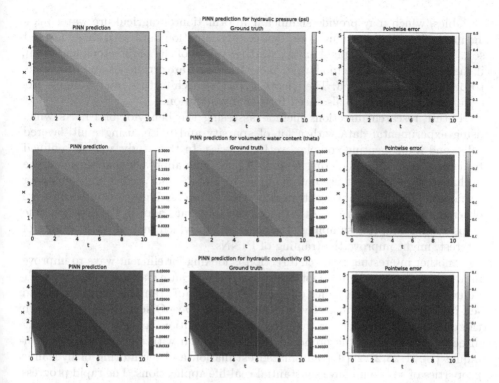

**Fig. 3.** Prediction of $\psi$, $\theta$, and K using PINN for the inverse problem of RRE using van-Genuchten relationships. The first column shows the PINN predictions, and the second column shows ground truth data from simulations [12]. The predictions are close to the ground truth, and the corresponding pointwise errors are shown in the third column.

## 4   Discussion and Conclusion

In this paper, we presented an interesting application of Physics-Informed neural networks to solve the inverse problem of the Richardson-Richards equation. We showed how it's useful for estimating the parameters ($\alpha$, $K_s$, $\theta_s$) and variables (Volumetric water content $\theta$ and hydraulic conductivity $K$) of the complete system from observations of a single variable (hydraulic potential $\psi$) across space-time domain. This is achieved by a multi-objective loss containing physics-based (RRE, parametric relationships) and data-based ($\psi$ values) objectives highlighting the powerful characteristic of PINNs to combine physical knowledge into deep learning systems. We further used simple transformations for outputs to restrict their ranges, which helped in stable training and robust estimation of parameters. We show that PINNS accurately solve the inverse problem for different parametric relationships, namely, Gardner and van-Genuchten relationships.

This inverse problem setting could be useful for knowing the hydraulic properties of some unknown soil. This can also be used to regularly monitor hydraulic variables, which may provide significant information on agriculture water management, flooding, etc. This work focuses on simulation data with single-layered soil. The estimations depend on the starting values of the parameters, and often, if the starting values are too far off, the estimations are erroneous and this needs to be further investigated. The effects of various architectures and training procedures on performance also need further investigation.

Future research directions include assessing the behavior of PINNs when using experimental data with a lot of sparsity and noise, using multi-layered soils, and implementing complex wetting cycles. In this study, we are confined to using $\psi$ data and predicting $K$ and $\theta$. This same approach is applicable if we exchange variables, i.e., use data samples of $\theta$ or $K$ and predict the remaining variables. The data sampling strategy is also an interesting question because we observe, for example, in Fig. 2 that most variables vary a lot in the first few time steps. Using adaptive sampling strategies [8,34], which include more points of interest, might improve the training of PINNs.

Another interesting research direction is looking for efficient ways to improve multi-objective optimization for inverse settings because classical weighting techniques [20,33,35] might suffer due to the presence of trainable parameters and self-referential loss terms. New algorithms like DeepONets [17], which learn the operator of PDE rather than the singular solution, are also an interesting direction. Combining the proposed PINN approach with Remote sensing techniques [14,21,23,27,29] for automated estimation and monitoring of hydraulic properties of soil could have substantial real-life applications. The rapid progress in modern-day deep learning [1,4,11] ensures a unique and exciting future for studying PDEs using deep learning.

**Acknowledgements.** This research is funded by the Federal Ministry for Economic Affairs and Climate Actions (BMWK) for the project titled LuFo VI-2 Melodi 20E2115B. We also thank Bandai et al. [3] and Ireson et al. [12] for their open-source simulation codes.

**Disclosure of Interest.** The authors declare no conflicts of interest.

# References

1. Ahmad, W., Shadaydeh, M., Denzler, J.: Deep learning-based group causal inference in multivariate time-series. In: AAAI Workshop on AI for Time-series (2024). https://arxiv.org/abs/2401.08386, (accepted)
2. Al Safwan, A., Song, C., Waheed, U.: Is it time to swish? Comparing activation functions in solving the helmholtz equation using PINNs **2021**(1), 1–5 (2021). https://doi.org/10.3997/2214-4609.202113254, https://www.earthdoc.org/content/papers/10.3997/2214-4609.202113254
3. Bandai, T., Ghezzehei, T.A.: Forward and inverse modeling of water flow in unsaturated soils with discontinuous hydraulic conductivities using physics-informed neural networks with domain decomposition. Hydrol. Earth Syst. Sci. **26**(16), 4469–4495 (2022). https://doi.org/10.5194/hess-26-4469-2022, https://hess.copernicus.org/articles/26/4469/2022/

4. Büchner, T., Guntinas-Lichius, O., Denzler, J.: Improved obstructed facial feature reconstruction for emotion recognition with minimal change CycleGANs. In: Blanc-Talon, J., Delmas, P., Philips, W., Scheunders, P. (eds.) Advanced Concepts for Intelligent Vision Systems. ACIVS 2023. LNCS, vol. 14124. Springer, Cham (2023). https://doi.org/10.1007/978-3-031-45382-3_22

5. Chen, Y., Xu, Y., Wang, L., Li, T.: Modeling water flow in unsaturated soils through physics-informed neural network with principled loss function. Comput. Geotech. **161**, 105546 (2023). https://doi.org/10.1016/j.compgeo.2023.105546, https://www.sciencedirect.com/science/article/pii/S0266352X23003038

6. Eivazi, H., Tahani, M., Schlatter, P., Vinuesa, R.: Physics-informed neural networks for solving reynolds-averaged navier-stokes equations. Phys. Fluids **34**, 075117 (2022). https://doi.org/10.1063/5.0095270

7. Farthing, M.W., Ogden, F.L.: Numerical solution of Richards' equation: a review of advances and challenges. Soil Sci. Soc. Am. J. **81**(6), 1257–1269 (2017). https://doi.org/10.2136/sssaj2017.02.0058, https://acsess.onlinelibrary.wiley.com/doi/abs/10.2136/sssaj2017.02.0058

8. Gao, Z., Yan, L., Zhou, T.: Failure-informed adaptive sampling for PINNs. SIAM J. Sci. Comput. **45**(4), A1971–A1994 (2023). https://doi.org/10.1137/22M1527763

9. Gardner, G.H.F.: Formation velocity and DENSITY—the diagnostic basics for stratigraphic traps. Geophysics **39**(6), 770 (1974). https://doi.org/10.1190/1.1440465

10. van Genuchten, M.T.: A closed-form equation for predicting the hydraulic conductivity of unsaturated soils. Soil Sci. Soc. Am. J. **44**(5), 892–898 (1980). https://doi.org/10.2136/sssaj1980.03615995004400050002x, https://acsess.onlinelibrary.wiley.com/doi/abs/10.2136/sssaj1980.03615995004400050002x

11. Huang, Y., et al.: Enhanced stability of grassland soil temperature by plant diversity. Nat. Geosci. **17**, 44–50 (2023). https://doi.org/10.1038/s41561-023-01338-5

12. Ireson, A.M., Spiteri, R.J., Clark, M.P., Mathias, S.A.: A simple, efficient, mass-conservative approach to solving Richards' equation (openre, v1. 0). Geosci. Model Dev. **16**(2), 659–677 (2023)

13. Depina, I., Saket Jain, S.M.V., Gotovac, H.: Application of physics-informed neural networks to inverse problems in unsaturated groundwater flow. Georisk: Assess. Manag. Risk Eng. Syst. Geohazards **16**(1), 21–36 (2022). https://doi.org/10.1080/17499518.2021.1971251

14. Jonard, F., Weihermüller, L., Schwank, M., Jadoon, K.Z., Vereecken, H., Lambot, S.: Estimation of hydraulic properties of a sandy soil using ground-based active and passive microwave remote sensing. IEEE Trans. Geosci. Remote Sens. **53**(6), 3095–3109 (2015)

15. Karniadakis, G.E., Kevrekidis, I.G., Lu, L., Perdikaris, P., Wang, S., Yang, L.: Physics-informed machine learning. Nat. Rev. Phys. **3**(6), 422–440 (2021). https://doi.org/10.1038/s42254-021-00314-5

16. Kingma, D., Ba, J.: Adam: A method for stochastic optimization. In: International Conference on Learning Representations (ICLR). San Diega, CA, USA (2015)

17. Lu, L., Jin, P., Pang, G., Zhang, Z., Karniadakis, G.E.: Learning nonlinear operators via DeepONet based on the universal approximation theorem of operators. Nat. Mach. Intell. **3**(3), 218–229 (2021). https://doi.org/10.1038/s42256-021-00302-5

18. Lu, L., Meng, X., Mao, Z., Karniadakis, G.E.: DeepXDE: a deep learning library for solving differential equations. SIAM Rev. **63**(1), 208–228 (2021). https://doi.org/10.1137/19M1274067

19. Lu, L., Pestourie, R., Yao, W., Wang, Z., Verdugo, F., Johnson, S.G.: Physics-informed neural networks with hard constraints for inverse design. SIAM J. Sci. Comput. **43**(6), B1105–B1132 (2021). https://doi.org/10.1137/21M1397908

20. Maddu, S., Sturm, D., Müller, C.L., Sbalzarini, I.F.: Inverse Dirichlet weighting enables reliable training of physics informed neural networks. Mach. Learn. Sci. Technol. **3**(1), 015026 (2022). https://doi.org/10.1088/2632-2153/ac3712

21. Mohanty, B.P.: Soil hydraulic property estimation using remote sensing: a review. Vadose Zone J. **12**(4), 1–9 (2013)

22. Pandey, S., Schumacher, J., Sreenivasan, K.R.: A perspective on machine learning in turbulent flows. J. Turbul. **21**, 567–584 (2020). https://doi.org/10.1080/14685248.2020.1757685

23. Pauwels, V.R., Balenzano, A., Satalino, G., Skriver, H., Verhoest, N.E., Mattia, F.: Optimization of soil hydraulic model parameters using synthetic aperture radar data: an integrated multidisciplinary approach. IEEE Trans. Geosci. Remote Sens. **47**(2), 455–467 (2009)

24. Raissi, M., Perdikaris, P., Karniadakis, G.E.: Physics-informed neural networks: a deep learning framework for solving forward and inverse problems involving nonlinear partial differential equations. J. Comput. Phys. **378**, 686–707 (2019). https://doi.org/10.1016/j.jcp.2018.10.045

25. Raissi, M., Perdikaris, P., Karniadakis, G.E.: Physics informed deep learning (part II): Data-driven discovery of nonlinear partial differential equations (2017). http://arxiv.org/abs/1711.10566

26. Rasht-Behesht, M., Huber, C., Shukla, K., Karniadakis, G.E.: Physics-informed neural networks (PINNs) for wave propagation and full waveform inversions. J. Geophys. Res. Solid Earth **127**(5), e2021JB023120 (2022). https://doi.org/10.1029/2021JB023120, https://agupubs.onlinelibrary.wiley.com/doi/abs/10.1029/2021JB023120

27. Rezaei, M., et al.: Incorporating machine learning models and remote sensing to assess the spatial distribution of saturated hydraulic conductivity in a light-textured soil. Comput. Electron. Agric. **209**, 107821 (2023). https://doi.org/10.1016/j.compag.2023.107821, https://www.sciencedirect.com/science/article/pii/S0168169923002090

28. Richards, L.A.: Capillary conduction of liquids through porous mediums. Physics **1**(5), 318–333 (2004). https://doi.org/10.1063/1.1745010

29. Santanello, J.A., Jr., et al.: Using remotely-sensed estimates of soil moisture to infer soil texture and hydraulic properties across a semi-arid watershed. Remote Sens. Environ. **110**(1), 79–97 (2007)

30. Song, C., Alkhalifah, T., Waheed, U.: Solving the frequency-domain acoustic VTI wave equation using physics-informed neural networks. Geophys. J. Int. **225**(2), 846–859 (2021). https://doi.org/10.1093/gji/ggab010, publisher Copyright: 2021 The Author(s) 2021. Published by Oxford University Press on behalf of The Royal Astronomical Society

31. Srivastava, R., Yeh, T.C.J.: Analytical solutions for one-dimensional, transient infiltration toward the water table in homogeneous and layered soils. Water Resour. Res. **27**(5), 753–762 (1991). https://doi.org/10.1029/90WR02772, https://agupubs.onlinelibrary.wiley.com/doi/abs/10.1029/90WR02772

32. Tartakovsky, A.M., Marrero, C.O., Perdikaris, P., Tartakovsky, G.D., Barajas-Solano, D.: Physics-informed deep neural networks for learning parameters and constitutive relationships in subsurface flow problems. Water Resour. Res. **56**(5), e2019WR026731 (2020). https://doi.org/10.1029/2019WR026731, https://agupubs.onlinelibrary.wiley.com/doi/abs/10.1029/2019WR026731, e2019WR026731 10.1029/2019WR026731

33. Vemuri, S.K., Denzler, J.: Gradient statistics-based multi-objective optimization in physics-informed neural networks. Sensors **23**(21), 8665 (2023). https://doi.org/10.3390/s23218665, https://www.mdpi.com/1424-8220/23/21/8665

34. Wu, C., Zhu, M., Tan, Q., Kartha, Y., Lu, L.: A comprehensive study of non-adaptive and residual-based adaptive sampling for physics-informed neural networks. Comput. Methods Appl. Mech. Eng. **403**, 115671 (2023). https://doi.org/10.1016/j.cma.2022.115671, https://www.sciencedirect.com/science/article/pii/S0045782522006260

35. Wu, W., Daneker, M., Jolley, M., Turner, K., Lu, L.: Effective data sampling strategies and boundary condition constraints of physics-informed neural networks for identifying material properties in solid mechanics. Appl. Math. Mech. **44**, 1039–1068 (2023). https://doi.org/10.1007/s10483-023-2995-8

# Accelerating Training of Physics Informed Neural Network for 1D PDEs with Hierarchical Matrices

Mateusz Dobija[1,3] , Anna Paszyńska[1,2(✉)] , Carlos Uriarte[4] ,
and Maciej Paszyński[2]

[1] Jagiellonian University, Krakow, Poland
mateusz.dobija@doctoral.uj.edu.pl
[2] AGH University of Krakow, Kraków, Poland
anna.paszynska@uj.edu.pl, maciej.paszynski@agh.edu.pl
[3] Doctoral School of Exact and Natural Sciences, Jagiellonian University,
Kraków, Poland
[4] Basque Center for Applied Mathematics, Bilbao, Spain
curiarte@bcamath.org

**Abstract.** In this paper, we consider a training of Physics Informed Neural Networks with fully connected neural networks for approximation of solutions of one-dimensional advection-diffusion problem. In this context, the neural network is interpreted as a non-linear function of one spatial variable, approximating the solution scalar field, namely $y = PINN(x) = A_n\sigma(A_{n-1}...A_2\sigma(A_1 + b1) + b2) + ... + b_{n-1}) + b_n$. In the standard PINN approach, the $A_i$ denotes dense matrices, $b_i$ denotes bias vectors, and $\sigma$ is the non-linear activation function (sigmoid in our case). In our paper, we consider a case when $A_i$ are hierarchical matrices $A_i = \mathcal{H}_i$. We assume a structure of our hierarchical matrices approximating the structure of finite difference matrices employed to solve analogous PDEs. In this sense, we propose a hierarchical neural network for training and approximation of PDEs using the PINN method. We verify our method on the example of a one-dimensional advection-diffusion problem.

**Keywords:** Partial Differential Equations · Physics Informed Neural Networks · Hierarchical Matrices

## 1 Introduction

Physics Informed Neural Networks (PINN) was introduced in 2019 by George Karniadakis [17]. PINNs have several applications from fluid mechanics [2,12,15, 20,21], wave propagation [7,14,18], phase-filed modeling [8], biomechanics [1,11], and inverse problems [5,13,16]. In this paper, we focus on a one-dimensional advection-diffusion problem. Its extension to a two-dimensional problem, known as the Eriksson-Johnson model problem [6] can be a subject of our future work.

L. Franco et al. (Eds.): ICCS 2024, LNCS 14834, pp. 352–362, 2024.
https://doi.org/10.1007/978-3-031-63759-9_38

Both one and two-dimensional model problems are often employed for testing the convergence of finite element method solvers [3,4]. Following the idea of PINN, we represent the solution of a one-dimensional advection-diffusion problem as the neural network. In order to speed up the training process, the neural network layers are represented by hierarchically compressed matrices.

## 1.1  One-Dimensional Advection-Diffusion Problem

One-dimensional advection-diffusion problem can be defined as:
Find $u \in C^2(0,1)$:

$$\underbrace{-\epsilon \frac{d^2 u(x)}{dx^2}}_{\text{diffusion}=\epsilon} \underbrace{+\beta \frac{du(x)}{dx}}_{\text{advection "wind"}=1} = 0, x \in (0,1). \tag{1}$$

The problem is augmented with boundary conditions

$$-\epsilon \frac{du}{dx}(0) + u(0) = 1.0, \ u(1) = 0. \tag{2}$$

The solution of this problem has the boundary layer of thickness $\epsilon$ at the right corner of the domain, as it is illustrated in Fig. 1.

## 1.2  PINN for One-Dimensional Advection-Diffusion Problem

Following the idea of PINN, we represent the solution as the neural network:

$$u(x) = PINN(x) = A_n \sigma \left( A_{n-1} \sigma(\ldots \sigma(A_1 x + B_1) \ldots + B_{n-1}) + B_n \right. \tag{3}$$

We define the loss function for the residual of the PDE

$$LOSS_{PDE}(x) = \left( -\epsilon \frac{d^2 PINN(x)}{dx^2} - \beta \frac{dPINN(x)}{dx} - 1 \right)^2 \tag{4}$$

We also define the loss function for the left boundary condition

$$LOSS_{BC0} = \left( -\epsilon \frac{dPINN}{dx}(0) + PINN(0) - 1.0 \right)^2, \tag{5}$$

and the loss function for the right boundary condition

$$LOSS_{BC1} = (PINN(1))^2, \tag{6}$$

The total loss function is defined by combining a weighted sum

$$LOSS = w_{PDE} \sum_{x \in (0,1)} (LOSS_{PDE}(x))^2 \tag{7}$$

$$+ w_{BC0} (LOSS_{BC0}(0))^2 \tag{8}$$

$$+ w_{BC1} (LOSS_{BC1}(1))^2. \tag{9}$$

The PINN methods can succesfully solve the advection-diffusion problem as it is shown in [19].

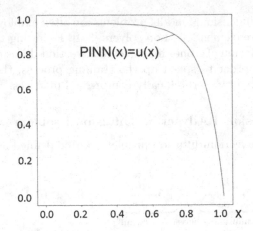

**Fig. 1.** PINN solution of the advection-diffusion problem for $\epsilon = 0.1$.

### 1.3 Neural Network with Hierarchical Matrices

In this paper, we investigate the fully connected neural networks with hierarchical matrices [9,10], see Fig. 2, namely $y = PINN(x) = \mathcal{H}_n \sigma(\mathcal{H}_{n-1}...\mathcal{H}_2 \sigma(\mathcal{H}_1 + b1) + b2) + ... + b_{n-1}) + b_n$, where $\mathcal{H}_i$ are hierarchical matrices, and $b_i$ are normal vectors. We do not compress the dense matrices of layers from the traditional fully connected neural network. We rather assume the structure of the compressed matrix for a layer, and we train it from the very beginning in the compressed form. We implemented our own training kernel for the matrices of layers compressed in a hierarchical manner. We have assumed that each layer has a structure of hierarchical matrix of rank one. The matrix is "refined" towards the diagonal, and the off-diagonal blocks are rank 1. In other words, each off-diagonal block is represented as a multiplication of 1 column and 1 row.

The main benefit of hierarchical matrices is that they enable linear computational cost matrix-vector multiplication, see Fig. 4.

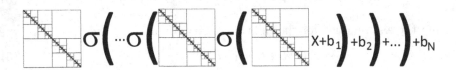

**Fig. 2.** Neural network with hierarchical matrices.

## 2   Matrix Compression

The main idea of speeding up the training process is storing the weight matrices $A_i$ in a compressed form. The matrix compression used in the paper is based on

Recursive Singular Value Decomposition, which is the key idea behind the hierarchical matrices [9,10]. In the Recursive Singular Value Decomposition Compression algorithm, the matrix is recursively divided into four smaller submatrices, and selected submatrices (for which so-called admissibility condition is fulfilled) are approximated using the singular value decomposition algorithm (SVD). The remaining submatrices, for which the admissibility condition is not fulfilled, are recursively divided into smaller matrices. The SVD algorithm decomposes matrix $A$ into three matrices $A = UDV$, where $D$ is diagonal matrix with singular values sorted in descending order, $U$ is the matrix of columns, and $V$ is the matrix of rows. In the approximate SVD, the singular values smaller than the predefined threshold are removed, together with the corresponding rows from $V$ and columns from $U$.

The compressed matrix can be stored in a tree-like structure, where the root node corresponds to the whole matrix, each node can have four sons corresponding to submatrices of the matrix or can be a leaf representing the corresponding matrix in the SVD compressed form. Figure 3 shows an exemplary tree representing the hierarchically compressed matrix.

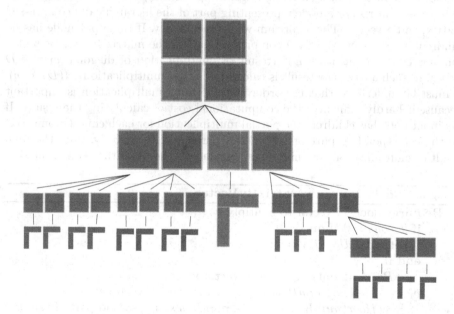

**Fig. 3.** An exemplary tree representing the hierarchically compressed matrix.

## 3   The Compressed Matrix-Vector Multiplication

The SVD compressed submatrices are stored as multi-columns $U$ multiplied by multi-rows $(DV)$. The SVD compression of a matrix allows to speed up the matrix-vector multiplication algorithm - the time complexity is $(O(Nr))$, where

$N$ is the size of the uncompressed matrix and $r$ is the number of singular values bigger than a given threshold (rank of the matrix). Figure 4 presents the idea of SVD compressed matrix-vector multiplication.

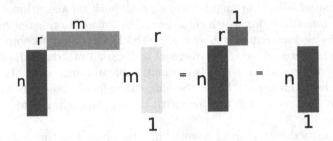

**Fig. 4.** The idea of SVD compressed matrix by vector multiplication.

The recursive algorithm for multiplication of the hierarchically compressed matrix by a vector takes as input a node representing the tree (hierarchically compressed matrix) or a node representing part of the hierarchically compressed matrix, and a vector. The algorithm works recursively. If the input node has no children, it represents the SVD compressed part of the matrix (stored as multi-columns $U$ and multi-rows $DV$ (result of multiplication of diagonal matrix $D$ by $V$). In such a case, the result is calculated as the multiplication $U((DV) * v)$. It must be underlined that the order of performing multiplication is important because it hardly influences the computational cost of calculating the results. If the input node has children, the partial multiplication for each child (submatrix) by the corresponding part of vector $v$ is performed recursively, then the final result of multiplications of children is calculated. The algorithm is presented in Algorithm 1.

---

**Algorithm 1. MultiplyMatrixByVector**

---

**Require:** node $T$, vector to multiply $v$
    **if** $T.sons = \emptyset$ **then**
        **return** $T.U * (T.V * v)$;
    **end if**
    $numRows$ =number of rows of vector v;
    $v_1 = v(1 : floor(numRows/2), :)$ //first part of vector $v$
    $v2 = v(floor(numRows/2 + 1) : numRows, :)$ //second part of vector $v$
    res1=MultiplyMatrixByVector(T.children(1),v1)
    res2=MultiplyMatrixByVector(T.children(2),v2)
    res3=MultiplyMatrixByVector(T.children(3),v1)
    res4=MultiplyMatrixByVector(T.children(4),v2)
    //calculate the final result of multiplication
    res1res2=res1+res2
    res3res4=res3+res4
    **return** result=[res1res2;res3res4]

---

## 4   Using the Algorithm of Hierarchically Compressed Matrix-Vector Multiplication to Speed up Neural Network Training

The main idea is to store the weight matrix $A$ in the hierarchically compressed form. The matrix of size $n \times n$ can be represented as the hierarchically compressed matrix of rank 1, where on each level of hierarchy the off-diagonal blocks are represented by SVD compressed blocks, and the remaining blocks are divided into smaller blocks. On this lower level of the compression, again, the off-diagonal blocks are represented by SVD compressed blocks, and the remaining blocks are divided into smaller blocks. The process of "refinement" of the matrix is stopped if the if the submatrix has size 1. An exemplary hierarchical compressed matrix of size $8 \times 8$ is presented in Fig. 5. The number of entries of the compressed form of the matrix of size $n \times n$ is equal to $2 * n * log_2(2 * n)$. In our neural network the compressed matrix of weights is represented as the vector of size $2 * n * log_2(2 * n)$, where the first entries are entries corresponding to the second submatrix, then entries for third submatrix, then entries for the first and finally entries for the fourth submatrix. The numbering of submatrices of the matrix as well as the vector form of hierarchical compressed matrix of size $8x8$ is presented in Fig. 5. In our neural network the iterative version of matrix-vector multiplication algorithm (see algorithm 1), where the $n \times n$ matrix is stored in form of a vector of size $2 * n * log_2(2 * n)$ was used.

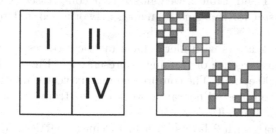

**Fig. 5.** Compressed matrix and its vector representation.

## 5   Results

Our test shows that an acceptable solution of the problem defined in equations (1) and (2) was obtained for the $LOSS$ of the order of 0.001. The tests were performed with a learning rate 0.02 and a number of epochs equal to 1000. We have used a neural network with 2 internal layers. In our tests in the classical approach (uncompressed full matrix) each internal layer has number of neurons equal to: $sizeofmatrix = 32, 64, 128, 256, 512$. So, the number of entries in uncompressed matrices were equal to: $32^2 = 1024$, $64^2 = 4096$, $128^2 = 16384$, $256^2 = 65536$, $512^2 = 262144$.

The number of entries in compressed matrices, corresponding to uncompressed matrices of sizes $32^2$, $64^2$, $128^2$, $256^2$, $512^2$ where equal to: 384, 896, 2048, 4608, 10240. The convergence of training is presented in Figs. 6a, 7a, 8a and 9a for uncompressed matrix with classic matrix-vector multiplication. For the fully connected neural network with full matrices and classical matrix-vector multiplication algorithm, the training finds correct solutions for matrices of sizes $32^2$, $64^2$, and $128^2$. The final loss was of order 0.001, and the shape of solution was correct. For example, Fig. 6a presents the successful training of neural network with two layers represented by full matrices of size $32^2$, where the loss function goes below 0.001.

However, the training does not find a correct solution after 2000 epochs for matrices of size $256^2$ and $512^2$ (the final loss was higher than 0.001 and the solution shape was wrong). For example Fig. 9a presents the training for neural network with two layers represented by full matrices of size $256^2$.

The convergence of training for compressed matrix with compressed matrix-vector multiplication is presented in Figs. 6b, 7b, 8b and 9b. For a neural network with 2 layers with hierarchical matrices corresponding to matrices of size $32^2$, with total of 384 non-zero entries to train, the convergence of training is presented in Fig. 6b. The loss value reaches $10^{-3}$ after 600 epochs. This accuracy and convergence rate is similar to the classical training presented in Fig. 6a. On top of that, the matrix-vector multiplication is way cheaper in the compressed matrix NN. Comparing Table 1 and Table 2, we can see that compressed matrices can be trained 3 times faster for this smaller neural network (see first rows in Table 1 and Table 2.

For a neural network with 2 layers with hierarchical matrices corresponding to matrices of size $64^2$, the convergence of training is presented in Fig. 7b. The loss value reaches $10^{-3}$ after 600 epochs. The compressed matrix-vector multiplications are cheaper, and the compressed neural network can be trained 2 times faster (see second row in Table 1 and Table 2).

Finally, for a neural network with 2 layers with hierarchical matrices corresponding to matrices of size $128^2$, the convergence of training is presented in Fig. 8b. The loss value reaches $10^{-3}$ after 600 epochs. The compressed matrix-

**Table 1.** Number of epochs and number of FLOPs of classic multiplication, learning rate 0.02, LOSS 0.001

| matrix size | number of FLOPs - classic multiplication | number of epochs - classic multiplication |
|---|---|---|
| 32 | 620 | 257,761,280 |
| 64 | 252 | 419,069,952 |
| 128 | 246 | 1,629,716,480 |
| 256 | - | - |
| 512 | - | - |

**Table 2.** Number of epochs and number of FLOPs of hierarchical multiplication, learning rate 0.02, LOSS 0.001

| matrix size | number of epochs hierarchical multiplication | number of FLOOPs hierarchical multiplication |
|---|---|---|
| 32 | 539 | 77,172,480 |
| 64 | 658 | 222,604,928 |
| 128 | 427 | 333,634,560 |
| 256 | 836 | 1,478,905,344 |
| 512 | 382 | 1,512,684,544 |

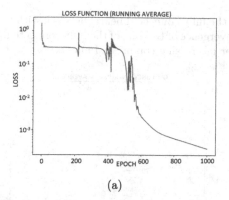
(a)

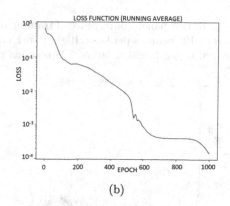
(b)

**Fig. 6.** Left panel: convergence of training of the fully connected neural network with 2 layers, 32 neurons per layer. Right panel: convergence of training of the fully connected neural network with 2 layers, 32 neurons per layer using compressed matrix.

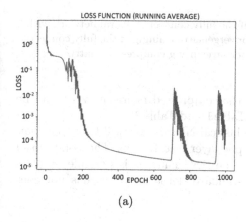
(a)

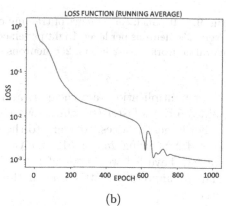
(b)

**Fig. 7.** Left panel: convergence of training of the fully connected neural network with 2 layers, 64 neurons per layer. Right panel: convergence of training of the fully connected neural network with 2 layers, 64 neurons per layer using compressed matrix.

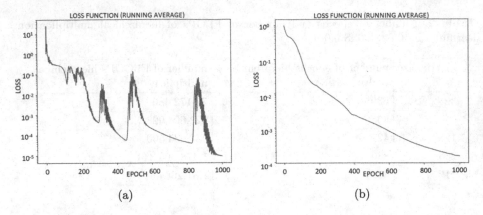

**Fig. 8.** Left panel: convergence of training of the fully connected neural network with 2 layers, 128 neurons per layer. Right panel: convergence of training of the fully connected neural network with 2 layers, 128 neurons per layer using compressed matrix.

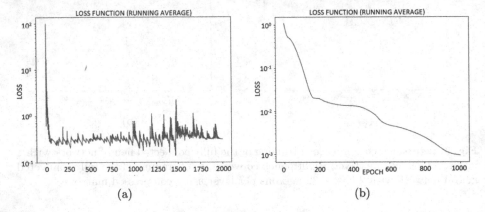

**Fig. 9.** Left panel: convergence of training of the fully connected neural network with 2 layers, 256 neurons per layer. Right panel: convergence of training of the fully connected neural network with 2 layers, 256 neurons per layer using compressed matrix.

vector multiplications are cheaper, and the compressed neural network can be trained 5 times faster (see third row in Table 1 and Table 2).

For larger matrices, contrary to the classical NN, we can still train the compressed NN, using 256 or 512 neurons per layer, see Fig. 9b. The compressed matrices have a lower order of trainable entries. Thus, it is possible to train a large compressed matrix even if the standard dense approach has not converged yet.

# 6    Conclusions

In this paper, we proposed a Physics Informed Neural Network with hierarchical matrices for approximation of one-dimensional advection-diffusion problems. The neural network represented a solution of one-dimensional PDE, namely $y = PINN(x) = \mathcal{H}_n\sigma(\mathcal{H}_{n-1}...\mathcal{H}_2\sigma(\mathcal{H}_1 + b1) + b2) + ... + b_{n-1}) + b_n$, where $\mathcal{H}$ are the hierarchical matrices. We have verified our method and showed that it allows to speed up the training process between 2–5 times (compare rows in Tables 1 and 2), while reducing the memory storage up to 3–20 times. The future work will involve generalization of this method for more complex PDEs.

**Acknowledgements.** This work was partially supported by the program "Excellence initiative - research university" for the AGH University of Krakow.

# References

1. Alber, M., et al.: Integrating machine learning and multiscale modeling-perspectives, challenges, and opportunities in the biologica biomedical, and behavioral sciences. NPJ Digit. Med. **2** (2019). https://doi.org/10.1038/s41746-019-0193-y
2. Cai, S., Mao, Z., Wang, Z., Yin, M., Karniadakis, G.E.: Physics-informed neural networks (PINNs) for fluid mechanics: a review. Acta. Mech. Sin. **37**(12), 1727–1738 (2021)
3. Calo, V., Łoś, M., Deng, Q., Muga, I., Paszyński, M.: Isogeometric Residual Minimization Method (iGRM) with direction splitting preconditioner for stationary advection-dominated diffusion problems. Comput. Methods Appl. Mech. Eng. **373**, 113214 (2021). https://doi.org/10.1016/j.cma.2020.113214
4. Chan, J., Evans, J.: A minimal-residual finite element method for the convection-diffusion equations. ICES-REPORT **13**(12) (2013). https://oden.utexas.edu/media/reports/2013/1312.pdf
5. Chen, Y., Lu, L., Karniadakis, G.E., Dal Negro, L.: Physics-informed neural networks for inverse problems in nano-optics and metamaterials. Opt. Exp. **28**(8), 11618–11633 (2020)
6. Eriksson, K., Johnson, C.: Adaptive finite element methods for parabolic problems I: a linear model problem. SIAM J. Numer. Anal. **28**(1), 43–77 (1991). http://www.jstor.org/stable/2157933
7. Geneva, N., Zabaras, N.: Modeling the dynamics of PDE systems with physics-constrained deep auto-regressive networks. J. Comput. Phys. **403** (2020). https://doi.org/10.1016/j.jcp.2019.109056
8. Goswami, S., Anitescu, C., Chakraborty, S., Rabczuk, T.: Transfer learning enhanced physics informed neural network for phase-field modeling of fracture. Theor. Appl. Fract. Mach. **106** (2020). https://doi.org/10.1016/j.tafmec.2019.102447
9. Hackbusch, W.: A sparse matrix arithmetic based on h-matrices. Part i: introduction to h-matrices. Computing **62**, 89–108 (1999)
10. Hackbusch, W.: Hierarchical Matrices: Algorithms and Analysis, vol. 49. Springer, Heidelberg (2015)

11. Kissas, G., Yang, Y., Hwuang, E., Witschey, W.R., Detre, J.A., Perdikaris, P.: Machine learning in cardiovascular flows modeling: predicting arterial blood pressure from non-invasive 4D flow MRI data using physics-informed neural networks. Comput. Methods Appl. Mech. Eng. **358** (2020). https://doi.org/10.1016/j.cma.2019.112623

12. Ling, J., Kurzawski, A., Templeton, J.: Reynolds averaged turbulence modelling using deep neural networks with embedded invariance. J. Fuild Mech. **807**, 155–166 (2016). https://doi.org/10.1017/jfm.2016.615

13. Lu, L., Pestourie, R., Yao, W., Wang, Z., Verdugo, F., Johnson, S.G.: Physics-informed neural networks with hard constraints for inverse design. SIAM J. Sci. Comput. **43**(6), B1105–B1132 (2021). https://doi.org/10.1137/21M1397908

14. Maczuga, P., Paszyński, M.: Influence of activation functions on the convergence of physics-informed neural networks for 1D wave equation. In: Mikyška, J., de Mulatier, C., Paszynski, M., Krzhizhanovskaya, V.V., Dongarra, J.J., Sloot, P.M. (eds.) ICCS 2023. LNCS, vol. 14073, pp. 74–88. Springer, Cham (2023). https://doi.org/10.1007/978-3-031-35995-8_6

15. Mao, Z., Jagtap, A.D., Karniadakis, G.E.: Physics-informed neural networks for high-speed flows. Comput. Methods Appl. Mech. Eng. **360**, 112789 (2020)

16. Mishra, S., Molinaro, R.: Estimates on the generalization error of physics-informed neural networks for approximating a class of inverse problems for PDEs. IMA J. Numer. Anal. **42**(2), 981–1022 (2022)

17. Raissi, M., Perdikaris, P., Karniadakis, G.: Physics-informed neural networks: a deep learning framework for solving forward and inverse problems involving nonlinear partial differential equations. J. Comput. Phys. **378**, 686–707 (2019). https://doi.org/10.1016/j.jcp.2018.10.045

18. Rasht-Behesht, M., Huber, C., Shukla, K., Karniadakis, G.E.: Physics-informed neural networks (PINNs) for wave propagation and full waveform inversions. J. Geophys. Res. Solid Earth **127**(5), e2021JB023120 (2022)

19. Sikora, M., Krukowski, P., Paszynska, A., Paszynski, M.: Physics informed neural networks with strong and weak residuals for advection-dominated diffusion problems (2023)

20. Sun, L., Gao, H., Pan, S., Wang, J.X.: Surrogate modeling for fluid flows based on physics-constrained deep learning without simulation data. Comput. Methods Appl. Mech. Eng. **361** (2020). https://doi.org/10.1016/j.cma.2019.112732

21. Wandel, N., Weinmann, M., Neidlin, M., Klein, R.: Spline-PINN: approaching PDEs without data using fast, physics-informed hermite-spline CNNs. In: Proceedings of the AAAI Conference on Artificial Intelligence, vol. 36, no. 8, pp. 8529–8538 (2022)

# On the Training Efficiency of Shallow Architectures for Physics Informed Neural Networks

J. Rishi[✉], Azhar Gafoor, Sumanth Kumar, and Deepak Subramani

Department of Computational and Data Sciences, Indian Institute of Science - Bengaluru, Bengaluru, India
{rishij,azharctp,ksumanth,deepakns}@iisc.ac.in

**Abstract.** Physics-informed Neural Networks (PINNs), a class of neural models that are trained by minimizing a combination of the residual of the governing partial differential equation and the initial and boundary data, have gained immense popularity in the natural and engineering sciences. Despite their observed empirical success, an analysis of the training efficiency of residual-driven PINNs at different architecture depths is poorly documented. Usually, neural models used for machine learning tasks such as computer vision and natural language processing have deep architectures, that is, a larger number of hidden layers. In PINNs, we show that for a given trainable parameter count (model size), a shallow network (less layers) converges faster than a deep network (more layers) for the same error characteristics. To illustrate this, we examine the one-dimensional Poisson's equation and evaluate the gradient for residual and boundary loss terms. We show that the characteristics of the gradient of the loss function are such that for residual loss, shallow architectures converge faster. Empirically, we show the implications of our theory through various experiments.

**Keywords:** PINNs · Partial Differential Equations · Deep Learning

## 1 Introduction

Neural networks have gained popularity in various fields such as computer vision, machine translation, and weather prediction [8,12,18,25]. In recent years, a type of deep neural model called Physics-Informed Neural Networks (PINNs) has emerged as a method for solving partial differential equations (PDEs) in a semi-supervised manner [20]. PINNs are trained using a loss function that includes the L2 norm of the residual of the governing PDEs, as well as losses for initial and boundary data. What makes PINNs special is their use of automatic differentiation to evaluate the residual rather than relying on simulation data from traditional PDE solvers. This makes the approach semi-supervised, as only boundary condition data are provided for supervision, while other points utilize the PDE residual loss.

© The Author(s), under exclusive license to Springer Nature Switzerland AG 2024
L. Franco et al. (Eds.): ICCS 2024, LNCS 14834, pp. 363–377, 2024.
https://doi.org/10.1007/978-3-031-63759-9_39

Despite the observed empirical success, the training efficiency of residual-driven PINNs at different network depths (number of layers) has not been well documented. In AI tasks such as computer vision and natural language processing, deep architectures with a larger number of hidden layers are commonly used in neural models. For these tasks, having an insufficient number of layers is shown to require a significantly higher number of parameters compared to deeper networks [2]. However, most PINN models use a shallow network with a modest number of layers (around 5) compared to deep networks that are typically used in AI tasks. Additionally, there is limited research that provides theoretical explanations for hyperparameter tuning in PINNs [6, 29].

Our objective is to answer the question "Why do PINN models work well with fewer layers?". To answer this question, we study the properties of the loss function through an analytical solution, a simple network with an analytically tractable number of parameters, and with numerical test cases to show how the gradient of PINN loss behaves for relatively shallower shallow and deep networks with the same parameter count. We find that shallower PINN models have steeper gradients that help them achieve faster convergence, compared to deeper networks with the same parameter count.

In what follows, we first provide a background of PINNs. Next, we analyze the gradient of the residual and boundary losses by considering an analytical solution of Poisson's equation, for a one-layer and two-layer neural network. Finally, we show the implications of our theory through simulations of the Kovasznay flow, Lid-driven cavity flow, and atmospheric boundary layer flow experiments.

## 2    Background of PINNs

Consider a general nonlinear PDE,

$$\mathcal{N}(u, x) = f, x \in \Omega, \tag{1}$$

$$\mathcal{B}(u(x)) = b, x \in \partial\Omega, \tag{2}$$

where $\mathcal{N}$ is a generic nonlinear function, $\mathcal{B}$ are the initial and boundary condition operator, $x \in \Omega$ is the coordinates in the domain $\Omega$, $\partial\Omega$ is the boundary of the domain, $b$ is the boundary condition value and $u \in \mathcal{U}$ is the solution field in a solution space $\mathcal{U}$ that satisfies the above PDE. As a specific example, for 2D fluid flow problems, the vector $u$ consists of the flow velocities (horizontal, vertical) and the pressure. The domain coordinates are $(x, y, t)$ for unsteady flow and $(x, y)$ for steady flows.

Traditionally, finite discretization-based numerical schemes or spectral methods have been used to solve the above PDE. The PINNs approach uses and trains a parametrized neural model $\mathcal{H}$ that approximates the functional map between the input domain $\Omega$ and the solution domain $\mathcal{U}$, i.e., $\mathcal{H} : x \times \theta \mapsto u$, such that the differential operator with the initial and boundary conditions are satisfied. Here $\theta$ is the set of parameters of the neural network $\mathcal{H}$ that is learned by minimizing a loss function $\mathcal{L}_{PINN}$ that comprises of the PDE-residual term and boundary data loss term, i.e.,

$$\mathcal{L}_{PINN} = \mathcal{L}_{residual} + \mathcal{L}_{boundary} \tag{3}$$

$$\mathcal{L}_{residual} = \frac{1}{N_c} \left[ \sum_{i=1}^{N_c} (\mathcal{N}(u_i, x_i) - f)^2 \right] \tag{4}$$

$$\mathcal{L}_{boundary} = \frac{1}{N_{bp}} \left[ \sum_{i=1}^{N_{bp}} (\mathcal{B}(u(x_i)) - b)^2 \right] . \tag{5}$$

Here, $N_c$ represents the total count of location points within the domain and $N_{bp}$ denotes the total number of boundary points. The PDE loss term is computed using the automatic differentiation approach, while the boundary loss term is determined using the mean squared error with provided boundary data for the Dirichlet boundary condition. If the Neumann/Robin condition is specified, the boundary loss is computed with the help of automatic differentiation. The use of automatic differentiation allows to solve the PDE without relying on simulation data from numerical solvers as is done in neural operator methods such as DeepONets [17], and Fourier Neural Operators [15]. Practically, a PINN model is a dense neural network with input as the coordinates and the output is the solution variable of the PDE. For neural models, the words *shallow* and *deep* are relative and context-dependent. Our usage is as follows. For the same number of trainable parameters, a network with fewer layers is called a shallow network, and a network with more layers is called a deep network. Most PINN models in the literature use a shallow dense neural network with about 5 layers [4, 8, 16, 18, 19, 21, 30] compared to typical deep networks used in AI tasks of computer vision and natural language processing.

## 3    Analysis of the Gradient of the Residual Loss

The key distinguishing factor between PINNs and neural models used for AI tasks is the objective function and training strategy. In PINNs, the goal is to minimize a combined loss function with PDE residual and boundary condition data. Additionally, the residual loss of the PDE is calculated by automatic differentiation. In contrast, traditional machine learning tasks typically involve working with existing datasets, and the objective is to generalize from that data. In PINNs minimizing the objective function is of paramount importance, whereas in usual machine learning tasks typically involve working with existing datasets, and the objective is to generalize from that data.

Gradient descent based algorithms are typically used to train neural networks. The nature of the gradient determines the convergence of the training. As PINNs have two components (residual and boundary loss), the gradients of each term of the loss affect the convergence. We examine the magnitude of the gradients of these two terms with respect to the parameters of the neural network to determine how the training will progress.

To conduct this study, we take a progressive approach. First, we use the analytical solution of the 1D Poisson's equation to gain insights into the difference between the gradients of these two terms with respect to parameters of the PINN model.

The 1D Poisson's equation is

$$\frac{\partial^2 u}{\partial x^2} = -\pi^2 \sin(\pi x), x \in [-0.5, 0.5]$$
$$u(-0.5) = -1,$$
$$u(0.5) = 1,$$

with an analytical solution $u(x) = \sin(\pi x)$.

Consider a trained neural network $f_\theta(x)$ that accurately approximates the solution $u(x)$. Without loss of generality, we may express the approximate solution as $f_\theta(x) = u(x)\epsilon_\theta(x)$, where $\epsilon_\theta(x)$ is defined for $x \in [-0.5, 0.5]$ and $\frac{\partial \epsilon_\theta(x)}{\partial x} < \epsilon$, for some $\epsilon > 0$. The gradient of residual loss and boundary loss of $f_\theta$ is

$$\frac{\partial \mathcal{L}_{residual}}{\partial \theta} \approx \pi^4(\epsilon_\theta(x) - 1)\frac{\partial \epsilon_\theta(x)}{\partial \theta},$$
$$\frac{\partial \mathcal{L}_{boundary}}{\partial \theta} \approx 4(\epsilon_\theta(x) - 1)\frac{\partial \epsilon_\theta(x)}{\partial \theta}.$$

The full derivation of the above is in Appendix 6.1, attached in supplementary material. We see that the gradient of the residual term with respect to the parameters of the neural network is approximately $\mathcal{O}(10) - -\mathcal{O}(100)$ times the gradient of the boundary loss term with respect to the parameters [28]. This observation holds for a variety of equations as reported in other PINN studies [24, 28].

Next, we consider two networks with two neurons and only two trainable parameters $w_1, w_2$. In the first network, we use only one layer (Fig. 1a); in the second, we use two layers (Fig. 1b). The weights on the remaining edges are assumed to be one for analysis. All hidden neurons use the swish activation function as commonly used in PINNs [22]. An input $(x)$ gives the output $(U)$ for the shallow (one-layer) and deep (two-layer) networks as follows.

$$U_{Shallow} = \frac{w_1 x}{1 + e^{(-w_1 x)}} + \frac{w_2 x}{1 + e^{(-w_2 x)}}, \tag{6}$$

$$U_{Deep} = \frac{\dfrac{w_2 w_1 x}{1 + e^{(-w_1 x)}}}{1 + e^{-\left(\dfrac{w_2 w_1 x}{1 + e^{(-w_1 x)}}\right)}}. \tag{7}$$

The efficiency of learning for a neural network through backpropagation depends on the characteristics of the loss function. From our analysis above,

**Table 1.** Hyperparameters used in all experiments to study the effect of network depth.

| Hyperparameters | | | | | | | | | |
|---|---|---|---|---|---|---|---|---|---|
| Flows | Kovaszny flow | | | Lid-driven cavity | | | ABL | | |
| Hidden layers | 5 | 10 | 15 | 5 | 10 | 15 | 5+1[a] | 10+1[a] | 15+1[a] |
| Neurons per layer | 10 | 6(5),7(10) | 5(10),6(5) | 300 | 200 | 161(5), 160(10) | 64(5),8(1) | 36(10),36(1) | 32(15),24(1) |
| Learning rate | 0.001 | | | | | | | | |
| Epochs | 2,000 | | | 30,000 | | | 25,000 | | |
| Optimizer | Adam | | | | | | | | |
| Trainable-parameters | 500 | 494 | 495 | 363,000 | 363,000 | 363,051 | 19,472 | 18,936 | 18,960 |
| Grid points | 64*64 | | | 50*50 | | | 50*500 | | |
| Activation function | Swish | | | | | | Swish + ReLU | | |
| Initializer | Glorot Uniform | | | Glorot Normal | | | Glorot Uniform | | |

[a] One layer of neurons for each output separately. More details in Appendix 6.2.

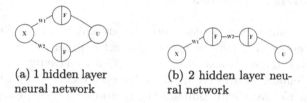

(a) 1 hidden layer neural network

(b) 2 hidden layer neural network

**Fig. 1.** The architectures used to analyse the residual loss for the Poisson's equation. Here, the 1 layer network is the "shallow" network and the 2 layer network is the "deep" network.

we know that the gradient of the residual loss term is greater. Therefore, we plot the residual loss in terms of $w_1$ and $w_2$ to analyze the difference in the loss landscape for a 1-layer (called shallow) and 2-layer (called deep) PINN. Approximately the residual term of the loss function for shallow and deep network is

$$\mathcal{L}_{Shallow} \approx \frac{w_1^2 e^{-w_1 x}}{(1 + e^{-w_1 x})^2} + \frac{w_2^2 e^{-w_2 x}}{(1 + e^{-w_2 x})^2},$$

$$\mathcal{L}_{Deep} \approx \frac{w_1 w_2}{(1 + e^{-w_1 x})^2 (1 + e^{\frac{-w_1 w_2 x}{1 + e^{-w_1 x}}})^2}.$$

This loss is evaluated for a randomly chosen $x$ in the input range and with a grid of weights and visualized in Fig. 2. We see that the loss landscape of the shallow network has a prominent valley that can easily be reached by the gradient descent algorithm, whereas the deep network has a shallow valley that is harder to optimize. Hence, we expect the shallow network to converge faster than the deep network during optimization [7,14].

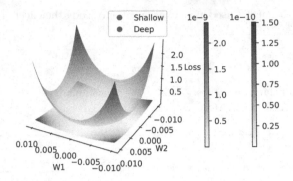

**Fig. 2.** The residual loss plotted against the parameters of the neural network for shallow (1 layer) and deep (2 layers) networks.

## 4    Experimental Results

To empirically confirm that shallow networks are indeed more efficient to train than deep networks for PINN tasks, we conduct experiments with these three test cases, viz., (i) the Kovasznay flow, (ii) the lid-driven cavity flow, and (iii) the Atmospheric Boundary Layer (ABL) flow. For each case, we train PINN models with 5, 10 and 15 layers and compare the solutions of each to a ground truth obtained either analytically or numerically.

### 4.1    Kovasznay Flow

The Kovasznay flow [10,27] is governed by the steady state Navier-Stokes equations

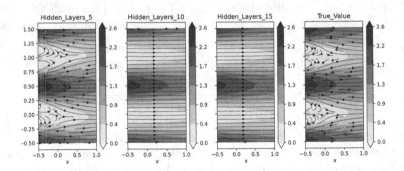

**Fig. 3.** The solution of PINNs trained with 5, 10 and 15 hidden layers is shown next to the true solution of Kovasznay flow with Re=40. Flow streamlines are overlaid on a background of velocity magnitude.

$$(\boldsymbol{V}.\nabla)\boldsymbol{V} = -\nabla p + \frac{1}{Re}(\nabla^2 \boldsymbol{V})$$
$$\nabla.\boldsymbol{V} = 0,$$
$$\boldsymbol{V}|_{bdry} = g(x_b, y_b),$$

(8)

where $\boldsymbol{V}$ is the velocity vector, $p$ is the pressure, $Re$ is the Reynold's number, $\nabla$ is the gradient operator, $\nabla^2$ is the Laplace operator and $g$ is the boundary data function. The Kovasznay flow has an analytical solution given by

$$\zeta = \frac{0.5}{\mu} - \sqrt{\frac{1}{4\mu^2} + 4\pi^2},$$
$$u = 1 - \exp(\zeta x)\cos(2\pi y),$$
$$v = \frac{\zeta}{2\pi}\exp(\zeta x)\sin(2\pi y),$$
$$p = 0.5(1 - \exp(2\zeta x)),$$

where $\mu$ is the viscosity, $x$ and $y$ are the cartesian coordinate system, $u$ and $v$ are horizontal and vertical components of the velocity, and $p$ is pressure. The boundary condition of the Navier-Stokes equation is given by the analytical solution evaluated at the boundary points.

We solve the Kovasznay flow using the PINN approach in a square grid with $x \in [-0.5, 1.0]$, $y \in [-0.5, 1.5]$ and $\mu = 0.025$. The loss function consists of the residual of the PDE in eq. (8) and the boundary loss.

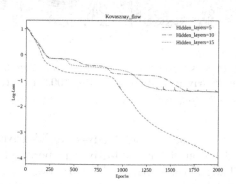

**Fig. 4.** Log-loss plot during training of the PINN with 5, 10, and 15 hidden layers for the Kovasznay flow test case.

We conducted experiments using PINN models with 5, 10, and 15 hidden layers. To ensure a fair comparison, we use approximately the same number of trainable parameters for these three models and keep all other training algorithm hyperparameters fixed. Consequently, shallower networks have more neurons in each layer compared to deeper networks, as the total number of trainable parameters remains approximately the same. The main hyperparameters used are presented in Table 1. The training process consists of 2000 epochs with a learning rate of 0.001. The regular training of the PINNs is completed and the trained model is then evaluated on a 64 × 64 regular grid for comparison with the ground truth obtained from the analytical solution. Figure 3 illustrates the solution for the PINN models with 5, 10, and 15 hidden layers, together with the reference ground truth solution. It is observed that only the shallow model with 5 hidden layers successfully trained, while the deeper models failed to accurately capture the physics of the flow. Figure 4 shows the log-loss plot for all three models. The reduction in loss is significant for the model with 5 hidden layers, whereas the loss plateaus for the other two models. This observation directly confirms our hypothesis that shallower networks have a favorable loss landscape for the optimization method to find a minimum for PINNs training dominated by the residual loss. The residual loss gradient is higher compared to the boundary loss gradient with respect to the network parameters for most of the epochs in this case as well, which is consistent with our analysis (Sect. 3) (See also Appendix 6.3)

## 4.2   Lid-Driven Cavity Flow

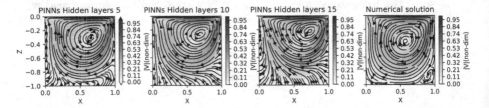

**Fig. 5.** The solution of PINNs trained with 5, 10 and 15 hidden layers is shown next to the true solution of the lid-driven cavity flow for Re=300. Flow streamlines are overlaid on a background of velocity magnitude.

The second test case chosen is the steady-state lid-driven cavity flow problem, which is an idealization of the wind-driven response of the ocean and is of practical relevance. The rightmost panel of Fig. 5 shows the flow setup. A square fluid-filled cavity in the $x - z$ plane is subject to a horizontal motion on the top surface, corresponding to the lid moving to the right at a specified velocity of 1 non-dimensional unit. The PINNs simulation employs the Navier-Stokes equation (Eq. 8) along with the following boundary conditions: $u, v = 0$ for left, bottom, and right, and $u = 1, v = 0$ on top surface. Simulations are compared with the solution obtained from a numerical CFD solver [26].

We train PINNs with 5, 10, and 15 hidden layers, all with approximately the same number of trainable parameters, and other hyperparameters are fixed, similar to the previous test case. The main hyperparameters are shown in Table 1.

Figure 5 shows the flow streamlines and magnitude of the lid-driven cavity problem for Re = 300 for different numbers of hidden layers. All models were trained for 30,000 epochs. The flow simulation using the network with five hidden layers agrees better with the flow simulation using the CFD tool, while those with more layers do not. Figure 6a shows the learning curve during training, i.e., the plot of log-loss versus the number of epochs. The network with five hidden layers has the lowest loss compared to the other networks. Table 2 documents the mean relative error between the PINN and CFD simulations for the different output variables. The relative error and the confidence interval are the lowest for the network with five hidden layers. A noticeable difference in the number of epochs to reach a certain threshold log-loss can also be observed between the models. For example, to reach a log-loss value less than −5.1, models with 5, 10, and 15 hidden layers took nearly 14000, 16000, and 17000 epochs, respectively. Figure 6b shows that the residual loss gradient is higher compared to the boundary loss gradient with respect to network parameters, which aligns with our analysis in Sect. 3.

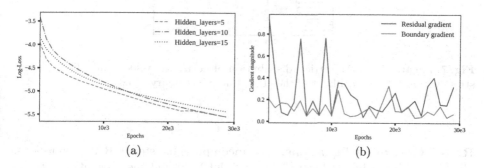

(a)                                                    (b)

**Fig. 6.** (a) Log-loss plot for 5, 10 and 15 hidden layers PINN models to simulate the lid-driven cavity flow. (b) L2 norm of the gradient vector of residual loss and boundary loss with respect to neural network parameters for the lid-driven cavity flow problem for 5 hidden layers.

**Table 2.** Relative error in $u$, $v$, and $p$ with their confidence interval for lid-driven cavity flow test case.

| Relative error | | | |
|---|---|---|---|
| Layers | U Error | V Error | P Error |
| 5 | $0.03 \pm 0.001$ | $0.04 \pm 0.001$ | $0.015 \pm 0.0001$ |
| 10 | $0.05 \pm 0.002$ | $0.06 \pm 0.002$ | $0.02 \pm 0.0002$ |
| 15 | $0.06 \pm 0.003$ | $0.07 \pm 0.003$ | $0.04 \pm 0.0006$ |

## 4.3   Atmospheric Boundary Layer (ABL)

Our third test case is the turbulent atmospheric boundary layer flow simulation, a much more complex test case than the previous two. A horizontally homogeneous atmospheric boundary layer flow, as shown in Fig. 7, is simulated in a uniformly rough, flat terrain under neutral stratification. The steady Reynolds-averaged Navier-Stokes (RANS) framework with two-equation turbulence models has been shown to be effective for the simulation of the ABL flow [1,11]. In the present work, we use the standard $k - \epsilon$ model [9] for turbulence closure along with the standard wall functions [13] modified to be consistent with the inlet profiles [23].

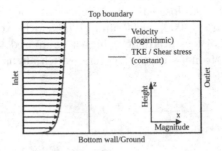

**Fig. 7.** Illustration of the vertical distribution of horizontal velocity and TKE / shear stress within the Prandtl layer of the atmospheric boundary layer.

**RANS Governing Equations.** The incompressible steady RANS equations, along with the turbulent kinetic energy (TKE) $(k)$ and the dissipation rate $(\epsilon)$ transport equations for the closure in the Cartesian coordinates, are as follows.

$$\text{Continuity equation: } \frac{\partial u_i}{\partial x_i} = 0 \,, \tag{9}$$

$$\text{Momentum equation: } \frac{\partial u_i u_j}{\partial x_j} = \frac{\partial}{\partial x_j}\left[(\nu + \nu_t)\left(\frac{\partial u_i}{\partial x_j} + \frac{\partial u_j}{\partial x_i}\right)\right] - \frac{1}{\rho}\frac{\partial P}{\partial x_i} \,, \tag{10}$$

$$k \text{ transport equation: } \frac{\partial k u_j}{\partial x_j} = \frac{\partial}{\partial x_j}\left[\left(\nu + \frac{\nu_t}{\sigma_k}\right)\frac{\partial k}{\partial x_j}\right] + P_k - \epsilon \,, \tag{11}$$

$$\epsilon \text{ transport equation: } \frac{\partial \epsilon u_j}{\partial x_j} = \frac{\partial}{\partial x_j}\left[\left(\nu + \frac{\nu_t}{\sigma_\epsilon}\right)\frac{\partial \epsilon}{\partial x_j}\right] + C_{1\epsilon}\frac{\epsilon}{k}P_k - C_{2\epsilon}\frac{\epsilon^2}{k} \,, \tag{12}$$

where $x_i$ is the spatial coordinate, $u_i$ is the time-averaged velocity vector ($u, w$ for 2D $(x, z)$ and $u, v, w$ for 3D $(x, y, z)$), $P$ is the mean pressure, $\nu$ is the kinematic viscosity, $\nu_t$ is the turbulent viscosity, $P_k$ is the production of turbulent kinetic energy. $P_k$ and $\nu_t$ are given by the equations,

$$P_k = \nu_t \left( \frac{\partial u_i}{\partial x_j} + \frac{\partial u_j}{\partial x_i} \right) \frac{\partial u_i}{\partial x_j}, \qquad\qquad \nu_t = C_\mu \frac{k^2}{\epsilon}.$$

Here, the values of the five constants $C_\mu, \sigma_k, \sigma_\epsilon, C_{1\epsilon}$ and $C_{2\epsilon}$ used for simulation of neutral atmospheric flow are 0.033, 1.0, 1.3, 1.176, and 1.92, respectively [5].

**Boundary Conditions.** For homogeneous ABL flow simulations in the stream-wise direction, we use a fully developed inlet velocity profile given by [23],

$$u = \frac{u_\tau}{\kappa} \ln \left( \frac{z + z_0}{z_0} \right), \tag{13}$$

where $u_\tau = \sqrt{\frac{\tau_w}{\rho}}$ is the frictional velocity (with $\tau_w$ is the wall shear stress, and $\rho$ is the fluid density), $\kappa \approx 0.418$ is the von Kármán constant, $z$ is the height co-ordinate and $z_0$ is the aerodynamic roughness height. Wall functions are used to achieve sufficiently precise solutions in the region close to the wall/ground, which helps reduce the computational cost and allows the inclusion of empirical information in special cases, such as rough wall conditions, to be used in ABL [13]. Turbulent kinetic energy and dissipation are specified on the bottom wall/ground using the equations,

$$k = \frac{u_\tau^2}{\sqrt{C_\mu}}, \qquad\qquad \epsilon = \frac{u_\tau^3}{\kappa(z + z_0)},$$

with $z = z_p$, where $z_p$ is the height from the wall such that $90 < \frac{z_p u_\tau}{\nu} < 500$ [3]. The velocity is also specified at the wall using eq. 13. At the top boundary, the symmetry boundary condition is used for velocity, pressure, and other turbulent quantities. A pressure outlet boundary condition is specified for the outlet with vanishing stream-wise gradients of other quantities.

**ABL Simulation Result.** Figure 8 shows plots of the stream-wise velocity, turbulent kinetic energy, and dissipation rate profiles against height obtained from the PINN models with 5, 10 and 15 hidden layers. The inlet profile is also plotted for reference. The hyperparameters used for the PINN models are shown in Table 1. Here, we use a reference velocity ($u_{ref}$) of $8m/s$ at a reference height ($z_{ref}$) of $70m$ and at a turbulence intensity of 7%. The domain is $100m$ high and $200m$ long in the stream-wise direction. The frictional velocity, calculated using Eq. 13 with 0.001 as the aerodynamic roughness height, is used to obtain values of other turbulent quantities. A successful simulation will have obtained horizontally homogeneous ABL flow under neutral stratification, that is, the

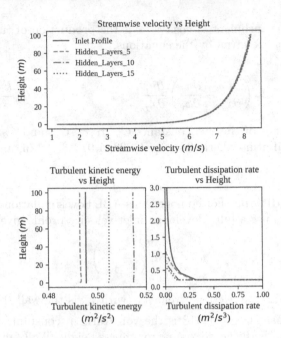

**Fig. 8.** Stream-wise velocity, turbulent kinetic energy and turbulent dissipation rate profiles of the ABL flow simulated from PINN models with 5, 10 and 15 hidden layers are compared to the reference (inlet profile). For the turbulent dissipation rate, the height is shown only until 3 m for highlighting the differences.

PINN simulations and reference inlet profiles would be identical. The flow is also characterized by constant shear stress over the height, providing a constant profile for the turbulent kinetic energy as seen in Fig. 8 panel for TKE.

From Fig. 8, it can be seen that the velocity profiles simulated by all networks are in good agreement with the inlet profile. However, the deviations in the TKE and dissipation rate profiles from the reference are more evident in models with greater number of hidden layers. The greater difference for 10 and 15 layer PINN models can be attributed to the fact that for TKE and dissipation rate, the Dirichlet condition was provided only at the bottom wall, and PDE losses were used to solve within the domain. This approach made the loss function residual dominated and as per our analysis (Sect. 3), we expect shallower networks to perform better here as evidenced by our experimental results. The absolute error between the inlet and outlet profiles simulated using the different PINN models is listed in Table 3. The TKE error in the 5 hidden layer model is almost one order less compared to other models. The velocity values were provided at the inlet and the bottom wall. Thus, all models were able to simulate this profile well, which shows that all models perform well in supervised learning tasks. From the log-loss plot (Fig. 9), we see that for the same number of epochs, the model with 5 hidden layers ended the training with the least loss. In our experiments, we have fixed the number of parameters and epochs to be equal for models with different hidden layers. A noticeable difference in the number of epochs to reach a certain threshold log-loss can also be observed between the models. For example, to reach a log-loss value less than 4, models with 5, 10, and 15 hidden layers took nearly 9000, 12000, and 16000 epochs, respectively. Also, it should be noted that the computational time for the model with 5 hidden layers is $\frac{4}{3}$ and $\frac{5}{3}$ times lesser than the models with 10 and 15 hidden layers. The residual loss gradient is higher compared to the boundary loss gradient with respect to the network parameters in this case as well, which is consistent with our analysis (Sect. 3) (See also Appendix 6.4)

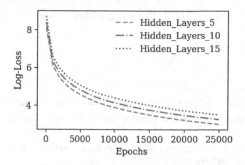

**Fig. 9.** Log-loss plot for 5, 10, and 15 hidden layers for the ABL flow.

**Table 3.** Mean absolute error in $u, k$, and $\epsilon$ with their confidence interval for ABL flow.

| Mean absolute error | | | |
|---|---|---|---|
| Hidden Layers | $u$ error | $k$ error | $\epsilon$ error |
| 5 | $0.016\pm5\times10^{-5}$ | $0.002\pm2\times10^{-5}$ | $0.003\pm4\times10^{-5}$ |
| 10 | $0.019\pm5\times10^{-5}$ | $0.020\pm8\times10^{-8}$ | $0.005\pm1\times10^{-3}$ |
| 15 | $0.024\pm2\times10^{-4}$ | $0.010\pm1\times10^{-7}$ | $0.004\pm2\times10^{-4}$ |

# 5   Conclusion

The results of our study suggest that shallow neural networks are more beneficial than deeper networks in the training of PINNs. We also provide an explanation for why shallow networks tend to converge better than deep networks, based on an analysis of the gradient of the loss function. To illustrate our findings, we present various test cases. Initially, these results may seem counter-intuitive, particularly when compared to the prevailing knowledge and empirical evidence in the fields of computer vision and natural language processing in AI. However, our analysis of the loss function for PDE residual-driven training clearly demonstrates why relatively shallow networks perform well in PINN model training. In all instances, it was observed that the gradient of the residual loss is higher when compared to the gradient of the boundary loss. While this holds true in the majority of cases, in the future we explore scenarios where this observation may not be accurate.

Supplementary material: Appendix and Codes

**Acknowledgment.** This research is partially supported by research grants from the Ministry of Earth Sciences (MoES/36/OOIS/Extra/84/2022) and Ministry of Education (MHRD STARS 1/388 SPADE). We are grateful for the PMRF fellowship support to Azhar Gafoor.

# References

1. Apsley, D.D., Castro, I.P.: A limited-length-scale k-$\varepsilon$ model for the neutral and stably-stratified atmospheric boundary layer. Bound.-Layer Meteorol. **83**, 75–98 (1997)
2. Bengio, Y., et al.: Learning deep architectures for AI. Found. Trends® Mach. Learn. **2**(1), 1–127 (2009)
3. Blocken, B., Stathopoulos, T., Carmeliet, J.: CFD simulation of the atmospheric boundary layer: wall function problems. Atmos. Environ. **41**(2), 238–252 (2007)
4. Cai, S., Wang, Z., Wang, S., Perdikaris, P., Karniadakis, G.E.: Physics-informed neural networks for heat transfer problems. J. Heat Transfer **143**(6), 060801 (2021)
5. El Kasmi, A., Masson, C.: An extended k-$\epsilon$ model for turbulent flow through horizontal-axis wind turbines. J. Wind Eng. Ind. Aerodyn. **96**(1), 103–122 (2008). https://doi.org/10.1016/j.jweia.2007.03.007
6. Elsken, T., Metzen, J.H., Hutter, F.: Neural architecture search: a survey. J. Mach. Learn. Res. **20**(1), 1997–2017 (2019)
7. Glorot, X., Bengio, Y.: Understanding the difficulty of training deep feedforward neural networks. In: Proceedings of the Thirteenth International Conference on Artificial Intelligence and Statistics, pp. 249–256. JMLR Workshop and Conference Proceedings (2010)
8. Jin, X., Cai, S., Li, H., Karniadakis, G.E.: NSFnets (Navier-stokes flow nets): physics-informed neural networks for the incompressible Navier-stokes equations. J. Comput. Phys. **426**, 109951 (2021)
9. Jones, W.P., Launder, B.E.: The prediction of laminarization with a two-equation model of turbulence. Int. J. Heat Mass Transf. **15**(2), 301–314 (1972). https://doi.org/10.1016/0017-9310(72)90076-2
10. Kovasznay, L.I.G.: Laminar flow behind a two-dimensional grid. Math. Proc. Cambridge Philos. Soc. **44**(1), 58–62 (1948). https://doi.org/10.1017/S0305004100023999
11. van der Laan, M.P., Kelly, M., Floors, R., Peña, A.: Rossby number similarity of an atmospheric rans model using limited-length-scale turbulence closures extended to unstable stratification. Wind Energy Sci. **5**(1), 355–374 (2020)
12. Lam, R., et al.: Learning skillful medium-range global weather forecasting. Science **382**(6677), eadi2336 (2023)

13. Launder, B.E., Spalding, D.B.: The numerical computation of turbulent flows. Comput. Methods Appl. Mech. Eng. **3**(2), 269–289 (1974). https://doi.org/10.1016/0045-7825(74)90029-2

14. Levin, E., Fleisher, M.: Accelerated learning in layered neural networks. Complex Syst. **2**(625–640), 3 (1988)

15. Li, Z., et al.: Fourier neural operator for parametric partial differential equations (2020). arXiv preprint arXiv:2010.08895

16. Lim, K.L., Dutta, R., Rotaru, M.: Physics informed neural network using finite difference method. In: 2022 IEEE International Conference on Systems, Man, and Cybernetics (SMC), pp. 1828–1833. IEEE (2022)

17. Lu, L., Jin, P., Pang, G., Zhang, Z., Karniadakis, G.E.: Learning nonlinear operators via DeepONet based on the universal approximation theorem of operators. Nat. Mach. Intell. **3**(3), 218–229 (2021)

18. Mao, Z., Jagtap, A.D., Karniadakis, G.E.: Physics-informed neural networks for high-speed flows. Comput. Methods Appl. Mech. Eng. **360**, 112789 (2020)

19. Pang, G., Lu, L., Karniadakis, G.E.: fPINNs: fractional physics-informed neural networks. SIAM J. Sci. Comput. **41**(4), A2603–A2626 (2019)

20. Raissi, M., Perdikaris, P., Karniadakis, G.: Physics-informed neural networks: a deep learning framework for solving forward and inverse problems involving nonlinear partial differential equations. J. Comput. Phys. **378**, 686–707 (2019)

21. Raissi, M., Perdikaris, P., Karniadakis, G.E.: Physics-informed neural networks: a deep learning framework for solving forward and inverse problems involving nonlinear partial differential equations. J. Comput. Phys. **378**, 686–707 (2019)

22. Ramachandran, P., Zoph, B., Le, Q.V.: Searching for activation functions (2017). arXiv preprint arXiv:1710.05941

23. Richards, P.J., Hoxey, R.P.: Appropriate boundary conditions for computational wind engineering models using the k-ε turbulence model. J. Wind Eng. Ind. Aerodyn. **46–47**, 145–153 (1993). https://doi.org/10.1016/0167-6105(93)90124-7

24. Sankaran, S., Wang, H., Guilhoto, L.F., Perdikaris, P.: On the impact of larger batch size in the training of physics informed neural networks. In: The Symbiosis of Deep Learning and Differential Equations II (2022)

25. Sun, L., Gao, H., Pan, S., Wang, J.X.: Surrogate modeling for fluid flows based on physics-constrained deep learning without simulation data. Comput. Methods Appl. Mech. Eng. **361**, 112732 (2020)

26. Ueckermann, M.P., Lermusiaux, P.F.: 2.29 finite volume MATLAB framework documentation. MSEAS report **14** (2012)

27. Wang, C.: Exact solutions of the steady-state Navier-stokes equations. Annu. Rev. Fluid Mech. **23**(1), 159–177 (1991)

28. Wang, S., Teng, Y., Perdikaris, P.: Understanding and mitigating gradient flow pathologies in physics-informed neural networks. SIAM J. Sci. Comput. **43**(5), A3055–A3081 (2021)

29. Wang, Y., Han, X., Chang, C.Y., Zha, D., Braga-Neto, U., Hu, X.: Auto-PINN: understanding and optimizing physics-informed neural architecture (2022). arXiv preprint arXiv:2205.13748

30. Yang, L., Meng, X., Karniadakis, G.E.: B-PINNs: Bayesian physics-informed neural networks for forward and inverse PDE problems with noisy data. J. Comput. Phys. **425**, 109913 (2021)

# Generative Modeling of Sparse Approximate Inverse Preconditioners

Mou Li[1] , He Wang[2(✉)] , and Peter K. Jimack[1(✉)]

[1] University of Leeds, Leeds LS2 9JT, UK
{scmli,p.k.jimack}@leeds.ac.uk
[2] University College London, London WC1E 6BT, UK
he_wang@ucl.ac.uk
http://drhewang.com

**Abstract.** We present a new deep learning paradigm for the generation of sparse approximate inverse (SPAI) preconditioners for matrix
systems arising from the mesh-based discretization of elliptic differential
operators. Our approach is based upon the observation that matrices
generated in this manner are not arbitrary, but inherit properties from
differential operators that they discretize. Consequently, we seek to represent a learnable distribution of high-performance preconditioners from
a low-dimensional subspace through a carefully-designed autoencoder,
which is able to generate SPAI preconditioners for these systems. The
concept has been implemented on a variety of finite element discretizations of second- and fourth-order elliptic partial differential equations
with highly promising results.

**Keywords:** Deep learning · Sparse matrices · Preconditioning ·
Elliptic partial differential equations · Finite element methods

## 1 Introduction

Finding the solution of systems of linear algebraic equations,

$$Ax = b \,, \tag{1}$$

has been a core topic in the field of scientific computation for many decades. Such
systems arise naturally in a wide range of algorithms and applications, including
from the discretization of systems of partial differential equations (PDEs). Not
only are the equations (1) the main product of the discretization process for
linear PDEs but such systems must be solved during every nonlinear iteration
when solving nonlinear PDE systems [4]. Furthermore, this step is often the most
computationally expensive in any given numerical simulation.

Direct methods, based on the factorisation of the matrix $A$ into easily invertible triangular matrices are designed to be highly robust. However they generally
scale as $O(n^3)$ as the system size, $n$, increases. Hence as $n$ approaches values in

© The Author(s), under exclusive license to Springer Nature Switzerland AG 2024
L. Franco et al. (Eds.): ICCS 2024, LNCS 14834, pp. 378–392, 2024.
https://doi.org/10.1007/978-3-031-63759-9_40

the millions the computational cost and memory requirements quickly become unacceptable. Fortunately, when the matrix $A$ is sparse, it is possible to optimize the factorization process to minimize the fill-in and avoid storing (or computing with) the zero entries in $A$ [21]. Such sparse direct methods significantly reduce the computational complexity but can still suffer from unacceptable memory and CPU requirements when $n$ becomes sufficiently large. Consequently iterative solution methods are a popular choice for large sparse systems (1).

Many iterative methods have been developed as an alternative approach to direct methods, offering a flexible level of accuracy as the trade-off for memory storage and computational cost. The Conjugate Gradient (CG) algorithm is arguably the most well-developed and efficient iterative method for solving large, sparse, symmetric positive definite (SPD) systems, whilst other Krylov-subspace algorithms are widely-used for indefinite or non-symmetric systems [11]. In the SPD case, the convergence of the CG method is dependent on the distribution of the eigenvalues of the stiffness matrix $A$ [33]. Assuming that these eigenvalues are widely spread along the real line, the CG algorithm will suffer from slow convergence when the problem is ill-conditioned. That is, the condition number $\kappa(A) \gg 1$, which is the ratio of the maximum, $\lambda_{\max}(A)$, over the minimum, $\lambda_{\min}(A)$, eigenvalues of $A$. Consequently, a longstanding topic of research has been the development of preconditioning techniques to reduce $\kappa(A)$ (strictly $\kappa(AP)$) via a transformation of the stiffness matrix $A$ through multiplication with a carefully selected non-singular matrix $P$ on both sides of the equation: $APx = bP$.

Selection of a "good" preconditioner, $P$ should take account of the following two requirements:

1  It should be easy to construct and cheap to implement (i.e. to solve $Px = y$ for a given vector $y$).
2  It should significantly reduce the condition number of the modified problem, $\kappa(AP)$.

In general these requirements are in conflict with each other, so frequently we seek to use prior knowledge about the linear system in order to manage the trade-off between point one and two. If little is known about $A$ then one of the simplest choices for $P$ is the diagonal of $A$ (frequently referred to as Jacobi preconditioning [11]), which clearly satisfies the first requirement above but typically satisfies the second requirement less well. More complex preconditioners, which aim to do better on the second requirement, include incomplete factorization approaches such as incomplete Cholesky (IC) and incomplete LU (ILU) decompositions [7,11], multigrid (MG) and algebraic multigrid (AMG) approaches [5,15], domain decomposition (DD) methods [27], and sparse approximate inverse (SPAI) methods [2] which are the main topic of this paper. In recent developments there are also some attempts to utilise data-driven methods to generate preconditioners [3,17,20,23,24,31].

If the exact inverse of $A$ were known then setting $P = A^{-1}$ would perfectly satisfy condition 2 above. However, the cost of constructing and implementing this $P$ would mean that condition 1 would fail to be satisfied. Sparse approximate inverse (SPAI) methods seek to balance these requirements more evenly by constructing a cheaper approximation to $A^{-1}$ ($P \approx A^{-1}$), where $P$ is a sparse matrix so as to ensure that its application, $y = Px$, has a cost of $O(n)$ rather than $O(n^2)$. Designing a high performance preconditioner by the conventional numerical approaches mentioned above requires prior knowledge and past experience of specific families of PDE systems. The final product also varies with chosen parameters, this suggests that there is a distribution of high-performance preconditioners for $A$, which naturally justifies using generative models. In this paper, we propose a deep learning based generative model for constructing a SPAI preconditioner, $P$, for a given SPD matrix $A$ arising from the finite element discretization of a self-adjoint PDE or system. Our method uses a conditioned variational auto-encoder to map the conditioned distribution of $p(A^{-1}|A)$ into a lower dimensional latent space. After training, it is able to generate high-performance preconditioners for SPD matrices arising from the discretization of unseen self-adjoint problems under the same mesh density.

Whilst this is not the first research to propose the use of machine learning (ML) techniques to generate SPAI preconditioners (see [31] for example), we believe that this is the first to consider the use of a graph representation of the sparse matrix $A$ and the first to consider modeling the distribution of $A$ and $A^{-1}$ for such a task. This permits the representation of $A$ in a smaller latent space through our use of a variational auto-encoder. Consequently, since our model only cares about the latent data distribution of the inverse of the stiffness matrix A, it is easily generalisable to other self-adjoint problems. Furthermore, we propose a controllable sparsity pattern for the preconditioner to allow a trade-off between the performance and the computational cost.

## 2    Related Work

In this section we provide a brief overview of the most common approaches to preconditioning that are in widespread use in numerical codes today. This is then followed by a short subsection on existing research into the use of data-driven methods for generating preconditioners. As part of that section we also highlight some of the proposed techniques for using ML to solve PDE systems since these could also be used to motivate new preconditioners (i.e. by applying as a preconditioner rather than a solver).

### 2.1    Numerical Preconditioning

Researchers have studied different methods for generating effective preconditioners over many decades, leading to a vast body of work on this topic [34]. For the purposes of this paper we restrict our attention to sparse SPD matrices, $A$, arising from mesh-based discretizatons of self-adjoint PDEs. As already noted, the simplest approach that is in widespread use is Jacobi preconditioning, where $P = \mathrm{diag}(A)^{-1}$. This is easy to construct and cheap to implement, but not sufficiently effective for many important problems [23]. Consequently more sophisticated techniques are frequently required, such as those based upon incomplete factorizations of $A$. Incomplete Cholesky (IC) factorization is most appropriate for SPD matrices, where $A \approx LL^T$. The sparsity pattern of the lower triangular matrix $L$ can be chosen to be identical to that of the lower triangle of $A$ or can be slightly more generous based upon a drop tolerance that is used to decide whether to allow any "fill-in" during the factorization process [34]. Note that this technique requires a forward and a backward substitution in order to apply the preconditioner, which is not generally well-suited to efficient parallel implementation. Furthermore, this approach requires careful design to balance the computational cost and accuracy associated with different levels of fill-in (complete fill-in leads to perfect factorization, which means that $\kappa(AP) = 1$ but at prohibitive cost, whereas insufficient fill-in can lead to a sub-optimal condition number).

The other class of preconditioners that we discuss in detail here are sparse approximate inverse (SPAI) methods, which are the main topic of this paper. This variant evaluates the preconditioner $P$ to be a sparse approximation to the inverse of the stiffness matrix $A$. Frobenius norm minimisation and incomplete bi-conjugation are the two most widely implemented frameworks to compute the SPAI preconditioners [4,34]. One major difficulty of SPAI preconditioning is the choice of the sparsity pattern. Since the inverse of an irreducible sparse matrix is proven to be a structurally dense matrix [9], when the sparsity pattern is predefined, such preconditioners may not work well if there exist entries with large magnitude outside the defined pattern. Attempts to address this by automatically capturing a pseudo-optimal sparsity pattern include the use of a drop tolerance [8] or of a "profitability factor" via a residual reduction process [13,16]. In this work we consider only predefined sparsity patterns based around the sparsity of the family of matrices, $A$, being considered.

## 2.2 Data Driven Preconditioning

In recent years a number of approaches have been proposed for the direct solution of systems of PDEs using both supervised and unsupervised ML methods. Noteworthy, and highly influential, early examples include the deep Ritz method [10], the deep Galerkin method [32] and physics-informed neural networks (PINNs) [28–30]. The latter approach has led to a substantial, and rapidly increasing, body of research into the unsupervised learning of PDE solutions and related inverse problems. However, as forward solvers, PINNs are not generally competitive with classical numerical approaches based upon efficient preconditioning. It is for this reason, and inspired by the successes of PINNs, that we seek to utilise the power of machine learning to generate high-performance preconditioners (as opposed to complete solvers).

There is relatively little prior work on the use of machine learning to develop preconditioners for use within conventional numerical algorithms. The first attempt to generate a SPAI preconditioner appears in [31], where a convolutional neural network (CNN) is used to derive an approximate triangular factorization of the inverse matrix. More recent research, such as [17,23] has focused on incomplete LU and Cholesky preconditioners, with the former targeting the non-self-adjoint Navier-Stokes equations and the latter two considering SPD systems: the work of [23] being most similar to that considered in this paper due to their use of a GNN. Other approaches that have been taken to develop novel preconditioners include [3], which mimics a multigrid approach through a combination of a CNN-based smoother and coarse-grid solvers, and [20], which builds a preconditioner directly upon a DeepONet approximation of the operator that represents the solution of the underlying PDE [24].

## 3   Methodology

In this initial investigation we focus on distributions of sparse matrices, $A$, arising from the finite element discretization of elliptic PDEs: the precise sparsity pattern depending upon the differential operator, the finite element spaces used, and the mesh on which the discretization occurs. Even though the inverse matrices, $A^{-1} \in \mathbb{R}^{n \times n}$ are generally dense in structure, other preconditioning methods are able to approximate cheaper versions of $A^{-1}$, which led us to assume that there exists a learnable distribution of high performance preconditioners within a lower dimensional subspace of $\mathbb{R}^{n \times n}$ with a prescribed sparsity pattern. We then propose a graph-conditioned variational autoencoder (GCVAE) architecture which is able to generate such SPAI preconditioners for these SPD linear systems.

### 3.1   Problem Setup

Inspired by the traditional Frobenius norm minimization approach of SPAI preconditioning, our starting point is the following assumption:

$$\forall A \in S_A \quad \exists R \in S_M : \left\| I - R^T A R \right\|_F \approx 0 , \tag{2}$$

where $S_A$ is our set of possible sparse SPD matrices and $S_M \subset \mathbb{R}^{n \times n}$ with a prescribed sparsity pattern defined by a selected mask, $M$. Let us consider the dataset $\mathcal{A} = \{A^{(i)}\}_{i=1}^{N}$ having $N$ independently and identically distributed samples and drawn from an unknown distribution $p(\mathcal{A})$, generated from a given family of linear PDE problems. Let $\mathcal{A}^{-1} = g(\mathcal{A})$ where $g(\cdot)$ is the inverse matrix transformation function. We assume that the conditional distribution of $p(\mathcal{A}^{-1}|\mathcal{A})$ lies within a parametric family of distributions, such as Gaussian, on a lower-dimensional latent space such that $p_\theta(z|\mathcal{A}, \mathcal{A}^{-1}) = \mathcal{N}(\mu, \sigma^2)$, where $\theta$ represents the learnable parameters of the neural networks. We also assume that for every given $A^{(i)}$ there exists a set of high-performance preconditioners $\mathcal{R}^{(i)}$ which satisfies Eq. 2 with a given sparsity pattern $M$ and $\mathcal{R} \sim p_\theta(z|\mathcal{A}, \mathcal{A}^{-1})$. Then the SPAI preconditioners $\mathcal{R}$ can be generated from the generative distribution $p_\theta(\mathcal{R}|z, \mathcal{A})$ conditioned on input $\mathcal{A}$, the prescribed sparsity pattern of $\mathcal{R}$, and latent variable $z \sim \mathcal{N}(\mu, \sigma^2)$.

The application of variational auto-encoders (VAEs) has shown a great potential in learning the probability distribution of a given dataset [18] and, when coupled with a condition, can ensure the robustness of the generative model [26]. VAEs have also shown great potential in modeling the distribution of the utility matrix for a recommender system [1] and for a gene expression matrix [25], both based upon large and sparse matrices. Therefore, we propose a conditional VAE generative model to approximate SPAI preconditioners.

## 3.2 Architecture Design

As noted above, the type of architecture we now consider is a conditioned variational auto-encoder (cVAE) [14]. The encoder part consists of two different encoders, which we now discuss in turn.

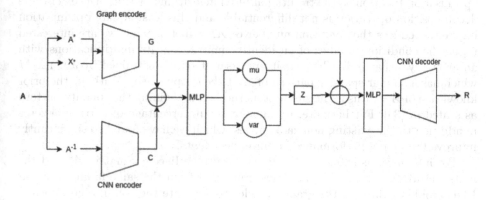

**Fig. 1.** Schematic diagram of the proposed graph-conditioned variational autoencoder

As shown in Fig. 1, the first is a conditional part (a graph encoder) that consists of several layers of a graph neural network (GNN). The input data $A$ is a sparse matrix which is represented as a graph, where the diagonal entries and off-diagonal entries are the nodes and edges of the graph respectively. GNNs have two clear advantages over CNNs under our setting: 1) a GNN can perform the feature mapping with a full "view" of the input data through its message passing mechanism, 2) a GNN is only interested in the entries that are non-zero therefore it is a sparse implementation, which significantly reduces the memory requirement and the training cost. The graph VAE is first proposed by [19], where it is used to learn latent representations of unweighted undirected graphs with multidimensional node features for link prediction in citation networks. In our case, the system matrix $A$ can be treated as a weighted and directed graph with single dimensional node features. Therefore we replace the graph convolutional layer (GCN) with the graph attention layer (GATv2) [6]. This graph encoder takes $A^*$ and $X^*$ as the input, where $A^*$ is the self-looped adjacency matrix of

$A$ and $X^*$ is the node feature matrix (where $X^* \in \mathbb{R}^{n \times 1}$ and $n$ is the number of nodes).

The second encoder is chosen to be a CNN encoder since it takes $A^{-1}$, the exact inverse of the system matrix $A$, as the input. CNNs are known to be powerful and efficient feature extractors for image data [12], where grey-scale images are fundamentally dense matrices. During training, we assume the latent distribution is a Gaussian, i.e. $z \sim \mathcal{N}(\mu, \sigma^2)$, both encoders' outputs are concatenated and then passed through a multilayer perceptron (MLP) to approximate the mean($\mu$) and the log variance($\phi$) of the latent distribution. The latent variable $z$ is then reconstructed deterministically through the reparameterization trick [18] i.e. $z = \mu + \sigma \epsilon$, where $\sigma = e^{\frac{1}{2}\phi}$, $\epsilon$ is an auxiliary noise variable sampled from standard Gaussian such that $\epsilon \sim \mathcal{N}(0, 1)$. This technique makes the stochastic estimation of the latent variable $z$ differentiable. The decoder reconstructs $R \sim A^{-1}$ from $z$ such that $\left\| I - R^T A R \right\|_F \approx 0$. We implemented the same optimisation function as the traditional SPAI algorithms, because the eigenvalue decomposition operation is not differentiable and this least-square optimisation implicitly reduces the condition number of $AR$ which is what we are interested in, as the condition number of an identity matrix and its multiplications with an arbitrary scalar is 1. The sparsity pattern of $R$ is constrained by a mask $M$ which has a similar sparsity pattern of $A$ at the output layer. Without the prior knowledge of $A^{-1}$, this is the simplest method of predefining the sparsity pattern as stated in [34]. Furthermore, we allow a small percentage of extra non-zeros in addition to the existing non-zero entries, which we have found to significantly improve the model performance in our experiments.

For inference, as shown in Fig. 2 we pass the adjacency matrix $A^*$ and the node feature $X^*$ of an unseen stiffness matrix $A$ from the same family of linear PDE problems through the graph encoder to generate the conditional information $G$, then concatenate with $z$ sampled from $z \sim \mathcal{N}(\mu, \sigma^2)$ and pass it through the decoder to generate the preconditioner $R$. Due to the nature of our model, multiple different $R$ can be generated for a single $A$.

### 3.3    Loss Function

In general, the family of VAE networks optimise the well-known evidence lower bound (ELBO) [18]:

$$L = \mathbb{E}[\log p(R|Z)] - \alpha D_{KL}[q(Z|X) \| p(Z)], \tag{3}$$

where the prior $Z \sim \mathcal{N}(0, I)$ s.t. $p(Z) = \prod p(z_i) = \prod \mathcal{N}(z_i|0, I)$. This type of loss function contains a reconstruction term, which is the first term that calculates the expected negative reconstruction error. The second term is the KL-divergence term, and it can be regarded as a regulariser which encourages the posterior distribution to be close to the prior, in this case the prior is assumed as the standard Gaussian. In our case, we replace the reconstruction term with $\left\| I - R^T A R \right\|_F$ and minimise the following loss function:

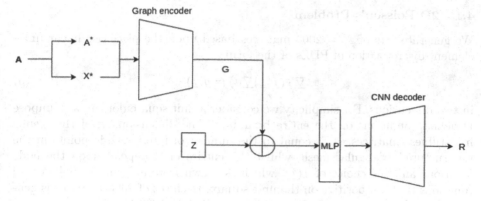

**Fig. 2.** Schematic diagram of the inference process

$$L = \mathbb{E}[\|I - R^T A R\|_F^2] - \alpha D_{KL}[q(Z|X)\|p(Z)]$$
$$\textbf{\textit{for}} \quad X = GNN_\theta(A^*, X^*) \oplus CNN_\theta(A^{-1}), \tag{4}$$

where $\alpha$ is the regularising parameter, which is chosen to be 0.1. $GNN_\theta(\cdot)$ and $CNN_\theta(\cdot)$ represent the GNN encoder and CNN encoder as shown in Fig. 1, and $\oplus$ is the aggregation operator.

## 4   Experiments

In the first set of computational experiments that we consider (Subsects. 4.1 and 4.2) the dataset, $\mathcal{A}$, is drawn from piecewise linear finite element discretizations of a family of second order elliptic PDEs in two dimensions. The second set of experiments considers a more challenging dataset, $\mathcal{A}$, that is drawn from discontinuous (piecewise quadratic) Galerkin discretizations of a family of two-dimensional biharmonic problems (Subsects. 4.3 and 4.4). Such problems are known to lead to highly ill-conditioned matrix systems upon discretization (as illustrated below). Both experiments are trained on a single NVDIA RTX 3090 GPU with 24 Gigabyte of VRAM. The dataset preparation of both problems takes less than 30 mins for the largest problem size. For the largest problem size of 2D Poisson's problem $1873 \times 1873$, the training converges at 200 epochs and around 6 mins per epoch. Similarly, for the largest biharmonic problem, of size $1089 \times 1089$, the training converges at 260 epochs and around 2 mins per epoch. The training cost is expected to grow linearly with the total number of hyper-parameters, which depends on many different factors, such as the discretized mesh density, parallelisation, depth of the model, number of the channels of the CNN layers, etc. For practical implementation of the model, fine tuning of the hyperparameters is required to find the optimal balance between the training cost and model performance, though this is not the focus of this paper.

## 4.1  2D Poisson's Problem

We generate sets of $N = 2000$ matrices based upon the piecewise linear finite element discretization of PDEs of the form

$$\underline{\nabla} \cdot (f(\underline{x})\underline{\nabla}u) = g(\underline{x}), \tag{5}$$

in two dimensions. For simplicity we consider a unit square domain and impose Dirichlet conditions on the entire boundary. The dimension, $n$, of the resulting stiffness matrices, $A$, is equal to the number of interior node points in the unstructured triangular mesh, whilst the entries of $A$ depend upon the node locations and the choice of $f(\underline{x})$, which is drawn from a family of polynomial functions that are positive on the unit square. Each set of 2000 matrices is generated on the same mesh and is randomly split into 1600 training samples and 400 for testing.

## 4.2  Results and Discussion - Poisson Family

This section demonstrates the performance of the preconditioners generated by our GCVAE model when applied with a CG solver using a relative convergence criterion of 1.0e-5. Included within our results is a comparison with two baseline methods: Jacobi preconditioning and Super Nodal incomplete LU factorisation (SPILU) [21] (using its symmetric mode to obtain IC preconditioning). Note that the performance of the latter comparator depends critically on the value of a "drop tolerance" parameter, which controls the amount of fill-in that is permitted during the incomplete factorization of $A$. When this is very small the IC preconditioner is highly effective but at the expense of significant additional computational cost (reducing the overall efficiency of the solver); when the drop-tolerance is larger the cost of computing and applying the preconditioner goes down but at the expense of a much larger number of CG iterations.

Figure 3 (left) shows the average estimated condition numbers for preconditioned systems with each choice of preconditioner. For the SPILU case we have artificially selected the drop-tolerance for each problem size so as to match the number of iterations taken using the GCVAE preconditioner. Consequently, by design, the equivalent curve in Fig. 3 (right), which shows average CG iterations, completely overlays the GCVAE curve. Also in this graph are the average iteration counts for SPILU applied in its more conventional form, with a constant drop-tolerance (in this case 0.12). It is clear from these examples that both the condition number and the CG iteration counts with Jacobi preconditioning grow at a much faster rate than for the GCVAE approach, demonstrating that our model will out-perform the Jacobi method on total execution time for sufficiently large problems.

Comparison against SPILU is less straightforward due to the trade-offs in the choice of drop-tolerance described above. The SPILU algorithm uses $(A^T + A)$ based column permutation [22] which causes the condition number to increase rapidly with the problem size as shown in Fig. 3 (left). This illustrates the limitation of using the condition number as the measure of quality, which is why we

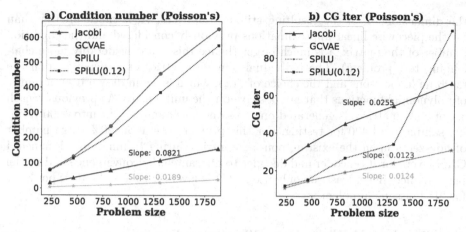

**Fig. 3.** Comparison of the condition number and CG algorithm iteration count for Jacobi preconditioning, SPILU preconditioning and our proposed method (GCVAE) when applied to discretizations of a family of second order operators

prefer to use iteration count. Nevertheless, even with this measure, Fig. 3 (right) shows that the parameters of the SPILU method need to be carefully tuned to achieve optimal performance. By carefully reducing drop-tolerance as $n$ increases we are able to match the iteration counts of the GCVAE preconditioner, though the latter has significantly lower condition numbers. In Fig. 5 (left) we also compare the density of non-zero entries in these two preconditioners for different choices of $n$, in order to give an indication of their computational costs. It can be seen that for sufficiently large systems the GCVAE preconditioner will be expected to have fewer non-zeros. Furthermore, to apply the SPILU preconditioner requires backward and forward substitution to be applied which cannot naturally be done in parallel, whereas the SPAI preconditioning is ideally suited to parallel implementation (since it just requires a sparse matrix-vector multiplication). The execution time benchmark test has not been carried out as our model is currently implemented in a non-optimised manner in a Python environment, whereas the state-of-art numerical software SPILU is implemented in a highly optimised C-language environment. Nevertheless, the advantage of the data driven approach is that, once it is tuned and trained, it can be used as a black box tool with execution time complexity close to $\mathcal{O}(n)$.

### 4.3  Biharmonic Problem

We again generate sets of $N = 2000$ matrices, this time based upon the piecewise quadratic discontinuous Galerkin discretization of fourth order PDEs of the form

$$\nabla^2(f(\underline{x})\nabla^2 u) = g(\underline{x}) \tag{6}$$

in two dimensions. We consider a unit square domain and impose Dirichlet conditions on both $u$ and $\nabla^2 u$ on the entire boundary. For a given triangular mesh, the dimension, $n$, of the resulting stiffness matrices, $A$, is much greater than for the piecewise linear approximations previously considered and the condition number of the matrix $A$ is much larger (hence this is a considerably more challenging test problem). The individual non-zero entries of $A$ depend upon the mesh node locations and the choice of $f(\underline{x})$, which is again drawn from a family of polynomial functions that are positive on the unit square. As previously, each set of 2000 matrices is generated on the same mesh and is split into 1600 training samples and 400 for testing. For this problem, we allow 20% extra number of non-zeros upon the existing non-zero entries during training and designed 4 CNN layers for its encoder and decoder to guarantee the convergence and model performance. In contrast, we only used 3 CNN layers with less channels per layer for 2D Poisson's problem.

## 4.4    Results and Discussion - Biharmonic Family

Similarly, in this subsection, we compare our method against two baseline methods: Jacobi and SPILU. As shown in Fig. 4, for this ill-conditioned problem, as the problem size increases our method significantly outperforms Jacobi preconditioning in terms of both the condition number reduction and the CG convergence rate.

Comparison against SPILU shows similar features as for the previous test case. For the reasons described previously we do not find condition number to be a useful metric in this case and therefore focus on iteration counts. Figure 4 (right) shows that it is possible to tune the value of drop-tolerance for each problem size in order to match the number of CG iterations obtained with the

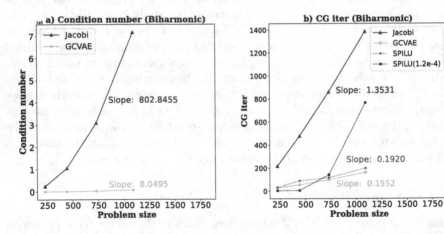

**Fig. 4.** Comparison of the condition number (in the unit of $1.0 \times 10^5$) and CG algorithm iteration count for Jacobi preconditioning, SPILU preconditioning and our proposed method (GCVAE) when applied to discretizations of a family of fourth order operators. The condition numbers of the SPILU approaches are not illustrated in **a)** as they are too large to fit into the plot.

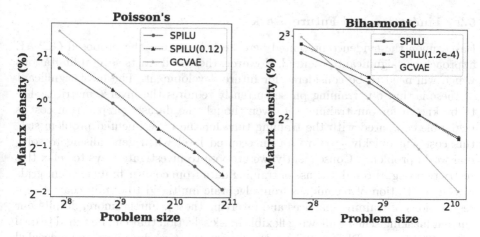

**Fig. 5.** Comparison of the proportion of non-zero entries in the preconditioners generated via the SPILU and GCVAE methods for the second order (left) and fourth order (right) problems

GCVAE preconditioner, as well as showing the growth in iterations when a constant drop-tolerance is used (in this case $1.2 \times 10^{-4}$). Equally significantly however, we see from Fig. 5 (right) that the number of non-zeros required for the GCVAE preconditioner is significantly fewer than for the SPILU preconditioner, even when drop-tolerance is tuned to match the number of iterations. Combined with the fact that application of SPILU requires backward and forward substitution, we observe that this preconditioner is much more expensive to apply than our proposed approach.

## 5   Conclusions and Future Work

### 5.1   Conclusions

This paper proposes a novel generative modelling framework to generate sparse approximate inverse preconditioners for matrix systems arising from the mesh-based discretization of families of self-adjoint elliptic partial differential equations. The approach has shown excellent potential in terms of its ability to reduce the condition number of the systems considered and to increase the convergence rate of the conjugate gradient solver. This performance has been demonstrated for two different families of partial differential equations: of second and fourth order respectively. Furthermore, comparing against a state-of-art factorisation-based preconditioning technique we observe that, for sufficiently large systems, our proposed approach should require fewer floating point operations to apply and that it is better suited to future parallel implementation.

## 5.2 Limitation and Future Work

Based upon the evidence presented here we believe that the proposed GCVAE approach has significant potential. However, there still exists several limitations which will need to be considered for future development. The most significant of these is that our training phase currently requires the inverse matrices $A^{-1}$ to be known for our training set, even though the dataset preparation cost is negligible compared with the training time for the experimented problem size, this cost will quickly grow to an unaccepted level when generalising to solve real world problems. Consequently, we propose to investigate ways to relax this restriction (e.g. through the use of training data from coarser finite element grids or the application of techniques from algebraic multigrid to obtain coarser representations of training matrices and inverting these). Furthermore, despite our current masking scheme allowing flexible mask selection that can be tuned to deal with more complex PDE problems, the sparsity pattern does need to be decided prior to training which may restrict the generalisation capability of the model. A learning-based masking approach could be more flexible and efficient, so will be attempted in further development. Finally, the work in this paper is limited to self-adjoint elliptic problem, which result in symmetric positive-definite stiffness matrices. We also intend to generalise our model to other families of PDE problems, which lead to non-symmetric and/or indefinite linear systems.

**Acknowledgments.** The first author gratefully acknowledges receipt of a doctoral training award from the UK Engineering and Physical Sciences Research Council.

**Disclosure of Interests.** The authors have no competing interests to declare that are relevant to the content of this article.

# References

1. Ali, A., Yu, L., Farida, M., Majjed, A.Q.: Variational autoencoder Bayesian matrix factorization (VABMF) for collaborative filtering. Appl. Intell. **51**(7), 5132–5145 (2021). https://doi.org/10.1007/s10489-020-02049-9
2. Anzt, H., Huckle, T.K., Bräckle, J., Dongarra, J.: Incomplete sparse approximate inverses for parallel preconditioning. Parallel Comput. **71**, 1–22 (2018)
3. Azulay, Y., Treister, E.: Multigrid-augmented deep learning preconditioners for the Helmholtz equation. SIAM J. Sci. Comput. **45**(3), S127–S151 (2023). https://doi.org/10.1137/21M1433514
4. Benzi, M.: Preconditioning techniques for large linear systems: a survey. J. Comput. Phys. **182**(2), 418–477 (2002). https://doi.org/10.1006/jcph.2002.7176
5. Bramble, J.H.J.H.: Multigrid methods. Pitman Research Notes in Mathematics Series, 294, Longman Scientific and Technical, Harlow (1993)
6. Brody, S., Alon, U., Yahav, E.: How Attentive are Graph Attention Networks? (2022). arXiv:2105.14491v3, https://doi.org/10.48550/arXiv.2105.14491
7. Chan, T.F., Van der Vorst, H.A.: Approximate and Incomplete Factorizations, pp. 167–202. Springer Netherlands, Dordrecht (1997). https://doi.org/10.1007/978-94-011-5412-3_6

8. Chow, E., Saad, Y.: Approximate inverse preconditioners via sparse-sparse iterations. SIAM J. Sci. Comput. **19**(3), 29 (1998). https://doi.org/10.1137/S1064827594270415

9. Duff, I.S., Erisman, A.M., Gear, C.W., Reid, J.K.: Sparsity structure and Gaussian elimination. ACM SIGNUM Newsletter **23**(2), 2–8 (1988). https://doi.org/10.1145/47917.47918

10. E, W., Yu, B.: The deep ritz method: a deep learning-based numerical algorithm for solving variational problems. Commun. Math. Stat. **6**(1), 1–12 (2018). https://doi.org/10.1007/s40304-018-0127-z

11. Golub, G.H.G.H., Van Loan, C.F.: Matrix Computations. Johns Hopkins Studies in the Mathematical Sciences, The Johns Hopkins University Press, Baltimore, 4 th edn. (2013)

12. Goodfellow, I., Bengio, Y., Courville, A.: Deep Learning. MIT Press, Cumberland, United States (2016)

13. Grote, M.J., Huckle, T.: Parallel preconditioning with sparse approximate inverses. SIAM J. Sci. Comput. **18**(3), 16 (1997). https://doi.org/10.1137/S1064827594276552

14. Gu, C., Zhao, S., Zhang, C.: Diversity-promoting human motion interpolation via conditional variational auto-encoder (2021). arXiv:2111.06762v1, http://arxiv.org/abs/2111.06762

15. Henson, V.E., Yang, U.M.: BoomerAMG: a parallel algebraic multigrid solver and preconditioner. Appl. Numer. Math. **41**(1), 155–177 (2002). https://doi.org/10.1016/S0168-9274(01)00115-5

16. Huckle, T.: Factorized sparse approximate inverses for preconditioning. J. Supercomput. **25**(2), 109–117 (2003). https://doi.org/10.1023/A:1023988426844

17. Häusner, P., Öktem, O., Sjölund, J.: Neural incomplete factorization: learning preconditioners for the conjugate gradient method (2024). arXiv:2305.16368v2. https://doi.org/10.48550/arXiv.2305.16368

18. Kingma, D.P., Welling, M.: Auto-encoding variational bayes (2014). arXiv:1312.6114, http://arxiv.org/abs/1312.6114

19. Kipf, T.N., Welling, M.: Variational graph auto-encoders (2016). arXiv:1611.07308, http://arxiv.org/abs/1611.07308

20. Kopaničáková, A., Karniadakis, G.E.: DeepOnet based preconditioning strategies for solving parametric linear systems of equations (2024). arXiv:2401.02016v2, https://doi.org/10.48550/arXiv.2401.02016

21. Li, X.S.: An overview of SuperLU: algorithms, implementation, and user interface. ACM Trans. Math. Software **31**(3), 302–325 (2005)

22. Li, X., Demmel, J., Gilbert, J., iL. Grigori, Shao, M., Yamazaki, I.: SuperLU users' guide. Tech. Rep. LBNL-44289, Lawrence Berkeley National Laboratory (1999). https://portal.nersc.gov/project/sparse/superlu/ug.pdf Last update: June 2018

23. Li, Y., Chen, P.Y., Du, T., Matusik, W.: Learning preconditioner for conjugate gradient PDE solvers (2023). arXiv:2305.16432v2, https://doi.org/10.48550/arXiv.2305.16432

24. Lu, L., Jin, P., Karniadakis, G.E.: DeepONet: learning nonlinear operators for identifying differential equations based on the universal approximation theorem of operators. Nat. Mach. Intell. **3**(3), 218–229 (2021). https://doi.org/10.1038/s42256-021-00302-5

25. Lukassen, S., Ten, F.W., Adam, L., Eils, R., Conrad, C.: Gene set inference from single-cell sequencing data using a hybrid of matrix factorization and variational autoencoders. Nat. Mach. Intell. **2**(12), 800–809 (2020). https://doi.org/10.1038/s42256-020-00269-9

26. Mirza, M., Osindero, S.: Conditional generative adversarial nets (2014). arXiv:1411.1784v1, https://arxiv.org/abs/1411.1784v1
27. Quarteroni, A., Valli, A.A.: Domain Decomposition Methods for Partial Differential Equations. Clarendon Press, Oxford, Numerical mathematics and scientific computation (1999)
28. Raissi, M., Perdikaris, P., Karniadakis, G.: Physics-informed neural networks: a deep learning framework for solving forward and inverse problems involving nonlinear partial differential equations. J. Comput. Phys. **378**, 686–707 (2019)
29. Raissi, M., Perdikaris, P., Karniadakis, G.E.: Physics informed deep learning (part I): data-driven solutions of nonlinear partial differential equations (2017). arXiv:1711.10561, http://arxiv.org/abs/1711.10561
30. Raissi, M., Perdikaris, P., Karniadakis, G.E.: Physics informed deep learning (Part II): data-driven discovery of nonlinear partial differential equations (2017). arXiv:1711.10566, http://arxiv.org/abs/1711.10566
31. Sappl, J., Seiler, L., Harders, M., Rauch, W.: Deep learning of preconditioners for conjugate gradient solvers in urban water related problems (2019). arXiv:1906.06925, http://arxiv.org/abs/1906.06925
32. Sirignano, J., Spiliopoulos, K.: DGM: a deep learning algorithm for solving partial differential equations. J. Comput. Phys. **375**, 1339–1364 (2018)
33. Tadmor, E.: A review of numerical methods for nonlinear partial differential equations. Bull. (New Series) Am. Math. Soc. **49** (2012). https://doi.org/10.1090/S0273-0979-2012-01379-4
34. Wathen, A.J.: Preconditioning. Acta Numer **24**, 329–376 (2015)

# Solving Sparse Linear Systems on Large Unstructured Grids with Graph Neural Networks: Application to Solve the Poisson's Equation in Hall-Effect Thrusters Simulations

Gabriel Vigot[✉], Bénédicte Cuenot, and Olivier Vermorel

Centre Européen de Recherche et de Formation Avancée en Calcul Scientifique,
42 Av. Gaspard Coriolis, 31100 Toulouse, France
{vigot,cuenot,vermorel}@cerfacs.fr

**Abstract.** The following work presents a new method to solve Poisson's equation and, more generally, sparse linear systems using graph neural networks. We propose a supervised approach to solve the discretized representation of Poisson's equation at every time step of a simulation. This new method will be applied to plasma physics simulations for Hall-Effect Thruster's modeling, where the electric potential gradient must be computed to get the electric field necessary to model the plasma's behavior. Solving Poisson's equation using classical iterative methods represents a major part of the computational costs in this setting. This is even more critical for unstructured meshes, increasing the problem's complexity. To accelerate the computational process, we propose a graph neural network to give an initial guess of Poisson's equation solution. The new method introduced in this article has been designed to handle any meshing structure, including structured and unstructured grids and sparse linear systems. Once trained, the neural network would be used inside a numerical simulation in inference to give an initial guess of the solution for each simulation time step for all right-hand sides of the linear system and all previous time step solutions. In most industrial cases, Hall-Effect thrusters' modeling requires a large unstructured mesh that one single processor cannot hold regarding memory capacity. We then propose a partitioning strategy to tackle the challenge of solving linear systems on large unstructured grids when they cannot be on a single processor.

**Keywords:** Plasma Physics · Partial differential equations · Graph Neural Networks · Sparse Linear systems · Partitioning

## 1 Introduction

Sparse linear systems may arise when discretizing Partial Differential Equations (PDEs) for numerical simulations. For elliptic problems like Laplace, Poisson, or Helmholtz equations, these problems are even more important when they

© The Author(s), under exclusive license to Springer Nature Switzerland AG 2024
L. Franco et al. (Eds.): ICCS 2024, LNCS 14834, pp. 393–407, 2024.
https://doi.org/10.1007/978-3-031-63759-9_41

need to be solved on a large unstructured mesh. Besides, the mesh itself could be partitioned into multiple subgraphs and distributed in parallel over multiple processors to compute the solution of the discretized PDEs. The linear system, when ill-conditioned, becomes harder to solve with a low convergence rate. Decades of research have been led to reduce the computational cost of solving a sparse linear system with efficiency with preconditioning techniques (Chen et al. [9]), or multi-level approach (Saad et al. [8]). Since the introduction of Physics Informed Neural Networks by Raissi et al. [7], multiple efforts have been made to help different kinds of solvers, find the solution to elliptic problems, applied to unstructured data that could be modeled as graph neural networks (Pfaff et al. [6], Sanchez-Gonzalez, Godwin et al. [13]). Other groups have focused more of their attention on solving sparse linear systems using neural network operators (Jiang et al. [2], Schäfer et al. [15]), whether with a supervised approach or an unsupervised approach (Stanziola et al. [18]). While some other groups dedicated their research to finding the right preconditioning matrix to help the iterative solvers converge faster (Li et al. [17], Sappl et al. [14], or Luz et al. [1], or Kopanicáková et al. [16]).

We propose a model to find the update $\Delta\Phi^t$ of the solution $\Phi^t = \Phi^{t-1} + \Delta\Phi^t$ for each time step $t$ of a simulation. In the end, we would like to demonstrate the ability of a neural network to solve elliptic PDEs discretized as sparse linear systems $\mathbf{A}\,\Phi^t = b^t$ where $\mathbf{A}$ represents the Laplacian operator and $b^t$ the right-hand side of Poisson's equation at a given time step $t$. In the context of industrial applications, we extend this work to a large unstructured mesh where the memory size of the graph is too large to fit onto a single graphics processing unit (GPU). The final goal is to have a graph neural network capable of predicting an initial guess $\Phi^t$ for each time step $t$ of a simulation based on the solution of the previous time step $\Phi^{t-1}$ and the right-hand side of the linear system $b^t$. The neural neural should also be able to understand the dynamics of the data represented in the graph, even if it is partitioned.

The following will first present the model and the neural network architecture. Then, two applications will be shown in the context of plasma simulations for Hall-effect Thruster design: the first one is a 2D simulation with triangular mesh, and the second one is a plasma discharge in 3D with irregular tetrahedral mesh, and the whole partitioned in multiple subgraphs.

## 2   Hybrid Model

The main objective of the method is to conciliate the solving speed of a linear system and the precision that the model has to reach at the end of each time step of a simulation. A particular focus is given to solving Poisson's equation, which is one fundamental step in plasma physics modeling. The expectations are that the neural network will be able to find a good approximation of the solution update for each time step of a simulation. During its training phase, the neural network is trained using a solution of Poisson's equation at a given simulation time step. Ultimately, the neural network should provide an initial guess of Poisson's equation that is close to the physical solutions given in its training phase.

The computational cost would decrease because a neural network's inference phase is less computationally expensive than a traditional iterative solver.

## 2.1 Poisson's Equation

Several discretization methods exist in numerical simulation to represent Poisson's equation on an unstructured grid. In this present work, Poisson's equation will be computed using the finite volume method. We describe our Poisson's problem on a given discretized domain $\mathring{\Omega}$ delimited by Dirichlet boundary conditions $\partial\Omega_D$:

$$\begin{cases} \nabla^2\Phi & = -\dfrac{q}{\epsilon_0}(n_i - n_e) \text{ on } \mathring{\Omega} \\ \Phi & = \Phi_D \quad \text{ on } \partial\Omega_D \end{cases} \tag{1}$$

where $-q(n_i - n_e)$ represents the right-hand side of our Poisson's equation with $q$ the electric charge of an electron, $\epsilon_0$ the vacuum permittivity, $n_i$ the charge density of ions constituting the plasma, and $n_e$ the electron charge density. The numerical scheme based on the AVBP solver created at CERFACS [12] defines the discretization of Poisson's equation with the Green-Ostrogradski theorem where for a nodal volume $V_i$ of a node $i$ belonging to the computational domain $E(i)$ for each cell $\tau$:

$$\int_{V_i} \nabla^2\Phi dV = \int_{\partial V_i} \nabla\Phi \cdot n dS = \sum_{\tau \in E(i)} \int_{\partial V_i \cap \tau} \nabla\Phi \cdot n \, dS, \tag{2}$$

where $n \, dS$ represents the orthogonal vector to the surface of the cell $\tau$. So that it is possible to build the discretized Laplacian operator over an unstructured mesh with primal cell volume $V_i$, which gives us the following linear system to solve:

$$\mathbf{A}\,\Phi^t = \mathbf{b}^t, \tag{3}$$

where for $t$ the designated time step of a simulation, $\mathbf{A}$ is the discretized Laplacian operator, $\Phi^t$ the solution of the linear system at time step $t$ and $\mathbf{b}^t$ the right-hand side of the linear system at time step $t$ since $\Phi$ and $\mathbf{b}$ will evolve over time.

## 2.2 Graph Convolution Network Model

We use a non-structured approach to harness a graph convolutional network that leverages node information stored on large unstructured meshes affiliated with the physical problem. By learning the node features on the graph, the neural network should have a global understanding of the nature of the data represented on the graph. We use the SAGEConv operator proposed by Hamilton *et al.* [19]. It consists of sampling the nodes of a graph, aggregating the features of the sampled node to a certain level of the node's local neighborhood, and repeating the process until the network has a complete representation of the node features of the graph.

SAGEConv is a mean aggregator which could be summarized with the following equation where for a designated graph convolutional layer $k$:

$$\mathbf{h}_v^k \leftarrow \sigma(\mathbf{W} \cdot \text{MEAN}(\{\mathbf{h}_v^{k-1}\}) \cup \{\mathbf{h}_u^{k-1}, \forall u \in \mathcal{N}(v)\}) \qquad (4)$$

where

- $\mathbf{h}_v^k$: represents the new node representation,
- $v$: the concatenation of the current node propagation,
- $\{\mathbf{h}_v^{k-1}\}$ : the current node information propagation,
- $\{\mathbf{h}_u^{k-1}, \forall u \in \mathcal{N}(v)\}$ means the information, aggregation from its immediate neighborhood,
- $u$ the selected neighborhood nodes,
- $\mathbf{W}$ is the weight matrix of the trainable parameters,
- $\sigma$ the nonlinear activation.

This strategy makes it possible to pass through a large graph and learn its dynamics in terms of node feature information.

### 2.3    Neural Network Architecture

As noted in Sect. 2.2, the SAGEConv model requires several layers to fully increase the capacity of the neural network to analyze globally the node features, mostly in the case where the graph is very large. As the neural network's main architecture, we suggest stacking several SAGEConv convolution layers. Each layer will be associated with PReLU as a nonlinear activation layer since this type of activation layer has a trainable parameter to adapt the output of the convolution layer [20]. Finally, the activation layer will be followed by a LayerNorm to stabilize the neural network's learning process [21]. The chosen neural network architecture has five layers with a linear operator at the end of it. Each layer's structure comprises a sequential operator of the SAGEConv model, a PReLU activation layer, and a LayerNorm. The neural network was optimized using the Adam optimizer [22] with $\beta_1 = 0.95$ and $\beta_2 = 0.90$ with all the trainable parameters initialized by default with the Glorot Normal distribution [23]. The learning rate is fixed at $1 \times 10^{-3}$. All training methods are written using PyTorch [3] and PyTorch Geometric [5] to model graph neural networks and monitor their training. Every training was conducted using the Distributed Data-Parallel paradigm from PyTorch [4] and ran on 4 NVIDIA A30 GPUs for the 2D case and 4 NVIDIA V100 for the 3D case.

### 2.4    Hybridization Method

The benefit of using a neural network inside a numerical simulation is that it decreases the computation time necessary for the discretized Poisson problem. In plasma physics simulation where we have to solve Poisson's equation at each time step of a simulation, the solution at the current time step $\Phi^t$ could be found incrementally such as:

$$\Phi^t = \Phi^{t-1} + \Delta\Phi^t, \tag{5}$$

where $\Delta\Phi^t$ represents the solution update for the current time step, in the correction of what we had in the previous simulation time step. At first, the neural network is built using the same connectivity as the unstructured mesh from the numerical simulation. Then, the neural network is employed to find the solution update $\Delta\Phi^t$ so that:

$$\Phi^t = \Phi^{t-1} + f_\theta(\Phi^{t-1}, b^t), \tag{6}$$

where $f_\theta(\Phi^{t-1}, b^t)$ represents the output of the neural network, with hyperparameters $\theta$. The output will update the current time step solution $\Delta\Phi^t$. As explained in Eq. (6), the neural network will have as input the solution of the previous time step $\Phi^{t-1}$, the current right-hand side of the discretized PDE $b^t$, source term as defined in Eq. (3). The supervised approach uses the neural network to provide an initial guess for each simulation time step. The initial guess is then refined by an iterative solver, which is supposed to perform a few iterations to converge.

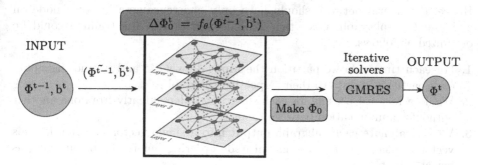

**Fig. 1.** Inference sketch for solving linear system $\mathbf{A}\Phi^t = b^t$ throughout a simulation

As demonstrated in Fig. 1, we take inputs from the solution of the previous time step $\Phi^{t-1}$, and the right-hand side $b^t$ of the linear system. We normalize $b^t$ and $\Phi^{t-1}$ with their respective $L_2$ norm giving $\tilde{\Phi}^{t-1}$ and $\tilde{b}^t$. Inside a simulation, the neural network will provide an update $\Delta\Phi_0^t$ that will give an initial guess $x_0 = \Phi^{t-1} + \Delta\Phi_0^t$ for an iterative solver such as a GMRES solver [8]. The iterative solver in that method is here to ensure good accuracy for the next step to avoid the divergence of the simulation throughout time iterations.

## 2.5   Training and Partitioning Algorithm

As a training method, we propose a supervised approach where the neural network should make a prediction close to a solution of reference obtained with an iterative solver for each simulation time step. The supervised learning objective will be the root-mean-squared error (RMSE) between the output of the neural network and the update of reference for the designated time step:

$$\mathcal{L}_\theta = \sqrt{\sum_{j=1}^{n} \frac{(f_{\theta,j}^t - \Delta\Phi_{j,\text{ref}}^t)^2}{n}}, \tag{7}$$

where $j$ corresponds to the node index of the update vector. The update of reference $\Delta\Phi_{j,\text{ref}}^t = \Phi_{\text{ref}}^t - \Phi_{\text{ref}}^{t-1}$ is also expressed with the same node index as the prediction of the neural network for all solution node $n$ of the unstructured mesh. In the major cases of industrial numerical simulations for plasma physics, it is necessary to study a discretized domain that largely exceeds the memory capacity of one single processor. This issue remains consistent for large graphs where the training phase and the feature node information linked to the graph structure exceeds the memory capacity of one single GPU.

In 2019, Chiang *et al.* [24] suggested partitioning the graph into several clusters and making predictions for each cluster separately. Using the library METIS [25], the clustering is computed by subdividing the initial graph $\bar{G}$ into $k$ groups of nodes $G_1, G_2, ..., G_k$, where $G_k := \{\mathcal{V}_i, \mathcal{E}_i\} \; \forall i \in [\![1, k]\!]$. $\mathcal{E}_i$ represents the edge connectivity of subgraph $G_i$. The partitioning is realized without overlapping. Hence, the neural network should have only one representation of a node on a designated subgraph. Once the graph is partitioned, the training could be organized as follows:

1. For each time step we partition the input data $\Phi^{t-1}$ and $b^t$ and distribute each partitioned data to their respective subgraph,
2. We do a forward passing for each subgraph independently from one another without communication between them,
3. We concatenate each subgraph output into a single vector and reorder this vector the same way as the global node ordering before the partitioning as for $\Phi^{t-1}$ or $b^t$,
4. This final vector will be used to compute the RMSE loss for each time step of a simulation,
5. We repeat the process for each time step of a simulation.

The process is repeated for each training iteration, giving the neural network an overview of the whole dynamics of the simulation. Despite the missing communication between the subgraph of the problem, we expect, with the backpropagation, a global understanding of the problem. Each subgraph's output vector has gradient values, which will be concatenated for backpropagation. Even though the forward pass of the neural network is done locally for each subgraph, the update of the neural network weights is done globally by backpropagating the final vector that regroups all the gradients of each subgraph output. In the end, the optimization process will be global, and the neural network must be able to learn the global patterns of the main graph before its partitioning.

# 3    Test Cases and Training Datasets

The present work has a special application to electric space propulsion. Hence, the datasets that constitute the training process for our method are based on the specific application field.

## 3.1    2D MTSI in a Radial-Azimuthal $(r, \theta)$ Plan

As a first example, we propose a 2D Particle-In-Cell (PIC) simulation from Petronio et al. [26] where we study the interaction of two opposite flows called Modified-Two-Stream-Instability (MTSI), meaning two species of charged particles going in opposite directions: one constituted of ions and the other, of electrons.

In the context of this simulation, the domain is discretized with regular squares, each split in half into triangles. A static electromagnetic field $\mathbf{E} \times \mathbf{B}$ is imposed on the virtual axis of the simulation (axial direction for $\mathbf{E}_x$ and azimuthal direction $\mathbf{B}_y$) so that the plasma instability will be correctly presented. This simulation aims to demonstrate the feasibility of solving Poisson's equation for a regular grid enough to fit in the memory of one single GPU (33,153 nodes). In this simulation, when a particle leaves on the right side of the domain, it will be reinjected on the left side, which is assimilated as a periodic boundary condition. The remaining boundaries marked in green are considered Dirichlet boundary conditions where the electric field is forced to zero.

## 3.2    3D PIC Simulation for Electron Drift Instability (EDI)

The second case is the study led by Villafana et al. [27], where 3D Particle-In-Cell simulation was made to describe the plasma instabilities due to the electrons which are very difficult to track inside a Hall-Thruster. A 3D unstructured mesh comprising 2,739,491 nodes was set to capture this physical phenomenon. The mesh is a section of the Hall Thruster channel in the azimuthal direction. Periodicity is added for the faces belonging to the azimuthal direction to guarantee an electric equilibrium of the plasma. The mesh itself is fully unstructured with a refined region inside the chamber of a Hall-Thruster. On this complex large geometry with different cell sizes, using a graph neural network becomes advantageous when capturing the dynamics of the physical fields without interpolating the latter. Due to the size of the graph, this graph will be split into 100 partitions to ensure every partition will fit in the memory capacity of one GPU for the training sequence.

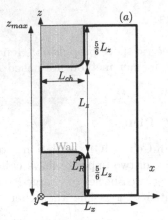

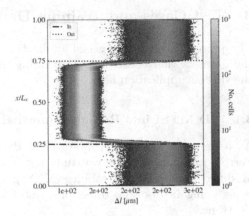

(a) 2D radial-axial ($z - x$) view
of the 3D geometry

(b) Radial distribution of the cell size
normalized on the radial length $L_z$

**Fig. 2.** 3D global view of 3D PIC simulation

## 3.3  Datasets Overview

In Fig. (2), the simulations constitute the datasets for the neural network's training. Throughout the simulation of the 2D case, every linear system solution is computed using a GMRES solver from PETSc [10] with stopping criteria of $1 \times 10^{-8}$ for the residual norm value without preconditioner. For the 3D case, an iterative solver MAPHYS [11] was used to obtain the solution with a similar threshold convergence.

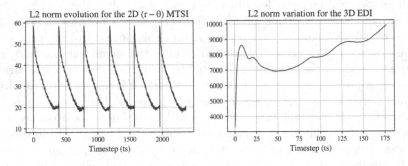

**Fig. 3.** Variation of $L_2$ norm $||\mathbf{A}\,\Phi_{\text{ref}}^{t-\text{delay}} - b^t||_2/||b^t||_2$ for MTSI 2D simulation (with delay = 1) and the 3D EDI simulation on the right (with delay = 5,000)

In the 2D case, the solutions that compose the dataset of our neural network are saved every 100 time steps. For every 100 time steps, we store the current solution, the solution of the previous time step, and the right-hand side of the current time step. Overall when computing the mean $L_2$ norm of the whole dataset $||\mathbf{A}\Phi^t - b^t||_2$, the $L_2$ norm of the simulation turns approximately around 30. In the 3D case, the solutions to save are more computationally expensive to store. The solutions are then stored every 5,000-time step without saving the solution of the previous time step for the current time step solution we want to store. Consequently, the mean $L_2$ norm of the 3D case is higher than the 2D case, with approximately 7,900.

## 4    Numerical Experiments

In this section, we report the results of our neural networks in training and inference. For both cases, we use the simulations reduced in PyTorch format datasets presented in Sect. 3.3. We split each simulation into training and validation sets. The validation represents the last 5% of the simulation itself, where we check the learning evolution of the model.

### 4.1    Training and Inference for the 2D Case

We present a synthesis of our training and inference results and the training and validation set for both cases. For the training process, we plot a map of neural network predictions along the time steps of the simulation and compare them to the solution of reference, which has a precision error of $1 \times 10-8$. The timeline results for the $300^{\text{th}}$ epochs of the training process are shown in Fig. (6a) in the appendix (6).

**Table 1.** Mean and standard deviation of the RMSE of the training set during the training phase, the validation set during inference, compared to the reference

| Training | Validation | Reference |
|---|---|---|
| $\approx 0.07 \pm 3 \times 10^{-4}$ | $\approx 0.10 \pm 3 \times 10^{-4}$ | $\approx 0.19 \pm 0.04$ |

From Fig. (6a), the predictions made by the neural network during the training process show a prediction with an absolute error that does not exceed 1.0. As we can notice in Table 1, the training phase shows an RMSE that does not exceed the mean level of 0.10. With a standard deviation lesser than 0.001, the neural network approximates the solution update well for all simulation time steps for the training and validation phase. The RMSE of reference refers to the RMSE between the previous time step and the current time step solution, meaning the error of the initial guess before the neural network correction giving $\Delta\Phi_0^t$ as mentioned in Sect. 2.4.

The validation set in inference presented in Fig. (4) below seems to follow the same trend as for the training set during the training phase. Indeed, supervised learning is expected to reach a certain level of accuracy for both training and validation sets since the difference $b^t$ and $\Phi^t$ have the same profile for both sets.

If the solution of reference is an acceptable target for the user with a known precision error, then optimizing the neural network with a supervised approach is recommended.

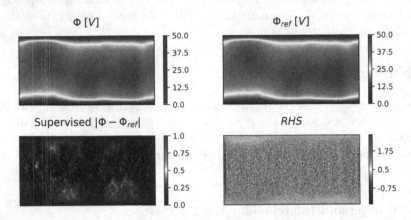

**Fig. 4.** Comparison of the solution updated by the neural network $\Phi$, the reference $\Phi_{ref}$, the absolute difference between them $|\Phi - \Phi_{ref}|$ and the right-hand side of the Poisson's equation at time step $t_s = 3.55\ \mu s$ for the validation set

## 4.2  Training and Inference for the 3D Case

Contrary to Sect. 4.1, the inputs given to the neural network for the current time step are not the solution from the previous time step but from the $5,000^{th}$ before the current one. Thus, the learning process would become harder, explaining the difference in resolution between the RMSE of the 2D case's RMSE and the 3D case. As in Sect. 4.1, we present a synthesis of our results for training and inference, training and validation set combined. For the training process, we plot in the appendix (6) on the Fig. (6b) a map of neural network predictions along the time steps of the simulation and compare them to the solution of reference, which has a precision error of $1 \times 10^{-12}$.

**Table 2.** 3D Case: Mean and standard deviation of the RMSE for all the training set during the training, the validation set during inference, compared to the reference

| Training | Validation | Reference |
|---|---|---|
| $\approx 3.49 \pm 8 \times 10^{-3}$ | $\approx 4.92 \pm 0.02$ | $\approx 7.02 \pm 0.59$ |

Overall, since the predictions made by the neural network are based on the previous time step with a delay of 5,000-time steps and the current right-hand side $b^t$, it becomes more difficult for the neural network to capture the update of the current time step properly. This could be explained by the plasma being stabilized progressively during the simulation, giving the same trend for all time steps at this phase. But it is not at the beginning of the simulation where we have a high variation of the Poisson's equation solution every time step. Consequently, the training process shows a prediction with an absolute error of 100 V amplitude at the training dataset's beginning. It ends with an absolute error of approximately 20 V of amplitude. As we can notice in Table 2, the training phase shows an RMSE with a mean level of 3.49, significantly higher than the results given in Table 1. But with a standard deviation lesser than 0.05, the neural network can provide for all time steps a good approximation of the solution update without a high dispersion degree of the solution update's precision for all time steps. The standard deviation of RMSE for the reference dataset is significantly higher. This could be explained by the heterogeneity of the dataset, which results in a higher dispersion of solution accuracy throughout the simulation.

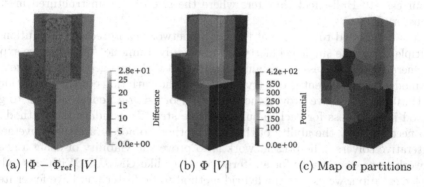

(a) $|\Phi - \Phi_{\text{ref}}|$ [V]          (b) $\Phi$ [V]          (c) Map of partitions

**Fig. 5.** Absolute difference between the solution updated by the neural network and the solution of reference (a) $|\Phi - \Phi_{\text{ref}}|$, the solution updated by the neural network (b) $\Phi$, at time step $t_s = 4.25\ \mu s$ for the validation set, and the partition map of the graph using METIS [25]

For the validation set, with the partitioning as for the training set, the supervised approach remains closer to the reference solution. Even with a partitioned process, the prediction does not seem to present the same discontinuities as the partitions presented in Fig. (5c). As shown in Fig. (5a), the amplitude of the difference does not exceed 15 V for a solution of reference where the amplitude exceeds 400 V despite that the neural network has an input of $\Phi^{t-5000}$. Nonetheless, for training and inference, the neural network seems to capture the global pattern of the solution even though the forward passing is realized locally for each subgraph.

# 5   Conclusion and Perspectives

In this work, we proposed a method capable of giving an initial guess for each time step of a simulation to speed up the solving process of a linear system. This method has been applied to plasma physics problems to obtain the electric potential from Poisson's equation. The method proposed in this study is also applied to large graph problems concerning unstructured mesh, which cannot be treated with one GPU processor. The partitioning process explained in this article proposed an approach to tackle this challenge. By summing all the subgraphs' outputs in one single vector indexed in global graph node ordering, we proceed with a global backpropagation of all the subgraphs' outputs to let the neural network learn about the problem globally. In the end, the neural network should be able to make an initial guess of our current linear system for large graph problems, which could be an asset for industrial applications, specifically in our case to Hall-effect thrusters where the size of the unstructured mesh is often very large.

As we intended to prove that the neural network can predict the solution for multiple $b^t$ for one single geometry, meaning only a unique $A$ matrix, we expect to generalize the process by training a neural network on a geometry then testing the model on a different geometry to observe the variations of the predictions.

Until now, we have proven the possibility of using a neural network to give a good initial guess for each simulation time step. To validate the method, we also need to show the ability of this new method to accelerate the convergence of iterative solvers. The present work is to prove the viability of using a neural network as an accelerator for an iterative solver like GMRES. With this hybrid implementation, we expect the hybrid method to be faster and have fewer iterations to converge. Future results will be presented at the conference with a fully coupled neural network with an iterative solver. In the end, once the neural network is trained, the latter will be used in inference inside a numerical simulation to provide a good approximation of the solution that helps the iterative solver to converge faster.

**Acknowledgments.** This research has been realized under the supervision of Bénédicte Cuenot and Olivier Vermorel (senior researchers at CERFACS). Gabriel Vigot acknowledges the support of Luciano Drozda (senior researcher at CERFACS) in conceiving and validating this present article for the machine learning part. Gabriel Vigot also acknowledges the support of Luc Giraud (senior researcher at INRIA) in conceiving and validating this present article for the part concerning linear algebra.

**Disclosure of Interests.** The present authors have no competing interests with the participants of this article. Gabriel Vigot acknowledges the financial support from Safran Spacecraft Propulsion, under the supervision of Benjamin Laurent, and the French Space Agency (CNES: Centre National d'Études Spatiales), under the supervision of Ulysse Weller, under the EPIC convention. This work is introduced within the CHEOPS project.

# 6  Appendix

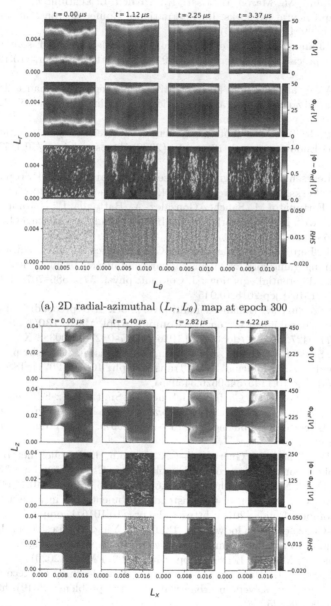

(a) 2D radial-azimuthal $(L_r, L_\theta)$ map at epoch 300

(b) 3D map in the centered axial-radial $(L_x, L_z)$ plane at epoch 300

**Fig. 6.** (a) 2D map along the simulation time line $t$ in microseconds. From the first to the fourth row is the update $\Phi$, the solution of reference $\Phi_{\mathrm{ref}}$, the absolute difference $|\Phi - \Phi_{\mathrm{ref}}|$, and RHS the right-hand side of the Poisson's equation.
(b) Neural network predictions $\Phi$ with the same variable order as in Fig. (6a)

# References

1. Luz, I., Galun, M., Maron, H., Basri, R., Yavneh, I.: Learning algebraic multigrid using graph neural networks. In: International Conference on Machine Learning, pp. 6489–6499. PMLR (2020)
2. Jiang, Z., Jiang, J., Yao, Q., et al.: A neural network-based PDE solving algorithm with high precision. Sci. Rep. **13**, 4479 (2023). https://doi.org/10.1038/s41598-023-31236-0
3. Paszke, A., et al.: PyTorch: an imperative style, high-performance deep learning library. In Advances in Neural Information Processing Systems, vol. 32, pp. 8024–8035. Curran Associates, Inc. (2019)
4. Li, S. et al.: PyTorch distributed: experiences on accelerating data parallel training, In: VLDB Endowment, vol. 13, no. 12 (2020). https://doi.org/10.14778/3415478.3415530
5. Fey, M., Lenssen, J.E.: Fast graph representation learning with PyTorch geometric. In: ICLR Workshop on Representation Learning on Graphs and Manifolds (2019)
6. Pfaff, T., Fortunato, M., Sanchez-Gonzalez, A., Battaglia, P.: Learning mesh-based simulation with graph networks. In: the International Conference on Learning Representations (2020)
7. Raissi, M., Perdikaris, P., Karniadakis, G.E.: Physics informed neural networks: a deep learning framework for solving forward and inverse problems involving nonlinear partial differential equations. J. Comput. physi. **378**, 686–707 (2019). https://doi.org/10.1016/j.jcp.2018.10.045
8. Saad, Y., Zhang, J.: Enhanced multi-level block ILU preconditioning strategies for general sparse linear systems. J. Comput. Appl. Math. **130**(1), 99–118 (2001). ISSN 0377-0427, https://doi.org/10.1016/S0377-0427(99)00388-X
9. Chen, J., Schäfer, F., Huang, J., Desbrun, M.: Multiscale Cholesky preconditioning for ill-conditioned problems. ACM Trans. Graph. **40**(4), (2021). ISSN 0730-0301, https://doi.org/10.1145/3450626.3459851
10. Balay, S., Gropp, W.D., Curfman McInnes, L., Smith, B.F.: Efficient management of parallelism in object oriented numerical software libraries, In: Modern Software Tools in Scientific Computing, pp. 163–202 (1997). https://doi.org/10.1007/978-1-4612-1986-6_8
11. Agullo, E., Giraud, L., Guermouche, A., Roman, J.: Parallel hierarchical hybrid linear solvers for emerging computing platforms. C. R. Mécanique **339**(2–3), 96–103 (2011). ISSN 1631-072, https://doi.org/10.1016/j.crme.2010.11.005
12. Gourdain, N.: Prediction of the unsteady turbulent flow in an axial compressor stage. Comput. Fluids (2015). https://doi.org/10.1016/j.compfluid.2014.09.052
13. Sanchez-Gonzalez, A., Godwin, J., Pfaff, T., Ying, R., Leskovec, J., Battaglia, P.: Learning to simulate complex physics with graph networks. In: International Conference on Machine Learning, pp. 8459–8468. PMLR (2020)
14. Sappl, J., Seiler, L., Harders, M., Rauch, W.: Deep learning of preconditioners for conjugate gradient solvers in urban water related problems (2019). https://arxiv.org/abs/1906.06925
15. Schäfer, F., Katzfuss, M., Owhadi, H.: Sparse Cholesky factorization by Kullback-Leibler minimization. SIAM J. Sci. Comput. **43**(3), A2019–A2046 (2021). https://doi.org/10.1137/20M1336254
16. Kopaničáková, A., Karniadakis, G.E.: DeepOnet based preconditioning strategies for solving parametric linear systems of equations (2024). arXiv:2401.02016

17. Li, Y., Chen, P.Y., Du, T., Matusik, W.: Learning preconditioner for conjugate gradient PDE solvers. In: International Conference on Machine Learning Computer Science, Mathematics, Engineering (2023). https://proceedings.mlr.press/v202/li23e.html

18. Stanziola, A., Arridge, S.R., Cox, B.T., Treeby, B.E.: A Helmholtz equation solver using unsupervised learning: application to transcranial ultrasound. J. Comput. Phys. **441** (2021). https://doi.org/10.1016/j.jcp.2021.110430

19. Hamilton, W.L., Ying, R., Leskovec, J.: Inductive representation learning on Large graphs. In: 31st International Conference on Neural Information Processing Systems (NIPS) (2017)

20. He, K., Zhang, X., Ren, S., Sun, J.: Delving deep into rectifiers: surpassing human-level performance on ImageNet classification. In: International Conference on Computer Vision (2015). https://doi.org/10.1109/ICCV.2015.123

21. Ba, J.L., Kiros, J.R., Hinton, G.E.: Layer normalization. In: Neural Information Processing System (2016)

22. Kingma, D.P., Ba, J.: Adam: a method for stochastic optimization, arXiv preprint arXiv:1412.6980 (2014)

23. Glorot, X., Bengio, Y.: Understanding the difficulty of training deep feedforward neural networks, In: Proceedings of the thirteenth international conference on artificial intelligence and statistics, JMLR Workshop and Conference Proceedings, pp. 249–256 (2010)

24. Chiang, W.L., Liu, X., Si, S., Li, Y., Bengio, S., Hsieh, C.J.: Cluster-GCN: an efficient algorithm for training deep and large graph convolutional networks, In: Proceedings of the 25th ACM SIGKDD International Conference on Knowledge Discovery & Data Mining, pp. 257–266 (July 2019). https://doi.org/10.1145/3292500.3330925

25. Karypis, G., Kumar, V.: A fast and high-quality multilevel scheme for partitioning irregular graphs. SIAMJ. Sci. Comput. **20**(1), 359–392 (1998). https://doi.org/10.1137/S1064827595287997

26. Villafana, W., et al.: 2D radial-azimuthal particle-in-cell benchmark for E×B discharges, In: Plasma Sources Science and Technology, vol. 30, no. 7 (2021). https://doi.org/10.1088/1361-6595/ac0a4a

27. Villafana, W., Fubiani, G., Garrigues, L., Vigot, G., Cuenot, B., Vermorel, O.: 3D Particle-In-Cell modeling of anomalous transport driven by the electron drift instability in hall thrusters, In: 37$^{th}$ International Electric Propulsion Conference, MIT (2022)

# Solving Coverage Problem by Self-organizing Wireless Sensor Networks: (ϵ,h)-Learning Automata Collective Behavior Approach

Franciszek Seredyński, Miroslaw Szaban$^{(\boxtimes)}$, Jaroslaw Skaruz, Piotr Świtalski, and Michal Seredyński

Institute of Computer Science, University of Siedlce, Siedlce, Poland
{franciszek.seredynski,miroslaw.szaban,jaroslaw.skaruz,piotr.switalski,
michal.seredynski}@uws.edu.pl

**Abstract.** We propose a novel multi-agent system approach to solve a coverage problem in Wireless Sensor Networks (WSN) based on the collective behavior of (ϵ,h)-Learning Automata (LAs). The coverage problem can be stated as a request to find a minimal number of sensors spending energy of their batteries to provide the requested level of coverage of the whole monitored area. We propose a distributed self-organizing algorithm based on the participation of LAs in an iterated Spatial Prisoner's Dilemma game. We show that agents achieve a solution corresponding to Nash equilibrium, which provides maximization of not known for agents a global criterion related to the requested level of the coverage with a minimal number of sensors which turn ON their batteries.

**Keywords:** Collective behavior · Learning automata · Network coverage problem · Self-organization · Sensor networks · Spatial Prisoner's Dilemma

## 1 Introduction

Wireless Sensor Networks (WSN) is a fast-developing technology belonging to a broader group of information and communication technologies [8] applied today in the Internet of Things. They are composed of a large number of tiny computers- communication devices called sensors deployed in some areas- that can sense a local environment and send related information to a remote user who can make an appropriate decision. Designing such systems is a complex task demanding solving a number of issues on different levels related to a single sensor node, a network of nodes, applications, etc. (see, e.g., [7]).

This paper focuses on some issues related to designing fault-tolerant WSN-based mission-critical systems. We assume that monitoring is performed in a remote and difficult-to-access area, and sensors are equipped with single-use batteries that cannot be recharged. From the Quality of Service (QoS) point of

L. Franco et al. (Eds.): ICCS 2024, LNCS 14834, pp. 408–422, 2024.
https://doi.org/10.1007/978-3-031-63759-9_42

view of such WSN, two closely related important issues exist: how to perform effective monitoring (coverage) of an area and how to maximize an operational lifetime. After deploying sensors, they should recognize their nearest neighbors to communicate and start making local decisions about turning ON or OFF their batteries to monitor events. These decisions will directly influence the level of area coverage, the amount of spending on sensors' battery energy, and the network's lifetime. One can notice that the lifetime maximization problem is closely related to the coverage problem. A group of sensors monitoring some area is usually redundant, i.e., more than one sensor covers the monitored targets, which forms some redundancy that can be exploited. Solving the coverage problem is crucial to solving the problem of maximization of the WSN's lifetime. In the paper, we will focus on the problem of the coverage.

This paper presents a novel approach to the problem of coverage, based on self-organization with the use of $(\epsilon,h)$- Learning Automata (LAs) [10,14]. Sensors do not have knowledge concerning a current level of WSN coverage or a total number of sensors. They have only information concerning their neighbors. Nevertheless, we expect that the system we will be able to self-optimize, i.e. to find, in a fully distributed way, a minimal number of sensors turned ON providing a necessary value of the level of coverage. Our approach is based on a multi-agent interpretation of the coverage problem and applies a recently proposed [11] methodology of game–theoretic interactions between players participating in the Spatial Prisoner's Dilemma (SPD) game. The paper extends the work [12], where a self-organizing system based on the application of a second-order Cellular Automata (CA) has been proposed. We show that the presented approach significantly outperforms the approach based on the application of CA.

Our approach contrasts with the others, currently used to solve the problem of the coverage. Because the problem is known to be NP-complete [2], centralized algorithms are oriented either on the delivery of exact solutions for specific cases (see, e.g. [1]) or applying metaheuristics to find approximate solutions (see, e.g. [5]). The main drawback of centralized algorithms is that they assume the availability of complete information about the problem and require an adequte computing power to find a schedule of sensors' activities. It means practically that it can be done only on a site of the remote user, and a solution must be delivered to WSN before starting the operation. As a result, distributed algorithms with only different forms of partial information about the problem have become more and more popular (see, e.g. [3]). Computational power independence of WSN from a remote user can provide scalability, and the entire operation of a WSN in real-time can be achieved only by self-optimizing systems. The need for such systems has been recognized in recent years in many industrial systems (see, e.g., [4]).

The structure of the paper is as follows. The next section states the problem of a coverage optimization in WSN. Section 3 presents a multi-agent approach to distributed online solving the considered problem. The concept of $(\epsilon,h)$-LA used as players in the game is proposed in Sect. 4. Section 5 discusses relations between concepts of a global solution and Nash equilibria. Section 6 presents results of the experimental study, and the last section contains conclusions.

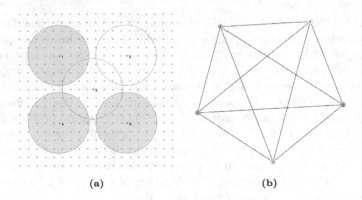

**Fig. 1.** WSN 5: (a) an area monitored by WSN consisting of 5 sensors with $R_s = 18$ m, (b) a WSN graph for $R_s = 35$ m.

## 2    Coverage Problem in Wireless Sensor Networks

We assume that an area of the size of $L_1 \times L_2$ $m^2$ should be monitored by a WSN consisting of $N$ sensors $(s_1, s_2, ..., s_i, ..., s_N)$ deployed over this area. More specifically, the area is represented by $M$ Points of Interest (PoI) which regularly cover the area. Each sensor has a non–rechargeable battery and can monitor PoIs in a sensing range of $R_s$ if its battery is turned ON. Figure 1a shows an example of such an area with $L_1 = L_2 = 100$ $m$ with $M = 441$ PoI (in orange), where WSN 5 consisting of $N = 5$ sensors with $R_s = 18$ m was deployed. Some sensors of WSN are currently turned ON and monitor the corresponding area.

It assumed that a QoS measure exists evaluating the quality of a WSN performing monitoring. As such a measure, we accept the coverage value $q$ defined as the ratio of the number of PoIs covered by active sensors to the whole number $M$ of PoIs, i.e. $q = \frac{M_{obs}}{M}$. A desirable objective is to preserve the complete area coverage, but sometimes, it may be more practical to achieve a predefined c o v e r age rate that is just high enough. Therefore, we assume that this ratio should not be lower than some predefined requested value $q_r$ ($0 < q_r \leq 1$).

The coverage problem can be stated in the following way. Find a solution $s = (s_1, s_2, ..., s_i, ..., s_N)$, where $s_i = \{0, 1\}$ with a corresponding value of the coverage $q(s)$ and a number $n\_ON(s)$ of sensors turned ON such that it fulfills the following requirements: a) a number $n\_ON$ of sensors turned ON is minimal, b) $q(s) \geq q_r$, and c) $q(s)$ has the largest possible value.

We have to deal with a combinatorial optimization problem, which can be described by a proposed function that should be maximized:

$$f(q(s), n\_ON(s), q_r) = \begin{cases} N - n\_ON(s) + q(s), \ if \ q(s) \geq q_r \\ q(s), \ if \ q(s) < q_r. \end{cases} \quad (1)$$

This function assigns univocally values to solutions in such a way that a maximal value of it corresponds to a solution (or solutions) which provides a maximal value of $q$ fulfilling the requirement $q \geq q_r$ under the minimal value of $n\_ON$ of sensors turned ON.

We want to solve the coverage problem online in real-time using only a tiny amount of computational power and the communication possibilities of WSN sensors. A potential solution algorithm should work in a monitored area in real-time and react quickly to changes in values of parameters of sensors. Therefore, we will focus in this paper on working out a variant of a distributed algorithm to solve the coverage problem by self-optimization.

The first step in this direction is converting a WSN instance into a WSN interaction graph. Such a graph will be used as a core of a multi-agent system oriented on solving a coverage problem. The conversion is based on the principle saying [12] that two nodes of a WSN graph are connected if they have at least one common PoI within their sensing range $R_s$ in a corresponding WSN.

Figure 1b shows an interaction graph for WSN 5 presented in Fig. 1a under an assumption that $R_s = 35$. The interaction graph contains five nodes, each corresponding to a sensor from the instance. The degree of each node corresponds to a number of neighbors which depends on the value of $R_s$, and in this case, each sensor has four neighbors.

## 3   Multi–agent System for Online Coverage Optimization

### 3.1   Agents and Their Actions

We assume that each node of a WSN interaction graph is controlled by an agent $A_i$ of a multi-agent system consisting of $N$ agents. Each agent has two alternative decisions (actions): $\alpha_i = 0$ (battery is turned OFF) and $\alpha_i = 1$ (battery is turned ON) and a decision unit is responsible for making a decision about turning ON and OFF the sensor. Different ways of coordination of agent decisions exist to reach a common goal. In our approach, we assume that the interaction of agents is based on a game-theoretic model, which is a variant of SPD game [11] related to the WSN coverage optimization problem [12]. In this game-theoretic model, we assume that all agents make discrete-time decisions regarding the activation of their batteries using specific rules/strategies.

We assume the following set of rules is available by agent-players:

- *all C*: always cooperate (C) what corresponds turning ON battery ($\alpha_i = 1$),
- *all D*: always defect (D) what corresponds to turning OFF battery ($\alpha_i = 0$),
- *k–D*: cooperate until not more than $k$ neighbors defect, otherwise defect,
- *k–C*: cooperate until not more than $k$ neighbors cooperate, otherwise defect,
- *k–DC*: defect until not more than $k$ neighbors defect, otherwise cooperate.

One can notice that the first two rules do not consider player-neighbor decisions. The remaining three rules of a given agent-player take into account the decisions of player-neighbors from a previous round. Selecting a rule to turn ON/OFF a battery depends on an algorithm of a player decision unit. While in our previous study [12] we were using a second order CA, in this paper, we propose to use a new original reinforcement learning algorithm called $(\epsilon,h)$-LA.

## 3.2 Payoff Function of a Game

It is assumed that each agent-player knows the value of a requested coverage $q_r$ and considers it as a local value $q_r^i = q_r$ which must be fulfilled. He participates in an iterated SPD-like game [12] consisted of some number of rounds (iterations) conducted in moments of time $t = 1, 2, ..., T$, where $T$ is known only for an organizer (a remote user of WSN) of the iterated game. An agent decides whether to turn on his battery (i.e. cooperate (C)) or turn off it (i.e. defect (D)) and obtain a payoff that depends on his decisions (C/D) and of only his neighbors decisions. Neighbors of a given player can be considered as a virtual player-opponent. A value of a local coverage $q_{curr}^i$ is a result of a game of the $i - th$ player with his virtual opponent. His payoff in a game depends on whether his current $q_{curr}^i$ is below or above the requested $q_r^i$. The payoff function of a player is given in Table 1.

**Table 1.** Payoff function of SPD-like game for coverage problem

| $i$–th agent's action | fulfilment of $q_r^i$ | |
|---|---|---|
| turn ON battery (C) | $q_{curr}^{i-off} \geq q_r^i$ | |
| | no | yes |
| | $payoff_i^{on+} = d$ | $payoff_i^{on-} = c$ |
| turn OFF battery (D) | $q_{curr}^i \geq q_r^i$ | |
| | no | yes |
| | $payoff_i^{off-} = a$ | $payoff_i^{off+} = b$ |

The payoff function assigns values to the $i$–th player in the following way:

• if he "turns OFF battery" then he calculates his local value of coverage $q_{curr}^i$; if this value $q_{curr}^i \geq q_r^i$ then he receives a payoff equal to $b$, otherwise a payoff equal to $a$,

• if he "turns ON battery" then he calculates what would be his value of $q_{curr}^i$ (denoted as $q_{curr}^{i-off}$) if in fact he would have "turned OFF" his battery; if $q_{curr}^{i-off} < q_r^i$ then he receives a payoff equal to $d$, otherwise a payoff equal to $c$.

The proposed payoff function transforms the global optimization criterion (see Eq. 1) stated in Sect. 2 for the coverage problem into local optimization goals of players.

We assume that players are rational and act in such a way to maximize their payoff defined by the payoff function. However, we are not be interested in players' payoffs but in the evaluation of the level of collective behavior of the system. As a measure of the collective behavior of the system, we use an external criterion (not known for players) - the average total payoff (ATP) $\bar{u}()$:

$$\bar{u}(s_1, s_2, ..., s_i, ..., s_N) = \frac{1}{N} \sum_{i=1}^{N} u_i(A_i(s_i), A_i^{virtual}(s_{i_1}, s_{i_2}, ..., s_{i_r})), \qquad (2)$$

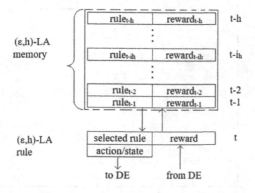

**Fig. 2.** Proposed ($\epsilon$,h)-Learning Automaton.

where $u_i()$ is a payoff of an agent-player $A_i$ in a game with a virtual player $A_i^{virtual}$ and $i_r$ is the number of neighbors of a player $A_i$ corresponding to his opponent, the player $A_i^{virtual}$.

Game theory predicts that players' behavior in noncooperating games is oriented towards achieving a Nash equilibrium (NE). We call a value of ATP corresponding to given NE the price of this NE. The game can have many NE points with different ATP. We call NE with the highest ATP the *maximal price point* (MPP). The question is whether we can expect of such a behavior of players that while they attempt to reach a NE at the same time ATP of the whole set of players is maximized, i.e. MPP is reached. Such a behavior depends on many factors, and one of them is a model of a player making decisions. In this paper, we examine the collective behavior of players modeled by $(\epsilon, h)$-LA.

## 4  $(\epsilon, h)$-Learning Automata

LAs are reinforcement learning algorithms proposed in 1969 by Tsetlin (see, [13]) and further extended and studied e.g. by [6,9,14] and many others. LAs distinctive feature is working in random environments and their ability to adapt in them. An idea of a Learning Automaton (LA) working in a deterministic environment and called $\epsilon$-automaton was presented in [14] with a comment saying that such an idea seems to have a sense only if a deterministic environment can be randomized. However, it was shown in [10] that the concept of $\epsilon$-LA is useful in game-theoretic models related to variants of PD game with deterministic environments. In this paper we extend this idea and propose a concept of $(\epsilon, h)$-LA suitable for an SPD-like game model oriented on solving coverage problems in WSN.

Figure 2 presents the proposed $(\epsilon, h)$-LA construction. Each LA uses a subset of $m$ rules $(0 < m \leq 5))$ from the whole set of available 5 rules (see, Subsect. 3.1). LA has a memory of a length $h$, where it stores pairs $(rule_t - i_h, reward_t - i_h)$ from last $h$ moments of a discrete time $t - 1, t - 2, ..., t - i_h, ..., t - h$ of its

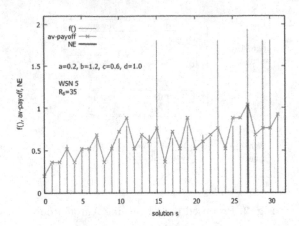

**Fig. 3.** Landscape of a global criterion function $f()$ and external criterion ATP for WSN 5.

operation. Each pair contains an information about a rule selected by LA at a discrete time $t - i_h$ and corresponding rewards for executing this rule.

At time $t$ LA selects a new rule in the following way: with a probability $1 - \epsilon$ $(0 < \epsilon < 1)$ it selects from the memory a rule with the highest reward, or with a probability $\epsilon$ selects randomly a rule from a set of $m$ rules. The selected rule is executed, i.e., the state of a corresponding sensor battery is changed by performing an action turn ON or turn OFF. Selected rule and a corresponding action (C or D) are sent to DE, which returns a reward for this action. Records of the memory are shifted in such a way that the oldest record is removed, and a new record with the latest strategy and corresponding reward is added to the memory.

## 5   Nash Equilibria and Global Solutions

Our approach based on self-organization differs significantly from the classical one. We replace the problem of global optimization (see, Eq. 1) by the problem of reaching NE by the agent-players. The level of self-organization, which we call a level of collective behavior, is measured by the value of ATP at NE.

ATP is closely related to two global characteristics of WSN represented by parameters $n\_ON$ and $q_r$, which are also not known to the players. The payoff function presented in Table 1 was designed in such a way to express global parameters $n\_ON$ and $q_r$ through local goals of the players. A correctly designed game should correctly link the global optimization criterion with a value of ATP at NE. We focus in this section on showing relations between global solutions offered by Eq. 1 and distributed solutions corresponding to ATP (Eq. 2) at NE.

Due to the exponential increase in the number of solutions, an analysis of relations between NE and a global optimization criterion can be performed only for small instances of the coverage problem. Therefore, in this Section, we will use WSN 5 shown in Fig. 1a under the assumption that $R_s = 35$ m. WSN graph of this instance is shown in Fig. 1b. For the analysis, the payoff function with the settings: $a = 0.2, b = 1.2, c = 0.6, d = 1.0$ will be used, and $q_r = 0.8$.

Figure 3 presents landscapes of both functions: the global function f() (see, Eq. 1) and ATP (see, Eq. 2) for WSN 5. The space of solutions $s$ of the coverage problem consists of 32 solutions. Figure 3 presents values of f() (in red) and ATP (in blue) for all solutions. One can notice that both functions indicate $s_{27} = (1, 1, 0, 1, 1)$ as an optimal solution with corresponding values $f(s_{27}) = 1.94$ and $\text{ATP}(s_{27}) = 1.04$, respectively. The solution shows that the number of sensors turned ON is equal to 4, with the corresponding value of $q = 0.94$.

It can be shown that $s_{27}$ is NE, and at the same time it is *maximal price point* (MPP). It is a unique NE for WSN 5 and is marked in violet in Fig. 3.

# 6 Experimental Results

A number of simulation experiments have been conducted to learn the performance of the proposed methodology. The monitored area was of the size $L_1 = L_2 = 100$ $m$ with $M = 441$ PoI. We use the game settings $a = 0.2, b = 1.2, c = 0.6, d = 1.0$, the request $q_r = 0.8$ and LAs with parameters $h = 8$ and $\epsilon = 0.005$. Batteries capacity is unlimited. Each experiment lasted 1000 iterations. When necessary, experimental results were averaged over 30–50 runs.

## 6.1 The Instance: WSN 5

The purpose of this set of experiments was to get some inside into the work of the self-organizing algorithm when a small instance of the problem WSN 5 is used. The experiments have been conducted under an assumption that $R_s = 35$ $m$. The WSN interaction graph of the multi-agent system is presented in Fig. 1b and a theoretical analysis of the game was presented in Sect. 5. Experiments have been conducted under the assumption that the whole set of 5 rules is used.

Figure 4 presents results of a single run of the algorithm. Figure 4a and Fig. 4b show changes of global parameters $q$ and $n_{ON}$, respectively. One can see that a suboptimal solution characterized by $q = 0.78$ and $n_{ON} = 3$ was reached very quickly at the iteration 7 and finally an optimal solution characterized by $q = 0.94$ and $n_{ON} = 4$ was found at the iteration 169. This solution was expected from the analysis provided in Sect. 5. It is stable till the end of the iterated game, because it corresponds to a unique NE in this game.

Figure 4c shows moments of time when some agents use $\epsilon$ alternative of the LA algorithm to select rules. The first time this alternative is used at iter=161 (see, the line in violet) by the agent 3 but it does not result in any changes in the system. The second time it is used at iter=169 (see, line in blue) by the agent 1 and it causes shifting a solution to the optimal one. Indeed, the agent 1 changed his rule *all D* used until now into the rule *all C* what resulted in changing the state of the sensor 1 from 0 to 1.

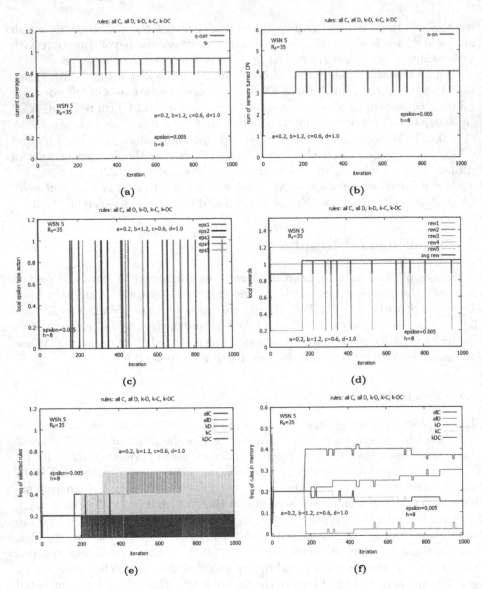

**Fig. 4.** Single run for WSN 5: (a) coverage $q$, (b) a number of sensors $n_{ON}$, (c) moments of taking actions caused by $\epsilon$ alternative of LA, (d) local rewards of agents, (e) frequencies of rules selected by agents, (f) structure of LA memories storing rules. (Color figure online)

Figure 4d shows how rewards of agents change in the game. One can see that before moving from the suboptimal solution to the optimal one the reward of the player 1 was equal to 0.2 (see, the line in blue), the reward of the player 3 was equal to 1.2 (see, the line in violet), and rewards of the remaining three players

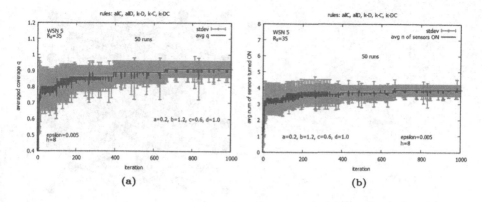

**Fig. 5.** WSN 5: averaged value of (a) coverage $q$, (b) a number $n_{ON}$ of sensors turned on.

were equal to 1.0. Changing by the player 1 at iter=169 the state of his battery into 1, moved, as we already noticed, the suboptimal solution to the optimal one, and moved his personal reward to 1.0, while rewards of remaining players did not change. The optimal solution reached by the players is characterized by the highest average payoff (see, the line in red) of the game (ATP) equal to 1.04 corresponding to a specific NE called MPP. We can see that the solution is stable during the remaining rounds despite the attempts of changing the course of the game caused mostly by the $\epsilon$ alternative of LAs.

Figures 4e,f give some insight into the process of managing rules of LAs which collectively influence on the global performance of the system. Figure 4e shows how frequencies of rules selected by agents change in time. Till iter=169 all 5 rules are used with the same frequency equal to 0.2, but later we can observe a complex dynamics of different use of rules. Dominating rules are $k-C$ (in orange) and *all C* (in red) which are used with frequencies from the range $(0,0.6)$ and $(0.2,0.6)$, respectively. The remaining three rules are used with the frequency from the range $(0,0.2)$.

Figure 4f gives some insight into the contents of LAs memories. One can see that while till the iter=169 all rules are equally distributed with the frequency equal to around 0.2, in further iterations we can observe the process of significant changes of LA memories structure. The rule *all C* becomes dominating occupying around 40% of the total LA memory, and *all D* occupies only around 5%. We can observe an increasing number of $k-D$ rules reaching finally around 30% of the whole population of rules, and at the same time populations of $k-C$ and $k-DC$ are decreasing to around 15%.

Figure 5 presents averaged over 50 runs results concerning of WSN 5. The requested coverage $q_r$ is reached in the average after around 100 iterations (see, Fig. 5a) but the process of searching a solution is accompanied by noticeable value of standard deviation (*std*). Improving the average values of $q$ and $n_{ON}$ (see, Fig. 5b)) can be observed in next iterations, with decreasing value of *std*.

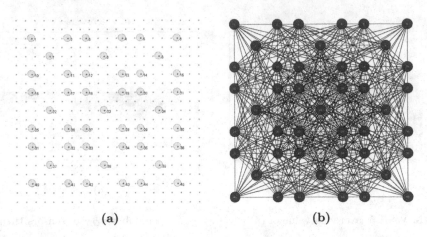

**Fig. 6.** WSN 45: (a) sensors localization, (b) WSN interaction graph for $R_s = 30$ m.

## 6.2   The Instance: WSN 45

To verify a general performance of the proposed approach we use in this subsection a realistic instance called WSN 45 consisting of 45 sensors with $R_s = 30\ m$. Figure 6a shows a deployment of sensors in the monitored area and Fig. 6b presents the corresponding WSN graph of the multi-agent system.

In the first set of experiments with WSN 45 we applied genetic algorithm (GA) (not shown here) to find an optimal or a suboptimal solution, and for this purpose the global function (see, Eq.(1)) was used. Conducted experiments indicated that there exists a solution providing the requested coverage $q \geq q_r$ with $n_{ON} = 4$, and it is used as a reference to results presented in this subsection.

Figures 7 and 8 show averaged over 30 runs values of $q$ and $n_{ON}$ for different subsets of rules. Figures 7a,b shows values of $q$ and $n_{ON}$ for the subset $\{all\ C, all\ D\}$ and Figs. 7c,d for the whole set of 5 rules. For both versions the algorithm achieves the requested level of $q_r$. With the use of 2 rules goals of self-optimization are reached in the average at the 7-th iteration providing in the average $q = 0.83$ (see, Fig. 7a) with the average $n_{ON}$ around 8.9 (see, Fig. 7b). The algorithm finds quickly a solution providing the average $q = 0.94$ which is maintained during the whole game close to this value. We can notice the decreasing number of sensors turned on, which reaches in the average $n_{ON} = 7.13$ in the final iterations.

The behavior of the system with the whole set of 5 rules is slightly different. The algorithm reaches at iter=14 the requested level of the coverage $q = 0.83$ (Fig. 7c) with low number of $n_{ON} = 6.16$ (Fig. 7d) and continues to improve the coverage until reaching $q = 0.94$ around iter=305 and maintains stable this value till the final iteration achieving $n_{ON}$ equal around 7.0. The main difference between both versions is that the algorithm with 2 rules requires at the beginning slightly more sensors turned on than the algorithm with the 5 rules, but finally

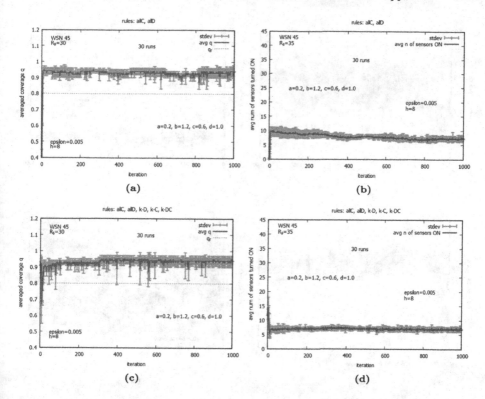

**Fig. 7.** WSN 45: averaged value of (a) coverage $q$ for a subset of 2 rules, (b) a number of sensors $n_{ON}$ for a subset of 2 rules, (c) $q$ for the whole set of 5 rules, (d) $n_{ON}$ for the whole set of 5 rules.

both algorithms tend to reach a solution with the average $n_{ON}$ close to around 7.00. Both versions are characterized by significantly reduced values of *std*.

Figure 8 presents behavior of the system with other rules from the whole set of rules. Figures 8a,b show changes of the average values of $q$ and $n_{ON}$, respectively for the rule $k$–$D$. One can see that this rule alone is able to provide a distributed search of a solution. However, it is accompanied by a relatively large *std* of $q$. At the iter=452 the average $q$ becomes equal to 0.8 and it is accompanied by $n_{ON} = 4.33$, what means that current solutions are very close to one offered by GA. However, the solutions are not stable and they are lost. Only from the iter=663 the values of $q$ fulfil the requirement of $q_r$ with corresponding values of $n_{ON} = 6.93$. During the last 100 iteration a solution with low *std* of both $q$ and $n_{ON}$ is maintained with values $q = 0.92$ and $n_{ON} = 6.10$.

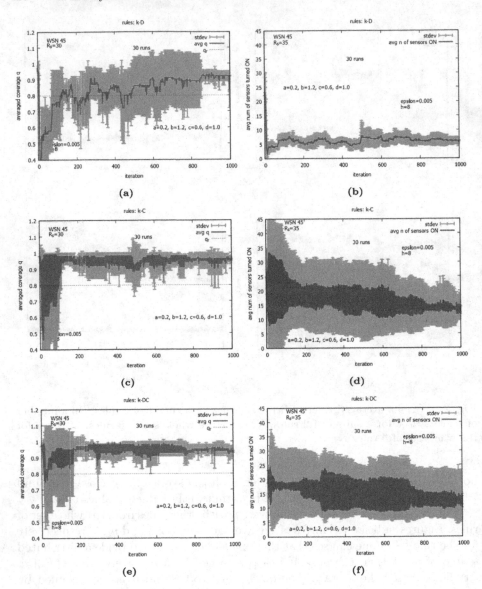

**Fig. 8.** WSN 45: averaged value of (a) $q$ for rule $k$–$D$, (b) $n_{ON}$ for rule $k$–$D$, (c) $q$ for $k$–$C$, (d) $n_{ON}$ for $k$–$C$, (c) $q$ for $k$–$DC$, (d) $n_{ON}$ for $k$–$DC$.

Figures 8c,d and Figs. 8e,f show the behavior of the algorithm with rules $k$–$C$ and $k$–$DC$, respectively. Both versions of the algorithm have self-organizing features. However, it is observed a relatively large number of sensors which are turned on in the beginning of a searching process. The convergence of both versions in terms of a number of active sensors with batteries turned on is slow. The algorithms are characterized by large values of $std$, and a phenomenon of

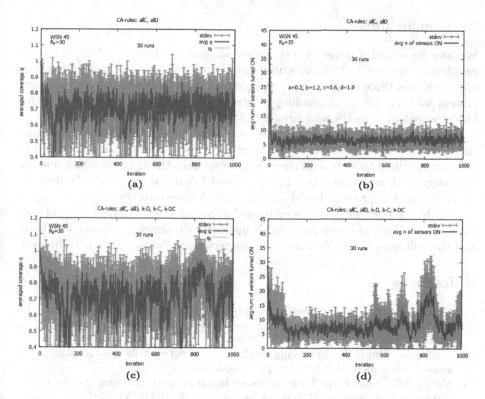

**Fig. 9.** WSN 45 (CA approach): averaged value of (a) $q$ for 2 rules, (b) $n_{ON}$ for 2 rules, (c) $q$ for 5 rules, (d) $n_{ON}$ 5 rules.

oscillation of values $q$ and $n_{ON}$ can be observed. We can see it in Figs. 8d,f, where values $n_{ON}$ oscillate what results in large average values of $n_{ON}$ (in blue).

### 6.3  The Instance: WSN 45 - Cellular Automata Approach

In [12] we proposed an approach to solve the coverage problem with the use of second-order CA. Figure 9 presents simulation results for WSN 45 with the use of CA for two options: with the use of two rules (see, Figs. 9a,b) and with the use of the whole set of 5 rules (see, Figs. 9c,d).

Results show that for both options we observe a relatively large *std* of $q$. We can see also that the average values of $q$ are mostly below the requested $q_r$. The option with the 2 rules is slightly better and more promising than the option with the 5 rules. It provides much more stable values of both $q$ and $n_{ON}$. The comparison of both approaches based either on the use of LA-based agents or CA-based agents shows unambiguously that the LA-based approach significantly outperforms the CA-based approach.

# 7   Conclusions

We have proposed a novel self-organizing algorithm which is able to solve the problem of coverage in WSN in a fully distributed way by self-optimization. The approach uses three components: (a) a graph model of WSN, (b) a multi-agent system with agents corresponding to nodes of a WSN graph, and (c) agents implemented as ($\epsilon$,h)-LA participating in the spatial PD game.

The system is able to self-optimize, i.e. agents act in such a way to maximize their own rewards but at the same time they reach unknown for them global solution specified by two objectives: a coverage and the corresponding number of sensors that are turned ON. The proposed LA-based approach significantly outperforms the approach based on using of CA-based agents.

Our future work will be oriented toward a more detailed study of the proposed approach with use of larger set of WSN instances to consider issues of scalability and the influence of specific model parameters on the algorithm's performance.

# References

1. Berman, P., et al.: Power efficient monitoring management in sensor networks. In: 2004 IEEE Wireless Communication and Networking Conference, pp. 2329–2334 (2004)
2. Cardei, M., Du, D.-Z.: Improving wireless sensor network lifetime through power aware organization. Wirel. Netw. **11**(3), 333–340 (2005)
3. Manju, S.C., Kumar, B.: Target coverage heuristic based on learning automata in wireless sensor networks. IET Wirel. Sensor Syst. **8**, 109–115 (2018)
4. Jaschke, J., Cao, Y., Kariwala, V.: Self-optimizaing control - a survey. Ann. Rev. Control. **43**, 199–223 (2017)
5. Manju, S.C., Kumar, B.: Genetic algorithm-based meta-heuristic for target coverage problem. IET Wirel. Sensor Syst. **8**(4), 170–175 (2018)
6. Narendra, K.S., Thathachar, M.A.L.: Learning Automata Theory: An Introduction. Prentice Hall, Englewood Cliffs (1989)
7. Ojeda, F., et al.: On wireless sensor network models: a cross-layer systematic review. J. Sens. Actuator Netw. **12**, 50 (2023)
8. Östberg, P., Byrne, J., et al.: Reliable capacity provisioning for distributed cloud/edge/fog computing applications. In: European Conference on Networks and Communications (EuCNC 2017), pp. 1–6 (2017)
9. Oommen, B.J., Omslandseter, R.O., Jiao, L.: Learning automata-based partitioning algorithm for Stochasting grouping problems with non-equal partition sizes. Pattern Anal. App. **26**(2), 751–772 (2023)
10. Seredyński, F.: Competetive coevolutionary multi-agent systems: the application to mapping and scheduling problems. J. Paral. Distr. Comp. **47**(1), 39–57 (1997)
11. Seredyński, F., Kulpa, T., Hoffmann, R.: Evolutionary self-optimization of large CA-based multi-agent systems. J. Comput. Sci. **68**, 101994 (2023)
12. Seredyński, F., Kulpa, T., Hoffmann, R., Désérable, D.: Coverage and lifetime optimization by self-optimizing sensor networks. Sensors **23**(8), 3930 (2023)
13. Tsetlin, M.L.: Automata Theory and Modeling of Biological Systems. Elsevier, Amsterdam (1973)
14. Warschawski, W.I.: Collective Behavior of Automata. Nauka, Moscow (1973). (in Russian)

# Enhancing a Hierarchical Evolutionary Strategy Using the Nearest-Better Clustering

Hubert Guzowski[1]([✉])[iD], Maciej Smołka[1][iD], and Libor Pekař[2][iD]

[1] AGH University of Krakow, Kraków, Poland
{guzowski,smolka}@agh.edu.pl
[2] Tomas Bata University in Zlín, Zlín, Czech Republic
pekar@utb.cz
https://www.informatyka.agh.edu.pl/ , https://fai.utb.cz/

**Abstract.** A straightforward way of solving global optimization problems is to find all local optima of the objective function. Therefore, the ability of detecting multiple local optima is a key feature of a practically usable global optimization method. One of such methods is a multi-population evolutionary strategy called the Hierarchic Memetic Strategy (HMS). Although HMS has already proven its global optimization capabilities there is an area for improvement. In this paper we show such an enhancement resulting from the application of the Nearest-Better Clustering. Results of experiments consisting both of curated benchmarks and a real-world inverse problem show that on average the performance is indeed improved compared to the baseline HMS and remains on par with state-of-the-art evolutionary global optimization methods.

**Keywords:** evolutionary algorithm · global optimization · continuous domain · Nearest-Better Clustering

## 1 Introduction

Real-world engineering applications often require solving a global optimization problem. A vast part of them features various kinds of multimodality. Hill climber optimization methods are naturally ill-suited to this type of task, but even single-population stochastic methods having the theoretical asymptotic guarantee of success tend to locate only one solution in practically available time. To tackle this challenge, various niching techniques are used which, by reducing information exchange or competition between groups of individuals, isolate

---

The research presented in this paper was partially supported by the Polish National Science Center under grant No. 2020/39/I/ST7/02285, by the funds of the Polish Ministry of Science and Education assigned to the AGH University of Krakow, by The Czech Science Foundation under grant No. GAČR 21-45465L, and the internal grant No. RVO/CEBIA/2021/001 by TBU in Zlín.

L. Franco et al. (Eds.): ICCS 2024, LNCS 14834, pp. 423–437, 2024.
https://doi.org/10.1007/978-3-031-63759-9_43

so-called species aimed at exploiting different optima of the objective function. Examples include various augmentations of Particle Swarm Optimization and Differential Evolution algorithms [16,24,26]. The Hierarchic Memetic Strategy (HMS) [23] is another optimization algorithm aimed at the multimodal optimization. It achieves the isolation between groups of individuals by running multiple optimization processes in parallel. In order to prevent several populations from converging to the same optima, they are organized in a tree structure, where higher-level processes decide whether and where to start lower-level processes. This operation is called sprouting.

HMS already proved its capabilities in finding multiple optima of the objective for both benchmarks and real-world inverse problems [22,25]. The method was also equipped with special mechanisms addressing the problem of optima located in the flat regions of the objective landscape [21]. These are not used in this paper. Here we present a technique based on the Nearest Better Clustering (NBC) aimed at improving the HMS sprouting. NBC is a well-established niching technique used both in multimodal optimization, dynamic optimization and even in Exploratory Landscape Analysis (ELA) [15]. We conjecture that by combining the unique structure of HMS with the ability of NBC to locate funnel structures in the target function landscape, we obtain a highly effective multimodal optimization method. The extensive tests we present later in this article are intended to test the new mechanism against several criteria:

- Does the application of NBC lead to better location of several target function optima?
- Does the application of NBC improve performance in terms of the best quality solution found?
- Are the parameters of the new mechanism better at adapting to the target problem?

## 2    Background

In the field of optimization, it has become customary to talk about exploration and exploitation features of various algorithms. The exploration is tied to discovering promising areas of the objective function domain where we expect to find an optimum. These areas are called peaks, funnels or basins of attraction, depending on the nature of problem solved or algorithm used. The exploitation refers to locating the solution as accurately as possible in an already defined area. For a single population of individuals that exchange knowledge between each other, either tendency for exploration or exploitation will at some point outweigh the other [23]. In the case of exploitation, convergence occurs and in the case of exploration we arrive at a method similar to the pure random search. Niche methods partition the population so that its parts running in parallel can converge to detected optima without depriving the algorithm of its ability to explore. This behavior is particularly beneficial in multimodal optimization problems when we can distinguish multiple equally good global optima, but it also has benefits for standard single-solution optimization problems [11]. To date, a number of niche techniques have been presented that differ in the way the division is achieved. One of these is a clustering based on the distance between individuals, exemplified by the NBC method [15]. The main observation behind the NBC is that the distance between an individual and its nearest neighbor with a better fitness value will be greater for the best individual in a given basin of attraction than for its surrounding individuals. Thus, a spanning tree where edges connect nearest better individuals with the best solution overall at its root is constructed. Then longer edges are removed creating subtrees representing species meant to occupy different niches.

The same observation about the inability to preserve exploration and exploitation for optimization algorithms using a single population inspired the creation of HMS. However, its answer is not the same as of typical niching methods. Instead of separating the population operating under a single algorithm into subspecies, HMS combines multiple populations operating under different algorithms into a hierarchical relationship [23] which operation is described in Algorithm 1. At the base of the tree hierarchy is the root algorithm, which can be a global optimization engine of user's choosing, but which has to be geared predominantly towards exploration. As a result, it would not be able to find the best solution with a decent accuracy in practically achievable time, but instead is responsible for detecting candidate basins of attraction. When it finds one, it launches a new process focused on exploiting the local optimum. In general this process is operating locally or semi-locally, thus its initial population is sampled from a Gaussian distribution as opposed to a uniform distribution over the whole search space which is used for the root process. Originally, the same evolutionary algorithm with different parameters was used at all levels, but the introduction of Covariance Matrix Adaptation Evolutionary Strategy (CMA-ES) as the algorithm responsible for exploitation has significantly improved performance [22]. HMS run consists of a loop executing a fixed number of iterations for each of its subprocesses. The subprocesses can be deactivated based on their specific stopping criterion e.g. no improvement over several epochs while the global stopping condition is checked for termination of the whole algorithm. An HMS user has control over the number of tree levels, which algorithms will be run on them, what will be their parameters and stopping conditions. This flexibility makes HMS adaptable to specific real-world tasks. One example would be the usage of multiwinner voting operators to promote a diversity of solutions for lower level algorithms, which made it possible to identify shapes and sizes of function niches [5]. To facilitate the use of HMS for non-expert users, an autoconfiguration mechanism was introduced based on objective function features calculated using ELA methods [7]. Although the HMS already proved to be a versatile and efficient global optimizer, there are still clear areas for improvement. In particular, the mechanism responsible for identifying the pools of attraction used to date could be a source of some practical problems. Its operation is very simple: consider the best current solution from every higher-level subprocess and if it is far enough from any centroid of lower-level population, start a new process at its location. Although intuitive and functional, this mechanism has clear dis-

**Algorithm 1:** High-level pseudocode of HMS

---

**Input**: Objective function $f$, *dimensions* and *bounds* of $f$, $hmsTreeConfig$
**Output**: $foundOptima$ of $f$

1  $root \leftarrow$ initPopulation($hmsTreeConfig[0]$, $\emptyset$)
2  $activePopulations \leftarrow \{root\}$
3  **while** *global stop condition is not satisfied* **do**
4      **foreach** $p$ *in* $activePopulations$ **do**
5          $isActive \leftarrow$ runMetaepoch($p$)
6          **if** *not* $isActive$ **then**
7              $activePopulations$.remove($p$)
8              **if** $p.engine = SEA$ **then**
9                  $p.best \leftarrow$ runLocalMethod($p.best$)
10      runSprout($activePopulations$, $hmsTreeConfig$)
11  $foundOptima \leftarrow []$
12  **foreach** $p$ *in all populations* **do**
13      $foundOptima$.add($p.best$)

---

advantages. Firstly, the value of the *far enough* distance parameter is highly problem dependent. This does not only mean that it should be scaled proportionally when working with a larger domain. Actually, the ideal value of the parameter should be exactly the radius of the basin of attraction so as not to allow more processes to run in it, but also not to cover the other basins. In practice, working with black box problems, we will certainly not set the exact value of this parameter correctly. So we can end up with a situation where the value is too high which makes it difficult to cover optima that are close to each other, or too low which results in a redundant running multiple processes in the same niche. In addition, the check approximates the radius of attraction pools to be perfectly spherical, which may not be an appropriate shape. Secondly, the best individual of a higher-level population can stagnate in the same peak for multiple iterations. This leads to underutilization of the information contained in the population of individuals. Even if solutions of similar quality to the best individual are found, they will remain ignored.

## 3    HMS with NBC Component

In the previous sections, we have already mentioned the idea and basics of the HMS algorithm. Apart from the new sprout mechanism, the core of the algorithm remains the same. Currently, we considered the Simple Evolutionary Algorithm (SEA) and CMA-ES as candidates for algorithm components. Additionally, if SEA is used as a lower-level method, then on its deactivation we execute an additional local optimizer. In particular, we use L-BFGS-B provided by Python SciPy library, but it can be replaced according to user's preference. The implementation used during experiments described in this paper can be found in [8], whereas the continuously developed Python implementation of the HMS is available on GitHub[1]. The core functionality is based on the LEAP library [4] while CMA-ES implementation is sourced from pycma package [10].

Our main contribution in this paper is the NBC-based sprout mechanism, which is presented in detail in Algorithm 2. Viewed from a high level, it consists of performing NBC clustering for each active population and then filtering resultant individuals based on quality and distance to active subprocesses at the lower level of the tree. Those distances are determined based on centroids calculated at the start of the process (line 2). At the start of the procedure we apply a simple elite selection mechanism introduced in line [14] which takes only the best fraction of the population given by $truncFactor$ (line 4). Afterward a standard Nearest Better Clustering is performed for each subprocess population separately. This consists of construction of a tree where individuals are connected to their nearest better neighbor and then cutting of edges that are $nbcCut$ times longer than mean edge length. Root individuals of each subtree are tested if they are far enough from active populations centroids mentioned above. Afterward a singular best individual that passes the distance criterion is selected per each subpopulation (line 20). The last filtering mechanism is taking the number of best candidates that will not exceed a subprocess limit imposed for each level (line 21). Resulting individuals serve as starting locations for new subprocesses.

The new mechanism addresses shortcomings described at the end of Sec. 2. Firstly, thanks to the capabilities of NBC, each parent population can identify multiple basins of attraction per metaepoch. The main benefit of this feature is that during consecutive metaepochs the parent population will sprout new subprocesses in different basins of attraction even if the best individual would not change. Secondly, we transfer the responsibility for determining the radii of the basins of attraction from the user to the method. Instead the user can operate on factors that influence the mechanism on the higher level. $nbcCut$ and $farEnough$ listed in pseudocode 2 both influence the granularity of sprouting process. $truncFactor$ can shift the behavior more towards exploration or exploitation, but our experiments suggest that the 0.8 can be safely assumed as its default value. $LevelLimit$ remains more as an emergency hard cap as the NBC tends to organically limit itself. The new mechanism therefore provides the user with more parameters than the previous one. On the one hand, this may allow the algorithm to be better tuned to the problem at hand, but in practice it is easier to use parameter-free algorithms.

## 4   Experimental Procedure

To answer the questions posed in the introduction, we prepared a series of comparisons. The first subsection below focuses on assessing the ability of the algorithm to locate multiple global optima. In the next one, the aim is to see what impact the newly applied mechanism has on the quality of the best solution found.

---

[1] https://github.com/agh-a2s/pyhms.

---

**Algorithm 2:** Pseudocode of NBC based HMS sprout

---

**Input**: *activePopulations, hmsTreeConfig*
**Parameters**: *truncFactor, nbcCut, farEnough, levelLimit*

1   *sproutCandidates* ← []
2   *centroids* ← calculatePopulationsCentroids(*activePopulations*)
3   **foreach** *p in activePopulations* **do**
4      *tp* ← takeBest(*p, truncFactor*)
5      *T* ← createEmpyTree()
6      *T*.addNode(best individual *i* from *tp*)
7      *distances* ← []
8      **foreach** *c in tp\\{i}* **do**
9          *nbi* ← findNearestBetterNeighbour(*c, tp*)
10         *T*.addNode(*c*)
11         *T*.addEdge(*c, nbi*)
12         *distances*.add(distance(*c, nbi*))
13      *cut_dist* ← *nbcCut* · mean(*distances*)
14      **foreach** *e in T.edges()* **do**
15         **if** *e.distance ¿ cut_dist* **then**
16            Cut off *e*
17            *c* ← newly created root node
18            **if** *all distances(c, centroids) ¿ farEnough* · mean(*distances*) **then**
19               *sproutCandidates*.add(*c*)
20   *sproutCandidates* ← TakeOnePerDeme(*sproutCandidates*)
21   *sproutCandidates* ← BestPerLevelUpToLimit(*sproutCandidates, levelLimit*)
22   **foreach** *c in sproutCandidates* **do**
23      *childPopulation* ← initPopulation(*treeConfig[level(c) + 1], c*)
24      *activePopulations*.add({*childPopulation*})

---

## 4.1 Global Optima Coverage

We compare HMS with base sprout against new NBC sprout mechanism on a multimodal testbed provided by the PyDDRBG multimodal optimization benchmark [1]. We use 10 default target functions with identifiers from 1 to 10 provided by the library in a static variant. All of those target functions are 10-dimensional with box constraints between -5.0 and 5.0 for every dimension. As a quality measure we provide how many of the function's global optima were located and what was the total evaluation budget of the algorithm. The decision whether a given optimum has been located is made by checking whether any of the best solutions found by each subprocess is at a distance no greater than the specified niche radius for the given optimum. Besides sprout mechanism both versions use the 2-levels HMS structure with root SEA and CMA-ES leaves with the same hyperparameters.

In the first part of this experiment, we investigated the effect of changing the values of the sprouting mechanism parameters on the number of global optima covered and sprouted subprocesses in general. For the baseline mechanism we had only 1 parameter to test. For NBC sprout, we can manipulate 3 parameters:

the subtree cut-off factor, the distance factor and the fraction of best individuals included. Each of these parameters can be used to limit the number of subprocesses sprouted and therefore, when testing the cut-off and distance factors, we set the second parameter to the liberal value of 1.0. When testing the *truncFactor* fraction value, we set the other two to 1.5. A run of each configuration was repeated 10 times.

In the second part, based on the previous results, we selected a single configuration for both sprouting mechanisms and ran such configured HMS 50 times for each target function to calculate the average Peak Ratio and the average function evaluation budget. For simple sprouting we set the value of the distance parameter as 4.0 while for NBC sprouting we set the cut-off value as 1.5, the distance factor as 1.0 and the fraction of solutions considered as 0.8.

## 4.2   Best Solution Quality

The tests we performed comparing the quality of the best found solution use the same procedures as the work done in the earlier article [7] and include the results achieved there for the base version of HMS. The comparison is performed on the Black Box Optimization Benchmark (BBOB) single objective continuous test suite [9]. To better compare sprout mechanisms, we keep the same two-level HMS architecture for the respective BBOB function classes consisting of SEA root with either CMA-ES or SEA leaves. We first run hyperparametric optimization for each of the five classes of 10-dimensional functions defined in BBOB by minimizing the sum of the quality of the solutions obtained after 9500 function evaluations for the first instances of every function in the class. As hyperparameter optimization tool we employed SMAC3 [12] and repeated a run 5 times for each class with a budget of 3000 evaluations in total. Details of the SMAC3 scenario and the ranges of hyperparameters considered in the optimization process can be found in linked repository [8]. Afterward we selected the best performing configuration out of the resulting 5 based on 50 runs per target function both in 10 and 20 dimensions with budget of 9500 and 19000 function evaluations respectively. We did not apply explicit parameter scaling when using the obtained configurations for 20-dimensional problems.

Our comparison focuses on two versions of HMS with base sprouting mechanisms and one based on NBC. In addition, we include the results of the BIPOP-CMA-ES [13] and iL-SHADE [3] which proved to be the most competitive and versatile algorithms in article [7]. In contrast to that work, we performed the same hyperparameter optimization and configuration selection procedure as for the HMS algorithm for a fair comparison. Although both of those algorithms are highly adaptive and do not require many parameters, they still gained in performance in the tuning process. We optimized population and archive size for iL-SHADE and population size, $\sigma_0$, population increment factor on restart and function value change tolerance for termination for BIPOP-CMA-ES. iL-SHADE implementation is sourced from PyADE library[2] while BIPOP-CMA-ES implementation is provided in pycma [10].

---

[2] https://github.com/xKuZz/pyade.

## 4.3   Real-World Inverse Problem

In addition to comparisons to the previous article, we also decided to include in the comparison a real-world parameter identification problem formulated as a global optimization task. It features a significantly different domain size relative to the problems included in the BBOB benchmark. Here we include its concise description.

Estimating the parameters of time-delay system (TDS) models constitutes a challenging task [6,27]. Its complexity arises mainly from the existence of infinite number of the so-called system modes in the TDS dynamics; however, the model is characterized via only a small set of parameters. Nevertheless, only a few dominant modes have a decisive role [18]. Hence, the identification problem can be formulated as searching for a parameter set yielding model dominant modes with the minimum distance to the actual TDS dominant modes.

Processes with interconnected heating-cooling loops (HCPs) are representatives of TDSs as they inherently incorporate heat and mass transfer, which causes latencies [28]. A stable laboratory HCP was investigated, e.g., in [19,20]. Despite its physical simplicity, the laboratory HCP has a relatively complex dynamics due to delays in the feedback loops. We herein focus on the relation between the input voltage to the heater $P_H(t) = u(t)$ and the outlet fluid temperature of the radiator $\vartheta_{CO}(t) = y(t)$ expressed by the delay-differential equation [19]

$$y^{(3)}(t) + a_2\ddot{y}(t) + a_1\dot{y}(t)a_0 y(t) + a_{0D}y(t-\theta) = b_0 u(t-\tau) + b_{0D}u(t-\tau-\tau_0) \quad (1)$$

where $b_0$, $b_{0D}$, $a_2$, $a_1$, $a_0$, and $a_{0D}$ are non-delay model parameters, while $\tau_0$, $\tau$, and $\theta$ are positive delays. The necessary stability conditions read

$$a_2 > 0, a_1 > 0, a_0 + a_{0D} > 0 \quad (2)$$

The equivalent model transfer function to (1) reads

$$G_m(s) = \frac{Y(s)}{U(s)} = \frac{b_0 + b_{0D}e^{-\tau_0 s}}{s^3 + a_2 s^2 + a_1 s + a_0 + a_{0D}e^{-\vartheta s}}e^{-\tau s} \quad (3)$$

where $Y$ and $U$ are the Laplace transforms of $y$ and $u$.

Among the large number of methods and techniques to determine unknown model parameters, a specific family of identification approaches utilizing a nonlinear element in the feedback loop exists. A subset of these approaches enables to estimate parameter values of (3) in the frequency domain [2,17]. Namely, one can obtain a set of experimentally obtained values $\widehat{G}_m(\imath\omega_j + a)$ for some suitable discrete values of $\omega_j, a \geq 0, j = 1, 2, ..., N$, where $\imath$ is the imaginary unit satisfying $\imath^2 = -1$. Hence, the parameter optimization task related to model (3) and with respect to (2) can be formulated as searching the closest distance of estimated $\widehat{G}_m(.)$ to the modeled ones $G_m(.)$ for selected $\omega_j, a$ that are decisive for dominant modes behavior. Then, the optimization problem can be expressed as follows

$$\mathbf{p}^* = \arg\min f(\mathbf{p})$$

$$f(\mathbf{p}) = \sum_{j=1}^{N} \left| \widehat{G}_m(\imath\omega_j + a) - G_m(\imath\omega_j + a) \right|^2 \tag{4}$$

$$\mathbf{p} = [b_{0D}, \tau_0, \tau, a_2, a_1, a_0, a_{0D}, \theta]$$

$$\text{such that} : [\tau_0, \tau, \theta, a_2, a_1, a_0 + a_{0D}] > 0$$

It is worth noting that problem (4) is multimodal and constrained. It i.a. means that some results with even an excellent cost function value do not reflect physical reality. Therefore, it is necessary to remain close to the values obtained by physical modeling and analysis [19,20]. Using ELA we discovered that this problem is of the same type as the BBOB class 5 problems, i.e. multimodal with the weak global structure. For this reason, we chose the same algorithm configurations for this problem as for the class 5 function. We also set a budget of 9500 evaluations which is consistent with previous tests. The actual computational problem domain is specified by box constraints of $[0, 500]$, $[0, 500]$, $[0, 1000]$, $[0, 2500]$, $[0, 2500]$, $[-250, 250]$, $[-250, 250]$, $[0, 1000]$ for respective components of $\mathbf{p}$. As its size is much larger than in the benchmark case, we scaled the distance parameters of baseline sprouting, mutation standard deviation and $\sigma_0$ linearly. In the case of NBC sprout, we did not need to apply scaling. To tackle additional domain constraints, we apply penalization during optimization process and at the end consider only valid individuals for the comparison.

## 5 Results

This section presents the results of the experiments carried out in subsections which correspond to their descriptions in Sect. 4.

### 5.1 Global Optima Coverage

The results of tests examining the effect of changing parameter values on the coverage of function optima show a clear improvement after applying the new NBC-based sprout mechanism. Because of limited space available, we include the resultant graphs for 3 out of 10 target functions under examination. However, the same general trend can be observed for every target function besides the instance 9, for which both variants perform poorly. The complete results are included in Zenodo archive [8]. As can be seen in Fig. 1 for the base sprout mechanism, when reducing the distance parameter below some border value the number of subprocesses increases fast, but the raise does not correspond to a similar increase in number of covered optima. Meanwhile, for the NBC sprout, we observe an even more drastic increase in number of created subprocesses with a value decrease between 2.0 and 1.5. However, this time it translates into substantially higher optima coverage. It is important to note, that the recommended NBC cut value of 2.0 produces quite limited number of sprouted subprocesses and also a lower optima coverage. This may be related to the nature of the PyD-DRBG [1] benchmark, which contains problems with global optima clustered inside a single peak or some yet to be discovered peculiarity in our application of NBC. Although the use of the new mechanism brings clear benefits in the number of optima covered, the way in which the amount increases rapidly in response to small changes in the value of the parameter is not desirable. Our preliminary tests on other test beds suggest, that this problem is not as drastic in the case of more easily distinguishable peaks, but it remains as an area to be addressed in the future works.

The next comparison shows the average number of covered peaks and budget expended during that process for both sprout variants. Although both mechanisms can be easily tuned to be highly exploratory or to expend the least amount of evaluations, we had to choose configurations that maximize both of those criteria. This context is relevant to the interpretation of the results in Table 1. They show higher optima coverage for 8 out of 10 functions for the new sprout mechanism. We can look more closely at the instance 6, which is the only one with worse coverage achieved with the new mechanism. As we can see on Fig. 1, the steep jump in coverage for this particular instance occurs for relatively low value of parameters, which were also lower than the ones we chose for this experiment. However, if we look at the results from the perspective of coverage efficiency, the picture shown is even more favorable as the gain in exploration capabilities of the algorithm does not come at the cost of sacrificing performance in terms of evaluation budget used. On average around 1526 evaluations were required for each optimum covered in case of the base sprout while the NBC sprout required 1081 per optimum. This is a clear qualitative improvement in the performance of the algorithm under these criteria.

**Table 1.** Comparing the efficiency of finding global optima between HMS with base and NBC sprout with set parameters on PyDDRBG benchmark. Values are averaged over 50 runs.

| Instance | | HMS base sprout | | HMS NBC sprout | |
|---|---|---|---|---|---|
| Id | N optima | Optima covered | Evaluation budget | Optima covered | Evaluation budget |
| 1 | 4 | 2.32 | 3091.2 | 2.34 | 2797.2 |
| 2 | 2 | 1.88 | 4018.4 | 1.96 | 3715.8 |
| 3 | 3 | 2.04 | 3515.0 | 2.88 | 5161.6 |
| 4 | 3 | 1.62 | 3023.4 | 1.86 | 3170.8 |
| 5 | 4 | 2.14 | 3700.2 | 3.82 | 4976.8 |
| 6 | 16 | 3.34 | 3003.0 | 2.74 | 2704.8 |
| 7 | 8 | 3.5 | 3785.2 | 4.82 | 4442.6 |
| 8 | 9 | 2.48 | 3351.6 | 6.38 | 5367.6 |
| 9 | 18 | 0.42 | 2944.2 | 0.42 | 3099.6 |
| 10 | 16 | 2.42 | 3385.2 | 10.6 | 5430.6 |

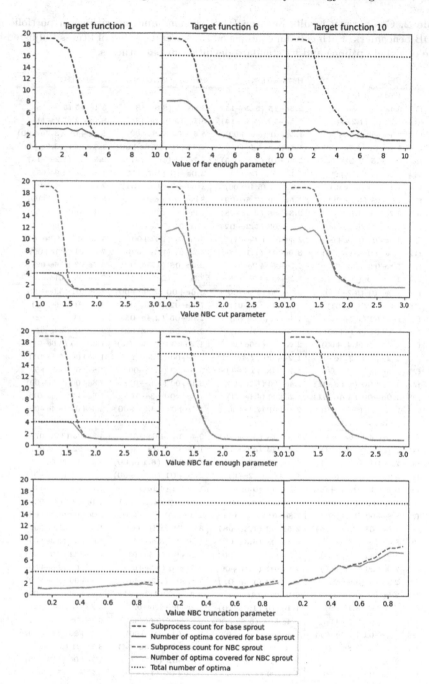

**Fig. 1.** Coverage difference for DDRB instance 5, 8 and 10. The first row corresponds to base sprout mechanism while rows 2–4 represent the change of one of the parameters in new NBC-based sprout mechanism. Results are averaged over 10 runs per value of a specific parameter.

**Table 2.** Comparison results between HMS versions and algorithms in portfolio on BBOB benchmark. Each column contains 50 runs average value of fitness (adjusted by value of global optima) and its standard deviation in parentheses.

|  | HMS | HMS with NBC | BIPOP-CMA-ES | iL-SHADE |
|---|---|---|---|---|
| | **10 dimensions** | | | |
| f1 | 9.4e−15 (7.8e−15) | **2.3e−15 (5.2e−15)** | 8.8e−13 (8.6e−13) | 3.1e−15 (5.9e−15) |
| f2 | 7.7e−01 (1.2e+00) | **1.1e−14 (2.3e−14)** | 8.1e−13 (6.2e−13) | 8.1e−12 (7.2e−11) |
| f3 | 8.4e+00 (3.1e+00) | 1.3e+01 (5.0e+00) | 8.6e+00 (3.9e+00) | **8.3e+00 (3.7e+00)** |
| f4 | 1.4e+01 (4.7e+00) | 1.8e+01 (7.4e+00) | **1.1e+01 (3.6e+00)** | **1.1e+01 (3.9e+00)** |
| f5 | 6.8e−15 (2.4e−14) | **2.8e−16 (2.8e−15)** | 6.7e−13 (4.0e−13) | 2.8e−15 (1.2e−14) |
| f6 | 7.2e−01 (3.7e+00) | 4.4e−03 (3.0e−02) | **2.0e−13 (1.7e−13)** | 1.7e−05 (4.9e−05) |
| f7 | 1.1e+00 (6.5e−01) | 1.3e+01 (5.4e+00) | **1.3e−02 (1.0e−01)** | 7.8e−02 (2.4e−01) |
| f8 | 1.2e−08 (1.1e−08) | **7.5e−09 (7.8e−09)** | 8.0e−02 (5.6e−01) | 3.1e+00 (1.5e+00) |
| f9 | 3.7e−08 (3.7e−08) | **2.3e−08 (2.5e−08)** | 1.9e−05 (1.9e−04) | 4.4e+00 (1.2e+00) |
| f10 | 1.0e+00 (6.8e+00) | **2.2e−03 (2.2e−02)** | 8.7e−01 (3.9e+00) | 7.0e−02 (1.9e−01) |
| f11 | 1.5e+00 (5.9e+00) | 2.3e−01 (6.6e−01) | 5.4e−01 (1.7e+00) | **1.2e−02 (8.3e−02)** |
| f12 | 1.8e+00 (2.6e+00) | **3.0e−01 (1.1e+00)** | 4.2e−01 (1.4e+00) | 7.8e−01 (1.1e+00) |
| f13 | 7.8e−03 (2.9e−02) | 1.2e−05 (1.2e−04) | **2.6e−08 (1.4e−07)** | 8.7e−03 (1.6e−02) |
| f14 | **1.1e−11 (9.4e−12)** | 1.2e−11 (8.9e−12) | 3.5e−10 (2.0e−10) | 2.2e−06 (3.1e−06) |
| f15 | 1.0e+01 (5.2e+00) | 9.5e+00 (4.5e+00) | **3.6e+00 (3.9e+00)** | 1.3e+01 (5.5e+00) |
| f16 | 3.0e−01 (3.7e−01) | 3.2e−01 (3.5e−01) | **7.4e−02 (1.5e−01)** | 8.7e−01 (9.6e−01) |
| f17 | 9.5e−02 (1.7e−01) | 1.4e−02 (4.0e−02) | **2.6e−05 (2.4e−05)** | 7.2e−03 (3.6e−02) |
| f18 | 6.4e−01 (1.1e+00) | 1.8e−01 (3.1e−01 | **2.9e−03 (2.8e−02)** | 1.4e−01 (3.4e−01) |
| f19 | 2.1e+00 (1.4e+00) | 2.4e+00 (1.6e+00) | **1.2e+00 (7.3e−01)** | 1.8e+00 (4.6e−01) |
| f20 | **1.2e+00 (2.3e−01)** | 1.2e+00 (2.4e−01) | 1.4e+00 (3.3e−01) | 1.3e+00 (2.8e−01) |
| f21 | 1.4e+00 (1.4e+00) | **1.2e+00 (1.6e+00)** | 3.5e+00 (4.4e+00) | 4.8e+00 (7.1e+00) |
| f22 | **1.3e+00 (8.6e−01)** | 1.4e+00 (1.2e+00) | 5.4e+00 (1.1e+01) | 2.8e+00 (1.7e+00) |
| f23 | **1.0e+00 (4.4e−01)** | 1.2e+00 (4.1e−01) | 1.3e+00 (6.3e−01) | 1.5e+00 (3.6e−01) |
| f24 | 2.0e+01 (5.8e+00) | 2.4e+01 (7.4e+00) | **1.9e+01 (8.5e+00)** | 2.6e+01 (8.4e+00) |
| | **20 dimensions** | | | |
| f1 | 1.6e−14 (4.6e−15) | **1.4e−14 (3.1e−15)** | 9.4e−13 (4.0e−13) | 4.3e−08 (3.9e−07) |
| f2 | 7.4e+01 (3.4e+01) | **9.3e−14 (4.0e−13)** | 1.0e−12 (5.1e−13) | 3.2e−02 (2.5e−01) |
| f3 | 2.7e+01 (7.0e+00) | 2.9e+01 (8.0e+00) | **1.9e+01 (6.1e+00)** | 3.2e+01 (1.0e+01) |
| f4 | 3.9e+01 (1.0e+01) | 4.3e+01 (1.2e+01) | **2.2e+01 (4.1e+00)** | 4.2e+01 (1.3e+01) |
| f5 | **8.5e−14 (0.0e+00)** | **8.5e−14 (0.0e+00)** | 1.0e−12 (4.1e−13) | 9.7e−14 (2.8e−14) |
| f6 | 7.8e+02 (5.4e+06) | 2.5e−03 (6.9e−03) | **4.7e−13 (3.1e−13)** | 2.1e−02 (6.5e−02) |
| f7 | 8.8e+00 (1.1e+01) | 7.8e+01 (2.3e+01) | **1.3e+00 (1.0e+00)** | 2.9e+00 (2.4e+00) |
| f8 | **2.5e−08 (2.4e−16)** | 2.5e−08 (1.8e−08) | 3.4e−01 (1.1e+00) | 1.5e+01 (1.1e+01) |
| f9 | 8.4e−08 (3.2e−15) | **8.1e−08 (5.3e−08)** | 1.3e−01 (6.8e−01) | 1.6e+01 (1.3e+00) |
| f10 | 7.2e+00 (1.8e+01) | **2.6e+00 (1.0e+01)** | 1.2e+01 (2.6e+01) | 1.9e+02 (1.9e+02) |
| f11 | 2.3e+00 (7.9e+00) | **4.6e−01 (2.1e+00)** | 2.7e+00 (9.1e+00) | 3.0e+00 (2.5e+00) |
| f12 | 2.2e−01 (8.6e−01) | **4.8e−02 (1.6e−01)** | 1.5e−01 (1.2e+00) | 1.0e+00 (2.5e+00) |
| f13 | 2.5e−01 (6.7e−01) | 6.1e−02 (2.1e−01) | **3.5e−02 (1.2e−01)** | 1.6e+00 (1.7e+00) |
| f14 | 3.9e−09 (3.1e−09) | **1.1e−10 (1.0e−10)** | 2.9e−09 (2.1e−09) | 3.2e−04 (1.9e−04) |
| f15 | 2.5e+01 (7.3e+00) | **2.4e+01 (6.8e+00)** | 8.6e+00 (3.0e+00) | 6.1e+01 (1.4e+01) |
| f16 | **5.1e−01 (4.0e−01)** | 1.3e+00 (3.8e+00) | 7.4e−01 (3.7e+00) | 3.7e+00 (2.8e+00) |
| f17 | 3.4e−02 (5.8e−02) | 2.5e−02 (6.6e−02) | **6.1e−05 (4.3e−04)** | 3.7e−01 (3.7e−01) |
| f18 | 4.6e−01 (7.2e−01) | 1.9e−01 (1.7e−01) | **1.6e−02 (6.7e−02)** | 1.7e+00 (1.7e+00) |
| f19 | 4.2e+00 (1.6e+00) | 4.1e+00 (2.0e+00) | **2.9e+00 (1.1e+00)** | 3.2e+00 (5.6e−01) |
| f20 | **1.5e+00 (1.5e−01)** | 1.5e+00 (2.1e−01) | 1.6e+00 (2.2e−01) | 1.7e+00 (2.5e−01) |
| f21 | **1.9e+00 (2.2e+00)** | 2.0e+00 (2.5e+00) | 6.1e+00 (6.3e+00) | 5.7e+00 (8.1e−01) |
| f22 | 3.3e+00 (4.4e+00) | 3.4e+00 (6.3e+00) | 2.1e+01 (2.1e+01) | **1.3e−02 (1.8e−02)** |
| f23 | **1.5e+00 (5.0e−01)** | 1.6e+00 (5.0e−01) | 2.3e+00 (6.9e−01) | 2.0e+00 (3.8e−01) |
| f24 | 6.9e+01 (1.3e+01) | 7.5e+01 (1.7e+01) | **5.6e+01 (3.2e+01)** | 7.5e+01 (1.5e+01) |

## 5.2 Best Solution Quality

Comparison of solution quality is included in Table 2. In 27 out of 48 cases the new variant of HMS with NBC sprout outperforms the previous version of the algorithm, while in the other 16 it performs worse. As the hyperparameter tuning produced different configurations with different performances on the same functions, some fluctuations in solution quality are to be expected. However, the number of improved results is substantially higher and also in 8 cases the improvement is more than tenfold. Compared to the original sprout mechanism the gains are most visible for classes of functions that are less multimodal (f1-14). We attribute this results to the improved filtering of unnecessary new subprocesses introduced in the new sprout mechanism. At the same time, when looking at the bigger picture of HMS in comparison to other algorithms, its purpose remains as the method to tackle highly multimodal problems (f20-24). Although the new mechanism certainly patched the performance on some of the previously problematic instances, making the algorithm more universal in usage (f2,f6).

## 5.3 Real-World Optimization Problem

Results included in Table 3 show improved results for the new version of HMS compared to the base one. Given that the real-world problem is similar in landscape to class 5 BBOB functions for which the new version have not performed better, these results hint at the better adaptability of the new mechanism. In the wider context of the HMS algorithm, we can see that the configurations trained for BBOB problems transferred very well to a real-world problem and remain competitive for BIPOP-CMA-ES and iL-SHADE. This is a very good sign for the use of default HMS configurations and the overall quality of the algorithm.

**Table 3.** Comparison results between HMS versions and algorithms in portfolio on real-world problem. Each column contains 50 runs average value of fitness and its standard deviation in brackets.

| HMS | HMS with NBC | BIPOP-CMA-ES | iL-SHADE |
|---|---|---|---|
| 6.5e–04 (3.0e–04) | 5.6e–04 (3.6e–04) | 4.0e–03 (8.7e–19) | 3.5e–04 (2.1e–04) |

# 6 Conclusion

In this paper we have introduced and tested a new Nearest Better Clustering based component for the Hierarchic Memetic Strategy. The new mechanism improves the algorithm's exploration capabilities by making a better use of the information contained in higher-order populations. At the same time, we have shown improvement in terms of the best solution found for a wide range of optimization problems included in the BBOB benchmark and for a real-world problem where HMS performed on a par with state-of-the-art global optimizers. The parameters of the new mechanism are less problem-dependent making the determination of universal default values simpler. However, the increase in the number of parameters and mechanisms complexity may make the new mechanism more difficult to use from the perspective of an unfamiliar user. Thus, in future works we plan to combine the new functionalities with configuration auto-selector proposed in [7]. Another promising area for improvement is the parameter adaptation during the runtime, which seems to be a natural direction of development for HMS, yet still remains unexplored.

# References

1. Ahrari, A., et al.: PyDDRBG: a python framework for benchmarking and evaluating static and dynamic multimodal optimization methods. SoftwareX **17**, 100961 (2022). https://doi.org/10.1016/j.softx.2021.100961
2. Bi, Q.A., Wang, Q.G., Hang, C.C.: Relay-based estimation of multiple points on process frequency response. Automatica **33**, 1753–1757 (1997)
3. Brest, J., Mau?ec, M.S., Bošković, B.: iL-SHADE: improved L-SHADE algorithm for single objective real-parameter optimization. In: 2016 IEEE Congress on Evolutionary Computation (CEC), pp. 1188–1195. IEEE (2016). https://doi.org/10.1109/CEC.2016.7743922
4. Coletti, M.A., Scott, E.O., Bassett, J.K.: Library for evolutionary algorithms in Python (LEAP). In: GECCO '20 Companion: Proceedings of the 2020 Genetic and Evolutionary Computation Conference Companion, pp. 1571–1579. ACM (2020). https://doi.org/10.1145/3377929.3398147
5. Faliszewski, P., et al.: Multiwinner voting in genetic algorithms. IEEE Intell. Syst. **32**(1), 40–48 (2017). https://doi.org/10.1109/MIS.2017.5
6. Gupta, S., Gupta, R., Padhee, S.: Parametric system identification and robust controller design for liquid-liquid heat exchanger system. IET Control Theory Appl. **12**, 1474–1482 (2018). https://doi.org/10.1049/iet-cta.2017.1128
7. Guzowski, H., Smołka, M.: Configuring a hierarchical evolutionary strategy using exploratory landscape analysis. In: GECCO 2023 Companion: Proceedings of the Companion Conference on Genetic and Evolutionary Computation, pp. 1785–1792. ACM (2023). https://doi.org/10.1145/3583133.3596403
8. Guzowski, H., Smo?ka, M., Pekař, L.: Experimental software used for Enhancing a Hierarchical Evolutionary Strategy Using the Nearest-Better Clustering article for ICCS 2024 conference. Zenodo (2024).https://doi.org/10.5281/zenodo.10730541
9. Hansen, N., et al.: Real-parameter black-box optimization benchmarking 2009: experimental setup. Research Report RR-6828, INRIA (2009). https://hal.inria.fr/inria-00362649
10. Hansen, N., et al.: CMA-ES/pycma. Zenodo (2023). https://doi.org/10.5281/zenodo.2559634
11. Li, X., et al.: Seeking multiple solutions: an updated survey on niching methods and their applications. IEEE Trans. Evol. Comput. **21**(4), 518–538 (2017). https://doi.org/10.1109/TEVC.2016.2638437
12. Lindauer, M., et al.: SMAC3: a versatile bayesian optimization package for hyperparameter optimization. J. Mach. Learn. Res. **23**(54), 1–9 (2022). http://jmlr.org/papers/v23/21-0888.html

13. Loshchilov, I., Schoenauer, M., Sèbag, M.: Bi-population CMA-ES algorithms with surrogate models and line searches. In: Proceedings of the 15th Annual Conference Companion on Genetic and Evolutionary Computation, GECCO 2013 Companion, pp. 1177–1184. ACM (2013). https://doi.org/10.1145/2464576.2482696

14. Luo, W., et al.: Identifying species for Particle Swarm Optimization under dynamic environments. In: 2018 IEEE Symposium Series on Computational Intelligence (SSCI), pp. 1921–1928 (2018). https://doi.org/10.1109/SSCI.2018.8628900

15. Luo, W., et al.: A survey of nearest-better clustering in swarm and evolutionary computation. In: 2021 IEEE Congress on Evolutionary Computation (CEC), pp. 1961–1967 (2021). https://doi.org/10.1109/CEC45853.2021.9505008

16. Luo, W., et al.: Hybridizing niching, particle swarm optimization, and evolution strategy for multimodal optimization. IEEE Trans. Cybern. **52**(7), 6707–6720 (2022). https://doi.org/10.1109/TCYB.2020.3032995

17. Miguel-Escrig, O., et al.: Multiple frequency response points identification through single asymmetric relay feedback experiment. Automatica **147**, 110749 (2023). https://doi.org/10.1016/j.automatica.2022.110749

18. Özer, M.S., Íftar, A.: Eigenvalue optimisation-based centralised and decentralised stabilisation of time-delay systems. Int. J. Control **95**, 2245–2266 (2022). https://doi.org/10.1080/00207179.2021.1906446

19. Pekař, L.: Modeling and identification of a time-delay heat exchanger plant. In: Pekař, L. (ed.) Advanced Analytic and Control Techniques for Thermal Systems with Heat Exchangers, pp. 23–48. Academic Press (Elsevier) (2020). https://doi.org/10.1016/B978-0-12-819422-5.00002-5

20. Pekař, L., et al.: Further experimental results on modelling and algebraic control of a delayed looped heating-cooling process under uncertainties. Heliyon **9**, e18445 (2023). https://doi.org/10.1016/j.heliyon.2023.e18445

21. Sawicki, J., et al.: Approximating landscape insensitivity regions in solving ill-conditioned inverse problems. Memetic Comput. **10**(3), 279–289 (2018). https://doi.org/10.1007/s12293-018-0258-5

22. Sawicki, J., et al.: Using covariance matrix adaptation evolutionary strategy to boost the search accuracy in hierarchic memetic computations. J. Comput. Sci. **34**, 48–54 (2019). https://doi.org/10.1016/j.jocs.2019.04.005

23. Sawicki, J., et al.: Understanding measure-driven algorithms solving irreversibly ill-conditioned problems. Nat. Comput. **21**, 289–315 (2022). https://doi.org/10.1007/s11047-020-09836-w

24. Sheng, W., et al.: Adaptive memetic differential evolution with niching competition and supporting archive strategies for multimodal optimization. Inf. Sci. **573**, 316–331 (2021). https://doi.org/10.1016/j.ins.2021.04.093

25. Smołka, M., et al.: An agent-oriented hierarchic strategy for solving inverse problems. Int. J. Appl. Math. Comput. Sci. **25**(3), 483–498 (2015). https://doi.org/10.1515/amcs-2015-0036

26. Wang, R., et al.: Adaptive niching particle swarm optimization with local search for multimodal optimization. Appl. Soft Comput. **133**, 109923 (2023). https://doi.org/10.1016/j.asoc.2022.109923

27. Wurm, J., Bachler, S., Woittennek, F.: On delay partial differential and delay differential thermal models for variable pipe flow. Int. J. Heat Mass Transf. **152**, 119403 (2022). https://doi.org/10.1016/j.ijheatmasstransfer.2020.119403

28. Zítek, P., Hlava, J.: An isochronic internal model control of time-delay systems. Control. Eng. Pract. **9**(5), 501–516 (2001). https://doi.org/10.1016/S0967-0661(01)00013-2

# Local Attention Augmentation
# for Chinese Spelling Correction

Shuo Wang[1,2], Chaodong Tong[1,2], Kun Peng[1,2], and Lei Jiang[1,2(✉)]

[1] Institute of Information Engineering, Chinese Academy of Sciences, BeiJing, China
{wangshuo4035,tongchaodong,pengkun,jianglei}@iie.ac.cn
[2] School of Cyberspace Security, University of Chinese Academy of Sciences,
BeiJing, China

**Abstract.** Chinese spelling correction (CSC) is an important task in
the field of natural language processing (NLP). While existing state-
of-the-art methods primarily leverage pre-trained language models and
incorporate external knowledge sources such as confusion sets, they often
fall short in fully leveraging local information that surrounds erroneous
words. In our research, we aim to bridge a crucial gap by introducing
a novel CSC model that is enhanced with a Gaussian attention mechanism. This integration allows the model to adeptly grasp and utilize
both contextual and local information. The model incorporates a Gaussian attention mechanism, which results in attention weights around erroneous words following a Gaussian distribution. This enables the model to
place more emphasis on the information from neighboring words. Additionally, the attention weights are dynamically adjusted using learnable
hyperparameters, allowing the model to adaptively allocate attention to
different parts of the input sequence. In the end, we adopt a homophonic
substitution masking strategy and fine-tune the BERT model on a large-
scale CSC corpus. Experimental results show that our proposed method
achieve a new state-of-the-art performance on the SIGHAN benchmarks.

**Keywords:** Gaussian Attention · Spelling Check · Chinese Spelling
Correction

## 1 Introduction

Spelling correction is an important task that detects and corrects language errors
made by humans. It finds wide applications in various fields, including text editors [1], machine translation [2,3], search engines [4–6], etc. Spelling errors in
Chinese can lead to ambiguities or misunderstandings, ultimately affecting the
readability and reliability of the content. Addressing spelling mistakes in Chinese
presents unique challenges due to the language's complex character combinations
and syllable structures. These complexities further compounds the challenge of
detecting and correcting errors. Moreover, CSC requires a deep understanding
of contextual information for accurate corrections, owing to the nuanced and

© The Author(s), under exclusive license to Springer Nature Switzerland AG 2024
L. Franco et al. (Eds.): ICCS 2024, LNCS 14834, pp. 438–452, 2024.
https://doi.org/10.1007/978-3-031-63759-9_44

context-dependent nature of Chinese. CSC involves two types of errors: phonological similarity errors and visual similarity errors. Phonological similarity errors are more prevalent, accounting for 83% of Chinese spelling error [7]. An example in Table 1 demonstrate phonological similarity errors in CSC. In the erroneous sentences, the word "绵" (soft) is misused, while the corrected sentences use the correct word "面" (surface), which has the same pronunciation but a different meaning. From the example, it can be observed that the error character "绵" (soft) has a stronger correlation with the surrounding characters "海" (sea) and "漂浮" (float) than the distant character "报告" (report). The example in Table 1 shows that characters closer to the mistaken character have a stronger association than those further. Therefore, the local information plays a crucial role in the CSC.

**Table 1.** An example of Chinese Spelling Correction.

| | |
|---|---|
| *Wrong*: | bao gao cheng quan qiu hai mian piao fu chao wu wan yi jian su liao la ji<br>报 告 称　全 球 海 绵　漂 浮 超　5　万 亿 件 塑 料 垃 圾<br>The report states that there are over 500 million pieces of plastic waste floating in the world's sponge |
| *Correct*: | bao gao cheng quan qiu hai mian piao fu chao wu wan yi jian su liao la ji<br>报 告 称　全 球 海 面　漂 浮 超　5　万 亿 件 塑 料 垃 圾<br>The report states that there are over 500 million pieces of plastic waste floating in the world's oceans |

Previous research has proposed various methods, primarily aimed at leveraging the contextual information within texts, to address challenges in CSC. Previous works proposed language models for CSC, using n-gram techniques to define corresponding rules for error detection and correction [7–10]. In recent years, BERT has been widely applied in the field of CSC. Soft-Masked BERT [11] proposed a novel masking strategy for CSC. Xu et al. (2020) [12] introduced a Chinese spell checker called REALISE, which directly leverages the multimodal information of Chinese characters. Additionally, FASPell [13], SpellGCN [14], DCN [15], PLOME [16], and others have also proposed BERT-based methods for the CSC task.

While recent research in CSC has mainly focused on utilizing contextual information from language models or neural network models, the local information of individual characters has often been overlooked. However, local information can play a crucial role in spelling correction tasks, especially in cases where errors occur at the character level. In this paper, we propose a novel CSC method architecture based on local attention enhancement. The CSC model utilizes the Gaussian attention mechanism, where the attention weights follow a Gaussian distribution, to integrate contextual and local information. The contributions of this paper are as follows: (1) We propose a CSC model that enhances local attention. We combine Gaussian attention with the neural network model (BiGRU) to effectively capture the contextual relationships between Chinese characters and focus on the specific regions where errors are likely to occur, thereby improving the accuracy of error correction. (2) In the Gaussian attention layer, we use learnable hyperparameters to adjust the attention weights of the characters in the sentence. Additionally, we adopt a homophonic substitution masking strategy and fine-tune the BERT model on a large-scale CSC corpus. (3) Our method achieves the best results and significant improvements compared to other methods in both character-level and sentence-level experiments. In terms of sentence-level precision, we achieve a substantial increase from 67.4% to 87.9% compared to the baseline in the SIGHAN2014 dataset.

## 2   Related Work

There is a significant need for efficient and accurate CSC methods to meet the diverse requirements of different applications. These methods should be capable of identifying and correcting a wide range of spelling errors, including typographical mistakes and word order errors. Additionally, they need to interpret contextual information to provide relevant correction suggestions. In recent years, there has been a surge in research methods focused on CSC. These innovations and advancements are continually evolving, reflecting the increasing complexity and demands of processing Chinese text in various technological applications.

The evolution of CSC methods marks substantial progress in NLP. Initially, the focus was on rule-based methods, exemplified by the approach [17]. Their technique involved segmenting sentences, identifying errors, and correcting them through dictionary lookups, which laid the foundation for future work in this area.

The field then shifted towards statistical and machine learning approaches. Researchers pioneered the use of n-gram language models for CSC [8,10,17]. This marked a significant step forward, leveraging statistical models to better grasp the intricacies of the Chinese language. More recently, the advent of neural networks has revolutionized CSC. Notable neural network architectures such as GRU, LSTM, and Transformer have been extensively applied. Soft-Masked BERT ch44zhang2020spelling and PLOME [16] utilized GRU models for error detection, demonstrating the efficacy of these models in identifying spelling errors [11,16]. Similarly, other works successfully employed LSTM models, showcasing their potential in correcting Chinese spelling errors [15,18].

The current state-of-the-art in Chinese Spelling Correction (CSC), primarily employing large-scale language models like BERT, marks a significant improvement over previous methods. These models tokenize input text and map each character to an embedding vector, leveraging contextual information to learn rich, nuanced representations. However, a notable limitation, particularly in the context of CSC, is their inadequate consideration of the relative importance of surrounding characters in relation to an erroneous character. Presently, pre-training of deep bidirectional transformers for language understanding (BERT)-[19] models have emerged as the most popular choice in this domain. They have been effectively used in various systems like FASPell [13], Confusionset [20],

SpellGCN [14], MLM-phonetics [21], PLOME [16], and uChecker [22]. These implementations demonstrate exceptional performance in CSC, capitalizing on the strengths of BERT models to understand and correct complex language structures and contextual nuances inherent in the Chinese language. The continuous evolution in this field underscores the dynamic nature of language processing technologies and their growing sophistication in handling the unique challenges presented by the Chinese language.

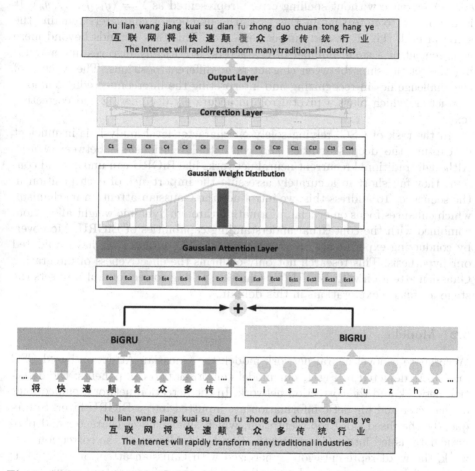

**Fig. 1.** Architecture of the CSC model. In the figure, $E_{ci}$ represents the vector representation combining character and phonetic information, while $C_i$ represents the vector representation of aggregated Gaussian weights.

# 3   Approach

## 3.1   Problem and Motivation

In the Chinese spelling correction task, we input a Chinese sentence comprising $n$ characters, denoted as $X = (x_1, x_2, \ldots, x_n)$. The core of this task involves using a model to detect and identify spelling errors within the sentence. Once errors are identified, the model undertakes necessary corrections to generate a correct sentence without spelling errors, represented as $Y = (y_1, y_2, \ldots, y_n)$. It is important to note that in this task, the number of characters remains the same in both the input and output sentences. This task extends beyond mere replacement of a small number of characters; it critically involves understanding the relationships between characters at different positions. The essence of the challenge lies in recognizing and interpreting the interconnectedness among characters, which plays a pivotal role in accurately deciphering and correcting text.

In the task of CSC, relying solely on character-level analysis is insufficient to capture the deep semantics and contextual relationships between words. Although traditional recurrent neural networks like BiGRU can understand context, they fall short in accurately assessing the importance of each position in the sequence. To address this, we introduce the Gaussian attention mechanism, which enhances focus on crucial information through dynamic weight allocation, combined with the contextual understanding capabilities of BiGRU. Moreover, by conducting experiments on a vast array of CSC datasets, we have validated our hypothesis. This research not only confirms the effectiveness of integrating Gaussian attention with BiGRU for Chinese spelling correction but also sets the stage for future explorations in this domain.

## 3.2   Model

We propose a novel neural network model: the Gaussian Attention-based CSC model. As depicted in Fig. 1, our model comprises two components: the detection network and the correction network. In the detection network, embeddings of character and phonetic information are inputted into a BiGRU layer. Subsequently, the fused word representation, which amalgemates character and phonetic data, is fed into a Gaussian attention layer. Moving to the correction network, the word representation, synergized with Gaussian attention weights, is channeled into a correction layer. Ultimately, the refined output is produced post-correction. For this task, we employ the BERT model, modifying its masking strategy to better suit CSC.

The detection network is engineered to identify potential spelling errors in the input text. We use a BiGRU model to capture the contextual information of the characters. Additionally, we incorporate a Gaussian attention mechanism, focusing on the characters surrounding the erroneous one. This enhancement enables more effective detection of Chinese spelling errors.

In the correction network, the BERT model generates suggestions for correcting detected errors. We adapt the BERT model's masking strategy to prioritize error characters during the correction process, enabling the model to provide more accurate and relevant suggestions for Chinese spelling errors.

### 3.3  Masking Strategy

The BERT model, known for its impressive performance across various tasks, is extensively applied in diverse domains. Its masking strategy, employed during training, involves randomly replacing a portion of the input sequence with the [MASK] token. The model then learns to predict these masked tokens. Typically, about 15% of the tokens in an input sequence are randomly substituted with the [MASK] token.

In our approach, we adjust the masking strategy of BERT to better suit CSC. Instead of randomly replacing 15% of the tokens with the [MASK] token, we replace 80% of the [MASK] tokens with characters that have the same pronunciation, and the remaining 20% with random characters. This modified masking strategy is more aligned with the requirements of the CSC task.

**Fig. 2.** An example of different masking strategies.

As shown in Fig. 2, for example, the character "坐 (sit)" would be masked using the original masking strategy with a random character "事 (matter)". In our approach, we replace it with a character that has the same pronunciation, such as "做 (do)". We maintain a separate file that contains characters with the same pronunciation for easy lookup and substitution by the model.

By adjusting the masking strategy to consider characters with the same pronunciation, our approach aims to improve the relevance and accuracy of the correction suggestions provided by the BERT model for CSC tasks.

### 3.4  Detection Network

In the detection module of Chinese Spelling Correction (CSC), our integration of the Gaussian Attention Layer with Bidirectional Gated Recurrent Units (BiGRU) not only significantly enhances the model's performance but also ensures that the model fully utilizes both contextual and local information within the text. This innovative blend capitalizes on the strengths of BiGRU and Gaussian attention mechanisms to achieve this dual focus.

**BIGRU Layer.** BiGRU is a bidirectional version of the Gated Recurrent Unit (GRU), consisting of two GRU layers. One processing the forward information in the time sequence, and the other processing the backward information. This allows the network to simultaneously consider past and future context information. In the context of spelling correction tasks, BiGRU effectively captures the contextual relationships between Chinese characters, helping the model understand and detect potential spelling errors.

The vector $\overrightarrow{h_t}$ represents the hidden state of the GRU in the forward direction, and $\overleftarrow{h_t}$ represents the hidden state of the GRU in the backward direction, as shown in the equations:

$$\overrightarrow{h_t} = \text{GRU}(\overrightarrow{h_{t-1}}, x_t) \tag{1}$$

$$\overleftarrow{h_t} = \text{GRU}(\overleftarrow{h_{t+1}}, x_t) \tag{2}$$

$$h_t = [\overrightarrow{h_t}; \overleftarrow{h_t}] \tag{3}$$

**Gaussian Attention Layer.** The Gaussian attention layer is a specially designed attention mechanism, aimed at enhancing the model's focus on the information surrounding specific characters in the text. It uses a Gaussian distribution to simulate the relative importance between characters, enabling the model to pay more attention to characters adjacent to erroneous ones. Applying the Gaussian attention layer to spelling detection can effectively capture long-range dependencies, allowing the model to focus more on the information surrounding the error character and improve the detection performance.

Let's assume that $a_t$ represents the attention weight of character $t$. Then, we can define it as:

$$a_t = \exp\left(-\frac{(t-\mu)^2}{2\sigma^2}\right) \tag{4}$$

where $\mu$ and $\sigma$ are the parameters of the Gaussian distribution, and they are learned by the network. Where $\mu$ and $\sigma$ are learnable hyperparameters used to dynamically adjust the mean and standard deviation of the attention weights. By using the Gaussian distribution form, the attention weight $a_t$ is computed as the exponential function of the distance between character position $t$ and the hyperparameters $\mu$ and $\sigma$. The normalization operation ensures that the sum of attention weights for all positions is equal to 1, making it a probability distribution.

The Gaussian attention layer finds a particularly compelling application in the realm of CSC. It empowers the model to focus keenly on the contextual information surrounding potential errors, leading to significant improvements in detection performance. This enhanced focus stems from the ability of the Gaussian attention layer to grasp intricate semantic relationships and spelling patterns within the surrounding characters.

The weighted hidden state with attention is represented as:

$$c_t = \sum_i a_i \cdot h_i \tag{5}$$

By combining the BiGRU's ability to capture contextual information with the focusing ability of the Gaussian attention mechanism, we aim to enhance the detection module of CSC, enabling the model to better understand and correct spelling errors.

### 3.5 Correction Network

In our correction module, we employ the BERT model, which is fundamentally based on the Transformer encoder architecture. BERT utilizes a self-attention mechanism to comprehend the contextual relationships within the text. Its bidirectional nature is especially crucial for understanding the context surrounding a given character in Chinese text, as the meaning of characters often relies on their neighboring characters. BERT is employed to identify and correct erroneous characters, and through fine-tuning, it learns to predict the most appropriate characters within a given context.

The self-attention mechanism used in the BERT model is represented by the following formula:

$$\text{Attention}(Q, K, V) = \text{softmax}\left(\frac{QK^T}{\sqrt{d_k}}\right) V \tag{6}$$

Here, $Q, K, V$ respectively represent the Query, Key, and Value, and $d_k$ denotes the dimensionality of the key.

For each input character $x_i$, BERT generates a contextually informed representation $h_i$:

$$h_i = \text{BERT}(x_i) \tag{7}$$

This representation $h_i$ contains the contextual information of the character $x_i$.

For characters identified as erroneous, a prediction is made using the Masked Language Model (MLM) task:

$$\hat{x}_i = \text{argmax}_x P(x|h_i) \tag{8}$$

In this equation, $\hat{x}_i$ is the character predicted by the model to be the correct replacement.

## 4  Experiments

In this section, we will introduce the dataset, experimental setup, and the performance of our model on the dataset. We evaluated the performance of our model using the SIGHAN dataset, which is a benchmark dataset for CSC task.

## 4.1  Datasets

For the CSC task, we utilized the SIGHAN dataset, a well-established benchmark for evaluating different methods. While relatively small in size, it offers valuable insights due to its diverse composition. SIGHAN comprises three key subsets: SIGHAN2013 [23], containing around 700 sentences with over 300 errors, and SIGHAN2014 [24] and SIGHAN2015 [25], each consisting of a few thousand sentences. The SIGHAN2013 subset is composed of a variety of text genres, including news articles, academic papers, and social media posts. This diversity helps to ensure that the dataset is representative of real-world spelling errors. The SIGHAN2014 and SIGHAN2015 subsets are both composed of news articles. This focus on a single genre allows for a more detailed analysis of the types of spelling errors that occur in this particular context.

For the training data, in our previous work employed a comprehensive dataset that strategically blends automatically annotated data with rigorously curated benchmarks. To ensure a dataset both ample and diverse, we leveraged an expansive collection of Chinese text automatically annotated with spelling errors. This dataset artfully simulates real-world errors by incorporating both visually similar and phonologically similar character substitutions. Visually similar characters mirror OCR-like errors, arising from visual resemblances between characters. Phonologically similar characters echo ASR-like errors, stemming from similarities in pronunciation [26]. To complement the breadth of the automated dataset, we judiciously incorporated training data from three smaller-sized, yet meticulously hand-annotated datasets: SIGHAN2013, SIGHAN2014, and SIGHAN2015. These datasets, widely recognized for their meticulous quality, offer a valuable source of precisely labeled spelling errors, fortifying the model's ability to discern and correct fine-grained distinctions.

**Table 2.** Experimental Data Statistics Information.

|  | Corpus Name | #Sentences | Avg. Length | #Errors |
|---|---|---|---|---|
| Training Set | Wang et al. (2018) | 271,329 | 44.4 | 382,704 |
|  | SIGHAN2013 | 700 | 41.8 | 343 |
|  | SIGHAN2014 | 3,437 | 49.6 | 5,122 |
|  | SIGHAN2015 | 2,338 | 31.3 | 3,037 |
|  | Total | 277,804 | 41.8 | 391,206 |
| Test Set | SIGHAN2013 | 1,000 | 74.3 | 1,224 |
|  | SIGHAN2014 | 1,062 | 50.0 | 771 |
|  | SIGHAN2015 | 1,100 | 30.6 | 703 |
|  | Total | 3,162 | 50.9 | 2,698 |

For the testing data, we assessed our model's performance using the SIGHAN-2013, SIGHAN2014, and SIGHAN2015 datasets. As these datasets originally contain traditional Chinese characters, we modified them to utilize simplified Chinese characters for validation and evaluation purposes. We also made necessary adjustments to the format of the datasets to ensure compatibility. The modified datasets consist of 1000 sentences for SIGHAN2013, 1062 sentences for SIGHAN2014, and 1100 sentences for SIGHAN2015.The specific data details are shown in Table 2.

## 4.2  Setup

We randomly divided the training data into a training set and a validation set, with the training set accounting for 90% of the data and the validation set accounting for 10%.

In real-world scenarios, 83% of Chinese spelling errors are caused by phonetic mistakes. Therefore, we randomly replaced 15% of the characters in the text with artificially generated errors. Among these errors, 80% were replaced with characters that have similar pronunciation, while 20% were replaced with random characters.

For the model's hyperparameter settings, we utilized the default parameters of the BERT model. Additionally, we fine-tuned the model using an optimizer. We set the learning rate of the model to 1e-4, the batch size to 16, and the loss weight to 0.5.

## 4.3  Baselines

We compared our method with existing methods in the field:

LMC [10] seamlessly blends bi-gram and trigram language models with Chinese word segmentation, leveraging dynamic programming and smoothing techniques to combat data sparsity and achieve robust accuracy. FASPell [13] based on a new paradigm which consists of a denoising autoencoder (DAE) and a decoder. Confusionset [20] utilizes the off-the-shelf confusionset for guiding the character generation. The Seq2Seq model jointly learns to copy a correct character from an input sentence through a pointer network, or generate a character from the confusionset rather than the entire vocabulary. SpellGCN [14] proposes to incorporate phonological and visual similarity knowledge into language models for CSC via a specialized graph convolutional network (SpellGCN). The model builds a graph over the characters, and SpellGCN is learned to map this graph into a set of inter-dependent character classifiers. MLM-phonetics [21] is a groundbreaking end-to-end CSC model that seamlessly integrates phonetic features within a unified framework for joint error detection and correction. PLOME [16] is a pre-trained masked language model with misspelled knowledge for CSC, which jointly learns how to understand language and correct spelling errors. To this end, PLOME masks the chosen tokens with similar characters according to a confusion set rather than the fixed token "[MASK]" as in BERT. DCN [15] generates the candidate Chinese characters via a Pinyin Enhanced Candidate Generator and then utilizes an attention-based network to model the dependencies between two adjacent Chinese characters. uChecker [22] is a Confusionset-guided masking strategy to fine-train the masked language model to further improve the performance of unsupervised detection and correction.

**Table 3.** The char-level performance of different models on SIGHAN test sets.

| Dataset | Model | Detection Level | | | Correction Level | | |
|---|---|---|---|---|---|---|---|
| | | Prec. | Rec. | F1 | Prec. | Rec. | F1 |
| SIGHAN2013 | LMC (Xie et al.,2015) | 79.8 | 50.0 | 61.5 | 77.6 | 22.7 | 35.1 |
| | FASPell (Hong et al.,2019) | 76.2 | 63.2 | 69.1 | 73.1 | 60.5 | 66.2 |
| | Confusionset (Wang et al.,2019) | 66.8 | 73.1 | 69.8 | 71.5 | 59.5 | 69.9 |
| | SpellGCN (Cheng et al.,2020) | 80.1 | 74.4 | 77.2 | 78.3 | 72.7 | 75.4 |
| | MLM-phonetics (Zhang et al.,2021) | **82.0** | 78.3 | 80.1 | 79.5 | 77.0 | 78.2 |
| | our method | 78.7 | **91.6** | **84.7** | **98.7** | **95.9** | **97.3** |
| SIGHAN2014 | LMC (Xie et al.,2015) | 56.4 | 34.8 | 43.0 | 71.1 | 50.2 | 58.8 |
| | FASPell (Hong et al.,2019) | 61.0 | 53.5 | 57.0 | 59.4 | 52.0 | 55.4 |
| | Confusionset (Wang et al.,2019) | 63.2 | **82.5** | 71.6 | 79.3 | 68.9 | 73.7 |
| | SpellGCN (Cheng et al.,2020) | 65.1 | 69.5 | 67.2 | 63.1 | 67.2 | 65.3 |
| | MLM-phonetics (Zhang et al.,2021) | 66.2 | 73.8 | 69.8 | 64.2 | 73.8 | 68.7 |
| | our method | **80.7** | 76.1 | **78.4** | **98.3** | **88.1** | **92.9** |
| SIGHAN2015 | LMC(Xie et al.,2015) | 83.8 | 26.2 | 40.0 | 67.6 | 31.8 | 43.2 |
| | FASPell (Hong et al.,2019) | 67.6 | 60.0 | 63.5 | 66.6 | 59.1 | 62.6 |
| | Confusionset (Wang et al.,2019) | 66.8 | 73.1 | 69.8 | 71.5 | 59.5 | 69.9 |
| | SpellGCN (Cheng et al.,2020) | 74.8 | 80.7 | 77.7 | 72.1 | 77.7 | 75.9 |
| | MLM-phonetics (Zhang et al.,2021) | 77.5 | **83.1** | 80.2 | 74.9 | 80.2 | 77.5 |
| | PLOME (Liu et al.,2021) | 77.4 | 81.5 | 79.4 | 75.3 | 79.3 | 77.2 |
| | uChecker (Piji Li.,2022) | 85.6 | 79.7 | 82.6 | 91.6 | 84.8 | 88.1 |
| | our method | **85.7** | 83.0 | **84.3** | **95.9** | **90.1** | **92.9** |

## 4.4   Results and Analysis

To evaluate the methods, we compared the previous methods and our method based on accuracy, recall, and F1 score at both the character-level and sentence-level. We assessed these metrics for both detection and correction tasks.

As shown in Table 3, we compared the performance of different methods on the character-level evaluation using the SIGHAN test set. We evaluated these methods on the SIGHAN2013, SIGHAN2014, and SIGHAN2015 datasets. In Table 3, we can observe that on the SIGHAN2013 dataset, our method performs slightly lower in terms of precision at the detection level compared to MLM-phonetics, but outperforms other methods in all other evaluation metrics. On the SIGHAN2014 and SIGHAN2015 datasets, compared to other methods, our method surpasses others in terms of precision at both the detection and correction levels.

As shown in Table 4, we present the performance of different methods on the sentence-level evaluation using the SIGHAN test set. Similarly, we compared these methods on the SIGHAN2013, SIGHAN2014, and SIGHAN2015 datasets. In Table 4, we can observe that our method outperforms other methods in both detection and correction levels on the SIGHAN2013, SIGHAN2014, and SIGHAN2015 datasets.

**Table 4.** The sentence-level performance of different models on SIGHAN test sets.

| Dataset | Model | Detection Level | | | Correction Level | | |
|---|---|---|---|---|---|---|---|
| | | Prec. | Rec. | F1 | Prec. | Rec. | F1 |
| SIGHAN2013 | LMC (Xie et al.,2015) | (−) | (−) | (−) | (−) | (−) | (−) |
| | FASPell (Hong et al.,2019) | 76.2 | 63.2 | 69.1 | 73.1 | 60.5 | 66.2 |
| | SpellGCN (Cheng et al.,2020) | 80.1 | 74.4 | 77.2 | 78.3 | 72.7 | 75.4 |
| | DCN (Wang et al.,2021) | 86.8 | 79.6 | 83.0 | 74.7 | **77.7** | 81.0 |
| | our method | **96.0** | **95.2** | **96.2** | **98.6** | 71.7 | **83.0** |
| SIGHAN2014 | LMC (Xie et al.,2015) | (−) | (−) | (−) | (−) | (−) | (−) |
| | FASPell (Hong et al.,2019) | 61.0 | 53.5 | 57.0 | 59.4 | 52.0 | 55.4 |
| | SpellGCN (Cheng et al.,2020) | 65.1 | 69.5 | 67.2 | 63.1 | 67.2 | 65.3 |
| | DCN (Wang et al.,2021) | 67.4 | 70.4 | 68.9 | 65.8 | 68.7 | 67.2 |
| | our method | **86.9** | **85.6** | **86.2** | **83.6** | **76.9** | **80.1** |
| SIGHAN2015 | LMC (Xie et al.,2015) | (−) | (−) | (−) | (−) | (−) | (−) |
| | FASPell (Hong et al.,2019) | 67.6 | 60.0 | 63.5 | 66.6 | 59.1 | 62.6 |
| | SpellGCN (Cheng et al.,2020) | 74.8 | 80.7 | 77.7 | 72.1 | 77.7 | 75.9 |
| | DCN (Wang et al.,2021) | 77.1 | 80.9 | 79.0 | 74.5 | 78.2 | 76.3 |
| | PLOME (Liu et al.,2021) | 77.4 | 81.5 | 79.4 | 75.3 | 79.3 | 77.2 |
| | uChecker (Piji Li.,2022) | 75.4 | 72.0 | 73.7 | 70.6 | 67.3 | 68.9 |
| | our method | **89.8** | **89.3** | **89.5** | **87.8** | **82.0** | **84.8** |

By conducting validation on the character-level and sentence-level using the three SIGHAN datasets, the experimental results demonstrate the effectiveness of our method, which is based on Gaussian distribution-enhanced local attention, for CSC. As expected, the local information around the error context in CSC carries more weight for error detection and correction than distant information. This fully validates the feasibility of our approach.

## 4.5  Ablation Study

In this subsection, we analyze the impact of several factors on the model, including Gaussian attention and masking strategy. We evaluate the effects of different factors through ablation experiments.

We conducted ablation experiments by individually removing different components to study their respective impacts on the model. First, we removed the Gaussian attention layer and compared the performance of the model without Gaussian attention. Next, we removed the masking strategy part of BERT and only used random masking. Finally, we performed experiments using only the BERT model. The experimental results, as shown in Table 5, indicate that the models without the Gaussian attention layer (Our-G) and the masking strategy (Our-M) were affected to varying degrees compared to our model. Furthermore, comparing with the BERT-only model highlights the effectiveness of our approach for the CSC task.

**Table 5.** The ablation result of our method.

| Dataset | Method | Detection Level | | | | Correction Level | | | |
|---|---|---|---|---|---|---|---|---|---|
| | | Acc. | Prec. | Rec. | F1 | Acc. | Pre. | Rec. | F1 |
| SIGHAN2013 | Bert | **77.0** | 74.2 | 83.2 | 78.6 | 75.2 | 83.0 | 75.2 | 78.9 |
| | Ours-G | 65.7 | 76.2 | 63.2 | 68.9 | 76.5 | 78.4 | 70.6 | 73.7 |
| | Ours-M | 68.2 | 76.6 | 82.5 | 79.6 | 82.6 | 76.3 | 72.8 | 74.4 |
| | Ours | 72.1 | **78.7** | **91.6** | **84.7** | **93.8** | **98.7** | **95.9** | **97.3** |
| SIGHAN2014 | Bert | 75.7 | 64.5 | 68.6 | 66.5 | 74.6 | 62.4 | 66.3 | 64.3 |
| | Ours-G | 72.3 | 76.2 | 70.5 | 73.2 | 75.2 | 85.4 | 80.9 | 82.9 |
| | Ours-M | 74.5 | 75.1 | 69.4 | 76.1 | 83.9 | 83.2 | 81.5 | 69.1 |
| | Ours | **77.4** | **80.7** | **76.1** | **78.4** | **91.7** | **98.3** | **88.1** | **92.9** |
| SIGHAN2015 | Bert | 82.4 | 74.2 | 78.0 | 76.1 | 81.0 | 71.6 | 75.3 | 73.4 |
| | Ours-G | 80.5 | 78.5 | 71.7 | 74.7 | 77.8 | 76.2 | 80.3 | 78.4 |
| | Ours-M | 81.2 | 82.3 | 80.7 | 81.5 | 84.2 | 75.8 | 71.5 | 73.8 |
| | Ours | **83.3** | **85.7** | **83.0** | **84.3** | **92.2** | **95.9** | **90.1** | **92.9** |

## 5   Conclusion

In this study, we introduce a novel Gaussian-based local attention enhancement approach specifically tailored for CSC tasks. Our method innovatively applies Gaussian attention to assign varying weights to characters, particularly emphasizing the information surrounding misspelled characters. This focus aids in extracting more pertinent information, crucial for both detecting and correcting errors. Furthermore, our approach includes adaptable hyperparameters, within the Gaussian attention layer, leading to significant performance enhancements. To refine our model further, we adopt a homophonic substitution masking strategy and fine-tune the BERT model on a large-scale CSC corpus. The efficacy of our proposed method is underscored by experimental results obtained from the SIGHAN benchmarks, where it surpasses previous methodologies. These findings confirm the effectiveness and potential of our Gaussian-based local attention enhancement method in addressing the challenges of CSC tasks.

**Acknowledgments.** This work is supported by Xinjiang Uygur Autonomous Region Key Researchand Development Program (No.2022B03010).

**Disclosure of Interests.** The authors have no competing interests to declare that are relevant to the content of this article.

## References

1. Hládek, D., Staš, J., Pleva, M.: Survey of automatic spelling correction. Electronics **9**(10), 1670 (2020)

2. Eger, S., vor der Brück, T., Mehler, A.: A comparison of four character-level string-to-string translation models for (OCR) spelling error correction. Prague Bull. Math. Linguist. **105**(1), 77 (2016)
3. Zhou, Y., Porwal, U., Konow, R.: Spelling correction as a foreign language. arXiv preprint arXiv:1705.07371 (2017)
4. Martins, B., Silva, M.J.: Spelling correction for search engine queries. In: Advances in Natural Language Processing: 4th International Conference, EsTAL 2004, Alicante, Spain, 20–22 October 2004. Proceedings, vol. 4, pp. 372–383 (2004)
5. Gao, J., Quirk, C.: A large scale ranker-based system for search query spelling correction. In: The 23rd International Conference on Computational Linguistics (2010)
6. Ye, D., et al.: Improving query correction using pre-train language model in search engines. In: Proceedings of the 32nd ACM International Conference on Information and Knowledge Management, pp. 2999–3008 (2023)
7. Liu, C.-L., Lai, M.-H., Chuang, Y.-H., Lee, C.-Y.: Visually and phonologically similar characters in incorrect simplified Chinese words. In: Coling 2010: Posters, pp. 739–747 (2010)
8. Wu, J.-C., Chiu, H.-W., Chang, J.S.: Integrating dictionary and web N-grams for Chinese spell checking. In: International Journal of Computational Linguistics Chinese Language Processing, Volume 18, Number 4, December 2013-Special Issue on Selected Papers from ROCLING XXV (2013)
9. Yu, J., Li, Z.: Chinese spelling error detection and correction based on language model, pronunciation, and shape. In: Proceedings of the Third CIPS-SIGHAN Joint Conference on Chinese Language Processing, pp. 220–223 (2014)
10. Xie, W., et al.: Chinese spelling check system based on n-gram model. In Proceedings of the Eighth SIGHAN Workshop on Chinese Language Processing, pp. 128–136 (2015)
11. Zhang, S., Huang, H., Liu, J., Li, H.: Spelling error correction with soft-masked BERT. arXiv preprint arXiv:2005.07421 (2020)
12. Xu, H.-D., et al.: Read, listen, and see: leveraging multimodal information helps Chinese spell checking. arXiv preprint arXiv:2105.12306 (2021)
13. Hong, Y., Yu, X., He, N., Liu, N., Liu, J.: FASPell: a fast, adaptable, simple, powerful Chinese spell checker based on DAE-decoder paradigm. In: Proceedings of the 5th Workshop on Noisy User-generated Text (W-NUT 2019), pp. 160–169 (2019)
14. Cheng, X., et al.: Spellgcn: incorporating phonological and visual similarities into language models for Chinese spelling check. arXiv preprint arXiv:2004.14166 (2020)
15. Wang, H., Wang, B., Duan, J., Zhang, J.: Chinese spelling error detection using a fusion lattice LSTM. Trans. Asian Low-Res. Lang. Inf. Process. **20**(2), 1–11 (2021)
16. Liu, S., Yang, T., Yue, T., Zhang, F., Wang, D.: PLOME: pre-training with misspelled knowledge for Chinese spelling correction. In: Proceedings of the 59th Annual Meeting of the Association for Computational Linguistics and the 11th International Joint Conference on Natural Language Processing, vol. 1: Long Papers, pp. 2991–3000 (2021)
17. Yeh, J.-F., Lu, Y.-Y., Lee, C.-H., Yu, Y.-H., Chen, Y.-T.: Chinese word spelling correction based on rule induction. In: Proceedings of The Third CIPS-SIGHAN Joint Conference on Chinese Language Processing, pp. 139–145 (2014)
18. Duan, J., Wang, B., Tan, Z., Wei, X., Wang, H.: Chinese spelling check via bidirectional lstm-crf. In: 2019 IEEE 8th Joint International Information Technology and Artificial Intelligence Conference (ITAIC), pp. 1333–1336 (2019)

19. Devlin, J., Chang, M.-W., Lee, K., Toutanova, K.: Bert: pre-training of deep bidirectional transformers for language understanding. arXiv preprint arXiv:1810.04805 (2018)
20. Wang, D., Tay, Y., Zhong, L.: Confusionset-guided pointer networks for Chinese spelling check. In Proceedings of the 57th Annual Meeting of the Association for Computational Linguistics, pp. 5780–5785 (2019)
21. Zhang, R., et al.: Correcting Chinese spelling errors with phonetic pre-training. Find. Assoc. Comput. Linguist. ACL-IJCNLP **2021**, 2250–2261 (2021)
22. Li, P.: uChecker: masked pretrained language models as unsupervised Chinese spelling checkers. arXiv preprint arXiv:2209.07068 (2022)
23. Wu, S.-H., Liu, C.-L., Lee, L.-H.: Chinese spelling check evaluation at SIGHAN bake-off 2013. In: Proceedings of the Seventh SIGHAN Workshop on Chinese Language Processing, pp. 35–42 (2013)
24. Yu, L.-C., Lee, L.-H., Tseng, Y.-H., Chen, H.-H.: Overview of SIGHAN 2014 bake-off for Chinese spelling check. In: Proceedings of The Third CIPS-SIGHAN Joint Conference on Chinese Language Processing, pp. 126–132 (2014)
25. Tseng, Y.-H., Lee, L.-H., Chang, L.-P., Chen, H.-H.: Introduction to SIGHAN 2015 bake-off for Chinese spelling check. In: Proceedings of the Eighth SIGHAN Workshop on Chinese Language Processing, pp. 32–37 (2015)
26. Wang, D., Song, Y., Li, J., Han, J., Zhang, H.: A hybrid approach to automatic corpus generation for Chinese spelling check. In: Proceedings of the 2018 Conference on Empirical Methods in Natural Language Processing, pp. 2517–2527 (2018)

# Fast Simulations in Augmented Reality

Mateusz Ksyta[1], Wojciech Kordylewski[1], Marcin Łoś[1], Piotr Gurgul[2],
Maciej Sikora[1]($\boxtimes$), and Maciej Paszyński[1]

[1] AGH University of Krakow, Krakow, Poland
maciejsikora2302@gmail.com, paszynsk@agh.edu.pl
[2] Snap Switzerland GmbH, Yverdon-les-Bains, Switzerland

**Abstract.** Augmented reality may soon revolutionize the world we live
in. The incorporation of computer simulations into augmented reality
glasses opens new perspectives for the perception of reality. In this paper,
we investigate the possibility of performing numerical simulations in real
time within augmented reality glasses. We present the technology that
can be successfully employed in the real-life simulations of the Partial Differential Equations (PDE) based phenomena. We designed and
implemented a two- and three-dimensional explicit dynamics solver in
Lens Studio using Finite Difference Method (FDM) on the augmented
reality glasses. We performed tests on the computational cost, memory
usage, and the capability of performing real-life simulations of advection-
diffusion and wave propagation problems.

**Keywords:** augmented reality · finite difference method · real time
simulations · three-dimensional advection-diffusion equations ·
two-dimensional wave equations · lens studio

## 1 Introduction

We consider a fast solver allowing for real-life simulations of physical phenomena described by Partial Differential Equations (PDEs) in augmented reality
of Lens studio [1]. The solver employs the explicit dynamics and Finite Difference Method (FDM) algorithm [6–8]. The Figs. 1 and 2 present screenshots from
simulations that have been rendered as Lenses in the Snapchat App. However,
these lenses can also run on Snap Spectacles, which are Snap's augmented reality glasses. Our goal is to develop algorithms that can perform simulations in
real-time using the SnapChat rendering engine of augmented reality glasses. The
proposed algorithm, a finite difference method with an Euler time integration
step scheme, has a linear computational complexity of $\mathcal{O}(N)$ where N is the
number of spatial points in which we recalculate the state of the modeled phenomenon. Similarly, the memory complexity of the proposed algorithm is linear
$\mathcal{O}(N)$ with respect to the number of spatial points processed, assuming that we
store the state of spatial points from the current ($N$ points) and previous ($N$
points) time instants. Specifically, the proposed algorithms have been tested on
two- and three-dimensional computational problems, such as

© The Author(s), under exclusive license to Springer Nature Switzerland AG 2024
L. Franco et al. (Eds.): ICCS 2024, LNCS 14834, pp. 453–460, 2024.
https://doi.org/10.1007/978-3-031-63759-9_45

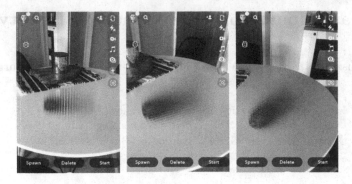

**Fig. 1.** Three-dimensional simulations of the advection-diffusion equations in augmented reality.

- the phenomenon of advection-diffusion in three dimensions (spreading of substances, e.g., smoke in the air by means of the phenomenon of diffusion and advection modeling air movements in the room)
- the phenomenon of wave propagation in two dimensions (to illustrate the possibility of visualizing the simulation on a virtual plane and the possibility of observing the course of the simulation by researchers using augmented reality glasses).

The work investigated the maximum size of the computational domain in two- and three-dimensions in which real-time simulations could be carried out using the fastest possible algorithm with linear complexity. The simulations performed by ANSYS with the use of the finite element method has been projected into the augmented reality [2]. The finite element method loading is based on the data provided by the sensors detecting the surrounding structures [2], and the computed results are displayed in the augmented reality. Another paper [3] projects the previously computed finite element method results onto real structures in augmented reality using the Microsoft Holo Lens. In our paper, for the first time, we present fast real-life computations with the finite difference method using a much lighter Snapchat Lens, allowing for integrating the lens with normal life activities without using any external sensors or servers. There are also some preliminary attempts to visualize the computational results of the finite element method computations performed offline on the real models using augmented reality [4,5].

## 2    Advection-Diffusion Problem

We start from the formulation of the advection-diffusion solver

$$
\begin{aligned}
&\frac{\partial c(x,y,z,t)}{\partial t} - \varepsilon_x \frac{\partial^2 c(x,y,z,t)}{\partial x^2} - \varepsilon_y \frac{\partial^2 c(x,y,z,t)}{\partial y^2} - \varepsilon_z \frac{\partial^2 c(x,y,z,t)}{\partial z^2} \\
&+ b_x \frac{\partial c(x,y,z,t)}{\partial x} + b_y \frac{\partial c(x,y,z,t)}{\partial y} + b_z \frac{\partial c(x,y,z,t)}{\partial z} = f(x,y,z,t)
\end{aligned}
\tag{1}
$$

where we seek the scalar concentration field $c(x, y, z, t)$; here, the vector $(b_x, b_y, b_z)$ is the advection vector, denoting the wind blowing in the domain, are the diffusion coefficients along the $x$, $y$, $z$-axis of the coordinate system. We introduce the explicit dynamics time integration scheme, where we compute the values of $c(x, y, z, t + \Delta t)$ based on the previous time step configuration $c(x, y, z, t)$:

$$\frac{c(x, y, z, t + \Delta t) - c(x, y, z, t)}{\Delta t} - \varepsilon_x \frac{\partial^2 c(x, y, z, t)}{\partial x^2} - \varepsilon_y \frac{\partial^2 c(x, y, z, t)}{\partial y^2}$$

$$-\varepsilon_z \frac{\partial^2 c(x, y, z, t)}{\partial z^2} + b_x \frac{\partial c(x, y, z, t)}{\partial x} + b_y \frac{\partial c(x, y, z, t)}{\partial y} + b_z \frac{\partial c(x, y, z, t)}{\partial z} \qquad (2)$$

$$= f(x, y, z, t)$$

$$c(x, y, z, t + \Delta t) = c(x, y, z, t)$$

$$+\Delta t \varepsilon_x \frac{\partial^2 c(x, y, z, t)}{\partial x^2} + \Delta t \varepsilon_y \frac{\partial^2 c(x, y, z, t)}{\partial y^2} + \Delta t \varepsilon_z \frac{\partial^2 c(x, y, z, t)}{\partial z^2}$$

$$-\Delta t b_x \frac{\partial c(x, y, z, t)}{\partial x} - \Delta t b_y \frac{\partial c(x, y, z, t)}{\partial y} - \Delta t b_z \frac{\partial c(x, y, z, t)}{\partial z} \qquad (3)$$

$$+\Delta t f(x, y, z, t)$$

We introduce the three-dimensional mesh with points $(x_i, y_j, z_k)$, $i = 1, \ldots, N_x$, $j = 1, \ldots, N_y$, $k = 1, \ldots, N_z$, and we write down the equations in the nodes of the mesh

$$c(x_i, y_j, z_k, t + \Delta t) = c(x_i, y_j, z_k, t)$$

$$+\Delta t \varepsilon_x \frac{\partial^2 c(x_i, y_j, z_k, t)}{\partial x^2} + \Delta t \varepsilon_y \frac{\partial^2 c(x_i, y_j, z_k, t)}{\partial y^2} + \Delta t \varepsilon_z \frac{\partial^2 c(x_i, y_j, z_k, t)}{\partial z^2}$$

$$-\Delta t b_x \frac{\partial c(x_i, y_j, z_k, t)}{\partial x} - \Delta t b_y \frac{\partial c(x_i, y_j, z_k, t)}{\partial y} - \Delta t b_z \frac{\partial c(x_i, y_j, z_k, t)}{\partial z} \qquad (4)$$

$$+\Delta t f(x_i, y_j, z_k, t)$$

We approximate the first and the second derivatives using the central finite differences

$$c(x_i, y_j, z_k, t + \Delta t) = c(x_i, y_j, z_k, t)$$

$$+\Delta t \varepsilon_x \frac{c(x_{i-1}, y_j, z_k, t) - 2c(x_i, y_j, z_k, t) + c(x_{i+1}, y_j, z_k, t)}{\Delta x^2}$$

$$+\Delta t \varepsilon_y \frac{c(x_i, y_{j-1}, z_k, t) - 2c(x_i, y_j, z_k, t) + c(x_i, y_{j+1}, z_k, t)}{\Delta y^2}$$

$$+\Delta t \varepsilon_z \frac{c(x_i, y_j, z_{k-1}, t) - 2c(x_i, y_j, z_k, t) + c(x_i, y_j, z_{k+1}, t)}{\Delta z^2} \qquad (5)$$

$$-\Delta t b_x \frac{c(x_{i+1}, y_j, z_k, t) - c(x_{i-1}, y_j, z_k, t)}{2\Delta x} - \Delta t b_y \frac{c(x_i, y_{j+1}, z_k, t) - c(x_i, y_{j-1}, z_k, t)}{2\Delta y}$$

$$-\Delta t b_z \frac{c(x_i, y_j, z_{k+1}, t) - c(x_i, y_j, z_{k-1}, t)}{2\Delta z} + \Delta t f(x_i, y_j, z_k, t)$$

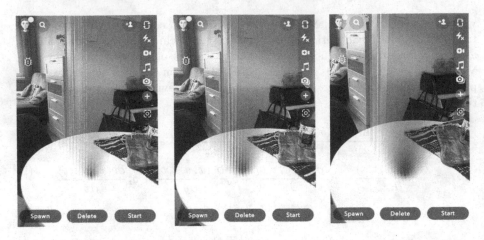

**Fig. 2.** Another simulation of three-dimensional advection-diffusion in augmented reality

The initial state is the zero concentration, the problem is driven by non-zero force component at a given point of the mesh, and the problem is modeled with free open boundary.

## 3   Wave Equations

We start from the formulation of the wave equation solver

$$
\frac{\partial^2 u(x,y,t)}{\partial t^2} - \frac{\partial}{\partial x}\left(g(u(x,y,t) - b(x,y))\frac{\partial u(x,y,t)}{\partial x}\right)
$$
$$
- \frac{\partial}{\partial y}\left(g(u(x,y,t) - b(x,y))\frac{\partial u(x,y,t)}{\partial y}\right) = f(x,y,t)
\tag{6}
$$

where we seek the water level $u(x,y,t)$. Here, $g$ denotes the acceleration due to gravity $g = 9.81$, $b(x,y)$ is given water bed. The initial condition is the shape of the initial wave, and the problem is modeled with free open boundary. We introduce the explicit dynamics time integration scheme, where we compute the values of $u(x,y,t + 2\Delta t)$ based on the two previous time step configurations $u(x,y,t + \Delta t)$, $u(x,y,t)$

$$
\frac{u(x,y,t + 2\Delta t) - 2u(x,y,t + \Delta t) + u(x,y,t)}{\Delta t^2}
$$
$$
- \frac{\partial}{\partial x}\left(g(u(x,y,t) - b(x,y))\frac{\partial u(x,y,t)}{\partial x}\right)
$$
$$
- \frac{\partial}{\partial y}\left(g(u(x,y,t) - b(x,y))\frac{\partial u(x,y,t)}{\partial y}\right) = f(x,y,t)
\tag{7}
$$

We rewrite the equation to emphasize that the new state $u(x, y, t+2\Delta t)$ is given by the update of the previous state $u(x, y, t + \Delta t)$, the estimation of the wave velocity computed based on the last two time steps $[u(x, y, t + \Delta t) - u(x, y, t)]$ and the physics of the wave propagation phenomena (all the remaining terms)

$$u(x, y, t + 2\Delta t) = u(x, y, t + \Delta t) + [u(x, y, t + \Delta t) - u(x, y, t)]$$
$$+\Delta t^2 \frac{\partial}{\partial x}\left(g(u(x, y, t) - b(x, y))\frac{\partial u(x, y, t)}{\partial x}\right)$$
$$+\Delta t^2 \frac{\partial}{\partial y}\left(g(u(x, y, t) - b(x, y))\frac{\partial u(x, y, t)}{\partial y}\right) + \Delta t^2 f(x, y, t) \tag{8}$$

We introduce the dumping constant in front of the difference $C(u(x, y, t+\Delta t) - u(x, y, t))$, to emphasize that the wave velocity is dumped due to internal forces

$$u(x, y, t + 2\Delta t) = u(x, y, t + \Delta t) + C(u(x, y, t + \Delta t) - u(x, y, t))$$
$$+\Delta t^2 \frac{\partial}{\partial x}\left(g(u(x, y, t) - b(x, y))\frac{\partial u(x, y, t)}{\partial x}\right)$$
$$+\Delta t^2 \frac{\partial}{\partial y}\left(g(u(x, y, t) - b(x, y))\frac{\partial u(x, y, t)}{\partial y}\right) + \Delta t^2 f(x, y, t) \tag{9}$$

We introduce the two-dimensional mesh with points $(x_i, y_j)$, $i = 1, \ldots, N_x$, $j = 1, \ldots, N_y$ and we write down the equations in the nodes of the mesh. We also assume a flat water bed $b(x, y, t) = 0$

$$u(x_i, y_j, t + 2\Delta t) = u(x_i, y_j, t + \Delta t) + C(u(x_i, y_j, t + \Delta t) - u(x_i, y_j, t))$$
$$+\Delta t^2 \frac{\partial}{\partial x}\left(gu(x_i, y_j, t)\frac{\partial u(x_i, y_j, t)}{\partial x}\right) + \Delta t^2 \frac{\partial}{\partial y}\left(gu(x_i, y_j, t)\frac{\partial u(x_i, y_j, t)}{\partial y}\right) \tag{10}$$
$$+\Delta t^2 f(x, y, t)$$

We expand the derivatives

$$u(x_i, y_j, t + 2\Delta t) = u(x_i, y_j, t + \Delta t) + C(u(x_i, y_j, t + \Delta t) - u(x_i, y_j, t))$$
$$+\Delta t^2 \left[\left(g\frac{\partial u(x_i, y_j, t)}{\partial x}\frac{\partial u(x_i, y_j, t)}{\partial x}\right) + \left(gu(x_i, y_j, t)\frac{\partial^2 u(x_i, y_j, t)}{\partial x^2}\right)\right]$$
$$+\Delta t^2 \left[\left(g\frac{\partial u(x_i, y_j, t)}{\partial y}\frac{\partial u(x_i, y_j, t)}{\partial y}\right) + \left(gu(x_i, y_j, t)\frac{\partial^2 u(x_i, y_j, t)}{\partial y^2}\right)\right] \tag{11}$$
$$+\Delta t^2 f(x, y, t)$$

and we approximate the first and the second derivatives using the finite differences,

$$u(x_i, y_j, t + 2\Delta t) = u(x_i, y_j, t + \Delta t) + C(u(x_i, y_j, t + \Delta t) - u(x_i, y_j, t))$$

$$+\Delta t^2 \left[ \begin{array}{c} \left(g\left(\frac{u(x_{i+1}, y_j, t+\Delta) - u(x_{i-1}, y_j, t+\Delta)}{2\Delta x}\right)^2\right) + \\ \left(gu(x_i, y_j, t + \Delta t)\frac{u(x_{i+1}, y_j, t+\Delta) - 2u(x_i, y_j, t+\Delta t) + u(x_{i-1}, y_j, t+\Delta)}{\Delta x^2}\right) \end{array} \right]$$

$$+\Delta t^2 \left[ \begin{array}{c} \left(g\left(\frac{u(x_i, y_{j+1}, t+\Delta) - u(x_i, y_{j-1}, t+\Delta)}{2\Delta y}\right)^2\right) + \\ \left(gu(x_i, y_j, t + \Delta t)\frac{u(x_i, y_{j+1}, t+\Delta) - 2u(x_i, y_j, t+\Delta t) + u(x_i, y_{j-1}, t+\Delta)}{\Delta y^2}\right) \end{array} \right] \tag{12}$$

$$+\Delta t^2 f(x, y, t)$$

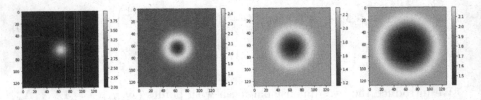

**Fig. 3.** Verification of the two-dimensional wave propagation simulations by Python code

## 4  Implementation

The whole project for implementing 2D or 3D simulation and visualization contains three files:

- WorldMeshController.js - the script which is responsible for controlling the simulation. It allows the definition of the dimensions of the computational mesh, the number of time steps, sources, the diffusion coefficients, and the advection vector.
- TweenColorChange_3D.js + Tween.js - these scripts perform the animation of the numerical results of the advection-diffusion model solved with explicit dynamics and finite difference method. They display the color values based on the concentration parameter in the following way:

```
StartColorValue = {        EndColorValue   = {
    r = 0,                     r = 255 * cellValue
    g = 0,                     g = 0,
    b = 255,                   b = 255 * (1 - cellValue)
    a = 0                      a = cellValue
}                          }
```

- Spawn3D.js - scripts that generate the matrix of the concentration values
  and the matrix of objects representing cells in the computational mesh. This
  script re-computes the values of the scalar concentration field via WorldMesh-
  Controller.js.

## 5  Numerical Results

We present two illustrative examples. The first concerns the advection-diffusion
simulation of a point-shape concentration scalar field source, with the assumed
advection and the constant diffusion coefficients. The snapshots from the simulations are presented in Figs. 1 and 2. The second one concerns the simulation and
visualization of the two-dimensional wave equation, executed over a flat surface,
with the initial states, the time step, the dumping constant, and the mesh size
of $100 \times 100$ elements. We first run the Python code on a laptop to verify our
simulation. The exemplary numerical results are presented in Fig. 3. They can
be compared to the simulations in Lens to be visualized on a floor in Fig. 4.

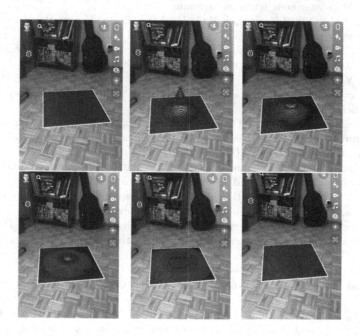

**Fig. 4.** Another two-dimensional simulations of the wave equation on a floor in augmented reality.

## 6  Conclusions

In this paper, we proposed a fast solver for performing finite difference explicit
dynamics simulations in augmented reality. For two-dimensional simulations, the

maximum possible size of the computational grid was $100 \times 100$ spatial points. For three-dimensional simulations, the maximum possible size of the computational grid was $30 \times 30 \times 30$ spatial points. Visualization of the results for the $30 \times 30 \times 30$ grid was characterized by a low frame rate (about five frames per second 5 FPS). Increasing the grid size resulted in an even more significant reduction in FPS. For a better numerical simulations we will need a computationally stronger hardware, or employ precomputed numerical results. Future work may involve experimenting with ARKit [9], and application of explicit dynamics solvers based on higher-order finite element method [10–12].

**Acknowledgements.** The funds the Polish Ministry of Science and Higher Education assigned to AGH University of Krakow. Research project partially supported by the program "Excellence initiative - research university" for the AGH University of Krakow.

# References

1. https://docs.snap.com/lens-studio/home
2. Huang, J.M., Ong, S.K., Nee, A.Y.C.: Real-time finite element structural analysis in augmented reality. Adv. Eng. Softw. **87**, 43–56 (2015)
3. Logg, A., Lundholm, C., Nordaas, M.: Finite element simulation of physical systems in augmented reality. Adv. Eng. Softw. **149**, 102902 (2020)
4. Erkek, Y., Erkek, S., Jamei, E., Seyedmahmoudian, M., Stojcevski, A., Horan, B.: Augmented reality visualization of modal analysis using the finite element method. Appl. Sci. **11**, 1310 (2021)
5. Lin, J., Cao, J., Zhang, J., Treeck, C., Frisch, J.: Visualization of indoor thermal environment on mobile devices based on augmented reality and computational fluid dynamics. Autom. Constr. **103**, 26–40 (2019)
6. Strikwerda, J.: Finite Difference Schemes and Partial Differential Equations, 2nd edn. Society of Industrial and Applied Mathematics (2004)
7. Smith, G.D.: Numerical Solution of Partial Differential Equations: Finite Difference Methods, 3rd edn. Oxford University Press, Oxford (1985)
8. LeVeque, R.J.: Finite Difference Methods for Ordinary and Partial Differential Equations. Society of Industrial and Applied Mathematics (2007)
9. ARKit, Apple Developer Documentation. https://developer.apple.com/documentation/arkit. Accessed 17 Apr 2024
10. Woźniak, M., Łoś, M., Paszyński, M., Dalcin, L., Calo, V.M.: Comput. Inf. **36**(2), 423–448 (2017)
11. Łoś, M., Munoz-Matute, J., Muga, I., Paszyński, M.: Isogeometric Residual Minimization Method (iGRM) with direction splitting for non-stationary advection-diffusion problems. Comput. Math. Appl. **79**(2), 213–229 (2020)
12. Łoś, M., Kłusek, A., Hassaan, M.A., Pingali, K., Dzwinel, W., Paszyński, M.: Parallel fast isogeometric L2 projection solver with GALOIS system for 3D tumor growth simulations. Comput. Methods Appl. Mech. Eng. **343**, 1–22 (2019)

# Towards Efficient Deep Autoencoders for Multivariate Time Series Anomaly Detection

Marcin Pietroń[1]([✉]), Dominik Żurek[1], Kamil Faber[1], and Roberto Corizzo[1,2]

[1] AGH University of Krakow, Krakow, Poland
{pietron,dzurek,kfaber}@agh.edu.pl
[2] American University, Washington, D.C., USA
rcorizzo@american.edu

**Abstract.** Multivariate time series anomaly detection is a crucial problem in many industrial and research applications. Timely detection of anomalies allows, for instance, to prevent defects in manufacturing processes and failures in cyberphysical systems. Deep learning methods are preferred among others for their accuracy and robustness for the analysis of complex multivariate data. However, a key aspect is being able to extract predictions in a timely manner, to accommodate real-time requirements in different applications. In the case of deep learning models, model reduction is extremely important to achieve optimal results in real-time systems with limited time and memory constraints. In this paper, we address this issue by proposing a compression method for deep autoencoders that involves three key factors. First, pruning reduces the number of weights, while preventing catastrophic drops in accuracy by means of a fast search process that identifies high sparsity levels. Second, linear and non-linear quantization reduces model complexity by reducing the number of bits for every single weight. The combined contribution of these three aspects allow the model size to be reduced, by removing a subset of the weights (pruning), and decreasing their bit-width (quantization). As a result, the compressed model is faster and easier to adopt in highly constrained hardware environments. Experiments performed on popular multivariate anomaly detection benchmarks, show that our method is capable of achieving significant model compression ratio (between 80% and 95%) without a significant reduction in the anomaly detection performance.

**Keywords:** Deep learning · Autoencoders · Anomaly detection · Pruning

## 1 Introduction

Multivariate time series anomaly detection is a very popular machine learning problem in many industry sectors. Therefore, many research works have been proposed in this field [2,4,9] Recent works highlight that the best results in terms of detection accuracy are achieved with deep autoencoders [4] based on convolutional layers. Other models with satisfactory results are autoencoders based on

© The Author(s), under exclusive license to Springer Nature Switzerland AG 2024
L. Franco et al. (Eds.): ICCS 2024, LNCS 14834, pp. 461–469, 2024.
https://doi.org/10.1007/978-3-031-63759-9_46

graph neural networks [3,13] and recurrent layers [7,9]. It is worth noting that their effectiveness depends heavily on the specific characteristics of the dataset they are assessed on. Neuroevolution provides a valuable way to address this issue, with the potential to extract optimized models for any given dataset. A notable example is the AD-NEv framework [11], which supports multiple layer types: CNN-based, LSTM based and GNN-based. One potential burden of deep autoencoder models is that each additional layer or channel inside the layer slows down the training and inference process, which negatively affects their efficiency in real-time or embedded systems. In fact, any delay in their inference can have a significant impact on the operation of the reliability of these systems. To this end, compression algorithms can significantly help in reducing the number of CPU cycles required to process input data. The second advantage of their adoption is the memory footprint reduction they provide. This aspect is extremely important in cases where models are exploited in dedicated hardware e.g. IoT, edge, etc. Many works focus on compression for deep learning [1,8,14], particularly for image-based data and natural language processing. The most efficient techniques are pruning [6,8,14] and quantization [1,8]. The pruning process presented in these works can be divided into structured pruning and unstructured pruning. The quantisation can be linear or non-linear. These works show that many deep learning models may present redundant weights which can be removed without any significant drop in detection accuracy. Additionally, given the robustness of these models to noise in input data, weights can be quantized to a lower bit format, further decreasing the memory footprint. However, studies focusing on reducing the complexity of deep learning models are still limited in the anomaly detection field. To the best of our knowledge, there is no work devoted to compressing models on multivariate anomaly detection benchmarks. To this end, in this paper we propose a compression workflow based on pruning and quantization. We adopt convolutional and graph autoencoders which have shown to be the most robust model architectures in anomaly detection tasks. In our work, pruning is incorporated in the training process, while linear and non-linear quantization based on nearest neighbour rounding is run on pruned and pre-trained models. Our experiments leveraging state-of-the-art base model architectures [4] show that compression techniques like pruning and quantization can significantly reduce the complexity of deep model architectures in multivariate anomaly detection tasks.

## 2   Method

In this section, we describe our proposed compression method for anomaly detection autoencoder models. Autoencoders learn a compressed representation, i.e., latent space $Z$ of raw input data $X$ in an unsupervised manner. Autoencoders are made of two parts: the encoder $E$, which transforms (encodes) the original input to the latent space, and the decoder $D$, which transforms (decodes) the latent space $Z$ to the original feature space:

$$Z = E_\Theta(X) = e_{\theta_L}(e_{\theta_{L-1}}...(e_{\theta_0}(X)))$$ (1)

$$AE_\Theta(X) = D_\Theta(Z) = d_{\theta_0'}(d_{\theta_1'}...(d_{\theta_L'}(Z))) \tag{2}$$

The objective of the training is the minimization of the *reconstruction loss*, which corresponds to the difference between the decoder's output and the original input data.

## 2.1 Pruning

The goal of the first stage is to carry out a pruning process, which allows the retention of the most relevant parameters, thus saving computational resources involved for model inference. The general idea is to identify a subset of weights that yield a similar anomaly detection performance to the full model. By doing so, it is possible to discard the remaining weights and reduce the model size, which facilitates its provisioning in resource-limited environments such as edge and IoT. To achieve this goal, in this stage we devise pruning algorithms that involve: *i)* identification of a separate sparsity level for each layer, and *ii)* pruning with model retraining, to foster a more effective identification of the sparsity levels [10].

The representation of the pruned model $AE_\Theta^p$ is a tuple $AE_\Theta^p = (AE_\Theta, M)$, where $AE_\Theta$ is the original model composed by convolutional, fully-connected or LSTM layers $e_{\theta_i}$ and $d_{\theta_i}$, arranged in the specified order. The weights for each layer are represented by the $\Theta$ tensor consisting of parameters from the encoder and decoder:

$$\Theta = \{\theta_0, \theta_1, ..., \theta_L, \theta_L', ..., \theta_1', \theta_0'\} \tag{3}$$

The mask tensor $M$ contains '0' and '1' entries, which denote, for a given layer, weights that are either pruned or retained, respectively:

$$M = \{M_{e_{\theta_0}}, M_{e_{\theta_1}}, ..., M_{e_{\theta_L}}, M_{d_{\theta_L}}, ..., M_{d_{\theta_1}}, M_{d_{\theta_0}}\}. \tag{4}$$

We note that, in our work, a *sparsity level* $v_i$ for a layer $i$ is defined as the ratio between the number of utilized weights and the total number of weights at that layer:

$$v_i = \frac{\sum_j M_{i,j}}{|M_i|} \quad , \quad v_{ws} = \sum_{j=0}^{L} \frac{|\theta_j| \cdot v_j}{|\Theta|} + \sum_{j=0}^{L} \frac{|\theta_j'| \cdot v_j}{|\Theta|}. \tag{5}$$

The weighted sparsity $v_{ws}$ is computed as a sum of two ratios which define the local sparsity for the encoder and decoder counterparts of the model.

We can find a threshold $\epsilon_i$ which ensures to retain the proper of weights (having a value greater than $x$) according to the sparsity $v_i$.

$$\epsilon_i = x \quad , \quad \text{where} \quad \frac{|abs(\theta_i) > x|}{|\theta_i|} = v_i. \tag{6}$$

We leverage $\epsilon_i$ to identify the strongest weights which should be retained for a given layer. It defines mask $M_i$ which is assigned to a specific layer and each entry $M_{i,j}$ can be regarded as binary value:

$$M_{i,j} = \begin{cases} 0 & \text{if } abs(\theta_{i,j}) < \epsilon_i \\ 1 & \text{if } abs(\theta_{i,j}) > \epsilon_i \end{cases} \tag{7}$$

---

**Algorithm 1.** Pruning - main algorithm

---

**Require:** $AE$, $P_S$, $V_{MIN}$, $V_{MAX}$, $N$, $N_{it}$, $O$, $B$, $P_S$
1: $\Lambda \leftarrow P_S$ copies of $AE$, $K \leftarrow \emptyset$
2: **for** $i = 0$ **to** $P_S$ **do**
3:     $M \leftarrow \emptyset$
4:     **for** $l = 0$ **to** $2 \cdot L$ **do**
5:         $min_l$, $max_l \leftarrow V_{MIN}[l]$, $V_{MAX}[l]$
6:         $mean_l$, $std_l \leftarrow min_l + \frac{max_l - min_l}{2}$, $mean_l + \frac{max_l - min_l}{6}$
7:         $v_l \leftarrow \text{bound}(\text{sample}(\mathcal{N}(mean_l, std_l)), min_l, max_l)$
8:         $\epsilon_l \leftarrow$ compute threshold based on $v_l$ (see Equation 6)
9:         $M_l \leftarrow$ generate mask based on $\epsilon_l$ (see Equation 7)
10:         $M \leftarrow M \cup M_l$
11:     **end for**
12:     **for** $epoch = 0$ **to** $N_{it}$ **do**
13:         **for** $b = 0$ **to** $B$ **do**
14:             $\Theta \leftarrow$ train batch $b$ with $O$
15:             **for** $l = 0$ **to** $2 \cdot L$ **do**
16:                 $\theta_l = \theta_l \odot M_l$     // Prune weights for layer $l$ according to mask
17:             **end for**
18:         **end for**
19:     **end for**
20:     $\Lambda \leftarrow \Lambda \cup (AE_\Theta, M, O)$
21:     $K \leftarrow K \cup (F_1 + \alpha \cdot v_{ws})$
22: **end for**
23: $i \leftarrow \text{argmax}(K)$
24: **for** $epoch = 0$ **to** $N$ **do**
25:     **for** $b = 0$ **to** $B$ **do**
26:         $\Theta_i \leftarrow$ train batch $b$ with $O_i$
27:         **for** $l = 0$ **to** $2 \cdot L$ **do**
28:             $\theta_l = \theta_l \odot M_l$     // Prune weights for layer $l$ according to mask
29:         **end for**
30:     **end for**
31: **end for**

---

Algorithm 1 starts with the initialization of the model population, and it sets up the empty list for pruned model quality metrics (line 1 and 2). Then, in a loop, the models are pruned following a retraining process (lines 2–22). At the beginning of the loop, the algorithm goes through the autoencoder layers (see internal loop: lines 4–10) and sets up the initial sparsity levels for each encoder and decoder layer. The sparsity levels are generated by normal distribution using specified mean and standard deviation (line 7). These parameters are computed based on predefined sparsity boundaries given as input parameters (line 6). Once the sparsity is computed, the algorithm uses Eq. 7 to define $\epsilon_l$ value for each layer (line 8). Afterwards, the layer mask is set (line 9). The next internal loop is responsible for short training with predefined $N_{it}$ epochs (lines from 12 to 19). The batch training is performed After each batch, the chosen layer weights are set to zero using element-wise multiplication with the layer mask (line 16). Then, the evaluation of the shortly pruned model is performed. The achieved F1 metric is added to the list (line 21). At the end, in lines from 24 to 31, the model with the best efficiency from the population is taken to the final long-term training stage with $N$ epochs (note: the model is taken with its mask tensor, $N \gg N_{it}$).

## 2.2 Quantization

The quantization process allows our models to be processed further, reducing their complexity. Quantization is a viable process to reduce a complete floating point representation of values to a format with fewer bits. In this paper, we present two types of quantization: linear and non-linear.

Linear quantization can be thought of as a mapping function from a floating-point value $x \in S$ to a fixed-point $q \in Q$ through a function $f_Q : S \to Q$:

$$q = f_Q(x) = \mu + \sigma \cdot \text{round}(\sigma^{-1} \cdot (x - \mu)), \tag{8}$$

where $\mu = 0$ and $\sigma = 2^{-\text{frac\_bits}}$, $\sigma$ is a scaling factor (shift up or down). Integer bit-width can be defined as:

$$\textbf{int\_bits} = \text{ceil}(\log_2(\max_{x \in S} |x|)). \tag{9}$$

The second type of quantization we present in this paper is the non-linear method inspired by [12]. At first, we cluster weights in each layer leveraging the k-Means algorithm according to the desired number of clusters $\omega$. Then, we assign an identifier of the cluster to each weight in a layer, selecting the closest cluster to the original value. The next step quantizes cluster centroids to $\psi$, which defines the bit-width format. Finally, we create a codebook, which contains a mapping between each original weight $w$ and it corresponding quantized cluster's centroid $w_q$. A similar approach has been adopted in [12] and showcased effective compression capabilities in the context of NLP models and GPU-based DL models acceleration, respectively.

## 3  Results and Discussion

The research questions posed by our paper are the following:

**RQ1.** How efficient dynamic pruning can be in anomaly detection autoencoder architectures?

**RQ2.** How effectively can deep state-of-the-art anomaly detection models be reduced by means of quantization?

**RQ3.** What is the efficiency of linear and non-linear quantization on a pre-trained autoencoder?

In our experiments, we consider state-of-the-art architectures in recent benchmarks for anomaly detection [5,11], i.e. convolutional autoencoders (CNN AE) and graph-based autoencoders (GDN). We adopt popular benchmark datasets: SWAT, WADI-2019, MSL, SMAP. The CNN AE for SWAT and WADI-2019 consist of 6 layers, for SMAP and WADI they have 12 layers. All our experiments were executed on a workstation equipped with Nvidia Tesla V100-SXM2-32GB GPUs using PyTorch framework. The $V_{MIN}$ and $V_{MAX}$ parameters were set to 0.2 and 0.8, respectively. The population size $P_S$ was set to 16. In the pruning experiments these models were trained from scratch by Algorithm 1. The linear and nonlinear quantization were run on pretrained models. In all quantization experiments the output neuron activations are in 16-bit format. The baseline results achieved by CNN AE are the best among all models tested on the analyzed datasets [11]. The GDN achieves the second result in the case of the WADI-2019 and SWAT [11]. The CNN AE achieves following point-wise F1: WADI-62.0, SWAT-82.0, MSL-77.0, SMAP-57.0. The GDN gives F1: WADI-57.0, SWAT-81.0, MSL-30.0, SMAP-33.0.

Results in Table 1 show the performance of models following the pruning stage of our proposed compression workflow with different Sparsity levels. We observe that with a sparsity level of 0.2 the anomaly detection performance drops slightly in SWAT: from 82.0 to 81.45 (CNN AE) and from 81.0 to 80.51 (GDN). The performance drop is more significant for WADI-2019: from 62.0 to 56.28 (CNN AE) and from 57.0 to 53.5 (GDN). These results show that is difficult to reduce significant number of weights for both models on WADI-2019. The drop for Sparity level 0.2 can be acceptable for SWAT (about 0.5). The higher Sparsity level increases the drop further for both datasets. In case of SMAP and MSL when Sparsity increases to 0.75, the performance is still at the same level as in baseline models (for both CNN AE and GDN). These surprising results can be motivated as pruning can, in some cases, provide a noise reduction capability in the presence of noisy data in multivariate time series datasets, resulting in a more robust model. Overall, our experimental result show that pruning can be an effective strategy to compress deep autoencoder models for anomaly detection, especially for MSL and SMAP datasets (**RQ1**).

Results in Table 2 show the performance of models following the quantization stage of our proposed compression workflow with different bit width configurations. Overall, our experimental results show that 16-bit and 8-bit quantization can be quite effective in reducing the complexity of deep autoencoder models used for anomaly detection tasks (**RQ2**). In case of 5-bit and 4-bit quantization there is significant drop on WADI-2019 and SWAT. Both models CNN AE and GDN are robust for 4-bit quantization and give the F1-score at the same level as the baseline counterparts. It can be observed that for nonlinear 16-bit and 8-bit there is no drop in accuracy for both models and datasets (**RQ2**). The drop in case of 4-bit quantization is acceptable only for MSL and SMAP. The research results presented in [12] show that sparse 1D convolutional layers can be speed up on GPU using sparse convolution. The sparsity above 70% guarantees that sparse convolution outperforms standard CuDnn implementation. Additionally, it shows that GPU can make usage from reduced precision format. These two aspects allows to improve models time efficiency on GPU (**RQ3**).

**Table 1.** Model performance in terms of F1-Score with proposed pruning workflow and different sparsity levels applied to each layer.

| Sparsity = 0.2 | | Sparsity = 0.75 | |
|---|---|---|---|
| CNN AE | GDN | CNN AE | GDN |
| 81.45 (SWAT) | 80.51 (SWAT) | 57.01 (MSL) | 30.2 (MSL) |
| 56.28 (WADI) | 53.5 (WADI) | 77.02 (SMAP) | 32.9 (SMAP) |

**Table 2.** Experimental results (F1-score) with linear (left|) and non-linear (|right) quantization and different bit-width configurations.

| | 16-bit | | 8-bit | |
|---|---|---|---|---|
| Datasets | CNN AE | GDN | CNN AE | GDN |
| SWAT | 81.90 \| 80.36 | 80.80 \| 80.70 | 81.98 \| 80.27 | 80.70 \| 80.75 |
| WADI | 62.13 \| 57.56 | 56.90 \| 55.01 | 62.06 \| 59.11 | 56.80 \| 55.50 |
| SMAP | 77.15 \| 77.02 | 32.85 \| 32.95 | 77.05 \| 76.81 | 32.82 \| 32.92 |
| MSL | 57.21 \| 56.97 | 29.95 \| 29.91 | 57.14 \| 56.98 | 29.91 \| 29.92 |
| | 5-bit | | 4-bit | |
| Datasets | CNN AE | GDN | CNN AE | GDN |
| SWAT | 80.70 \| 78.45 | 79.80 \| 79.54 | 16.44 \| 16.35 | 23.51 \| 21.43 |
| WADI | 54.52 \| 17.87 | 55.52 \| 31.52 | 48.20 \| 10.89 | 45.51 \| 24.45 |
| SMAP | 76.09 \| 76.21 | 32.65 \| 32.71 | 75.61 \| 75.89 | 32.49 \| 32.63 |
| MSL | 56.45 \| 56.15 | 29.69 \| 29.67 | 56.09 \| 56.91 | 29.55 \| 29.63 |

## 4   Conclusions and Future Works

In this paper we proposed a compression workflow leveraging pruning and quantization stages. While pruning is incorporated in the training process, linear and non-linear quantization is performed on pruned and pre-trained models. Our experiments leveraging state-of-the-art convolutional and graph autoencoder model architectures revealed the trade-off between model compression and anomaly detection performance that pruning and quantization techniques can achieve in benchmark multivariate anomaly detection settings. Among key findings, we observed that pruning can be quite effective with MSL and SMAP datasets, and 16-bit and 8-bit quantization only impacted in a small drop in terms of F1 score. Additionally, the 4-bit quantization gives the same accuracy levels as in floating point mode. On the other hand, pruning was not effective with the WADI dataset. The presented results show that anomaly detection autoencoders can be reduced from 80% (8-bit quantization and 20% sparsity level, WADI-2019 and SWAT) to about 95% (4-bit quantization and 75% sparsity level, MSL and SMAP). Future work will focus on more advanced quantization techniques based on model retraining, which could decrease the drop in F1-Score for lower bit-widths.

**Acknowledgments.** This paper was realized with funds of Polish Ministry of Science and Higher Education assigned to AGH University and it was supported by PLGrid Infrastructure.

# References

1. Al-Hami, M., Pietron, M., Casas, R., Wielgosz, M.: Methodologies of compressing a stable performance convolutional neural networks in image classification (2020)
2. Audibert, J., Michiardi, P., Guyard, F., Marti, S., Zuluaga, M.A.: USAD: unsupervised anomaly detection on multivariate time series. In: Proceedings of the 26th ACM SIGKDD, KDD 2020, New York, NY, USA, pp. 3395–3404 (2020)
3. Deng, A., Hooi, B.: Graph neural network-based anomaly detection in multivariate time series. In: AAAI International Conference on Artificial Intelligence (2021)
4. Faber, K., Pietron, M., Zurek, D.: Ensemble neuroevolution-based approach for multivariate time series anomaly detection. Entropy **23**(11) (2021). https://doi.org/10.3390/e23111466
5. Faber, K., Pietron, M., Zurek, D.: Ensemble neuroevolution-based approach for multivariate time series anomaly detection. Entropy **23**(11), 1466 (2021)
6. Frankle, J., Dziugaite, G., Roy, D., Carbin, M.: The Lottery Ticket Hypothesis at Scale (2019)
7. Garg, A., Zhang, W., Samaran, J., Savitha, R., Foo, C.S.: An evaluation of anomaly detection and diagnosis in multivariate time series. IEEE Trans. Neural Netw. Learn. Syst. 1–10 (2021). https://doi.org/10.1109/TNNLS.2021.3105827
8. Han, S., Pool, J., Tran, J., Dally, W.: Learning both weights and connections for efficient neural network (2015)
9. Hundman, K., Constantinou, V., Laporte, C., Colwell, I., Soderstrom, T.: Detecting spacecraft anomalies using LSTMs and nonparametric dynamic thresholding. In: Proceedings of the 24th ACM SIGKDD, pp. 387–395 (2018)
10. Pietron, M., Wielgosz, M.: Retrain or not retrain? – efficient pruning methods of deep CNN networks (2020)
11. Pietron, M., Zurek, D., Faber, K., Corizzo, R.: Ad-nev: A scalable multi-level neuroevolution framework for multivariate anomaly detection. arXiv preprint arXiv:2305.16497 (2023)
12. Pietron, M., Zurek, D., Sniezynski, B.: Speedup deep learning models on GPU by taking advantage of efficient unstructured pruning and bit-width reduction, vol. 67. Elsevier (2023). https://doi.org/10.1016/j.jocs.2023.101971

13. Ren, Z., et al.: Graph autoencoder with mirror temporal convolutional networks for traffic anomaly detection. Sci. Rep. **14**(1), 1247 (2024)
14. Renda, A., Frankle, J., Carbin, M.: Comparing fine-tuning and rewinding in neural network pruning (2020)

# Active Learning on Ensemble Machine-Learning Model to Retrofit Buildings Under Seismic Mainshock-Aftershock Sequence

Neda Asgarkhani[1,2], Farzin Kazemi[1,2(✉)], and Robert Jankowski[1]

[1] Faculty of Civil and Environmental Engineering, Gdańsk University of Technology, ul. Narutowicza 11/12, 80-233 Gdansk, Poland
{neda.asgarkhani,farzin.kazemi,jankowr}@pg.edu.pl
[2] Department of Structures for Engineering and Architecture, School of Polytechnic and Basic Sciences, University of Naples "Federico II", Naples, Italy

**Abstract.** This research presents an efficient computational method for retrofitting of buildings by employing an active learning-based ensemble machine learning (AL-Ensemble ML) approach developed in OpenSees, Python and MATLAB. The results of the study shows that the AL-Ensemble ML model provides the most accurate estimations of interstory drift (ID) and residual interstory drift (RID) for steel structures using a dataset of 2-, to 9-story steel structures considering four soil type effects. To prepare the dataset, 3584 incremental dynamic analysis (IDA) were performed on 64 structures. The research employs 6-, and 8-story structures to validate the AL-Ensemble ML model's effectiveness, showing it achieves the highest accuracy among conventional ML models, with an $R^2$ of 98.4%. Specifically, it accurately predicts the RID of floor levels in a 6-story structure with an accuracy exceeding 96.6%. Additionally, the programming code identifies the specific damaged floor level in a building, facilitating targeted local retrofitting instead of retrofitting the entire structure promising a reduction in retrofitting costs while enhancing prediction accuracy.

**Keywords:** Computational Method · Active Learning · Ensemble Machine-Learning Model · Retrofitting Structures · Mainshock-Aftershock Sequence

## 1 Introduction

The utilization of dissipation devices, such as viscous dampers, buckling-resisting braces (BRBs), and shape memory alloys (SMAs), constitutes an advanced and strategic approach in structural engineering and seismic retrofitting. Each of these dissipation devices serves a unique purpose in enhancing the resilience and performance of structures under dynamic loads, and avoiding intensive damages particularly in seismic-prone regions [1–3]. Viscous dampers are strategically placed within a structure or between structures to enhance its overall seismic performance and resilience [4–6], while knee braces and BRBs are implemented as bracing system [7]. By dissipating energy through controlled

© The Author(s), under exclusive license to Springer Nature Switzerland AG 2024
L. Franco et al. (Eds.): ICCS 2024, LNCS 14834, pp. 470–478, 2024.
https://doi.org/10.1007/978-3-031-63759-9_47

yielding or ductile behavior, BRBs contribute to the structure's resilience against seismic forces [8–10]. In addition, using infill walls also can be a reachable alternative for retrofitting of buildings [11, 12]. When incorporated into structural elements, SMAs contribute to damping vibrations and reducing the overall impact of seismic forces, enhancing the structure's performance and minimizing damage. Moreover, using SMA bolts can enhance the connections behavior and prevents large residual interstory drift (RID) that is crucial in the decision-making process for retrofitting procedures [13, 14]. Enhancing the safety and resilience of buildings in seismic challenges, this study aims to propose a novel retrofitting scheme.

Machine learning (ML) computational algorithms are widely used by researchers to provide prediction models on engineering problems such as seismic response and performance assessment [15–17], seismic risk assessment [18–20], and predicting concrete material strength [21] for steel and reinforced concrete (RC) structures. Meanwhile, there is still a gap for predicting the interstory drift (ID) and RID of floor levels of buildings for retrofitting purpose. Instead of relying on a single ML algorithm, ensemble methods combine predictions from several base models to make more robust and reliable predictions [22]. By iteratively selecting the most relevant data points for labeling, active learning can often achieve higher accuracy with fewer labeled instances compared to traditional supervised learning approaches. In addition, active learning allows retrofitting decisions to be based on the most informative data, optimizing the allocation of resources by strategically selecting dissipation devices that will benefit the most from retrofitting efforts.

This study aims to provide an active learning-based ensemble ML computational model to estimate the seismic response of ID and RID, which play a crucial role on illustrating the seismic behavior of building. Having these responses can help civil engineers to decide on retrofitting of building with recognizing the weak floor level and introduce it for retrofitting scheme. Moreover, by changing the structural members of the weak floor, it is possible to check the reliability of the retrofitted structure. This procedure can cut the complex modeling, time-consuming analysis, and need for a professional expert for modeling process. Since the structural conditions and seismic responses can vary widely among different buildings, active learning enables ensemble ML models to adaptively learn from the most relevant retrofitting cases (i.e., training dataset), allowing for the formulation of retrofitting strategies tailored to specific characteristics and vulnerabilities of each structure. Therefore, active learning can be a guidance to ensemble ML models to identify and prioritize the critical parameters influencing the effectiveness of retrofitting measures. Active learning helps reduce this uncertainty by iteratively refining the ensemble model based on the most relevant and informative data, leading to more robust retrofitting decisions that align with actual structural performance. Since the retrofitting of building after mainshock and before aftershock can be a challenging for its complex modeling of damaged building, the proposed method can be a useful strategy. It is noteworthy that using this procedure can widely reduce time of seismic evaluations and retrofitting of buildings.

## 2  Structural Modeling

For providing a dataset, buildings having 2-, to 9-story elevations have been designed according to ASCE 7-16 [23] (see also [15] for details of designing process) considering four soil types (i.e., B, C, D, and E). It should be noted that to improve the modeling quality of the structure, IMK hinges have been used for beams [4], fiber section has been used for columns [5], and P-Δ effect has been considered in models using a leaning column [24, 25] modeled in OpenSees [26] software. The modeling procedure and dataset used in this research has been provided by Kazemi et al. [15] and 3584 incremental dynamic analysis (IDA) were done on selected 64 structures based on the $S_a(T_1)$ (i.e., intensity measure) and ID and RID (i.e., demanding thresholds) [27, 28]. In addition, the dataset has been improved by adding the floor labels of structures to have the ID of each floors as output of ML model. Moreover, the dataset has been changed to include the sections of beams and columns of each floor levels into account. Therefore, it will be possible to estimate the ID or RID of each floor level of structure; and then, check the sections of structural elements related to that floor level. This ability provides information regarding the weak floor level that can be useful for retrofitting purpose. Figure 1 presents the IDA curves and median of IDA curves of the 6-, and 8-story structures subjected to pulse-like records.

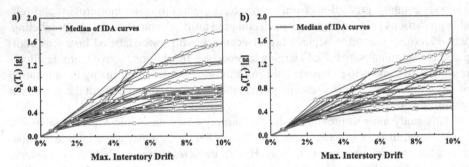

**Fig. 1.** IDA curves and median of IDA curves of the 6-, and 8-story structures subjected to pulse-like records.

## 3  Computational Method

Literature review show that many studies used dissipation devices such as viscous dampers, knee braces, BRBs, and SMAs to improve the seismic behavior of building under seismic excitations. Meanwhile, they used dissipation devices as structural member in all floor levels rather than implementing on the floor level with high possibility of weakness. Therefore, as alternative retrofitting scheme, this study proposes dissipation devices to be implemented on the weak floor level to control the ID and RID, which this floor has been selected by active learning-based ensemble ML model. Figure 2 illustrates the computational method based on active learning ensemble ML model for retrofitting of buildings. As it is presented, after modeling of structure in stage 1, the mainshock will be performed and ID and RID of the structure will be calculated in stage 2. According to the floor level with highest ID and RID, the designer can decide to use aforementioned dissipation devices as retrofitting scheme in stage 3. This process will be time-consuming since the modeling and performing the analysis need more complex modeling. To overcome this shortening, the active learning ensemble ML model can use the structural characteristics to estimate the ID and RID of structure and corresponding floor level.

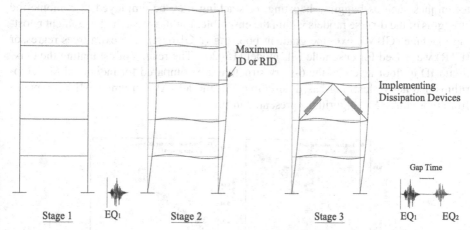

**Fig. 2.** Computational method based on active learning ensemble ML model for retrofitting of buildings.

It should be noted that modeling local damages of structural members and a damaged building is not an easy task due to differences in strength of each elements and different damage limitations. Therefore, total structural evaluations can ease the way of retrofitting by reducing modeling process and structure can be retrofitted before after-shock. To automate the procedure, a Tcl code has been developed in OpenSees [26] to model structure and provide a mainshock-aftershock analysis [29], then, the procedure has been controlled by MATLAB software to achieve results of analysis and prepare the dataset of each structures. Python programming code has been developed for labeling of dataset and performing active learning process on ensemble ML model. The results of ID and RID for floor levels have been illustrated on text file and can be used as source of retrofitting scheme. Although this study explores the procedure for steel structures, the procedure can be used for RC structures by providing related dataset.

## 4  Retrofitting of Building with ML Method

The application of active learning in ensemble ML models is a systematic and strategic approach to optimizing model performance through iterative data selection and labeling. A pool of unlabeled data serves as the starting point. The ensemble actively selects instances from this pool for labeling based on their perceived potential to improve model performance. Various active learning query strategies guide the ensemble ML model in selecting instances for labeling. Common strategies include uncertainty sampling, query by committee, expected model change, and other metrics that quantify the model's confidence or uncertainty in its predictions.

The active learning process unfolds iteratively, and then, in each iteration, the ensemble model makes predictions on the unlabeled instances, selects a subset based on the chosen query strategy, and queries a user for labels. After each iteration, the predictions of the individual base models are aggregated to form a collective prediction. Ensemble techniques such as bagging, boosting, or stacking may be employed to combine the strengths of the diverse models within the ensemble. In this research, the gradient boosting machine (GBM), extreme gradient boosting (XGBoost), and extra trees regressor (ETR) were used for ensemble ML model [15–20]. The results of estimating the maximum ID in floor levels of the 6-story structure are compared for individual ML algorithms and active learning-based ensemble ML model (AL-Ensemble ML) composed from those three ML algorithms presented in Fig. 3.

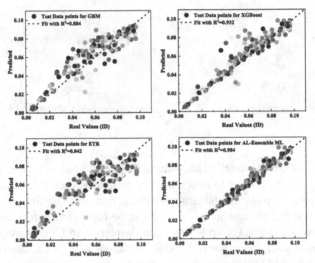

**Fig. 3.** Scatter results of the 6-story structure assuming soil D considering the conventional and AL-Ensemble ML models.

It can be seen that the conventional ML models has less ability to estimate the maximum ID of 6-story structure by large dispersion on x = y line. For instance, the XGBoost model achieved the accuracy of 93.2% and the result show that it has good ability to estimate the ID less than 0.04. Although using ensemble modeling can widely improve the performance of estimation model, active learning can enhance it further. The result confirm that AL-Ensemble ML is the best prediction model with accuracy of 98.4%. For brevity, only result of the 6-story structure has been plotted, while similar results has been achieved. Since the AL-Ensemble ML model has the best prediction among other models, it has been used to estimate RID of the 6-story structure and the results has been presented in Fig. 4. It is clear that having accuracy more than 96.6% allows the AL-Ensemble ML model to estimate the RID of floor levels of the 6-story structure that can be used for retrofitting scheme. According to results, maximum RID has been determined in the first and second floor levels of the 6-, and 8-story structures and these floors are introduced for retrofitting with viscous dampers. By adding dissipations devices, the RID of the structures has been compared to non-retrofitted structures in Fig. 5. It can be concluded that the precise predictions made by AL-Ensemble ML model helped to find those weak floor levels and reduce maximum RID of structures and the cost of retrofitting accordingly. Therefore, the procedure can be a useful tool for retrofitting structures under seismic sequences.

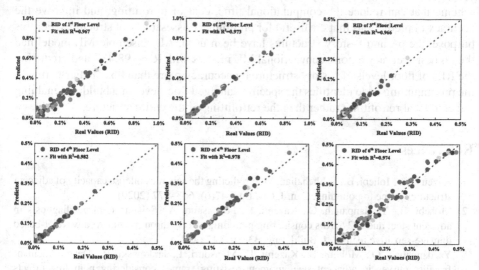

**Fig. 4.** Scatter RID prediction results of the 6-story structure assuming soil D considering the conventional and AL-Ensemble ML models.

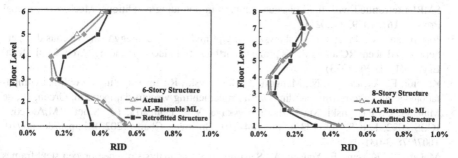

**Fig. 5.** Retrofitting first and second floor levels of the 6-, and 8-story structures according to prediction results of AL-Ensemble ML model.

## 5 Conclusions

This study introduces an efficient computational approach aimed at retrofitting of buildings affected by seismic mainshock-aftershock sequences using active learning-based ensemble ML method. The proposed method is versatile, capable of retrofitting both steel and reinforced concrete structures while accommodating various intensity measures and engineering requirements. It allows for the application of diverse retrofitting devices, including viscous dampers and BRBs. Results confirm that the AL-Ensemble ML model has the best estimations of ID and RID of steel structures that can be used for retrofitting of structures by implementing dissipation devices at floor levels with highest values of RID and ID. The proposed procedure introduces a novel ML-based retrofitting scheme that can reduce the computational time, cost of retrofitting, and improve the accuracy of predictions that are useful for preliminary assessment of structures. For this purpose, the 6-, and 8-story structures have been used. AL-Ensemble ML model had the highest accuracy among conventional ML models with $R^2 = 98.4\%$, and predicting the RID of floor levels of 6-story structure by accuracy more than 96.6%. Furthermore, the programming code identifies the specific damaged floor level of a building, enabling targeted local retrofitting rather than the retrofitting of the entire structure.

# References

1. Kazemi, F., Mohebi, B., Yakhchalian, M.: Predicting the seismic collapse capacity of adjacent structures prone to pounding. Can. J. Civ. Eng. **47**(6), 663–677 (2020)
2. Mohebi, B., Yazdanpanah, O., Kazemi, F., Formisano, A.: Seismic damage diagnosis in adjacent steel and RC MRFs considering pounding effects through improved wavelet-based damage-sensitive feature. J. Build. Eng. **33**, 101847 (2021)
3. Yazdanpanah, O., Mohebi, B., Kazemi, F., Mansouri, I., Jankowski, R.: Development of fragility curves in adjacent steel moment-resisting frames considering pounding effects through improved wavelet-based refined damage-sensitive feature. Mech. Syst. Signal Process. **173**, 109038 (2022)
4. Kazemi, F., Mohebi, B., Jankowski, R.: Predicting the seismic collapse capacity of adjacent SMRFs retrofitted with fluid viscous dampers in pounding conditions. Mech. Syst. Signal Process. **161**, 107939 (2021)
5. Asgarkhani, N., Kazemi, F., Jankowski, R.: Optimal retrofit strategy using viscous dampers between adjacent RC and SMRFs prone to earthquake-induced pounding. Arch. Civil Mech. Eng. **23**(1), 1–26 (2023)
6. Kazemi, F., Asgarkhani, N., Manguri, A., Jankowski, R.: Investigating an optimal computational strategy to retrofit buildings with implementing viscous dampers. In: Groen, D., de Mulatier, C., Paszynski, M., Krzhizhanovskaya, V.V., Dongarra, J.J., Sloot, P.M.A. (eds.) ICCS 2022. LNCS, vol. 13351, pp. 184–191. Springer, Cham: (2022). https://doi.org/10.1007/978-3-031-08754-7_25
7. Mohebi, B., Kazemi, F., Yousefi, A.: Seismic response analysis of knee-braced steel frames using Ni-Ti shape memory alloys (SMAs). In: Di Trapani, F., Demartino, C., Marano, G.C., Monti, G. (eds.) EOS 2022. LNCE, vol. 326, pp. 238–247. Springer, Cham (2023). https://doi.org/10.1007/978-3-031-30125-4_21
8. Mohebi, B., Sartipi, M., Kazemi, F.: Enhancing seismic performance of buckling-restrained brace frames equipped with innovative bracing systems. Arch. Civil Mech. Eng. **23**(4), 243 (2023)
9. Asgarkhani, N., Yakhchalian, M., Mohebi, B.: Evaluation of approximate methods for estimating residual drift demands in BRBFs. Eng. Struct. **224**, 110849 (2020)
10. Kazemi, F., Jankowski, R.: Seismic performance evaluation of steel buckling-restrained braced frames including SMA materials. J. Constr. Steel Res. **201**, 107750 (2023)

11. Kazemi, F., Asgarkhani, N., Jankowski, R.: Probabilistic assessment of SMRFs with infill masonry walls incorporating nonlinear soil-structure interaction. Bull. Earthq. Eng. **21**(1), 503–534 (2023)

12. Kazemi, F., Asgarkhani, N., Jankowski, R.: Enhancing seismic performance of steel buildings having semi-rigid connection with infill masonry walls considering soil type effects. Soil Dyn. Earthq. Eng. **177**, 108396 (2024)

13. Kazemi, F., Jankowski, R.: Enhancing seismic performance of rigid and semi-rigid connections equipped with SMA bolts incorporating nonlinear soil-structure interaction. Eng. Struct. **274**, 114896 (2023)

14. Mohebi, B., Kazemi, F., Yousefi, A.: Enhancing seismic performance of semi-rigid connection using shape memory alloy bolts considering nonlinear soil–structure interaction. In: Di Trapani, F., Demartino, C., Marano, G.C., Monti, G. (eds.) EOS 2022. LNCE, vol. 326, pp. 248–256. Springer, Cham (2023). https://doi.org/10.1007/978-3-031-30125-4_22

15. Kazemi, F., Asgarkhani, N., Jankowski, R.: Predicting seismic response of SMRFs founded on different soil types using machine learning techniques. Eng. Struct. **274**, 114953 (2023)

16. Kazemi, F., Asgarkhani, N., Jankowski, R.: Machine learning-based seismic response and performance assessment of reinforced concrete buildings. Arch. Civ. Mech. Eng. **23**(2), 94 (2023)

17. Asgarkhani, N., Kazemi, F., Jakubczyk-Gałczyńska, A., Mohebi, B., Jankowski, R.: Seismic response and performance prediction of steel buckling-restrained braced frames using machine-learning methods. Eng. Appl. Artif. Intell. **128**, 107388 (2024)

18. Kazemi, F., Jankowski, R.: Machine learning-based prediction of seismic limit-state capacity of steel moment-resisting frames considering soil-structure interaction. Comput. Struct. **274**, 106886 (2023)

19. Kazemi, F., Asgarkhani, N., Jankowski, R.: Machine learning-based seismic fragility and seismic vulnerability assessment of reinforced concrete structures. Soil Dyn. Earthq. Eng. **166**, 107761 (2023)

20. Asgarkhani, N., Kazemi, F., Jankowski, R.: Machine learning-based prediction of residual drift and seismic risk assessment of steel moment-resisting frames considering soil-structure interaction. Comput. Struct. **289**, 107181 (2023)

21. Kazemi, F., Shafighfard, T., Yoo, D.Y.: Data-driven modeling of mechanical properties of fiber-reinforced concrete: a critical review. Arch. Comput. Meth. Eng., 1–30 (2024)

22. Shafighfard, T., Kazemi, F., Bagherzadeh, F., Mieloszyk, M., Yoo, D. Y.: Chained machine learning model for predicting load capacity and ductility of steel fiber–reinforced concrete beams. Comput.-Aid. Civ. Infrastruct. Eng. (2024)

23. Minimum Design Loads for Buildings and Other Structures (ASCE/SEI 7-16), first, second, and third printings. Minimum Design Loads for Buildings and Other Structures (2017)

24. Mohebi, B., Asadi, N., Kazemi, F.: Effects of using gusset plate stiffeners on the seismic performance of concentrically braced frame. Int. J. Civ. Environ. Eng. **13**(12), 723–729 (2019)

25. Mohebi, B., Kazemi, F., Asgarkhani, N.: Retrofitting damaged buildings under seismic mainshock-aftershock sequence. In: 9th International Conference on Computational Methods in Structural Dynamics and Earthquake Engineering (COMPDYN) 21523 (2023)

26. McKenna, F., Fenves, G.L., Filippou, F.C., Scott, M.H.: Open system for earthquake engineering simulation (OpenSees). Pacific Earthquake Engineering Research Center, University of California, Berkeley (2016)

27. Mohebi, B., Kazemi, F., Asgarkhani, N., Ghasemnezhadsani, P., Mohebi, A.: Performance of vector-valued intensity measures for estimating residual drift of steel MRFs with viscous dampers. Int. J. Struct. Civ. Eng. Res. **11**(4), 79–83 (2022)

28. Kazemi, F., Mohebi, B., Asgarkhani, N., Yousefi, A.: Advanced scalar-valued intensity measures for residual drift prediction of SMRFs with fluid viscous dampers. Int. J. Struct. Integr. **12**, 20–25 (2023)

29. Kazemi, F., Asgarkhani, N., Manguri, A., Lasowicz, N., Jankowski, R.: Introducing a computational method to retrofit damaged buildings under seismic mainshock-aftershock sequence. In: Mikyška, J., de Mulatier, C., Paszynski, M., Krzhizhanovskaya, V.V., Dongarra, J.J., Sloot, P.M. (eds.) ICCS 2023. LNCS, vol. 14074, pp. 180–187. Springer, Cham (2023). https://doi.org/10.1007/978-3-031-36021-3_16

# Kernel-Based Learning with Guarantees for Multi-agent Applications

Krzysztof Kowalczyk[1]([✉])[iD], Paweł Wachel[1][iD], and Cristian R. Rojas[2][iD]

[1] Department of Control Systems and Mechatronics, Wrocław University of Science and Technology, Wrocław, Poland
{krzysztof.kowalczyk,pawel.wachel}@pwr.edu.pl
[2] School of Electrical Engineering and Computer Science, KTH Royal Institute of Technology, Stockholm, Sweden
crro@kth.se

**Abstract.** This paper addresses a kernel-based learning problem for a network of agents locally observing a latent multidimensional, nonlinear phenomenon in a noisy environment. We propose a learning algorithm that requires only mild *a priori* knowledge about the phenomenon under investigation and delivers a model with corresponding non-asymptotic high probability error bounds.

Both non-asymptotic analysis of the method and numerical simulation results are presented and discussed in the paper.

**Keywords:** Multi-agent systems · distributed learning

## 1 Introduction

A multi-agent system is a network of autonomous entities called agents that share information and collaborate to solve tasks usually beyond an individual agent's scope [11]. This broad description fits well in the recent research trends like cloud computing [10], or Industry 4.0 [9], and allows multi-agent systems to find applications in many other fields. In robotics, in scenarios including groups of mobile robots or swarms of drones, it is necessary to avoid collisions or obstacles and to navigate collaboratively [8]. We can also find numerous other examples, like analyzing the traffic flow [6] or modelling purchasing decisions [3].

Inspired by these multidisciplinary applications, we formally discuss the general problem of distributed learning, with a particular focus on the modelling of nonlinearities under limited information, *cf.* [4]. In the considered scenario, every agent (node) locally observes the outcome of some unknown global phenomenon of interest. Although the agents aim to provide a non-local comprehensive model of the phenomenon, this may be not possible for individual nodes due to the limited range of their own observations. Thus, collaboration is necessary. Nonetheless, we assume that the agents cannot communicate freely throughout the entire network, but a single agent can only interact with a narrow group of its neighbourhood nodes (*cf.* Fig. 1).

L. Franco et al. (Eds.): ICCS 2024, LNCS 14834, pp. 479–487, 2024.
https://doi.org/10.1007/978-3-031-63759-9_48

One can find numerous approaches related to this problem in the literature, among which Kalman-based filtering [2], diffusion [5], and consensus [1] techniques can be distinguished; see *e.g.* [7] for a more extensive discussion. While our approach is motivated by the abovementioned methods, we introduce, however, a few substantial modifications. In particular, regarding the investigated nonlinear phenomenon, we require only limited *a priori* knowledge, usually insufficient for many parametric estimation techniques proposed so far. We use kernel regression for efficient non-parametric modelling and provide corresponding error-bound guarantees that hold for a finite number of samples. The algorithm proposed in this paper is an extension of the method introduced in [13], suited for multivariate phenomenons.

**Fig. 1.** A network of distributed agents with highlighted neighbourhood of a selected node.

## 2   Problem Formulation

We investigate a problem of distributed learning, where a group of agents observes an unknown phenomenon in a noisy environment and aims to provide noise-free estimations with high probability guarantees for a given region of interest.

We consider a set of $M$ agents and model their cooperation via a connected and undirected graph $\mathcal{G} = (\mathcal{M}, \mathcal{E})$ with $\mathcal{M} = \{1, 2, \ldots, M\}$ nodes and a set of unweighted edges $\mathcal{E}$. To reflect possible restrictions and to reduce the communication burden, we assume that two nodes $i, j \in \mathcal{M}$ can exchange information if and only if they are directly connected, *i.e.*, if $\{i, j\} \in \mathcal{E}$. Thus, we define the neighbourhood of a node $i \in \mathcal{M}$ as the set $\mathcal{N}_i = \{j \colon \{i, j\} \in \mathcal{E}\}$.

In the considered setup, at every time step $t \in \mathbb{N}$, every agent $k \in \mathcal{M}$ obtains an explanatory data point $\xi_{k,t} \in \mathbb{R}^p$, for some fixed $p \in \mathbb{N}$, and observes a noisy outcome $y_{k,t}$ of the latent nonlinear phenomenon modelled by an unknown nonlinear mapping $m \colon \mathcal{D} \subset \mathbb{R}^p \to \mathbb{R}^d$,

$$y_{k,t} = m(\xi_{k,t}) + \eta_{k,t}, \quad k \in \mathcal{M}, \quad t \in \mathbb{N}, \tag{1}$$

where $\eta_{k,t}$ denotes an additive noise.

This paper aims to provide a distributed inference of $m$ under mild *a priori* knowledge about its structure. Hence, the following assumptions regarding the observed phenomenon and the additive noise have a general form. For simplicity of notation, we will use the symbol $a_{1:m}$ as a short for a sequence $a_1, \ldots, a_m$.

*Assumption* 1. The latent phenomenon of interest, $m\colon \mathcal{D} \subset \mathbb{R}^p \to \mathbb{R}^d$, is a Lipschitz continuous mapping, *i.e.*, for a known constant $0 \le L < \infty$,

$$\|m(\xi) - m(\xi')\|_2 \le L\|\xi - \xi'\|_2, \quad \forall \, \xi, \xi' \in \mathcal{D}. \tag{2}$$

*Assumption* 2. The explanatory sequence $\{\xi_t \in \mathbb{R}^p : t \in \mathbb{N}\}$ is an arbitrary stochastic process.

*Assumption* 3. The disturbance $\{\eta_t \in \mathbb{R}^d : t \in \mathbb{N}\}$ is a sub-Gaussian stochastic process, that is, there exists some $\sigma > 0$ such that, for every $\gamma_t \in \mathbb{R}^d$ (possibly a function of $\xi_t$), and every $t \in \mathbb{N}$, $\mathbb{E}\{\exp(\gamma_t^\top \eta_t)|\eta_{1:t-1}, \xi_{1:t}\} \le \exp\left(\gamma_t^\top \gamma_t \sigma^2/2\right)$.

The above requirements have a somewhat general character and are inspired by the real-world properties of many technical processes. Informally, Assumption 1 allows, in particular, any nonlinear function with a limited rate of increase (or decrease), and Assumption 3 admits any bounded *i.i.d.* disturbances with zero mean, independent of the explanatory data.

## 3  Local Agents' Modelling

To construct the proposed learning technique, we begin from a single-agent perspective. Given a fixed time instant $t$ and a set of local data measurements, we define for agent $k \in \mathcal{M}$ the following kernel regression estimator:

$$\hat{\mu}_{k,t}(x) := \sum_{n=1}^{t} \frac{K_h(x, \xi_{k,n})}{\kappa_{k,t}(x)} y_{k,n} =: \frac{\psi_{k,t}(x)}{\kappa_{k,t}(x)},$$

$$\kappa_{k,t}(x) := \kappa_{k,t}(x, h) = \sum_{n=1}^{t} K_h(x, \xi_{k,n}), \tag{3}$$

with $K_h(x, \xi) := K(\|x - \xi\|_2/h)$, and where $K$, $h$ are the kernel function and the bandwidth parameter, respectively. To ensure appropriate statistical properties of $\hat{\mu}_{k,t}$, we make the following assumption:

*Assumption* 4. The kernel $K\colon \mathbb{R} \to \mathbb{R}$ is such that $0 \le K(v) \le 1$ for all $v \in \mathbb{R}$. Also, $K(v) = 0$ for all $|v| > 1$.

We are now about to develop the main technical result, which is the basis for the network estimation algorithm introduced in the sequel (*cf.* [13]).

**Lemma 1.** *Let Assumptions 1–4 be in force. Consider the estimator $\hat{\mu}_{k,t} \in \mathbb{R}^d$ and fix a bandwidth parameter $h$. Let $x \in \mathcal{D} \subset \mathbb{R}^p$ be fixed or in general a measurable function of $\eta_{k,1:t-1}, \xi_{k,1:t}$ (e.g., $x = \xi_{k,t}$). Then, for every $0 < \delta < 1$, with probability at least $1 - \delta$, if $\kappa_{k,t}(x) \neq 0$,*

$$\|\hat{\mu}_{k,t}(x) - m(x)\|_2 \le \beta_{k,t}(x), \quad \text{where} \quad \beta_{k,t}(x) := Lh + 2\sigma \frac{\alpha_{k,t}(x, \delta)}{\kappa_{k,t}(x)}, \tag{4}$$

$$\alpha_{k,t}(x,\delta) := \begin{cases} \sqrt{\log(\delta^{-1}2^{d/2})}, & \text{for } \kappa_{k,t}(x) \leq 1 \\ \sqrt{\kappa_{k,t}(x)\log\left(\delta^{-1}\left(1+\kappa_{k,t}(x)\right)^{d/2}\right)}, & \text{for } \kappa_{k,t}(x) > 1. \end{cases} \quad (5)$$

*Proof.* See the Appendix[1]. □

In Lemma 1, we provide error bounds for local (single-agent) estimates that hold with probability $1 - \delta$. The Lipschitz constant $L$ and the noise proxy variance $\sigma$ are, however, required to be known (in practice, at least upper bounds on these quantities are needed).

Due to the fact, that the dimensionality of the output influences the bounds, for higher $d$'s, it may be worth considering techniques of MIMO system decompositions as *e.g.* [12].

## 4   Distributed Modelling – Data Aggregation

Having a single-agent estimator, we are now ready to introduce a distributed modelling procedure.

According to the considered approach, every agent $k$ spreads its local estimations by broadcasting tuples of essential data $T_{k,t}(x) = (\psi_{k,t}(x), \kappa_{k,t}(x), x)$, which contains locally computed numerator, denominator and the estimation point, to its neighbourhood $\mathcal{N}_k$. The acquired tuples are then stored in set $\mathbb{T}_k$. To avoid data repetition in a container of tuples, only a single tuple from a single agent and fixed estimation point $x$ is included in $\mathbb{T}_k$, *i.e.*, the newer (incoming) tuples overwrite the older ones.

---

**Algorithm 1.** Data exchange and aggregation ▷ Agent $k$

1: **input:** $\mathcal{X}$ ▷ Estimation points
2: **for** $t = 1, 2, \ldots$ **do**
3:     GET $(\xi_{k,t}, y_{k,t})$ ▷ Get local measurement
4:     **if** acquired_new_tuple **then**
5:         UPDATE $\mathbb{T}_k$
6:     **if** send_local_data **then**
7:         SELECT $x \in \mathcal{X}$ ▷ Select an estimation point
8:         EVALUATE $\psi_{k,t}(x), \kappa_{k,t}(x)$
9:         $T_{k,t}(x) \leftarrow (\psi_{k,t}(x), \kappa_{k,t}(x), x)$
10:     **if** send_acquired_data **then**
11:         SELECT $T_i(x) \in \mathbb{T}_k$
12:     BROADCAST selected tuple ▷ Send data to the neighbors
13: **end**

---

[1] For the full proofs we refer the reader to https://arxiv.org/pdf/2404.09708.pdf.

The proposed algorithm requires a few comments. We assume that all the agents work on the same set $\mathcal{X}$ (*i.e.*, $x \in \mathcal{X}$) and they can freely share their data. We do not specify here when the agents should transfer their local data and when they acquire information from their neighbourhoods. Currently, we leave this open for the user, by setting the flags *send_local_data* and *send_acquired_data* (in the experiments these flags were set randomly).

Following the data exchange and aggregation routine proposed in Algorithm 1, every agent builds a tuple set $\mathbb{T}_k$ that will be used next to construct a model of $m(\cdot)$. For every agent $k$ with $\mathbb{T}_k$, we define an estimator that combines all the acquired data as follows:

$$\hat{m}_{k,t}(x) = \frac{\sum_{i=1}^{M} \psi_i(x)}{\sum_{i=1}^{M} \kappa_i(x)} = \frac{\Psi_{k,t(x)}}{\mathcal{K}_{k,t(x)}}, \quad \psi_i(x), \kappa_i(x) \in T_i(x) \in \mathbb{T}_k. \tag{6}$$

For the estimator in (6), we provide non-asymptotic error bounds in Theorem 1 below.

**Theorem 1.** *Let Assumptions 1–4 be in force. Consider any agent $k \in M$ with data exchange and aggregation procedure as in Algorithm 1 and estimate $\hat{m}_{k,t}$. Then, for $x \in \mathcal{X}$ and any $0 < \delta < 1$, with probability $1 - \delta$,*

$$\|\hat{m}_{k,t}(x) - m(x)\|_2 \leq \beta_{k,t}(x), \tag{7}$$

*where $\beta_{k,t}(x)$ is given by Eqs. (4) and (5).*

*Proof.* We introduce a merged index $q$ that takes values from 1 to $\tau = \sum_{i=1}^{M} t_i$ and mappings $i_q$ and $n_q$, that transfer a single $q$ back to the original $i$ and $n$, respectively. Thus,

$$\hat{m}_{k,t} = \frac{\sum_{i=1}^{M} \sum_{n=1}^{t_i} K_h(x, \xi_{i,n}) y_{i,n}}{\sum_{i=1}^{M} \sum_{n=1}^{t_i} K_h(x, \xi_{i,n})} = \frac{\sum_{q=1}^{\tau} K_h(x, \xi_{i_q,n_q})}{\kappa_\tau(x)} y_{i_q,n_q} = \hat{\mu}_\tau(x). \tag{8}$$

This can be interpreted as the local estimator of an agent, that directly acquired all $\tau$ observations. Hence, we can apply the error bound from Lemma 1, which completes the proof. □

As we have shown, Theorem 1 can be proven by reinterpreting Lemma 1 since the final estimate combines the acquired numerators and denominators, and is, in fact, the same as the estimate calculated from raw data transferred to a single agent. This is however possible only if all the agents operate with the same upper bound of the noise proxy variance $\sigma$.

## 5 Numerical Experiments

In this section, we illustrate the main concept of the proposed approach[2]. To this end, we use a network of 25 agents with randomly selected topology, as shown in

---

[2] The Python code to obtain the numerical results is available at https://github.com/kkowalc/Kernel-based-learning-with-guarantees-for-multi-agent-applications.

484    K. Kowalczyk et al.

Fig. 2. In the experiments we consider a nonlinearity $m \colon \mathbb{R}^2 \to \mathbb{R}$ being a mixture of three Gaussian surfaces $\mathcal{N}([0,0], 0.5\mathbb{I})$, $\mathcal{N}([1,2], 0.55\mathbb{I})$, $\mathcal{N}([2,-2], 0.7\mathbb{I})$. The output noise sequences $\eta_{k,t}$ for every agent $k$ are sampled from a normal distribution $\mathcal{N}(0, 0.05)$. The total region of interest $\mathcal{D}$ is a set $[-2,2] \times [-2,2]$ and the estimation grid $\mathcal{X}$ is evenly spaced with a step 0.25. The explanatory data $\xi_{k,t}$ is generated from a normal distribution $\mathcal{N}(\mu_{\xi_k}, \sigma_{\xi_k})$. Both $\{\xi_{k,t}\}$ and $\{\eta_{k,t}\}$ are mutually independent. For simplicity of calculations and clarity of presentation, the parameters $\mu_{\xi,k}$ and $\sigma_{\xi,k}$ are selected to ensure that $\mathcal{D} \subset \mathcal{D}_1 \cup \mathcal{D}_2 \cup \cdots \cup \mathcal{D}_M$; otherwise, it would be necessary to propagate the bounds with the Lipschitz constant for the regions where measurements could not be obtained. The required parameters $L$ and $\delta$ are set to 0.3 and 0.001, respectively (Fig. 3).

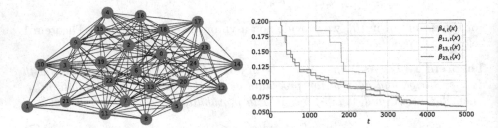

**Fig. 2.** Random topology network with 25 nodes.

**Fig. 3.** Bound evolution over time for a selected estimation point and a few selected agents.

In Fig. 4 we present the evolution of our confidence bound over time for a selected estimation point $x = (0,0)$. At the beginning the bound is high due to the lack of reacquired tuples, but with time more tuples for the estimation point

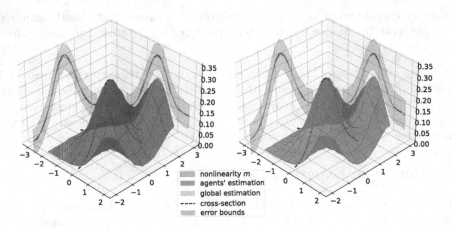

**Fig. 4.** Comparison of the model provided by a single agent, with a global model, that uses all the agents' data.

are obtained. This process however, slows down with time, since the number of agents in the network is finite, hence no new tuples are acquired and the improvement of the bounds is a result of updating existing tuples.

As we mentioned in the previous sections, transferring all the data to a single processing centre usually requires a significant communication cost. One of the main goals of distributed learning is to provide a result that is close to the centralised approach, and the proposed data exchange and aggregation algorithm has the possibility (under a proper number of connections between the nodes) to achieve it. In Fig. 4. we present a side-by-side comparison of the model provided by the single agent and the model calculated in a centralised way with the usage of all the agents' local data.

## 6   Conclusions

In this paper, we have proposed a new distributed learning algorithm, designed for learning multivariate phenomena. Following the data exchange and aggregation procedure as described in Algorithm 1 and using a distributed estimator, a single agent is able to model a phenomenon in regions that are far beyond its local scope. We have formally investigated, under rather mild assumptions, the non-asymptotic properties of the proposed method. Also, we have illustrated the obtained theoretical results via numerical simulations, which clearly show the advantages of the techniques described in this paper.

As future work, we plan to expand the proposed approach to a more general setting, where we allow the agents to have different bandwidth parameter $h$ and noise proxy variance upper bounds $\sigma$. Also we aim to investigate network topology properties in order to develop a technique for setting data sharing flags.

## Appendeix

*Proof (of Lemma 1).* We begin with the observation that

$$\left\| \sum_{n=1}^{t} \frac{K_h(x,\xi_n)}{\kappa_t(x)} y_n - m(x) \right\|_2 \leq \sum_{n=1}^{t} \theta_n \left\| m(\xi_n) - m(x) \right\|_2 + \left\| \sum_{n=1}^{t} \theta_n \eta_n \right\|_2, \quad (9)$$

where $\theta_n := K_h(x,\xi_n)/\kappa_t(x)$. Note that $\Sigma_{n=1}^t \theta_n = 1$. Due to Assumption 4, if $K_h(x,\xi_n) > 0$, then $\|x - \xi_n\|_2/h \leq 1$. Hence, *cf.* Assumption 1, $\sum_{n=1}^{t} \theta_n \|m(\xi_n) - m(x)\|_2 \leq \sum_{n=1}^{t} \theta_n L \|x - \xi_n\|_2 \leq Lh$. For the last term in (9), observe that

$$\left\| \sum_{n=1}^{t} \theta_n \eta_n \right\|_2 = \frac{1}{\kappa_t(x)} \left\| \sum_{n=1}^{t} K_h(x,\xi_n)\eta_n \right\|_2. \quad (10)$$

According to Lemma 2 and since $K_h(x,\xi_n) \leq 1$ (*cf.* Assumption 4), the right-hand side of Eq. (10) is upper bounded (with probability $1 - \delta$) by

$$\frac{1}{\kappa_t(x)} \left\| \sum\nolimits_{n=1}^{t} K_h(x,\xi_n)\eta_n \right\|_2 \leq \sigma\sqrt{2\log\left(\delta^{-1}\left(1+\kappa_t(x)\right)^{d/2}\right)} \frac{\sqrt{1+\kappa_t(x)}}{\kappa_t(x)}.$$

Observe next that, if $\kappa_t(x) > 1$, then $\sqrt{1+\kappa_t(x)}/\kappa_t(x) < \sqrt{2}/\sqrt{\kappa_t(x)}$. Therefore, with probability $1 - \delta$, for $\kappa_t(x) > 1$,

$$\frac{1}{\kappa_{k,t}(x)} \left\| \sum\nolimits_{n=1}^{t} K_h(x,\xi_n)\eta_n \right\|_2 \leq \frac{2\sigma}{\kappa_t(x)}\sqrt{\kappa_t(x)\log\left(\delta^{-1}\left(1+\kappa_t(x)\right)^{d/2}\right)},$$

whereas for $0 < \kappa_t \leq 1$,

$$\frac{1}{\kappa_t(x)} \left\| \sum\nolimits_{n=1}^{t} K_h(x,\xi_n)\eta_n \right\|_2 \leq \frac{2\sigma}{\kappa_t(x)}\sqrt{\log(\delta^{-1}2^{d/2})},$$

which completes the proof.

**Lemma 2.** *Let $\{v_t \in \mathbb{R}: t \in \mathbb{N}\}$ and $\{\eta_t \in \mathbb{R}^d: t \in \mathbb{N}\}$ be stochastic processes. Assume that there exists some $\sigma > 0$ such that, for every $\gamma_t \in \mathbb{R}^d$ (possibly a function of $v_t$), and every $t \in \mathbb{N}$, $\mathbb{E}[\exp(\gamma_t^\top \eta_t)|\eta_{1:t-1}, v_{1:t}] \leq \exp\left(\gamma_t^\top \gamma_t \sigma^2/2\right)$. Define $S_t := \sum_{n=1}^{t} v_n \eta_n$ and $V_t := \sum_{n=1}^{t} v_n^2$. Then, for every $t \in \mathbb{N}$ and $0 < \delta < 1$, with probability $1 - \delta$,*

$$S_t^\top S_t \leq 2\sigma^2 \log\left[(V_t + 1)^{d/2}/\delta\right](V_t + 1).$$

# References

1. Bertrand, A., Moonen, M.: Consensus-based distributed total least squares estimation in ad hoc wireless sensor networks. IEEE Trans. Signal Process. **59**(5), 2320–2330 (2011)
2. Cattivelli, F.S., Sayed, A.H.: Distributed nonlinear kalman filtering with applications to wireless localization. In: 2010 IEEE International Conference on Acoustics, Speech and Signal Processing, pp. 3522–3525. IEEE (2010)
3. Jedrzejewski, A., Sznajd-Weron, K., Pawłowski, J., Kowalska-Pyzalska, A.: In: Groen, D., de Mulatier, C., Paszynski, M., Krzhizhanovskaya, V.V., Dongarra, J.J., Sloot, P.M.A. (eds.) ICCS, pp. 719–726. Springer, Heidelberg (2022). https://doi.org/10.1007/978-3-031-08754-7_74
4. Łagosz, S., Śliwiński, P., Wachel, P.: Identification of Wiener-Hammerstein systems by ℓ1-constrained Volterra series. Eur. J. Control. **58**, 53–59 (2021)
5. Lopes, C.G., Sayed, A.H.: Diffusion least-mean squares over adaptive networks: formulation and performance analysis. IEEE Trans. Signal Process. **56**(7), 3122–3136 (2008)
6. Małecki, K., Górka, P., Gokieli, M.: Multi-agent cellular automaton model for traffic flow considering the heterogeneity of human delay and accelerations. In: Mikyska, J., de Mulatier, C., Paszynski, M., Krzhizhanovskaya, V.V., Dongarra, J.J., Sloot, P.M. (eds.) ICCS 2023, pp. 539–552. Springer, Heidelberg (2023). https://doi.org/10.1007/978-3-031-35995-8_38

7. Modalavalasa, S., Sahoo, U.K., Sahoo, A.K., Baraha, S.: A review of robust distributed estimation strategies over wireless sensor networks. Signal Process. **188**, 108150 (2021)
8. Rasheed, A.A.A., Abdullah, M.N., Al-Araji, A.S.: A review of multi-agent mobile robot systems applications. Int. J. Electr. Comput. Eng. (2088-8708) **12**(4) (2022)
9. Sakurada, L., Leitão, P.: Multi-agent systems to implement industry 4.0 components. In: 2020 IEEE Conference on Industrial Cyberphysical Systems (ICPS), vol. 1, pp. 21–26. IEEE (2020)
10. Kwang Mong Sim: Agent-based cloud computing. IEEE Trans. Serv. Comput. **5**(4), 564–577 (2011)
11. Srinivasan, D., Jain, L.C.: Innovations in Multi-Agent Systems and Applications-1. Springer, Heidelberg (2010). https://doi.org/10.1007/978-3-642-14435-6
12. Wachel, P., Tiels, K., Filiński, M.: Learning low-dimensional separable decompositions of mimo non-linear systems. Int. J. Control **96**(4), 900–906 (2023)
13. Wachel, P., Kowalczyk, K., Rojas, C.R.: Decentralized diffusion-based learning under non-parametric limited prior knowledge. Eur. J. Control **75**, 100912 (2024)

# Author Index

© The Editor(s) (if applicable) and The Author(s), under exclusive license
to Springer Nature Switzerland AG 2024
L. Franco et al. (Eds.): ICCS 2024, LNCS 14834, pp. 489–491, 2024.
https://doi.org/10.1007/978-3-031-63759-9

Printed in the United States
by Baker & Taylor Publisher Services

Printed in the United States
by Baker & Taylor Publisher Services